Thomas Jefferson

Other Works by Willard Sterne Randall

Ethan Allen: His Life and Times

Alexander Hamilton: A Life

George Washington: A Life

Benedict Arnold: Patriot and Traitor

A Little Revenge: Benjamin Franklin at War with His Son

Forgotten Americans
(with Nancy Nahra)

American Lives
(with Nancy Nahra)

Thomas Chittenden's Town
(with Nancy Nahra)

Building Six: The Tragedy at Bridesburg
(with Stephen D. Solomon)

The Founding City
(with David R. Boldt)

Thomas Jefferson

A Life

Willard Sterne Randall

HARPERPERENNIAL ★ POLITICALCLASSICS

NEW YORK • LONDON • TORONTO • SYDNEY • NEW DELHI • AUCKLAND

HARPERPERENNIAL ★ POLITICALCLASSICS

This book was originally published in 1993 by Henry Holt and Company. It is here reprinted by arrangement with Henry Holt and Company.

FIRST HARPER PERENNIAL EDITION PUBLISHED 1994. REISSUED IN HARPER PERENNIAL POLITICAL CLASSICS IN 2014.

Designed by Paula R. Szafranski

The Library of Congress has catalogued the previous edition as follows:
Randall, Willard Sterne.
Thomas Jefferson : a life / Willard Sterne Randall.
p. cm.
Previously published: New York : H. Holt, 1993.
Includes bibliographical references and index.
ISBN 0-06-097617-9
1. Jefferson, Thomas, 1743–1826. 2. Presidents—United States—Biography.
I. Title.
[E332.R196 1994]
973.4'092—dc20
[B] 94-14363

ISBN 978-0-06-097617-0 (reissue)

14 15 16 17 18 RRD 30 29 28 27

For Nan
and
for my children,
Christopher
Polly
Alice
and
Lucy

Contents

Acknowledgments

THE GENESIS OF this book came in France when, on breaks from writing an earlier book about Benjamin Franklin and his son, it became clear that Thomas Jefferson was involved when negotiations for the rights of French citizens were under way in the 1780s. Learning about Jefferson from French sources proved disappointing until it was my good fortune to meet Doreen Objois-Peel, for many years a member of the staff of the Benjamin Franklin Documentation Center of the American Embassy in Paris. For twenty-five years, Madame Objois-Peel had hoarded every Jefferson article available while helping to build a first-rate collection on the early American envoys at the embassy. She and her successors, Anne-Marie d'Attis and Christianne Laude, allowed me access to the embassy's "hanging files" on all past and present ambassadors. Because of their help I was able to present a paper on Jefferson before the Association France—Grande Bretagne at Nice whose president, Professor Pierre Marambaud, a leading authority on colonial America, made his library, along with original and valuable insights, available to me. His daughter, Yvette Salviati, professor of literature at the University of Nice and a Jefferson scholar in her own right, and her husband, Gérard, introduced me to the private collections of the Chevalier de Cessole at the Musée Massena. Paul Malaussena, Olivier Vernier, and Nadine Aimar enhanced my visits to that *bibliothèque*.

André and Florence Pinglier helped me to follow in Jefferson's footsteps through southern France, as did Jean-Pierre and Claudie Thibault, and Adelaide Kane, who drove my wife and me over a route that was no less hair-raising for us in a small car than it must have been for Jefferson on a mule.

At first, I planned to write a book about Jefferson's five years in France, but I was finally and wisely persuaded to expand the work to a full-dress biography by my editor, John Macrae, a man with as keen an eye for biography as anyone I have met. Jack's many helpful suggestions drawn from his apparently inexhaustible knowledge of American history, his generosity with his time, and his gentle yet firm editing more than once helped my flagging courage in taking on the daunting life of Thomas Jefferson.

At every turn, I have received kind assistance from the staffs of archives and libraries that held Jefferson's disparate and voluminous papers, especially at the Henry E. Huntington Library in San Marino, California, at the New York Public Library, at Princeton University Library, at the Thomas Jefferson Memorial Association, and at the Alderman Library of the University of Virginia, which made available microfilm and doctoral dissertations through interlibrary loans. The shared resources of cooperating libraries make works such as this one possible. My deep thanks to Patricia Mardeusz, head of interlibrary loans at the University of Vermont, for tracking down Jefferson materials available on loan from the Universities of Virginia, Minnesota, and Chicago, Mount Holyoke College, the Massachusetts Historical Society, and the New-York Historical Society. Special thanks to Joe Ryan for helping me to search databases to develop several bibliographies, to Nancy Crane and Monica Racine of the Bailey/Howe Library staff, and to Robert Dean, for his sharp-eyed vigilance.

Helpful colleagues at the University of Vermont include Dean Howard Ball of the College of Arts and Sciences, Associate Dean James Lubker, and Professor Patrick Henry Hutton, acting chairman of the History Department. Three research grants from the Dean's Fund and timely funds from the History Department helped to make my archival research and my travels in Jefferson's footsteps easier.

For help at every stage, not only in preparing numerous drafts of the manuscript but in coordinating my efforts to keep the book on schedule while I traveled, I owe great thanks once again to Diann Varricchione. For heroic labors in copyediting, my thanks to Margaret Wolf. Thanks, also, to Nancy Effron, Elizabeth O'Connell, Philip Fitzpatrick, M. Jerome Diamond, Gail Hampton, and Steven Hopkins. To Lawrence Boynton, especial thanks for rounding up hundreds of books. Two people I cannot thank enough: Ray Lincoln, my old friend and longtime literary agent, and my wife, Nancy Nahra, whose numerous suggestions, careful criticisms and extraordinary knowledge of the eighteenth century, its literature, and the classical origins of its thought, have added immeasurably to this effort.

For permission to quote from papers in their collections, I am grateful to the Henry E. Huntington Library; the Bibliothèque de Cessole of the Musée Massena, Nice, France; The Bailey/Howe Library of the University of Vermont; the William L. Clements Library of the University of Michigan; the Virginia State Library; the Newberry Library; the Massachusetts Historical Society; the New York Public Library; the New-York Historical Society; Princeton University Library; the Yale University Library for permission to quote from *The Papers of Benjamin Franklin,* 28 volumes, edited by Leonard W. Larrabee et al.; and from the *Papers of Thomas Jefferson,* 24 volumes, edited by Julian Boyd, Charles T. Cullen, John Catanzariti et al., published by (or in press from) Princeton University Press. To Professor Boyd and his successors, I and the American people owe an enormous debt.

Introduction

"JEFFERSON IS AN old, old subject," noted Merrill Peterson in 1960 in *The Jefferson Image in the American Mind,* "but the quest for the historical Jefferson, under the formal discipline of scholarly inquiry, is young."[1] Since 1950, when Julian Boyd began to gather some twenty-eight thousand Jefferson letters and documents at Princeton University and, with a small staff, commenced to painstakingly edit and publish them, experts have not only been subjecting each original Jefferson document to critical analysis but have been challenging earlier assumptions about Jefferson's life, thought, motives, and work.

A prime example of this laborious process is their research into Jefferson's *Literary Commonplace Book,* one of a series of notebooks Jefferson filled with passages that he copied from Greek, Latin, French, and English books throughout his life. They have proven that earlier profiles of Jefferson have been based on flawed translations of classical passages and on misinterpretations and halfhearted biographical analysis written between 1925 and the present day. Many earlier works have sidestepped the problem of dating Jefferson's three commonplace books. Dumas Malone, author of a compendious six-volume study of Jefferson, argued in 1948 that "the dates of the various entries in the notebooks cannot be determined with entire confidence; and, even if they could be, the works that he cited cannot be assumed to have been the source of his ideas. . . . It is a more important fact that he read widely and stored his mind with riches than that he abstracted particular writings."[2] Yet Malone, without carefully studying Jefferson's periodic handwriting, attempts in several instances—mistakenly—to link *Commonplace* entries with Jefferson's thinking on such major topics as Christianity and the law. Malone was struggling with an old problem. Paul Leicester Ford, turn-of-the-century editor of Jefferson's writings, misdated Jefferson's studies of the law by as much as fifteen years. He erroneously interpreted many events on his assumption that the *Literary Commonplace Book* dated from 1764, when Jefferson was twenty-two, whereas Douglas L. Wilson, editor of the new edition, published in 1989 by the Jefferson Papers project at Princeton, demonstrates that Jefferson began it eight years earlier, at age fourteen, while his father was still alive.[3]

Wilson, after patient decades of literary detective work, has discovered that, as an old retired statesman editing his personal papers, Thomas Jefferson cut up, rearranged, and then rebound the literary notebooks he had kept as a young man half a century earlier. Jefferson's edited and rebound version, long accepted by historians such as Chinard, Malone, and Fawn Brodie as being in its original order, created the impression that Jefferson had copied down seemingly misogynistic remarks from Milton, Pope, and from classical Greek and Latin writers after Jefferson flirted timidly and unsuccessfully with Rebecca Burwell when he was twenty-one. Aided by other scholars, historian Wilson has established the dating of watermarks in the paper of the *Literary Commonplace Book*, and has discovered that there were eleven different and often experimental styles of Jefferson's evolving handwriting. Wilson then matched the passages with Jefferson letters, whose dates were known, to establish further the period of each passage. He was able to conclude that Jefferson had copied out many of the supposedly antifeminine passages when he was a mere boy of fifteen or sixteen who was chafing under his mother's reimposed authority after his father's sudden death.

Thus, part of the problem has been that many books about Jefferson have built on a flawed foundation. The *Literary Commonplace Book* remained in Jefferson's family until 1915, when it was acquired by the Library of Congress. It was edited for the public in 1928 by Gilbert Chinard, who titled it *The Literary Bible of Thomas Jefferson*. Chinard considered the commonplace book "Jefferson self-revealed,"[4] but he translated and transcribed carelessly, noted only a "certain progress" in Jefferson's handwriting styles, and made no attempt to give even approximate dates for

Jefferson's entries, all undated in the original notebooks. Chinard attached little significance to Jefferson's varying handwriting styles, or to the fact that Jefferson used one style when copying and another when writing down his original thoughts. Biographer Marie Kimball made the first serious attempt to date the abstracts in *Jefferson: The Road to Glory* (1943),[5] but five years later Malone ignored her analysis when he launched his multivolume biography. Kimball's pioneering if amateur study led Wilson to his landmark work.

Misdating the entries in the *Literary Commonplace Book* has, for one thing, helped to feed the false impression that Thomas Jefferson was a misogynist. The attack on him as antifeminist then merged with criticisms of his private slaveholding throughout his adult life even as he, from time to time, publicly wrote documents and participated in debates that would have abolished or curtailed slavery. Among the progeny of this anti-Jefferson critique have been Fawn Brodie's bestselling *Thomas Jefferson: An Intimate History*[6] and the popular novel, *Sally Hemings*,[7] both based on an alleged sexual relationship between Jefferson and a beautiful slave who purportedly bore him an illicit second family at Monticello, his mountaintop home. Brodie's psychobiography has triggered a barrage of scholarly rebuttals little noticed by the public, even though the use of psychohistory has been largely discredited among scholars. The slave-mistress controversy has also raised many unanswered questions about Jefferson's personal life and the influence of his thoughts and beliefs on his public actions as the framer of the Declaration of Independence, the first Secretary of State, the founder of the Democratic-Republican Party, and two-term president of the United States. Many new questions have arisen that cry for reexamination and for a new interpretation of Jefferson from his youth to his old age.

Other recent scholarship reveals that there is considerable disagreement among writers such as Garry Wills over the sources from which Jefferson drew his thinking when he wrote the immortal Declaration of Independence in little more than two weeks in 1776 and, less than ten years later, the Northwest Ordinance, the pattern for most succeeding state governments, which banned all involuntary servitude in newly created states. It has long been held that, in creating both of these historic documents, the man who was the leading writer of the American Revolution, while attempting quite publicly to abolish slavery, allowed slave labor at Monticello to support him in high style.

As this book will attempt to demonstrate, much of Jefferson's revolutionary political thinking as well as his later diplomatic and presidential writing were influenced by a combination of his studies of Enlightenment *philosophes* and his nearly twenty years of studying the laws of biblical, Greek, and Roman eras up to much more recent writings by scholars of English Common Law and Continental law. His early legal studies were often the result of pioneering cases on slavery, divorce, and religious freedom that he took on without fee and argued before Virginia's highest court. His preeminence as a constitutional lawyer, therefore, helps to explain his selection as the leading legal spokesman of the American Revolution. These early law studies and the court records of them painstakingly exhumed by such legal historians as Frank L. Dewey, W. Hamilton Bryson, Stanley N. Katz, and Charles T. Cullen reveal that Jefferson the young lawyer clashed repeatedly with aristocratic Tidewater land speculators and slaveowners, setting him on the path that led him to write, as the revisor of revolutionary Virginia's statutes, the most sweeping reform of law in American history. His massive three-year overhauling of Virginia's statutes and the 133 new bills he personally wrote made concrete America's theoretical break with his Declaration of Independence.

Jefferson's previously underestimated years as a lawyer on the frontier and his Olympian labors in the law at Monticello between 1776 and 1779 help to explain his election as two-time Revolutionary War governor of Virginia. This brief period is one of the more controversial segments of his forty years of public life, a painful time that ended in his disgrace. Yet as a strong supporter of westward expansion and the advocate of voting rights for frontier smallholders, Jefferson worked secretly at this time to set up new counties in Kentucky and to write the secret orders that enabled George Rogers Clark to seize the northwest country for Virginia under cover of the Revolution. Then, personally breaking the congressional logjam over adoption of the Articles of Confederation, Jefferson paved the way for the diplomatic victory of American negotiators in Paris, which doubled American territory at the end of the Revolution.

A better understanding of Jefferson's lifelong conviction that western land was linked to American democracy, its freedoms, and the prosperity of a majority of its people helps to explain many of his greatest contributions. From writing the first draft of the Northwest Ordinance with its ban on slavery in the vast new federal territories to his (quite unconstitutional) Louisiana Purchase, which again doubled American territory, Jefferson always had his eyes on the prize of cheap land as the key to American freedom and economic independence from the Old World he so distrusted. All his life, Jefferson faced west, pushing the United States to the Pacific Rim nearly two centuries ago.

Yet his policies as Secretary of State and president had everything to do with a thorough grounding in the laws, customs, and trade of the European nations that dominated his world. No president before the twentieth century was so well traveled or so well prepared to create foreign policy. Earlier biographers have paid too little attention to Jefferson's years in Europe between the American and French revolutions. It is a mistake to take at face value his careful claims that he was only traveling for his health or merely to study the remains of Roman architecture. While, indeed, he had a discerning eye for anything that contributed to his efforts to make the new American republic look less like the mother country, England, which he so thoroughly detested, there was more to those travels—with vast consequences for American trade, diplomacy, and military preparedness—than met the prying eyes of the French spies always planted around him. The reexamination of Ambassador Jefferson's five years in Europe is absolutely necessary to understand his ensuing two decades as Secretary of State, vice president, and president.

Along the way, it has been necessary to study several seeming contradictions raised by opposing schools of historical thought. Did Jefferson's public versus his private views and actions on the haunting question of slavery make him nothing more than a thoroughgoing hypocrite? Was Jefferson so impecunious after his long public life that he had to sell his slaves rather than free them, as did Washington and Franklin? Was the spokesman of American freedom in the Revolution a tyrannical trampler of civil liberties when he came to power as president in what he was fond of calling the Revolution of 1800? Is his invention of the modern Democratic Party and the utter routing of the other Founding Fathers' political party, the Federalists, indeed a sweeping revolution that shattered the party of Washington and opened up American politics to the New Man from the frontier?

While much research remains to be done on the voluminous Jefferson papers, after forty years of study there are many new conclusions that can be drawn, enough at this stage to warrant a reinterpretation of the life of Thomas Jefferson.

Thomas Jefferson

1

"I Cannot Live Without Books"

My father's education had been quite neglected but, being of a strong mind, sound judgment and eager after information, he read much and improved himself.

—Thomas Jefferson, *Autobiography*

At the age of seventy-seven, when Thomas Jefferson sat down to write the story of his life, he said that his earliest memory was of being handed up to a slave on horseback and carried on a pillow from his father's frontier farm in Albemarle County, Virginia, to a far wealthier plantation fifty miles to the east near Richmond. Whenever Jefferson began to reminisce in letters or during visits to Monticello by his relatives and friends, he invariably included this first memory.

Seventy-five years later, he still remembered that long-ago, bone-jarring ride down a narrow, rutted, red-clay road past fields where tobacco leaves were maturing in the sun, as his father left behind his home and privacy to manage the estate of the Randolphs at Tuckahoe, there to fulfill a friend's deathbed request to raise his children. Through much of his own eighty-three years, Thomas Jefferson would be asked to leave his home to serve his country and, as he began to record his memories for public consumption, it evidently struck him that this was what had always been expected of him, no matter the personal cost. So much of what he had done had been influenced by his father, even if there had been so little time to get to know him.

Thomas Jefferson was born in a simple, one-and-a-half-story frame farmhouse in the middle of a horseshoe of outbuildings in a clearing on the edge of the Virginia wilderness on April 13, 1743. It was the season of dogwood and honeysuckle blossoms white-speckling the dark forests of the Piedmont. The Jefferson farm overlooked the still-clear Rivanna River, the north branch of the James. On a bright spring morning, from the veranda of Peter Jefferson's house, he could see through a gap in the Southwest Mountains to the distant Blue Ridge. No other habitation obscured his view. There had been only three or four other farms in the county when Peter Jefferson had arrived to survey his wilderness acres.

In the autobiography that Thomas Jefferson wrote nearly a century after his father, Peter, had staked out the family lands, the man who was the author of the Declaration of Independence made light of the genealogy that obsessed so many of his Virginia neighbors:

> The tradition in my father's family was that their ancestor came to this country from Wales, and from near the mountain of Snowdon, the highest in Great Britain . . . [My mother's family] trace their pedigree far back in England and Scotland, to which let every one ascribe the faith and merit he chooses.[1]

Thomas Jefferson poked fun at pedigrees and family connections. Shortly before he married, he wrote to his agent in London,

> I have what I have been told were the family arms, but on what authority I know not. It is possible there may be none. If so, I would with your assistance become a purchaser, having [Laurence] Sterne's word for it that a coat of arms may be purchased as cheap as any other coat.[2]

Jefferson, according to one recent biographer, was a direct descendant of Welsh kings, but if he knew it, he would have hated the disclosure of such a fact as much as he came to hate more recent kings of England. Yet he benefited mightily from connections and pedigree in a society that required them. Whether he knew it or not, he was descended, on his mother's side, from a lord chief justice of England; from one of the barons who signed the Magna Carta in 1215; King David I of Scotland; Hugh Magnus, leader of the First Crusade; Alfred the Great; and Charlemagne. Fortunately, too, for him, there had been Jeffersons in Virginia at least as early as 1619, when one of his name had sat with the first House of Burgesses, the earliest representative assembly in America. As a young student of the law, Jefferson had "noted once a case from Wales in the law reports where a person of our name was Secretary to the Virginia Company," which established the first permanent English settlement in the New World.

"The first particular information I have of any ancestor was of my grandfather,"[3] Jefferson wrote in his memoirs. Jefferson seems to have known nothing of his great-grandfather, Thomas Jefferson I, a yeoman farmer who was also the first of his family to move to the western frontier of Virginia. Thomas I was married to Mary Branch, the granddaughter of Christopher Branch, who had been a member of the House of Burgesses and had amassed at least 1,380 acres. In April 1682, Thomas Jefferson I purchased 167 acres in Henrico County near the falls of the James. A skilled hunter, he collected bounties for killing wolves. He also learned surveying, thus establishing a family tradition. Accumulating a modest fortune, he was nevertheless referred to as "mister," rather than "esquire," indicating he was less than prominent in the colony. Trusted by his neighbors, he was frequently asked to serve on juries and several times was executor of his neighbors' estates. The Henrico County Court appointed him surveyor of the public highways. His marriage to Mary Branch brought him a son, Thomas, a daughter, Martha, and one-fifth of his wife's father's property. By the time he died in 1697, Thomas I owned several slaves, enough to produce nearly two tons of tobacco a year. His estate included such hints of luxury as "an old silver dram cup, buttons and shoe buckles."[4]

By the beginning of the eighteenth century, his son, Thomas II, had acquired the status of a gentleman. One mark of his entry into the gentry class was ownership of a racing mare called Bony, who won at least one high-stakes race. Another was his appointment as a "gentleman justice" of Henrico County. His friends included aristocratic Randolphs. He became a captain of the county militia and, after musters, dined with the other militia officers. At one roast-beef dinner at his house, Jefferson played host to the wealthiest men in Virginia, including Col. William Byrd of Westover, owner of some 170,000 acres. Captain Jefferson supervised the building of a church long called "Jefferson's Church" in Bristol Parish. The captain was the first Jefferson to make a family connection with the land-rich Randolphs: he married Mary Field, the daughter of Maj. Peter Field and Judith Soane, herself daughter of a former Speaker of the House of Burgesses. Like many Virginia planters, he received large grants of the colony's open lands for importing laborers. At first they were white indentured servants who paid off their ship's passage to America and received small farms from the colony in exchange for seven years of hard labor. According to the quitrent rolls of 1704, Captain Jefferson owned 492 acres in Henrico County. In 1718, he received a one-fourth share of one such fifteen-hundred-acre grant of western Henrico land on Fine Creek from Governor Spotswood. The captain lived out his life at Jefferson's (later renamed Osborne's) Landing, adding his wife's inherited lands to his own and making more land purchases outright. By 1718, he had acquired enough social status to be elected sheriff of Henrico County.

His fourth child was Peter, the father of President Jefferson. Born February 29, 1708, Peter Jefferson was only eight when his mother died and, throughout his childhood, he had no expectation of inheriting his father's land. Under Virginia's laws of primogeniture and entail, all his father's lands would pass to his firstborn brother, Thomas III. One result was that his father wasted no education on him. But Thomas III died at sea at age twenty-three when Peter was fifteen. By then, his father had lost much of his wealth in a fire that left him temporarily threadbare, at least enough to petition the House of Burgesses for relief. Later, when he could not honor a debt of 6,480 pounds of tobacco, the specie of the colony, the Henrico County Court attached his plantation for debt. Fortunes based on land speculation were fragile. By the time Peter Jefferson inherited his father's lands, he could not stay on his home farm. He also inherited "two Negroes Farding and Pompey,"[5] a chest of clothes, six silver spoons, weapons, furniture, horses, cattle, sheep, and hogs.

As his father's executor, he was able to hold on to the undeveloped western tracts on Fine and Manikin creeks in the new frontier county of Goochland. He moved west above the fall line of the James River with his father's remaining possessions, including horses, cattle, and pigs, and built a modest house on Fine Creek. At the time there were only four or five other families living on hardscrabble farms hacked from the dense forests of what was to become Albemarle County. For ten years, Peter Jefferson, a large and powerful man, cleared land and built a house and planted crops and sought out and surveyed more choice land. He became one of the first justices of the peace and then sheriff of Goochland County. He ranged far and wide as a surveyor who was becoming legendary for his courage. Of his mapping and surveying expeditions, Sarah Randolph recounts, "Jefferson and his companions had often to defend themselves against the attacks of wild beasts during the day and at night found but a broken rest, sleeping, as they were obliged to do for safety—in trees."[6] In his travels he frequently called at Tuckahoe, the home of the county's first family, the William Randolphs. William was, as Colonel Byrd observed, "a pretty young man"[7] some five years younger than Peter Jefferson, who became his closest friend. (According to a Randolph family tradition, the Randolphs of Tuckahoe were descended from John Rolfe and Pocahontas.) For ten years William and Peter served as magistrates and militia officers together, exchanging lengthy visits.

Farther west up the James River, about forty miles above Richmond at the edge of settlement, was Dungeness, the plantation of William Randolph's uncle, Isham, a member of the House of Burgesses and adjutant general of the colony. His tall, slender daughter, Jane, was nineteen when Peter Jefferson, massive, freckled, red-haired, and somber, courted her in 1739. Isham Randolph, the maternal grandfather of President Jefferson and a graduate of the College of William and Mary, was not only a planter but a merchant sea captain who imported slaves and indentured servants and traded with England. As agent for Virginia in England, he had resided there for a few years when his close friend, Col. William Byrd, the first American nominated to the Royal Society, had made a London sojourn. Captain Randolph, like Byrd, had made friends in the amateur scientific circles of London. He met and married Jane Lilburne Rogers, a distant and wealthy relative of the seventeenth-century Puritan radical "Free-born John" Lilburne. By Virginia standards, Jane Randolph's family made the Jeffersons seem like rude countryfolk. Her father's British forebears had long been country squires in Northumberland and Warwickshire and were allied with the powerful Scottish clan of the Earls of Murray. An uncle of the first Randolph to emigrate to Virginia in 1660 had been the poet Thomas Randolph, who wrote "An Ode to Master Anthony Stafford" in praise of country life while an undergraduate at Cambridge and became a drinking crony of playwright Ben Jonson. Randolph had proposed that his philosophy of drinking be added to the university syllabus, a theme he amplified in *The Drinking Academy* (1626). His play, *The Muses' Looking-Glass* (1630), lampooned Puritan opposition to the theater.

Peter Jefferson and Capt. Isham Randolph admired each other. Captain Randolph, whose own father, the president's great-grandfather, had arrived a poor emigrant to America, had "the polished and courteous manners of a gentleman of the colonial days," according to one family historian, "with a well-cultivated intellect."[8] Several members of the Royal Academy paid tribute to Isham Randolph's abilities as a naturalist. As botanist John Bartram of Philadelphia, who was traveling all through British America gathering specimens, was preparing to visit Virginia, Peter Collinson of London urged Bartram to look up Randolph: "I know no person who will make thee more welcome . . . to take several days' [botanical] excursions all round."[9] Bartram followed Collinson's advice and, near Dungeness, discovered, as he reported:

> the Arbor Vitae which, if thee could procure some seed thereof, if it growed, would be a curious ornament in thy garden. I doubt not, if thee was to write or speak to Isham, he would procure thee some.[10]

Bartram later reported to Collinson that Isham Randolph had proven "generous" and "good-natured."[11]

Among the suitors for Jane Randolph's hand and dowry, Peter Jefferson no doubt stood out for his sheer magnetism, for his reputation as a rising young planter, and as an explorer, and mapmaker, on the frontier two vital empirical branches of the new science. He also benefited from the introduction of his best friend, Jane's cousin William. In 1735, four years before he courted Jane Randolph, Peter Jefferson had surveyed and patented for himself some one thousand acres of fertile bottomland ideal for tobacco cultivation on the Rivanna River southward from Secretary's Ford and just north of the South Mountains, including the future site of Monticello. He had also already earned for himself a reputation as something of a diplomat. He had his eye on a choice upland tract, across the river, owned by his friend William Randolph of Tuckahoe, as the ideal site for a house. They made a bargain: in return for Williamsburg's Raleigh Tavern owner "Henry Wedderburn's biggest bowl of arrack punch,"[12] Randolph on May 18, 1736, deeded Jefferson two hundred acres and then sold him another two hundred acres for cash. When Peter Jefferson, already a justice of the peace, stood for sheriff in 1738, Isham Randolph posted a £1,000 bond for him; a year later, Randolph agreed to his daughter Jane's marriage to this ambitious young planter. According to family lore passed down by his great-granddaughter, Thomas Jefferson "inherited his cheerful and hopeful temper and disposition" from his mother, "a woman of a clear and strong understanding."[13]

Peter Jefferson apparently never had been to school, but he had taught himself, beyond reading and writing, the fine points of accounting, surveying, and mapmaking. "My father's education had been quite neglected but, being of a strong mind, sound judgment and eager after information, he read much and improved himself."[14] His small library included well-worn copies of Shakespeare, the Bible, Swift, Addison and Steele. The men in his wife's family were all better educated: Jane's father and all of her uncles had gone to the College of William and Mary in Williamsburg.

From the time of Peter Jefferson's marriage to Jane Randolph, education, regardless of the expense, became important in his family. Jane Randolph brought not only refinement to Jefferson's farm, but more slaves from her father's estate: in the year of their marriage, her father became one of Virginia's leading slaveholders by sharing in a consignment of 380 black slaves. She also brought with her a promise of a large dowry of £200 sterling, which was never paid in her father's lifetime, as well as social connections with the Byrds, the Washingtons, and a plethora of Randolphs, the first families of Virginia. Peter Jefferson brought Jane Randolph to the Rivanna with her belongings and her servants in the fall of 1739 and he named his place Shadwell, after the parish in London where she had been born.

According to characterizations passed down by Thomas Jefferson's grandchildren to his nineteenth-century biographer, Henry Stephens Randall, Peter Jefferson was "grave, taciturn, slow to make and not overprompt to accept advances."[15] He was a giant not only in size but in his son's estimation. Jefferson's great-granddaughter, Sarah Randolph, depicted Jefferson's father as "a man of most extraordinary vigor both of mind and body."[16] That Peter Jefferson was in the vanguard of western settlement was a source of pride to his famous son, who was equally proud of his father as a noted explorer. Thomas Jefferson never penetrated the wilderness himself, never challenged his father on his own ground, but he always honored, even romanticized, western explorers. Perhaps they reminded him of Peter Jefferson's legendary strength and bravery. Jefferson was proud that his father had so

> much improved himself insomuch that he was chosen, with Joshua Fry, professor of mathematics in William and Mary College, to run the boundary line between Virginia and North Carolina, which had been begun by Colonel Byrd, and was afterwards employed with the same Mr. Fry to make the first map of Virginia which had ever been made.[17]

For half of their fourteen years together Thomas was in school or his father was away on land-speculation expeditions. When Thomas was born on April 13, 1743, he already had two sisters. In all, there were eight children who survived infancy, six girls, and two boys. Jefferson did not remember his first years at Shadwell. When he was two years old, his father's close friend, William Randolph, died. Under a codicil to his will, Randolph entrusted his "dear and loving friend"[18] Peter Jefferson as resident executor of Tuckahoe, where the three Jeffersons relocated so that Peter could administer the estate and raise the two motherless Randolph children. By this time, Peter had begun building his own clapboarded one-and-a-half-story house. In September 1744, the Virginia Assembly voted to divide vast Goochland County to make the administration of laws more convenient to settlers. Peter attended the organization meeting of Albemarle County and took his oath as justice of the peace and judge of the court of chancery, becoming one of the ranking gentlemen of the new county. A month later, he was appointed lieutenant colonel of the county militia.

But Peter Jefferson's political career was cut short by the death of his friend, Randolph. While Randolph was undoubtedly wealthier than Jefferson and the managerial post was lucrative, it was nonetheless a selfless gesture for him to leave his own home and lands to care for those of his friend. It was at this time that he moved his family the bone-jarring fifty miles to Tuckahoe with his two-year-old, Thomas, riding on a pillow. Leading his family entourage up the avenue of elms at Tuckahoe, Peter Jefferson took charge of the plantation's seven overseers and managed to coax some 667,336 pounds of tobacco from its rich bottomland during the six years of his charge. In the meantime he left the clearing and cultivation of his own Shadwell to an overseer. In return he seems to have accepted only in-kind payment of food, lodging, and schooling for his growing family.

There were five Jefferson children to feed by the time the family returned to Shadwell seven years later, in the autumn of 1752. But there were also skilled slave artisans working in shops at Tuckahoe. An inventory of the personal property in Peter Jefferson's will indicates he made good use of their services. His will included two desks (at a time and place where having one was rare), two bookcases (one cherry, one walnut), three large walnut tables (two of them five feet long and one four feet), six other tables, two chairs, two dressing tables and two dressing mirrors, a large mirror, five four-poster beds, and a chest of drawers. Of this, eight tables, the three dressing tables, and fourteen chairs were made in 1750 to furnish the Jefferson quarters at Tuckahoe.

At Tuckahoe, in the airy, high-ceilinged rooms of the capacious H-shaped white-frame Randolph mansion with its solid Flemish-bond brick ends, it appears that the Jeffersons took up residence in one gabled wing and the Randolphs in the other. Each two-story wing had four large rooms on a floor filled with beautifully carved pine-paneled woodwork; between the wings was a large black-walnut-paneled salon. In the evenings, porches front and back, overlooking lands and riverfront, beckoned with their breezes. It was here, as a child, that Thomas Jefferson first imbibed a taste for luxurious living; it was probably the most spacious house Jefferson ever lived in.

Tom's father was away much of the time, literally carving his reputation from the Virginia wilderness. Tom was only three years old when Peter Jefferson, now called "the Colonel," set off toward the Blue Ridge with forty men on horseback through mountains and forests to extend the Fairfax Line, which delineated the vast holdings of Lord Fairfax from the colony of Virginia, across the Northern Neck of Virginia. They drew their chains for seventy-six miles from the headspring of the Rappahannock across mountains and primeval forests to the headspring of the Potomac. According to the diary of Thomas Lewis, an Augusta County surveyor, the expedition set out on September 10, 1746, with Peter Jefferson, surveyor of adjoining Albemarle County, in the lead with Royal Surveyor Fry. Lord Fairfax and Col. William Beverly, the founder of Augusta County, between them held patents to most of the Shenandoah. The expedition took on the pace and trappings of a royal progress, Fairfax and Beverly holding court in an open field as "a great number of the neighboring gentlemen" paid their respects before the expedition set out. After a Sunday sermon, a hearty dinner, and some "very good cider," there was a furious brawl with fencerails. Hung over from too much "very good cider," some of the less hardy "took sick" at the sight of the mountains they had to climb. At last, heavily laden horses and hungover men struggled up and over New Market Gap. To these flatlanders from the Virginia Tidewater, the Blue Ridge seemed "exceedingly high and very rocky." They remained on horseback as much as possible over ranges of mountain "prodigiously full of fallen timber and ivy," down precipices where horses slithered and fell, and through tangled swamps, including one they called Purgatory, "that struck terror in any human creature." Here, one surveyor nearly drowned. All of the horses starved; many assistants, some much younger than Jefferson, dropped out or were left behind.

After five months of surveying and carousing, the party completed the survey of the Northern Neck, the surveyors returning with Peter Jefferson to Shadwell by February 22, 1747, to draw up their maps. Older than the rest at age thirty-nine, Colonel Jefferson survived the wilderness ordeal only by force of his personality and will, ending the combined 150-mile mountain trek "much indisposed."[19]

Tom Jefferson never saw the terminal point where his father carved "P.J." in a beech tree near the Fairfax Stone on November 15, 1746, but he would later stand on his father's shoulders as a naturalist. He learned from his father that the northern mountains were "not solitary and scattered confusedly over the face of the country," as he recounted forty years later when he wrote *Notes on the State of Virginia,* but were "disposed in ridges one behind another," adding that, "for the particular geography of our mountains I must refer to Fry and Jefferson's map of Virginia," which he had engraved and appended to his book.[20] The expedition, which had taken more than five months through a winter in the mountains, returned to Tuckahoe in February 1747, shortly before Thomas's fourth birthday. The new map of Virginia made possible by the expedition and sent to royal officials in England was drawn by Peter Jefferson and his partner, Robert Brooke, in 1751. Three years later, Colonel Jefferson and Joshua Fry set out again to extend the dividing line between Virginia and North Carolina for ninety miles to Steep Rock Creek in present-day Washington County. Peter Jefferson's explorations also added significantly to his wealth. He was paid £300 for the Virginia–North Carolina boundary survey plus all expenses and £150 for drawing the map, a large sum for the times.

When Colonel Jefferson returned from these dangerous missions, he evidently recounted them for his Tom and the other children. Years later, Peter Jefferson's great-grandchildren repeated them for Jefferson's biographer Henry Randall. There were stories of fighting off wild animals, battling and killing a family of bears, sleeping in trees, running out of food, and surviving on raw meat. The tales, stressing Peter Jefferson's courage and tenacity, must have been transmitted by his admiring son.

Tom was also proud that he had learned surveying from his father and he, in his turn, taught it to his nephew, Meriwether Lewis. The opening lines of Thomas Jefferson's only book, *Notes on the State of Virginia,* were written by a true surveyor:

> Virginia is bounded on the East by the Atlantic: on the North by a line of latitude crossing the Eastern Shore through Watkin's Point, being about 37° 57′ North latitude.[21]

In his book, he used a map based on the one drawn by his father. He showed it with pride to the Hessian officers who were prisoners near Monticello in 1779. According to William Peden, editor of a recent version of Jefferson's *Notes,* he "frequently declared that his map was of more value than the book in which it appeared."[22]

Peter Jefferson's explorations brought him into the inner circle of the new aristocrats of the Virginia Piedmont; he was fast becoming the wealthiest squire in Albemarle County. He joined the largest land-speculating group in the colony, the Loyal Company, which received an unprecedented eight-hundred-thousand-acre grant, mainly in present-day Kentucky. His cronies included his closest friend, Dr. Thomas Walker, a physician and surgeon practicing without a medical degree; John Lewis, representing the Shenandoah Valley; five Meriwethers, related by blood to almost everyone of importance in western Virginia; Francis Thornton, a member of several land schemes; Edmund Pendleton, Virginia attorney general and front man for powerful Williamsburg interests. The company, formed by Speaker of the House of Burgesses John Robinson, received the speedy approval of the royal governor and the Governor's Council. While the official purpose of the colony's lands was to be granted to settlers in fifty-acre freeholds, the Loyal Company was not required to settle any lands and was purely speculative. In the words of George Mercer, agent to London for the rival Ohio Company, Peter Jefferson and his friends were well-connected "private land-mongers who were incapable of making effectual settlements."[23] Over the years, Peter Jefferson drew ever closer to this western in-group.

Before he died, William Randolph left instructions that his children were not to be sent away to school to England or to Williamsburg but were to be educated at home. Colonel Jefferson was to hire a tutor to teach them in a little schoolhouse that still stands in the yard at Tuckahoe. The tutor apparently taught boys and girls together. By age five, Tom Jefferson had joined his two older sisters, the two young Randolphs, and several cousins in the classroom of what he afterward referred to as "the English school," typical of the one-room schoolhouses where reading, writing, and arithmetic were taught.

Few records remain of Jefferson's childhood, but it appears that he saw his father reading incessantly and laboring over account books and maps from Tuckahoe's library to supplement his own stock of books. Family legend has it that Tom read all of his father's books by the age of five, *before* his father sent him to school, including the Bible, Rapin's *History of England,* and *The Spectator.* Not only was Tom exposed to his father's strong example of self-improvement but he had religion drummed into him by his tutor. Years later, as governor of Virginia, Jefferson was to crusade against established state religion, but as a boy he had to learn his prayers by rote as if they were sums. One of the few images of him praying comes down through his grandchildren (which means he probably handed down the story himself). Desperately hungry as any five-year-old might be after a long school day, Tom ducked behind the schoolhouse and prayed that dinner might be hastened along.

By the age of five, Tom was certainly becoming aware of the enforced differences between blacks and whites without being formally taught. At Shadwell, there had been only a small number of black servants; at Tuckahoe there were more than one hundred slaves—household servants, artisans, and field hands—ten slaves for every free white. One of the inequities of the system was driven home when Tom's father permitted him to go to school on the plantation while his black playmates could not. Put another way, Tom *had* to go to school each day and was no longer free to play with his black friends. Even as a child, Tom was learning to rule over blacks, as he remembered when he wrote in *Notes on the State of Virginia* about the method of inculcating the tyranny of slavery in the white child:

> The whole commerce between master and slave is a perpetual exercise of the most boisterous passions, the most unremitting despotism on the one part, and degrading submissions on the other. Our children see this, and learn to imitate it; for man is an imitative animal . . . The parent storms, the child looks on, catches the lineaments of wrath, puts on the same airs in the circle of smaller slaves, gives a loose to the worst of his passions, and thus nursed, educated, and daily exercised in tyranny, cannot but be stamped by it with odious peculiarities.

He had just resigned as governor of Virginia when he began to write *Notes.* He outraged his white neighbors and relatives when he heretically proposed that all black youths be educated "to tillage, arts and sciences" till age twenty-one, all girls to eighteen "at public expense."[24]

In the small, white frame schoolhouse on the grounds at Tuckahoe, Tom and his older sisters, Jane and Mary, learned to read and pray, but Colonel Jefferson's idea of the proper training of a boy was not confined to learning from books. In the evenings he taught his son his own exquisite penmanship, his veneration of books, his love of mathematics. He also saw to it that Tom learned the delights of the naturalist familiar with animal and plant life in the forest. From his father, Tom learned to swim and ride, fish and stalk game throughout the Southwest Mountains.

There are some things that cannot be transmitted. Tom was not powerfully set like his father, whose strength was prodigious and legendary, but was instead thin like the Randolphs. Sarah Randolph recounts of the father:

> So great was his physical strength that, when standing between two hogsheads of tobacco lying on their sides, he could raise or "head" them both up at once. [According to Henry Randall, a hogshead weighed one thousand pounds at the time.][25]

When three slaves were unable to pull down a dilapidated shed, Colonel Jefferson took the rope and hauled hard, tearing the building down. Thomas Jefferson loved to tell his grandchildren this particular story. Colonel Jefferson gave his son a lifelong appreciation of exercise to balance his sedentary love of books, frequently telling his son, "It is the strong in body who are both the strong and free in mind." He also taught his son to be self-reliant: "Never ask another to do for you what you can do for yourself."[26]

But it was with his favorite sister, Jane, nearly three years older than he, that Tom enjoyed long walks without parental pressure. Together they wandered in the lowlands and searched for bluebells and picked their way through the woods, looking to see the wild violets open. More than anyone else, Jane encouraged her younger brother's reading, cultivated his taste for music, and taught him to sing the psalm tunes.

In his autobiography, however, there is scant mention of Jane or of his childhood, scant mention of his mother, only the dates of her birth, marriage, death. But Jefferson had a lifelong aversion to revealing his personal life except to members of his own family, and then only discreetly, and he was always silent about his grief at the time of death of people close to him: his father, his mother, his wife, his children. Most other traces of his childhood were obliterated when Shadwell burned in 1770, destroying legal papers, correspondence, records, accounts, most of the family library—almost all trace of the crucial formative period of Thomas Jefferson's childhood and youth and the development of his extraordinarily complex personality. Only the commonplace notebooks and a few Greek and Latin texts he had carted off to his lawyer's digs in Williamsburg survived the fire. But, as long as Thomas Jefferson lived, he believed that the most powerful influence on his young life had been undoubtedly his father, and he wrote and spoke of him with admiration.

Yet when Jefferson composed his autobiography as an old man, it was only as a mapmaker and surveyor that he honored his father to the public. He never mentioned his status as a fourth-generation member of the Virginia House of Burgesses, a justice of the peace and sheriff of Albemarle County as well as judge of the Court of Chancery and lieutenant colonel of the militia in a frontier county during a war with the French and Indians. Virtually all of these were accomplishments he would equal and surpass in his own lifetime. He stressed not Colonel Jefferson, emerging leader of the increasingly important Virginia back country and major land speculator, but Peter Jefferson, man of science, a man of "strong mind, sound judgment, and eager after information," one who "read much and improved himself,"[27] a father who was like himself.

Perhaps because he had the advantage of all his father's attributes, Thomas Jefferson emulated his father and improved himself prodigiously until he became America's most learned president, its best-read leader, one of its most distinguished men of science. But he also had a debt to his mother, who had brought education and refinement to the Jefferson family as well as physical grace and elegant manners. These were debts he gave no public hint of honoring. Indeed, he seems never to have come to grips with his coolness and ingratitude to his mother and, except for one brief period in his life, in France, seems never to have enjoyed a warm relationship with a woman, not even his wife.

Thomas Jefferson's years close to his father at Tuckahoe were over when he was nine. The family, with a sixth child on the way, returned to Shadwell. Here, his father set about expanding his plain Virginia farmhouse into a spacious country house to accommodate his brood. Colonel Jefferson's account books show a great deal of construction taking place by the time Thomas was ten: to live in the dust and tumult of constant construction came to seem a normal thing to this boy. There was the clatter of putting on a steep new gabled shingle roof after the years of neglect, the construction of a mill and of various small houses—probably for slaves—and other outbuildings. Peter Jefferson the surveyor undoubtedly laid out the foundations of each building and it is likely he designed the expansion of the main house, then supervised while a professional builder, John Biswell, worked on the house with his two slaves, Jupiter and Samson. They most certainly had an avid young spectator who watched his father watching over all the myriad details. All his life, Thomas Jefferson loved to be building, expanding, changing things.

According to architectural historian Jack McLaughlin, houses like Shadwell "were patterned after rural English dwellings."[28] A typical Virginia farmhouse was gabled, a one-and-a-half-story wooden house with a fireplace on each end and a central hallway, front to back, dividing the two main rooms on the ground floor. The passage, as it was called, provided much-needed cross-ventilation during the hot Virginia summers. The lower rooms were invariably a parlor for dining and entertaining and a master bedroom. The bedchambers on the second floor had dormer windows and sharply sloping ceilings with poor headroom.

By modern standards, this is not much space, particularly for such a large family of large people. At the time Peter Jefferson was beginning to expand his house in 1753–54, he and his wife had six children and, in 1755, as he finished the improvements, twins were born. From the remains of the house's foundations, it appears that Colonel Jefferson accommodated them by adding a room past one of the end chimneys and possibly by running a shed—without a foundation—the length of the building at the rear. The finished and more commodious house, then, was similar to the restored John Blair House in Colonial Williamsburg, with its outbuildings forming a balanced pair of U-shaped wings, probably the inspiration for the layout of Thomas Jefferson's own mountaintop home at nearby Monticello.

As the only son and an adolescent by the time the house was finished (his brother Randolph had not been born yet), Tom probably had his own room upstairs, away from his seven siblings, when he came home on weekends from school. But Tom saw little of his father now: Colonel Jefferson spent four days out of every week riding over to Castle Hill where he preferred the company of his closest friend, Dr. Thomas Walker. Together they had traced the Virginia boundary lines and now, as principals in the Loyal Land Company, they were staking out claims to eight hundred thousand acres of southwest Virginia wilderness.

Amid the hubbub of his growing home and family, Colonel Jefferson decided to send Tom away from home for the formal education he himself had been denied. The practice in rural Virginia was for a small group of gentlemen to support a tutor, often some recently ordained young clergymen who could not make a living by preaching alone, who took his pupils into his family, educated, fed, and housed them for a modest sixteen pounds a year. George Washington chose an Anglican clergyman, the Rev. Jonathan Boucher, to prepare his stepson, Parke Custis, for college; James Madison was instructed by a famous Scottish tutor, Donald Robertson. The young tutors often made desperate promises to attract pupils: one young clergyman advertised in the *Virginia Gazette* that he

> proposes to teach Ladies and Gentlemen the French, Latin, Greek and English languages, Book-keeping by double entry, Algebra, Geometry, Measuring, Surveying, Mechanicks, Fortification, Gunnery,

> Navigation, and the use of the Globes and Maps, after a natural, easy and concise Method, without Burthen to the Memory.[29]

But not every tutor was equally well qualified or fated to receive credit in later years for having molded the character and mind of a great man. In Thomas Jefferson's case, he had two tutors, one whom he detested, the other he honored. When Jefferson was nine, the year after the family moved back to Shadwell, his father sent him to begin five years of schooling with the Rev. William Douglas, at Dover Church, five miles beyond Tuckahoe, where he could be close to his Randolph cousins. Douglas had been tutor for a while for the Monroe family, and Peter Jefferson was the vestrymen in one church where he preached. Douglas was the first of several Scots who were to teach and influence Jefferson. The young Reverend Douglas taught him some French, Latin, and Greek—with a Scottish burr. Years later, as American ambassador to France, Jefferson still was embarrassed by his Scottish accent and worked hard to lose it. More years later, he recalled being served up dinners of moldy pies and lessons by a teacher who was "a superficial Latinist less instructed in Greek." Reverend Douglas somehow contrived, however, "with the rudiments of these languages [to teach] me the French."[30] By the age of fourteen, Tom could read Greek, Latin, and French literature in the original.

Young Tom Jefferson was on his way to developing his own complicated personality, including his own style of clever wit, which he displayed in his weekend visits home. When his father, who had insisted on turning Tom loose in the woods with a gun to learn self-reliance, insisted that he bring home a wild turkey one day for his dinner, Tom, unable to find any of the elusive birds on the wing, found a penned-up turkey, tied it to a tree with his garter, shot it, and hauled it home without comment (except, years later, to his grandchildren).

Young Tom had also learned to dance minuets, reels, and country dances like any other Virginia gentleman, if somewhat more awkwardly. His favorite amusement, his passion, was to play the violin, not only by ear but by sight reading. When he heard country tunes at dances and fairs, he copied them down in his music books. When his sister Jane sang from Purcell's *Psalms Set Full for the Organ or Harpsichord,* Tom transposed them for the violin. He taught himself music, probably at first from John Playford's *An Introduction to the Skill of Musick.* By the age of fourteen, he was already a competent fiddler and undoubtedly owned his first violin. He played without urging at school, and when he was home he accompanied Jane as she sang for him.

✧ ✧ ✧

Thomas Jefferson liked to hike in the woods, but not to hunt. He was not graceful on his feet, but he learned to ride boldly and wonderfully. He loved conversation, but he couldn't make a speech; mumbling, he became tongue-tied when he made extemporaneous remarks. Perhaps as one result, Jefferson lionized Indians, "a people with whom, in the early part of my life, I was very familiar"[31] who naturally excelled at both conversation and public speaking. In *Notes on the State of Virginia,* he praised and quoted the rhetoric of Indians. He always considered Indians at least the equals of white men. With his father now the colonel in charge of frontier Albemarle County's militia and its representative in the Assembly, it was not surprising that Indians en route to the capital at Williamsburg paid their respects at Shadwell. The most famous chief of the Cherokees was Outacity, who visited Shadwell with 165 other Indians on his way to Williamsburg before he embarked for England, where he created a sensation in London, was received by Oliver Goldsmith, and was painted by Joshua Reynolds.

Years later, in a letter to John Adams, Jefferson recounted his own visit to Outacity's camp outside Williamsburg the night before the famous Indian orator departed for England:

> The moon was in full splendor, and to her he seemed to address himself in his prayers for his own safety on the voyage, and for the welfare of his own people during his absence. His sounding voice, distinct articulation, animated action and the solemn silence of his people at their several fires filled me with awe and veneration, although I did not understand a word he uttered.[32]

In the summer of 1757, when Col. Peter Jefferson had not quite reached fifty, his prodigious strength gave out. Despite or perhaps because of all the eighteenth-century medicine administered by his own closest friend and neighbor, the self-taught and unlicensed Dr. Thomas Walker, who visited Shadwell professionally fifteen times in six weeks that summer, the Colonel died on August 17 of that year at age forty-nine. There is no surviving explanation: was he only slightly ill and then bled and purged to death like Washington? Colonel Jefferson left a thirty-seven-year-old widow and eight children. Tom was only fourteen. His father had left him a small fortune of half his seventy-five hundred acres, the choice of two plantations, his best body servant, another twenty-five slaves when he turned twenty-one, along with the residue of his estate. His father left an estate unencumbered by debt and he distributed his wealth carefully: he did not believe in primogeniture and entail, which would have left Tom far wealthier but the rest of his family and his dependents virtually destitute. Colonel Jefferson had accumulated a sizable estate in a single generation. In addition to the Rivanna River lands, he left personal property worth £316, "stock and slaves" amounting to £2,018, including sixty slaves, two hundred hogs, seventy cows and steers, eleven sheep, and twenty-five horses. His wealth compares favorably with the estate of Thomas Lee, one of the richest planters in Virginia, whose Stratford Hall inventory was filed one year later. Jefferson's house, which he left with its four hundred acres to his widow in her lifetime, was filled with fine furniture. His wife served endless rounds of guests from a considerable silver service: silver coffeepot, teapot and milk pot, silver punch ladle, teaspoons, salts, sugar tongs, and strainers. Colonel Jefferson arranged things so that his widow could attract a new husband, if that were her wish, before turning over the residue of Tom's share to him at twenty-one. In his will, Peter Jefferson did what most eighteenth-century Englishmen did: he refused to give his wife outright ownership of his landed estate. He gave her lifetime use of her widow's third of the estate, including the house and home farm, which she would lose if she remarried. He also guaranteed sizable two-hundred-pound dowries to all his six daughters as well as shares in the land and slaves. Even in his death, Peter Jefferson influenced his son. The nature of his bequests, uncommonly democratic among Virginia planters in the eighteenth century, was not lost on Tom, who was to lead the movement to abolish primogeniture and entail in Virginia as its revolutionary governor. Peter also left to Tom his forty-book library, his bookcase, and the cherry-wood writing desk where he had taught his son accounts, his surveying tools, and the mathematical instruments he had inherited from Joshua Fry. According to family lore, the Colonel's final wish was that his son should continue his classical education. It was for this that Tom was most grateful. If he had had to make a choice between the money and his education, he later told Joseph Priestley, he would have chosen "the sublime luxury" of his classical studies.[33]

But the youthful Jefferson, an awkward and shy boy, was devastated by his father's death. He missed his father terribly, and he transferred his grief into blaming his mother. He sank himself in his studies and in playing the violin. Many years later, in a letter to his eldest grandson, Thomas Jefferson Randolph, Jefferson gave a rare glimpse into his lingering painful memory of the loss of the person he had most loved as a boy:

Thrown on a wide world, among entire strangers, without a friend or guardian to advise, so young, too, and with so little experience of mankind, your dangers great, and still your safety must rest on yourself. . . . When I recollect that at fourteen years of age the whole care and direction of myself was thrown on myself entirely, without a relative or friend qualified to advise or guide me, and recollect the various sorts of bad company with which I associated from time to time, I am astonished that I did not turn off with some of them and become as worthless to society as they were.[34]

2

"I Am Surrounded With Enemies"

To live with them is far less sweet than to remember thee!

—Thomas Jefferson's translation of
an epitaph by William Shenstone

From the age of five Thomas Jefferson's life had been totally circumscribed by men—teachers, schoolmates, cousins, slaves, clerics, uncles, his father—and young Jefferson had enjoyed the privileged position of the firstborn boy surrounded by a horde of sisters in a male, patriarchal society. Now, that advantage was all but stripped away at Shadwell, and for the next seven years, until he turned twenty-one, Tom, as he preferred to call himself, would be subject to the primary control of his mother and the legal counsel of five executors. Jane Randolph Jefferson, left a widow at age thirty-seven with eight children, and fourteen slaves, controlled every aspect of his life at Shadwell, which he began to call "my mother's house."

According to his father's dying wish, Tom was packed off, in February 1758, just before he turned fifteen, to a small classical academy in a log house on the grounds of Edgeworth, the plantation of the Rev. Mr. James Maury. Accompanied by his father's trusted body servant, the mulatto Sawney, the young Jefferson moved the fourteen miles to the small plantation of the Reverend Mr. Maury, to be his surrogate father for the next two troubled years of Jefferson's adolescence. Shadwell was close enough, however, to this new school for Tom to be obliged by his mother to make the return journey every Saturday regardless of weather, to assist her, as the man of the house, in an endless round of southern hospitality that filled the place, at great expense, with the relatives and friends of this amiable young widow. Then it was back to Reverend Maury's for the weekdays under the tutelage of this highly respected Virginia-born Huguenot scholar.

A prickly, frazzled man of letters by the time Tom Jefferson met him, Reverend Maury, in part because he had eight children of his own, was obliged to take in eight would-be classicists—at twenty-two pounds sterling each per year paid in tobacco—from among his wealthier neighbors. To be a "compleat gentleman" in those days required learning Greek, Latin, and possibly French. In addition to the boys and girls he crammed into his house and lined up around his dinner table and taught in his log schoolhouse, Reverend Maury planted his own crops and cultivated the spiritual welfare of the largest, wildest, and farthest-west parish in Virginia, which consisted of three Church of England churches and a chapel scattered over the hilly backwoods of western Albemarle County. Reverend Maury, a cultivated Latin master, was also a slaveowner and a bit of a cynic, giving slaves classical names such as Cato, Clio, Ajax, Aggy, and Memnon. During Tom's two years with Reverend Maury, he was to learn his love for words. This schoolmaster was noted for stressing their precise meanings. Another Virginia dominie of the time, the Rev. Jonathan Boucher, testified to Reverend Maury's fine English prose style. Boucher described his colleague Maury as a "singularly ingenious and worthy man" who encouraged Jefferson to launch his literary career:

> Mr. Maury was of French parents; begotten, as he used to tell, in France, born at sea, reared in England and educated in America. His particular and great merit was the command of a fine style.[1]

The Reverend Mr. Maury wrote, added Boucher, with "force and elegance,"[2] two virtues he imparted to Jefferson, and Jefferson himself later lauded Maury's classical scholarship. Reverend Maury had attended the College of William and Mary and had taught there briefly in its grammar school, then had voyaged to England to receive Holy Orders. His surviving sermons indicate an intelligent, cultivated, almost mystical man who wrote and spoke with great vigor.

Years later, Jefferson acknowledged his "deep and lasting"[3] debt to Reverend Maury's scholarship, especially for introducing him to the study of natural philosophy. A few years before, typically, Reverend Maury, according to his brother-in-law, sent him a letter "with a box containing a piece of antediluvian mud, petrified with the perfect print of a cockle shell upon it, taken from the top of one of the [Blue Ridge] mountains."[4] There, Maury led his eager young students on field trips. Jefferson seems also to have had a natural talent for learning languages. What Reverend Maury did was to encourage Jefferson to read, in the original Latin and Greek, Cicero and Horace, Homer and Plato, as well as the modern French literature of the Enlightenment. He also taught him that his English mother tongue was, as Maury was to write in his *Dissertation on Education,* a language "as significant and expressive, as numerous and musical, nay, to my own ears as enchanting as any that was ever spoken."[5] He taught Jefferson well: Jefferson's prose style was decidedly classical, owing much to Cicero's periodic sentences.

Lanky, freckled, over six feet tall, with red hair and hazel eyes, young Tom Jefferson had two families through these two years, and he intermingled them. He brought home Reverend Maury's son, James, three years his junior, on weekends, and he lived all week as part of the Maurys' noisy, bookish household, relishing its well-stocked library of four hundred books, vast by frontier standards, ten times the size of Peter Jefferson's. He followed the advice Reverend Maury gave to his son as carefully as if it had been his own father's, no doubt listening gravely as the Reverend Maury intoned

> I would recommend it to you to reflect and remark on and digest what you read, to enter into the spirit and design of your author, to observe every step he takes to accomplish his end, and to dwell on any remarkable beauties of diction, justness or sublimity of sentiment or masterly strokes of true wit which may occur in the course of your reading.[6]

The long-suffering Reverend Maury also taught Thomas Jefferson three necessary qualities for either scholar or statesman: caution, self-discipline, patience. Jefferson survived the drudgery largely through the inspiration of Reverend Maury, whose obituary in the *Virginia Gazette* stated, "It might have been hard to say whether he was more admired as a learned man or reverenced as a good man."[7] As an old man, Jefferson still thought of those days, and he wrote of them to Maury's son, his schoolboy friend at Edgeworth, a letter full of longing to meet and "beguile our lingering hours with talking over our youthful exploits, our hunts on Peter's Mountain." Time and other trials had preserved his memories of Maury's schoolhouse and his happy years there. "Reviewing the course of a long and sufficiently successful life, I find no portion of it happier moments than these were."[8]

Tall, serious, and awkward even for his age, Tom Jefferson passed two intellectually exhilarating yet emotionally painful years being pulled between a demanding mother and the man who had become his surrogate father, at times outspokenly preferring the company of his learned teacher to that of his mother. In these two years, at ages fifteen and sixteen, Jefferson nursed a growing hostility toward his mother, one that did not end with adolescence. His only direct mention of her in a public document was in his aborted autobiography, which he was to abandon at age seventy-seven when he appeared to be on the brink of allowing himself to express his personal feelings about her. In his old age, nearly half a century after her death and after losing his own spouse and several children, he remained sarcastic about her family lineage, even though he profited greatly from it. The Randolphs remained a cold, distant "they," even if he, by birth, was one of "them." When, nineteen years after his father's death, his mother died in 1776, there was never any later hint of grieving over her, only a cold obligation, which seems to have had its roots in the months and years immediately following his father's death.

While Thomas Jefferson was an inveterate, even obsessive letter writer and saver, leaving behind some twenty-eight thousand known letters, he apparently burned at the time of her death any scrap of correspondence that may have passed between his mother and himself, those that had not already been destroyed in the fire at Shadwell. Even in his few surviving direct references to her, there is none of the open admiration toward her that he showed to his father. Much of what evidence remains, according to Jefferson scholar Douglas Wilson, is in his *Literary Commonplace Book*, where numerous extracts suggest that he resented his mother after so many years so close to his father. And, in his later writings, there is no recorded sympathy for her repeated severe losses: the deaths of her husband and of three children, the destruction by fire of Shadwell, the home her husband had so painstakingly built for her with all of her accumulated possessions in it. Young Jefferson would be away at the time, and there is never in all his surviving writings a note of pity for a youngish woman left with a brood of eight children and without adult companionship except for her servants and left virtually homeless for an entire winter after the fire, presumably living in outbuildings with five young daughters and a teenage son until travel to relatives became possible. From his fifteenth birthday, Thomas Jefferson was away at school or on business nearly all the time and, when the family home burned, his only superficial reaction was an expression of loss for his books (which included those left him by his father) and his own papers, his apparent chief concern being for his legal briefs and lawbooks.

Was young Thomas Jefferson uncommonly self-absorbed, self-centered, and unsympathetic where his mother was concerned, or was he no more so than many young men of the period? Is there some question in his case of his early establishment of a pattern of brittle behavior toward women, or was he fairly typical of men in the paternalistic eighteenth-century society in which he grew up? Other Founding Fathers give their mothers equally scant notice in their writings: Benjamin Franklin, in his famous *Autobiography*, after pages of anecdotes about his father, his uncle, his brothers, noted only two facts about his mother: she had ten children and she nursed all of them. Jefferson's father's sudden death and the terms of his will, which defied convention by not leaving everything immediately to his firstborn son, may have contributed to young Jefferson's resentment of his mother. His father's death had come at the outset of Jefferson's adolescence, during the period of his life when a boy normally struggles with the problems of dependence and independence and he begins both to break away from parental, especially maternal, control and to assert his masculinity. As if his struggles in Virginia's hierarchical and paternalistic society would not have been difficult enough in his dominant father's shadow, his father's death cast Thomas back under the complete day-to-day control of his mother, her routines, her needs, and her pretenses.

In the two and a half years between his father's death and his temporary escape from his mother to attend college, Jefferson went through a tortured period in which, if Douglas Wilson's analysis of his *Literary Commonplace Book* is correct, he was, by turns, morbid, hostile, gloomy, melancholy, antimaternal and, finally, stoical. Apparently in his years at Reverend Maury's school, young Jefferson could only find an outlet for his frustrations in copying from the readings his tutor provided him as he studied literature. In the pages of his *Commonplace Book*, he protests his mother's authority and he broods about his father's death and about death generally, in fact writing and reacting to passages about death to a remarkable degree. Even had his mother ever examined the thousands of lines Tom copied out under Master Maury's tutelage, she would not have paid much attention to them. Many of them were in Greek and Latin, which she could not understand anyway, and she apparently regarded them only as his assigned homework.

All through his long life, Jefferson was to resort to literary metaphor to grapple with his feelings toward women and often, during the Revolution and as a diplomat, he preferred to resort to secret codes and ciphers to communicate with other revolutionaries. Obliquely, secretly, he turned now to rebellion against his mother, finding the shafts for his bow in the writings of Milton, Shakespeare, Cicero, and Horace, as well as contemporaries such as Pope and Thomas Mallet. Jefferson's entries in his *Literary Commonplace Book* reveal the youthful tendencies of his inner life, as well as his fantasies and his attempts to find, in fictional characters, suitable models for himself. There is, from the *Odyssey of Homer*, Telemachus, son of Odysseus, watching patiently and resentfully for his father's return as his mother's suitors, hordes of them, eat and drink up the family's fortune. He longed for his lost father, quoting Cicero as he wished "the phantoms to speak." In the pages of the *Commonplace Book*, there are powerful struggles between Adam and the Devil, between Caesar and Cassius, Milton's *Samson Agonistes* and his manifold sufferings, friendless and alone, in deep despair and suffering at the hands of a woman.

Jefferson's early abstracts in Latin, in a slanted, italic hand, are the predictable assignments of his teacher, Reverend Maury. They are the most prominent authors in the standard classical curriculum—Horace, Virgil, Ovid, Livy, Cicero. The earliest English excerpts are from Pope, Milton, and Shakespeare. Jefferson also included two contemporary favorites, James Thomson, whose principal theme was rural landscape, and Edward Young, whose primary subject was death. While the books, plays, and poems Jefferson read were often no doubt chosen by his teacher, "the individual selections," as Wilson points out, were "decidedly his own."[9] He did not concern himself with copying the most famous passages or even the most characteristic, but often he chose the moral and philosophical cores of the works. Ignoring the humorous content of Samuel Butler's immensely popular *Hudibras*, for example, which was a particular favorite of Anglican Virginians because it satirized Puritans, Jefferson concentrated on passages about honor. He took them straight, but then, he was fifteen. Jefferson regarded the study of poetry as useful as well as pleasurable, a classical ideal learned from Horace. Poetry, by far his favorite form of excerpt throughout his life, afforded instruction in morals as well as literary style. The most persistent motif in the *Literary Commonplace Book* was death. It appeared all through it, in English, in Latin and Greek, in poetry and prose. In the earliest section, a passage from Horace's *Odes* invokes the specter of death appearing at the door of the rich as well as the poor:

> Pale Death with foot impartial knocks at the poor man's cottage and at princes' palaces.[10]

and Jefferson's final entry, written in his old age, is a kind of farewell:

> What shall betide
> Hereafter, care I not—yea, though this day Death's doom stand by my feet; no man may live For ever; each man's fate is foreordained.[11]

By far the most revealing excerpts about death come from Cicero's *Tusculan Disputations,* which, Douglas Wilson maintains, were entered not long after Jefferson's father died. Cicero had been writing in grief at the loss of his own daughter. As Jefferson faces Cicero's observation that

> All have to die: still there would have been an end of wretchedness in death[12]

he asks the question

> What satisfaction can there be in living, when day and night we have to reflect that at this or that moment we must die?[13]

Jefferson continues to trace Cicero's thoughts on such topics as immortality and death, on the afterlife, on believing in God, on suicide:

> Your true wise man will joyfully pass forthwith from the darkness here into the light beyond.[14]

Writings on suicide and premature death fascinated him:

> For if the final day brings, not annihilation, but a change of place, what more can be wished for? But if on the other hand that day brings total destruction and obliteration, what can be better than to fall asleep in the midst of the toils of life and so, closing one's eyes, be lulled in everlasting slumber?[15]

Jefferson already had developed a sure instinct for style, but it is Cicero's peaceful, accepting philosophy that he comes back to study again and again. A few years later, Jefferson held up an example of Cicero's stoicism to his closest friend, John Page, and in 1816, at age seventy-three, Jefferson opened his *Commonplace Book* again and copied out one of his favorite excerpts to send to Amos J. Cook, a schoolteacher in Maine, to read to students on the northern frontier:

> The most fortunate of us all in our journey through life frequently meet with calamities and misfortunes which may greatly afflict us . . . To fortify our minds against the attacks of these calamities and misfortunes should be one of the principal studies and endeavors of our lives. The only method of doing this is to assume a perfect resignation to the divine will . . .[16]

If Jefferson learned a stoical acceptance of death after his father died, he also studied rebellion, defiance, and bravery. His "passages voicing defiance and rebellion all belong to the early part of the book," writes Douglas Wilson. "No passage of this sort was entered after 1762,"[17] when Jefferson was nineteen and fully two years before his infatuation with Rebecca Burwell, his first flame. He no doubt introduced a note of misogyny when he excerpted Thomas Otway, but it could not have been directed toward a woman he had not met:

> *Wed her!*
> *No! were she all Desire could wish, as fair*
> *As would the vainest of her Sex be thought,*
> *With Wealth beyond what Woman's Pride could waste,*
> *She should not cheat me of my freedom. Marry!*
> *When I am old and weary of the World,*
> *I may grow desperate,*
> *And take a Wife to mortify withal.*[18]

Such defiant outbursts against authority—in this case almost certainly directed toward his mother—show Jefferson's early proclivity for rebellion more than any hatred of women. There is defiance again in his abstracting when he writes out lines from Nicholas Rowe's *Tamerlane:*

> *The fiery Seeds of Wrath are in my Temper,*
> *And may be blown up to so fierce a Blaze*
> *As Wisdom cannot rule.*[19]

But Jefferson was, as he wrote his young friend John Page, learning to control his emotions. In his notebook, he records the stoicism he was imbibing from Cicero: he was "not sprung from rock," but

> We must, however, with all our might and main resist these disturbances . . . if we wish to pass our allotted span in peace and quiet.[20]

He was gradually, deliberately disciplining himself, suppressing his passions, his grief, inducing pacifism:

> Therefore the man, whoever he is, whose soul is tranquilized by restraint and consistency and who is at peace with himself so that he neither pines away in distress nor is broken down by fear . . . he is the wise man of whom we are in quest, he is the happy man. . . .[21]

He kept his *Commonplace Book* to himself, copying out extracts in letters to friends, but never intending it for public scrutiny, least of all by his mother, against whom he fairly rants in its pages.

Young Jefferson learned to study long hours in his years with Reverend Maury. He was remembered by his schoolmates for preparing his lessons before he joined in any games, for keeping his Greek grammar book with him even as he played, and for practicing long hours, often three hours a day, on his violin. There is also a receipt that indicates Jefferson was one of five pupils who took dancing lessons beginning in October 1759, when he was sixteen.

Jefferson was occasionally capable of the unexpected bit of humor. When his friend, elegant young Dabney Carr, owner of a fine racing horse, challenged Tom to a race on his fat old mare, Jefferson accepted. They had in common a love of horses. The year after Jefferson's father died, Jefferson began to record the births and breedings of horses at Shadwell in his *Stud Book.* He was fifteen when his first mare, Allycrocker, was foaled. Sixteen years later, when he began to keep his *Farm Book,* he carefully recorded her genealogy. He noted that she was descended from fast racing stock, her grandsire, Spanker, imported from Andalusia in Spain. For twenty years he bred her, keeping careful notes. It was a few days before the end of the month before anyone noticed that Jefferson had set the racing date for February 30. Carr and Jefferson were to become the closest of friends, riding frequently together to Shadwell where, on long weekends, they kept company with Tom's sisters, Martha and Jane, singing, riding, hiking, and laughing together.

At sixteen, a young Virginia gentleman considered himself mature; at sixteen, George Washington was a full-fledged surveyor. At the end of a gay Christmas holiday spent at Shadwell, during which his sister Martha was married, Jefferson left and rode over to the plantation of Colonel Dandridge in Hanover County. There, he met Patrick Henry and together they "enjoyed a week together at the neighborhood's revelries and season."[22]

Jefferson had already determined to become a learned man, a philosopher. But his two-year isolation on the mountain at Shadwell with a houseful of women and slaves, was, Jefferson had found, a cause of concern to his male uncles and his father's executors. Jefferson wrote his earliest surviving letter on January 14, 1760, just before he turned seventeen, to John Harvie, his guardian, sketching his restlessness to have a life away from his mother and her endless round of social obligations:

> My schooling falling into discourse, [Colonel Randolph] said he thought it would be to my advantage to go to the College and [he] was desirous I should go, as indeed I am myself, for several reasons.[23]

Young Jefferson's classical preparation included the ability to marshal arguments. "In the first place, as long as I stay at the mountains, the loss of one fourth of my time is inevitable by company's coming here and detaining me from school." John Harvie could be counted on to respond to financial considerations.

> My absence will in a great measure put a stop to so much company and by that means lessen the expenses of the estate in housekeeping . . . By going to the College, I shall get a more universal acquaintance which may hereafter be serviceable to me. . . . I can pursue my studies in the Greek and Latin as well there as here, and likewise learn something of the Mathematics. . . .[24]

By age sixteen, Jefferson was already as well educated as all but the front rank of Virginia gentry—and had received as much schooling as many sons of English squires. Only a handful of Virginians went to England to receive either a grammar-school or university education. While most Virginia upper-class fathers clung to the value of a classical education, fewer than three dozen Virginians in the seventeenth and eighteenth centuries combined undertook the expense and the hazards of the journey. But to a young Virginia gentleman of the colonial period, ignorance was not only a disgrace but a handicap in his business affairs, making management of his estate difficult and the supporting of his obligations in the county and its society impossible. A young gentleman of the first rank was expected to attend William and Mary College, and he was expected to work hard for at least one year to learn enough Latin to sprinkle his conversation with classical allusions, to learn to speak and write good English and be a competent letter writer, and to learn enough mathematics to manage his accounts and carry out basic surveying. In general he was expected to be able to conduct himself intelligently in any company. To be otherwise, as Nathaniel Burwell complained about a lazy half brother, was to become "a scandalous person," "a shame to his relations," and a "blockhead" disgraced "before the face of the whole college." To save himself from such disgrace, a young Virginia gentleman was to read and interpret his own collection of lawbooks and then, prepared to settle land or labor disputes, take his place as a magistrate or elected official. Jefferson knew this as he rode over to his cousin's estate at Christmas 1759.

3

"A More Universal Acquaintance"

I had the good fortune to become acquainted very early with some characters of very high standing, and to feel the incessant wish that I could even become what they were.

—THOMAS JEFFERSON, *Autobiography*

SIX FEET TWO inches tall, sandy-haired Thomas Jefferson left Shadwell with his horses, a wagon and his slave, Jupiter, riding one hundred fifty miles from his hilltop home on the fringe of the Virginia frontier to the royal provincial capital at Williamsburg in March 1760. It was his first journey of any length from his home county. Even when his father had taken up his seat in the House of Burgesses, Tom had been left behind in school. Travel was expensive and time consuming and children did not customarily accompany the gentry lawmakers to the capital city of Williamsburg.

Jefferson's emergence from the foothills of Appalachia to become a "compleat gentleman" was part of his new education. As he traveled from the hill country with its rough newness through the generations-old plantations of the Tidewater, young Jefferson was becoming more aware of social class distinctions, which were personified during a stopover he made at the hospitable Captain Nathaniel West Dandridge's plantation in Hanover County. Another houseguest during Jefferson's lingering two-week visit was a big, red-haired upcountry youth, Patrick Henry, who was seven years his senior, had twice failed as a storekeeper, had helped his father-in-law as a bartender in his tavern, and was about to try his luck at law. To Jefferson, Henry had some admirable gifts, among them fiddling, dancing and, in private, easy, endless camaraderie. But there was a coarseness to his manners that endeared him to the rough backcountry hunters with whom he pursued white-tail Virginia deer by day and cracked crude jokes around the campfire at night. Though he came from a minor gentry family similar to Jefferson's, Henry preferred the company of plain country people. Patrick Henry lived nearby and had materialized one day without invitation: this was a perfectly acceptable country visit under the rules of Virginia hospitality. Jefferson thought him a bit uncouth at first but, like the younger boys and girls at Dandridge's, he found Henry's boisterous gaiety contagious, even charming, which may explain why Jefferson uncharacteristically dawdled so long on the way to Williamsburg. Tom lingered, rode, took walks, danced and flirted with the Dandridge girls, perhaps celebrating his temporary emancipation from his mother, and Tom Jefferson and Patrick Henry fiddled and laughed and were on their way to becoming fast friends. For the next several years, Henry would stay with young Jefferson whenever he visited Williamsburg.

Patrick Henry was more typical of the backcountry yeomen Jefferson's father's family had sprung from than was Jefferson himself, who was by now a member of the gentry class. The backcountry gentry lived the modest lives of yeomen farmers compared to the opulence Jefferson increasingly encountered as he made his way to Williamsburg. He wrote many years later of the often-subtle evidences of social cleavage around him. Even then, he found it difficult to spell out all "the difference between the classes of society and the lines of demarcation which separated them." By law, there were no social distinctions "except as to the twelve Councillors" in the upper House of Burgesses. Yet

> in a country isolated from the European world, insulated from its sister colonies, with whom there was scarcely any intercourse, little visited by foreigners . . . certain families had risen to splendor by wealth and the preservation of it from generation to generation by the law of entails [which] had produced [in some families] a series of men of talents. . . . Families in general had remained stationary on the grounds of their forefathers, for there was no emigration to the westward in those days.

Society, according to Jefferson, "scarcely admitting any change of station," had "settled itself down into several strata." There were "no marked lines" to demarcate degrees of segregation, but there was "shading off imperceptibly from top to bottom, nothing disturbing the order of their repose." There were, Jefferson found, "aristocrats, half-breeds, pretenders, a solid independent yeomanry looking askance at those above yet not venturing to jostle them." At the bottom of the white social scale were the slaves' overseers, "the most abject, degraded and unprincipled race," always "cap in hand to the dons" and notorious for "the exercise of their pride, insolence and spirit of domination."[1]

The well-beaten, winter-hardened road that led Jefferson along the James River to Williamsburg was better than most in colonial America. It followed the serpentine river through a landscape that one English traveler found primitive, "one continuous immense forest intercepted by openings where the trees had been cut down and the land cultivated."[2] Plantations were from one to four miles apart, "each having a dwelling-house in the middle with kitchens and outhouses all detailed"[3] because of the danger from cooking over great logs in large open fireplaces.

When he reached Williamsburg, young Jefferson rode down its "one handsome street"[4] just a mile long, the unpaved, crowded, catalpa-lined main thoroughfare, Duke of Gloucester Street. Each end was anchored by an elegant brick building, with the Capitol, at the western end, the sixty-year-old College of William and Mary, based on the design of famed Restoration architect Sir Christopher Wren after his plan for Chelsea Hospital in London, at the eastern end. Whatever Jefferson's initial reaction to the scene, two decades later, after, as revolutionary governor, he had moved the state capital to Richmond and was under the spell of building his own Palladian-style villa atop Monticello, he looked back somewhat askance at Williamsburg. Nearly twenty-five years later, the by-then ex-governor called its public buildings "rude, misshapen piles which, but that they have roofs, would be taken for brick kilns." The single exception to his sharp and scathing architectural critique in *Notes on Virginia* was the porticoed Capitol, "a light and airy structure."[5] The colonial capital was little more than a village of about one mile square and about two hundred houses "of wood chiefly, painted white," the population fluctuating with the legislative season. But by then he found the old Georgian English architecture part of a dead, prerevolutionary past: "The genius of architecture seems to have shed its maledictions over this land." For now, the fresh new collegian drank in an architectural landscape grander in scale than he had ever seen. Many landed gentlemen maintained townhouses in the capital. "The houses stand at convenient distances apart," wrote one German visitor, "and, on account of the general white paint, have a neat look."[6] The burgesses were in town in late March when Jefferson arrived and the streets were jammed with six-horse coaches utterly unseen in the backcountry.

When the Tidewater aristocracy came to Williamsburg for the legislative season, no town in British America could rival it for fashion, high life, or dissipation. Virginians thought it was like going to the Court of St. James in London. Indeed, the society matrons of Virginia arranged a lively social calendar revolving around the Governor's Palace. To visiting Englishmen, Williamsburg represented a familiar scene. Lord Adam Gordon, touring the colonies in 1764, called the Virginia capital reminiscent of "a good Country Town in England." Gentlemen with courtly manners escorted young ladies suitably outfitted at Madam Finette's in rich silks, velvets, laces, and ribbons, which had been spread across counters by obsequious mercers eager to tempt the ladies and their mothers. Gentlemen were no less sumptuary in their tastes. One dandy advertised in the *Virginia Gazette* that he would give a reward for "an elegant toothpick case lately imported from Paris, with a smelling bottle and gold stopper at one end."[7]

Race week, the highlight of the Williamsburg social season, began shortly after Jefferson arrived. The town became even more crowded for the spring thoroughbred races. Gentlemen and wealthy students wagered heavily not only at the track but at cards, dice, and cockfights. Gambling was the besetting vice in Williamsburg, reducing ancient families to the brink of ruin. At the races, purses were raised by subscription and went to the owner of the horse that won two out of three four-mile heats. Purses for matches and sweepstakes were high, and the horses, according to one English traveler, "were such as to make no despicable figure" in the best races in England, "their speed, bottom or blood [not] inferior to their appearance."[8] Jefferson, an experienced rider by age seventeen who loved good horses, was certainly among the spectators during race week.

From the day of his departure for college for the rest of his life, Jefferson kept meticulous records of his daily expenditures in a pocket expense account book. There are frequent notations for tickets to plays at the theater on Waller Street, where Jefferson kept constant attendance on the visiting companies from London and New York come to perform Shakespearean tragedies and Congreve comedies, and he was in the audience when the famed Hallams came from London on an American tour. The other diversion that drew young Jefferson was the Raleigh Tavern, where distinguished lawyers gathered and drank the strong arrack punch and where young people, including Jefferson, crowded into the Apollo Room to weave in Virginia reels and face off in quadrilles and minuets as musicians in the balcony provided accompaniment on flutes, fifes, and fiddles.

Jefferson's college days in Williamsburg began on May 25, 1760, surprisingly late in the term, and continued for two years and one month, during which time, except for a brief sojourn in rooms on Duke of Gloucester Street, he paid board, lived, and studied at the College of William and Mary. There was a new young king on the throne of England, British armies were winning an empire in a year of triumphs at Quebec and in Germany, in India, the Mediterranean, and the Caribbean. The American revolutionaries who would one day make up Jefferson's political orb were mostly older and already pursuing careers. Benjamin Franklin had already retired from printing and was the agent for the Pennsylvania Assembly in London. John Adams was five years out of Harvard College and practicing law in Boston. George Washington, a Virginia planter who never went to college, after seeing a British army defeated at the Monongahela and searching in vain for a military command, had married the richest widow in the Tidewater and settled down to life as a gentleman farmer at Mount Vernon.

For the next two years, as the French and their Indian allies were finally defeated after seventy years of intermittent wars along the Appalachian chain, Tom Jefferson began to contemplate Europe, its laws and literature. He lived a Spartan existence: long days of lectures, tutorials, and studies punctuated by bad dinners. The William and Mary faculty had recently petitioned the housekeeper to assure that the young men and their teachers receive matching meals of both fresh and salt meat for dinner and puddings and pies on Sundays and two other nights each week instead of the odd scraps they had been served. The college, built in 1694 and supported by customs duties, consisted of two main red-brick buildings, President's House, now known as the Wren Building, and the Brafferton Building, connected by a walled garden. The College of William and Mary of Jefferson's day comprised a preparatory grammar school; the Indian school (a handful of Native Americans were in residence); the divinity school, which prepared Anglican clergy for Virginia's official state Church of England; and the philosophy school, in which Jefferson enrolled. There were seven celibate and generally hard-drinking men on its faculty, several of them distinguished scholars from Oxford and Cambridge, and not more than one hundred students, most of them the rowdy sons of country gentry bent on having a good time more than on studying. Jefferson took up residence at first in the Wren Building; from its balcony he could enjoy the bustling mile-long vista of Duke of Gloucester Street to the Capitol with its throngs of strollers and riders, jostling carriages, oxcarts, covered wagons, and the occasional herd of livestock.

The faculty was racked by loud and acrimonious controversy and the college was in decline. All of the faculty except William Small were Anglican clergy. The Anglican clergy had become embroiled in the Parson's Cause, a public dispute over their pay rates, and were unpopular with the gentry, some of whom served on the college's Board of Visitors, which had recently sacked half the teachers. Amid all the controversy, Dr. Small had recently been brought from Scotland to teach physics, metaphysics, and mathematics. A graduate of the Scottish college at Aberdeen, he soon wound up teaching virtually everything else after Jefferson's other teacher, the Rev. John Rowe, was dismissed for his involvement, along with the master of the grammar school, in a drunken town-and-gown street brawl. (At about the same time, the president of the college was charged with public drunkenness: he soon thereafter died.) For a year, before new professors could be imported, William Small was Jefferson's tutor in mathematics, moral philosophy, and science as well as his surrogate father.

Few men had more influence over Thomas Jefferson's youthful character formation than William Small, yet remarkably little is known about him other than that he taught at William and Mary from 1758 to 1764 and then returned to England, where he numbered among his friends James Watt, inventor of the steam engine; Erasmus Darwin of Lichfield, the leader of England's North Country literati, who would one day write his elegy; Joseph Priestley, discoverer of oxygen and a political radical; and the highly successful industrial innovator and potter, Josiah Wedgwood. Jefferson, who began to absorb from Small the ideas and the attitudes of the Enlightenment, considered him a great man, the embodiment of the spirit of the Enlightenment. When Small died in England in 1775, his obituary listed him as an "M.D.," a title and a practice he had appropriated for himself in Virginia, but his students knew him, in the words of one, as an "illustrious" teacher, and Benjamin Franklin also hailed his contributions to science. Jefferson considered his mathematics studies with Small "peculiarly engaging and delightful." To Bernard Moore, the young son of one of Peter Jefferson's friends, Jefferson summed up the importance of his studies thus:

> Before you enter on the study of the law, a sufficient groundwork must be laid. For this purpose, an acquaintance with the Latin and French languages is absolutely necessary. . . . Mathematics and natural philosophy are so useful in the most familiar occurrences of life,

> and are so peculiarly engaging and delightful as would induce every person to wish an acquaintance with them. . . . Besides this, the faculties of the mind, like the members of the body, are strengthened and improved by exercise. Mathematical reasonings and deductions are therefore a fine preparation for investigating the abstruse speculations of the law.[9]

Writing his autobiography, Jefferson still considered it "my great good fortune" that Small's predecessor had been recently sacked and "the philosophical chair became vacant soon after my arrival at college and he was appointed to fill it and was the first who gave in that college regular lectures in ethics, rhetoric and belles-lettres."[10] He was particularly fond of his mathematics studies with Small. As president of the United States nearly fifty years later, Jefferson told his grandson that one thing that he liked about mathematics as it was then studied was that there were "no theories" and "no uncertainties [to] remain on the mind. . . . All is demonstration and satisfaction."[11] Jefferson attributed his successes as an architect, inventor, and man of science "to the good foundation laid at college by my old master and friend Small." Dr. Small, an arch-looking, all-inquiring man who never drank too much or joined in the student brawls, steadily and singlehandedly among the faculty denied the power of a master to punish a student, and Jefferson admired and emulated Small's "even and dignified line" of behavior:[12]

> It was my great good fortune, and what probably fixed the destinies of my life, that Dr. William Small of Scotland was the professor of mathematics, a man profound in most of the useful branches of science, with a happy talent of communication, correct gentlemanly manners, and an enlarged and liberal mind. He, most happily for me, became soon attached to me, and made me his daily companion when not engaged in school, and from his conversation I got my first views of the expansion of science, and of the system of things in which we are placed.[13]

To Joseph Priestley, inventor of modern physics, Jefferson would one day put it more simply: "When I was young, mathematics was the passion of my life."[14]

One of his lasting contributions was to put William and Mary in the lead of a race to teach the new science. Small obtained a leave of absence to bring from England the finest collection of scientific apparatus in America. One of the leading polymaths alive at the time, he taught scientific experimentation to the young Jefferson and scores of other young Virginians. Jefferson regarded him as a father, and he conversed with Small for many long hours, which he had never managed to do with his own laconic father. Jefferson would one day sharply criticize William and Mary, and eventually he designed, built, and administered the University of Virginia at Charlottesville in open opposition to his alma mater, but for the moment he was grateful that Small was introducing into its curriculum the spirit of philosophical skepticism. While Small met with faculty resistance for his efforts, he was the successful torchbearer of the Enlightenment in Virginia. From him, Jefferson first learned of Montesquieu and Molière, Voltaire, Rousseau, and Diderot. For nearly a quarter century, he read and worshipped Enlightenment thinkers from afar until he finally went to France as American minister plenipotentiary on the eve of the French Revolution. As Dumas Malone put it, Small "was responsible for the liberality of spirit which came to characterize William and Mary."[15] In 1760, however, the goals of the college were as notoriously divided as its faculty was generally inept, and many Virginians refused it their sons.

Under Small's guidance Jefferson adhered to a rigorous, self-disciplined scheme of studying. "I was a hard student,"[16] he later said of himself, but a contemporary admirer, Francis Walker Gilmer, spelled out the reasons for Jefferson's academic success:

> His mind must have been by nature one of uncommon capaciousness and retention, of wonderful clearness and as rapid as is consistent with accurate thoughts. His application from early youth [was] not only intense but unremitted. When young, he adopted a system, perhaps an entire plan of life from which neither the exigencies of business nor the allurements of pleasure could drive or seduce him. Much of his success is to be ascribed to methodical industry.

By his college days, Jefferson was studying fifteen out of every twenty-four hours, often long after midnight.[17] "It is while we are young that the habit of industry is formed," Jefferson later reflected. "If not then, it never is afterwards."[18]

Just how rigorous and omnivorous was his self-imposed regimen can be deduced from his answer to a request from young Robert Skipwith ten years later to suggest a list of books for him to study. The list included *Percy's Reliques of Ancient English Poetry,* Chaucer, Shakespeare, Milton, Dryden, Spenser, Thomson, Gray, Prior, Gay, Pope, the plays of Steele, Congreve, and Addison, the novels of Smollett, Richardson, Langhorne and Sterne, the works of Swift, Addison and Steele's *Spectator* and *Tatler,* the essays of John Locke, Bolingbroke's five volumes of political philosophy, Hume's two-volume *History of England,* plus the Bible, the works of Josephus, and his favorite Greek and Latin writers. To keep himself fit for these systematic sedentary sieges, Jefferson ran, took hikes, and swam back and forth across a pond. He habitually bathed his feet in cold water every morning, a practice to which he attributed a lifelong freedom from colds.

Young Thomas Jefferson, trying to make himself the embodiment of the Age of Reason with its belief in self-improvement, would have come dangerously close to making a fussy attempt at self-perfection were it not for his ability to make friends. He seems to have begun deliberately to cultivate friendships when he discovered that, despite all the dances and social introductions that came his way naturally, young women would not notice such a quiet, bookish loner. Still, it took quite an effort for him to stay away from his readings more than a few hours. Jefferson, said his college friend John Page, "could tear himself away from his dearest friends to fly to his studies."[19]

Offsetting his studiousness was his confession to his guardian at the end of his first year at college that he had been extravagant, overspending his budget by half, had yielded to youthful temptations, had even participated in the annual town-and-gown battles against local boys, and had spent a good deal of time with his friends, Dabney Carr, John Page, Jack Walker (all formerly of Reverend Maury's school), and John Tyler, just plain talking about girls. Jefferson, as he now preferred to be called, belonged to the Flat Hat Club, as did his friend Walker. It was a secret if silly student club made up of six members. Its membership certificates were composed of humorous Latin prose and its goals included that each member should be "a great ornament and pillar of things general and particular." Jefferson wrote about club meetings, which he attended even after college, in humorous "hog Latin." Dr. Small became "Parvi," Carr was "Currus".[20] Jefferson shared rooms with the studious John Tyler, who was to become a distinguished circuit court judge and the father of President Tyler, and for a short time with Frank Willis of Gloucester County, who was something of a wag and who kept his horse in the cellar. Jefferson evidently joined them in fox hunting but, while Willis pursued other forms of pleasure in the provincial capital, Jefferson and his friend Tyler read, wrote papers, and practiced on their violins. So good a violinist was Jefferson becoming that Tyler later would concede that, if he had had the bow arm of Jefferson, he "would yield the palm to no man living in excellence of performance."[21]

Jefferson's Randolph family connections and his friendship with John Tyler gave him access to the highest level of Williamsburg society, for which his descent from the Randolphs as well as two generations of Virginia burgesses already qualified him, but as something of a country cousin. John Tyler came from an old English family of dubious distinction; he believed (although modern genealogists have challenged this) that he was descended from Wat Tyler, the fourteenth-century blacksmith who rebelled against King Richard II. (Jefferson's roommate named one of his sons Wat Tyler.) A lifelong friend of Jefferson, Tyler was, as one biographer put it, "a congenital rebel and individualist" and, like his roommate Jefferson, "an intellectual child of the French Enlightenment."[22] Tyler's stature in Williamsburg was unquestionable: the Governor's Palace was built on his family's land, which had been sold to Virginia when his great-grandfather had served on the committee that had superintended the settling and building of this new capital at Williamsburg in 1699.

Of all his college chums, Jefferson singled out John Page from the Flat Hat Club to become his closest friend. Their friendship and correspondence lasted half a century; Jefferson's letters reveal his deep respect for Page. Page lacked Jefferson's stern determination but, partly because of his inherited position in Virginia society, he participated in every key political movement and eventually filled almost every high office—burgess, member of the Governor's Council, lieutenant governor, delegate to the Continental Congress and, finally, governor. Page boarded with the president of the college and he kept a journal of his school days. His father paid the president of the college "handsomely" to be "my private tutor." Two things Page and Jefferson had in common were excellent Latin and an interest in history, but Page refused to think of himself in the same academic breath as his studious friend, Jefferson:

> I never thought, however, that I had made any great proficiency in my study, for I was too sociable and fond of the conversation of my friends to study as Mr. Jefferson did.[23]

Page eventually became Jefferson's confidant on the subject of women. At this stage, however, to be Jefferson's friend meant to crack the books with him. Family lore holds that Dabney Carr, another close college friend and erstwhile victim of Jefferson's horse-racing hoax at Reverend Maury's, often visited Jefferson at Shadwell, where he became fond of Jefferson's older sister, Martha. Jefferson and Carr climbed Monticello and studied together beneath a great oak tree, agreeing that whoever should die first would be buried there by the other. All too soon, Jefferson would keep this promise

Amazingly, at the end of his first year at William and Mary, Jefferson believed he had squandered too much of his time:

> I was often thrown into the society of horse racers and players, fox hunters, scientific and professional men. . . . Many a time, I have asked myself, in the enthusiastic moment of the death of the fox, the victory of a favorite horse, the issue of a question eloquently argued at the bar . . . Well, which of these kinds of reputation should I prefer? That of a horse jockey? Or a fox hunter? An orator? Or the honest advocate of my country's rights?[24]

Examining his accounts for that first year as well as his conscience, Jefferson saw that he had charged fancy clothes and a fine horse and entertainment to his account and, writing to his guardian that he thought he had spent too much, he proposed that all of his expenditures should be charged against his separate share of the inheritance. His guardian declined, noting that the expenses had been harmless and that the estate could well afford to pay the bill.

Most if not all of Jefferson's activities at Williamsburg seem to have been quite innocent. He danced at the elegant, gray-blue-paneled assemblies in the Apollo Room at the Raleigh Tavern (under the gilded Latin motto that translates, "Jollity, offspring of wisdom and good living," and at the Governor's Palace. He visited plantation houses, made wagers with girls he met at dances, and gossiped about courtships and served as an escort at weddings. Stepping into a close-knit social group, he became one of its eligible, presentable fixtures. He took to calling the capital "Devilsburgh" but he missed it whenever he went home. Home, as it seems to almost all college students, compared adversely.

In college Thomas Jefferson was no humorless prig, but he was still haunted by the loss of his father. Time wasted was the insupportable sin. Devilsburgh and his friends tempted him. A passage Jefferson excerpted in his *Literary Commonplace Book* from Laurence Sterne's then popular novel, *Tristram Shandy*, shows how he was holding himself accountable for so many wasted hours:

> Time wastes too fast: every letter I trace tells me with what rapidity life follows my pen; the days and hours of it, more precious, my dear Jenny, than the rubies about thy neck. . . . Every time I kiss thy hand to bid adieu, and every absence which follows it, are preludes to that eternal separation which we are shortly to make . . .[25]

One result of his continuing sense of loss was that he turned to none of his father's friends as a mentor or intimate friend. From the beginning of his years at William and Mary until his death, Thomas Jefferson was obsessed by accountability. At eighteen, he was, if not rich, then financially self-sufficient, yet he was technically responsible in even the smallest financial matters to the executors of his father's estate. To be sure, he knew that he could always afford to pay from his share of his father's estate for anything of which they disapproved. In the ultimate sense he felt that no one, not his mother, not his guardians, *no one* was responsible for him and that he was really not responsible to anyone but himself. His father had left him the wealth and freedom to indulge his love of books and self-education, but he in turn had internalized the bookkeeping his father had taught him. He must account for every penny, every minute of his legacy to his internal auditor.

Jefferson spent the Christmas 1760 interterm holidays at Shadwell, where he seems to have studied fifteen hours a day in his room and shunned his mother's social activities as much as possible. Each morning as soon as it was light enough to make out the hands of his mantel clock, he was up and back at his studies. In the afternoon, after spending the minimum time possible in company, he went down to the Rivanna where he kept a small canoe ready and paddled across to Monticello, his favorite mountain, then trudged up through the woods to its nine-hundred-foot slope. By now, at age seventeen, he was already daydreaming of building his own house on land his father had left him atop his Monticello as soon as he became twenty-one. For the time being, he had to content himself with gazing out over the Blue Ridge, the western horizon of his world.

In his letters to his schoolmates, he complained of the monotony of home visits. To John Page, he bemoaned,

> All things here appear to me to trudge on in one and the same round: we rise in the morning that we may eat breakfast, dinner and supper, and we go to bed again that we may get up the next morning and go the same, so that you never saw two peas more alike than our yesterday and today.[26]

He had to weigh carefully the decision of whether to become a lawyer or remain merely a planter, sifting through questions of money and time. He was not rich, he knew, although his father had died land-rich. But by the time Peter Jefferson's estate was divided, half of the best lands would go to his brother and his young mother would have the use of the main house and farm. Jefferson would one day inherit the patrimonial farm at Shadwell, but in the meantime, the practice of law could augment his income from agriculture and combine his love for learning with his pragmatic love of independence. Jefferson had little question that the law was an honorable adjunct for a gentleman. His Randolph cousins and uncles were among Virginia's most distinguished attorneys. Thanks to his tutelage under Dr. Small, he was familiar with the injunctions of seventeenth-century Scottish philosophers of the importance of mastering the law. Jefferson bought John Locke's 1693 treatise, *Some Thoughts Concerning Education,* which advised young gentlemen to read works on international law and legal philosophy by Grotius and Pufendorf, two jurists very popular in Virginia. Jefferson owned Locke's *Education* with this much-read passage:

> It would be strange to suppose an English gentleman should be ignorant of the law of his country. This, whatever station he is in, is so requisite that, from a justice of the peace to a minister of state, I know no place he can well fill without it.

Jefferson also owned Gilbert Burnet's *History of His Own Time,* first published in 1734; which held that "a competent skill in this [the law] makes a man very useful in his country, both in conducting his own affairs and in giving good advice to those about him." The law would help Jefferson become a good justice of the peace, a tacit part of his Albemarle inheritance, help him to arbitrate disputes in his community and settle lawsuits and "which ought to be the top of an English gentleman's ambition, to be an able parliament man."

By early 1761, Jefferson had made up his mind to study law. One of the kindest things Dr. Small had done for his young protégé Jefferson was to hand him over to George Wythe, who was to become the first law professor in America and who brought the Enlightenment to the teaching of law by treating it as a science, with scientific methods by which it could be taught. Lawyers were to be "men of science." Jefferson as an old man remembered that Dr. Small had "filled up the measure of his goodness to me by procuring for me, from his most intimate friend" a willing "reception as a student of law under his direction."[27]

There were still no law schools in America, and would-be attorneys had to serve informal apprenticeships that lasted four or five years. It was to be another twenty years before then-Governor Jefferson could repay his teacher by initiating the movement to have Wythe appointed to the first professorship of law in America at William and Mary. In the meantime, most of the aspiring law students read law and acted as unpaid law clerks from a few months to several years if, like Patrick Henry, they only wanted to practice in the lower or county courts. Then they stood for written and oral bar examinations.

A relative handful of Virginians, including his cousins Edmund and Peyton Randolph, traveled to London to study at the Inns of Court to become English barristers, as did John Hancock and Benjamin Franklin's son, William, but this was terribly expensive and many families feared that their sons would be corrupted to an even greater extent than was possible in Williamsburg. In the century preceding the Revolution, only sixty Virginians chose this route and only twenty were called to the bar. Besides, Jefferson seems not to have seriously considered it possible to leave his widowed mother and his other family responsibilities at Shadwell so far behind. As a gentleman, he was supposed to eschew working for a living: he had read Daniel Defoe's definition of "the gentry," the class to which he belonged, as "such who live on estates and without the mechanism of employment, including the men of letters, such as clergy, lawyers, and physicians." In modern terms, Jefferson was supposed to live on the unearned income of his estates; managing his lands was a permissible way of exercising his authority. And like the ancient Roman landed aristocrats whom he was educated to emulate, he was expected to be active in all sorts of what historian Gordon S. Wood categorizes as "commercial and entrepreneurial activities—breeding their cattle, upgrading their soil, improving their fruit trees, speculating in land, or even trafficking or trading." Just so they engaged in any commercial activities "in something other than a pure money-making spirit."

The third possibility was to sign on with a distinguished lawyer practicing before the Virginia General Court, the highest court in Virginia, and, after reading law for an average of five years, joining the province's self-perpetuating legal elite. This pragmatic and scholarly approach appealed to Jefferson. Admission to the General Court bar did not require a bar examination. Normally a lawyer was admitted to the General Court after having distinguished himself at the county-court level. Virginia's General Court was bristling with the brilliant and the famous in the eighteenth century. It was to this lofty legal beginning that Jefferson set his sights.

One day in the spring of 1761, Jefferson had Jupiter saddle up his horse for a ride up sandy, catalpa-fanned Duke of Gloucester Street past Bruton Parish Church, where he was accustomed to sit in the college pews every Sunday for the inevitably dry sermon. One block farther past the Governor's Palace, he came to Palace Green and the stout brick mansion of George Wythe with its four massive chimneys, two with their stacks standing tall and free at each end. At thirty-five, the country-educated, self-effacing and self-taught legal scholar already had an imposing appearance, with a high, wide, overhanging forehead, penetrating dark gray eyes, a Roman nose, and a strong chin. A man of medium height known for his suave manners and deep, courtly bow, Wythe introduced young Jefferson to his wife, Elizabeth; a Taliaferro, she was the daughter of the colony's best architect and builder. Over the next five years, as student and master grew even closer, Jefferson became part of another surrogate family: the Wythes had no children of their own. In letters to friends, he was to refer to Wythe as "my second father."[28] In his autobiography, he described Wythe as "my faithful and beloved mentor in youth and my most affectionate friend through life."[29] He also called Wythe "my ancient master, my earliest and best friend."[30]

No doubt Wythe knew from Small what an uncommon young scholar was seeking his tutelage, and certainly Jefferson knew Wythe's reputation as the most learned lawyer practicing before the Virginia courts. In seeking Wythe as his preceptor, Jefferson was shunning the easier if expected course of signing on with one of his lawyerly relations. Both Peyton Randolph, the king's attorney for Virginia, and his brother, John, were among the colony's leading legal practitioners. In choosing Wythe, Jefferson sought to study with a man who not only had the best legal mind in Virginia but also was a self-taught classical scholar of rare ability. Years later, when he was asked to advise an in-law on how to choose a mentor, Jefferson was almost dismissive of the question—"the only help a youth wants is to be directed what books to read and in what order to read them,"[31] but, at age nineteen, he knew that Dr. Small had done him an exceptional favor. Like Small, George Wythe knew the importance of reading the right books in the right order. For the next five years, as Jefferson alternated his time between the capital and Shadwell, he would gratefully be under Wythe's erudite direction, plowing through the mounds of lawbooks from Edward Coke to Montesquieu.

In April 1762, Thomas Jefferson left the College of William and Mary to study law in Wythe's Williamsburg office, cementing one of the happier relationships of his long and turbulent life. He never tired of paying tribute to the "beloved mentor" of his youth. He rarely applied such terms of warmth to anyone, even his own parents. "No man ever left behind him a character more venerated," Jefferson wrote. His teacher possessed "virtue of the purest tint" with "inflexible" integrity and "exact" justice, "warm" patriotism, and devotion "to liberty and the natural and equal rights of man—a more disinterested person never lived."

Wythe "was not quick of apprehension but, in a little time, profound in penetration and sound in conclusion." Jefferson admired Wythe's tenacity. "In philosophy, he was firm,"—and he imbibed some of Wythe's agnosticism: "Neither troubling nor perhaps trusting anyone with his religious creed, he left the world to the conclusion that religion must be good which could produce a life of such exemplary virtue."[32]

Jefferson was not the only impressionable young law clerk to come under Wythe's spell and go on to individual greatness. John Marshall, fourth chief justice of the United States, and James Monroe, the fifth president, studied with him. Henry Clay became Wythe's clerk thirty years later. Clay, a frontiersman, commented on Wythe's courtly appearance and suave manners, including "the most graceful bow I have ever witnessed,"[33] something that must have set off in his young clerks hours of repetitive practice. Clay admired Wythe's dark wigs, his black velvet breeches fit for the hot summers in the Tidewater yet elegant enough for evenings at the Governor's Palace. Wythe's students, including Jefferson, imbibed their mentor's sartorial example along with their Coke.

But it was Wythe's preeminence at the Virginia bar that attracted bright young law students such as Jefferson and Clay, his "superior learning, correct elocution and logical style of reasoning." Wythe was the polished and learned father Jefferson's own inarticulate sire had not been, even as he embodied many of the homely qualities he had so respected in his own father. Wythe was "distinguished by correctness and purity of conduct in his profession as he was by his industry and fidelity to those who employed him." As Wythe directed Jefferson's studies and "led me into the practice of the law," he also became "my most affectionate friend."[34]

A member of the House of Burgesses and one of Virginia's three most successful trial lawyers, a populist and champion of religious liberty and toleration before such beliefs were fashionable, George Wythe was the first great man and probably the first man of genius young Jefferson met. One Virginia legal historian ranked him "almost alone . . . in the solid learning of the law."[35] An impressive speaker who riddled his discourse with Latin and Greek references and allusions, he always prepared his arguments with great care. Jefferson said that in pleading a case, Wythe "never indulged himself with a useless or declamatory word,"[36] a trait that Jefferson had already learned to admire from his earlier tutor, Reverend Maury. Wythe was meditative and methodical. In extemporaneous debate, some of his colleagues were far more facile at the bar, yet he could be witty and he could give as well as he took. His greatest rival was Edmund Pendleton. One day Pendleton's assistant did not show up in court and Wythe's did. Pendleton asked Lord Dunmore, who was presiding, for a continuance. When Lord Dunmore told him,

"Go on, Mr. Pendleton, for you will be a match for both of them," Wythe, bowing, retorted, "Yes, with your lordship's assistance."[37] A famed teacher, Wythe numbered among his students not only Chief Justice John Marshall, but also Henry Clay, the great card-playing master of compromise. But his favorite student was to be Jefferson: to him he ultimately willed his fine library.

4

"I Was Bold in the Pursuit of Knowledge"

Conscience is as much a part of man as his leg or arm. It is given to all human beings in a stronger or weaker degree . . . It may be strengthened by exercise . . . Therefore, read good books, as they will encourage as well as direct your feelings.

—Thomas Jefferson to Peter Carr

Each day that the Williamsburg courts were in session over the next five years, Jefferson walked over from his rented rooms to the solid red-brick Georgian house with the high chimneys on Palace Green. Admitted by a servant, he went to a first-floor back room, the clerks' study, and took a place at the round table with the other young clerks. Wythe taught Jefferson to treasure the morning, to tackle the toughest readings early in his day. Opening his assigned reading, he broke only long enough to drop another log on the embers in the corner fireplace in winter, to gaze briefly out the tall, narrow window toward the garden in summer.

It was Jefferson's good fortune that George Wythe had bitter memories of the sterile clerkship that he had served under an uncle, Stephen Dewey, a lawyer in Prince George County. Wythe refused to follow his uncle's bad example. He became an exemplary master, taking on several students free of charge at a time and taking the time to teach his students rather than exacting clerical chores from them. Young Jefferson was free, for the next five years, to read the law, drawing on his master's extensive law library.

He was expected to observe very closely every stage of every case in his master's office. Wythe explained his every legal step to Jefferson and, as Jefferson progressed, Wythe gave him more and more responsibility for legal research in the provincial records and from Wythe's lawbooks, eventually turning over to him more of the out-of-court preparation vital for his master's clients. Under Wythe's watchful eyes, Jefferson learned law by handling actual cases. On court days in Williamsburg, Jefferson carried George Wythe's books and notes into the Palace for him and stayed nearby at his beck and call. When Wythe went on the hustings to county courts, Jefferson rode with him.

As he began his guided reading of the law, Jefferson's old habit of commonplacing served him well. In a new blank book, he wrote at the top of each page, in alphabetical order, the major divisions of the common law. Under the various headings, he wrote comments on what he considered to be the essential points of law as he read treatises, reports, statutes, writing the individual points of law on the appropriate page. Beginning to keep his legal commonplace book in 1762, Jefferson entered cases in it and subsequent volumes for a dozen years. To a young would-be lawyer, he gave detailed instruction in his method and its underlying importance:

> In reading the Reporters, enter in a commonplace book every case of value, condensed into the narrowest compass possible, which will admit of presenting distinctly the principles of the case. . . . This operation is doubly useful, insomuch as it obliges the student to seek out the pith of the case, and habituates him to a condensation of thought, and to an acquisition of the most valuable of all talents, that of never using two words where one will do. It fixes the case, too, more indelibly in the mind.[1]

The foundation of Wythe's own legal learning was his mastery of all of Sir Edward Coke's *Commentaries upon Littleton,* what Coke's biographer, Catherine Drinker Bowen, once called the "greatest of lawbooks." Within its pages lay the records of some five hundred years of English law, all the way back to the Magna Carta. Sir Thomas Littleton had written a learned treatise on the law of real property in the fifteenth century; in the early seventeenth century, English attorney general Sir Edward Coke had brought it up to date and enlarged it. By the time Jefferson tackled it, it had, as legal historian Hamilton Bryson puts it, "acquired several layers of footnotes" and was "authoritative, erudite, complicated and thoroughly turgid." Before the Revolution, Coke's *Littleton* was, as Jefferson reminded James Madison some sixty years later, "the universal elementary textbook of law students and a sounder Whig never wrote, nor of profounder learning in the orthodox doctrines of the British Constitution, or what were called English liberties . . . You remember also that our lawyers were then all Whigs."[2]

Wythe, who believed in learning and teaching the fundamentals of the law, was soon to put all of Coke in eager young Jefferson's hands. After a year of slogging through it, Jefferson would feel as one of Coke's Inner Templar colleagues had put it: "A lawyer must have an iron head, a brazen face and a leaden breech."[3] Coke's tome required, as well, a mastery of Latin and French to decode its terms—*villein service, service de chevalier, fee simple,* and *fee tail.* Jefferson would find Coke's "black text" a work of "uncouth but cunning learning," and he was soon writing to his friend, John Page, complaining of Coke's "chaotic form": "I do wish the Devil had old Coke, for I am sure I never was so tired of an old dull scoundrel in my life."[4]

Over the years, Jefferson and Wythe found that they had many interests in common. Wythe so loved the classics that sometimes, when he was arguing a case, he quoted Horace forcefully and, when he became a judge, he gave free classes in Latin and Greek. He was just as conversant in Elizabethan and Restoration literature. Mentor and protégé habitually and interchangeably discussed natural philosophy, the law of nations, and the natural rights of man with Dr. Small and their equally learned friend, the royal governor, Francis Fauquier, at frequent intimate dinners at the Governor's Palace.

Fauquier was, in Jefferson's later estimation, "the ablest man who had ever filled that office."[5] About sixty years old, the scholarly son of a director of the Bank of England who had eschewed the countinghouse and preferred life as a Hertfordshire country squire, Fauquier was a good enough amateur scientist to have been elected a Fellow of the Royal Society shortly before he sailed for America. He excelled, in descending order, in music, scientific studies, and gambling, and he was considered somewhat mysterious. Virginia gossip had it that he had lost his entire inheritance in a single night at cards with Admiral Lord Anson, who had the decency as well as the right connections to see to it that Fauquier was appointed royal lieutenant governor of Virginia (the governorship was a position held by a nobleman who usually remained at home in England). This event coincided almost exactly with the publication of his tract on economics, *An Essay on Ways and Means of Raising Money for the Support of the Present War without Increasing the Public Debts.* Most royal officials in America were the recipients of patronage plums for past favors or proper family ties in England.

On his arrival in Williamsburg, Fauquier had been greeted by a hailstorm that had shattered every window on the north side of the Governor's Palace and left enough ice on the ground to chill his wine and freeze enough cream for the next day's dessert. An amateur scientist, Fauquier had a reaction typical of the breed: he measured the hailstones and sent an account back to England to be published in the *Philosophical Transactions* of the Royal Society. And then he began to keep a diary of Williamsburg weather, an idea emulated by Jefferson for long periods at Monticello and in Washington.

But, while extremely popular with the local river barons, the royal governor had become jaded by Tidewater society and increasingly eschewed its languid, gossipy, dissolute horse-breeding country gentry, preferring the literate badinage that was coming to characterize the Enlightenment. He was happy to recruit a gifted young fiddler to his weekly concerts in the Palace, whom he quickly added as a fourth fluent and well-read conversationalist. For young Jefferson, an incomparable period of happiness had begun. He imbibed the elegance of polished marble floors, of blazing chandelier light playing on rich walnut panels. He heard and engaged in, he later wrote, after knowing Franklin, Washington, Adams, and several French philosophers, "more good sense, more rational and philosophical conversation than in all my life besides."[6] As minister to France and as president of the United States, Jefferson would be at his best as he presided over literate, witty dinner-table discussions. An extraordinary friendship developed among the foursome, the worldly governor, the Enlightenment don, the pedantic lawyer, and the young yet learned Virginian. When he penned his autobiography nearly half a century later, Jefferson still remembered fondly those dinners. Fauquier, "Dr. Small and Mr. Wythe, his *amici omnium horarum* [friends of all hours] and myself formed a *partie quarée* [a foursome], and to the habitual conversations on these occasions I owed much instruction."[7]

His memories of the college were contradictory. As governor, he drew up a plan to revamp its curriculum, severing its church ties and chastising it for years of wasting money: "The experience of near an hundred years hath proved that the said college, thus amply endowed by the public, hath not answered their expectations."[8] Yet he advised his friends not to send their sons abroad when, except for medical training and languages, they could learn all they needed at the college, "the best place to go" for law, mathematics, and philosophy. "I know of no better place in the world while the present professors remain."[9] They included George Wythe, "the pride of the institution," and "one of the greatest men of the age."[10]

When the years of his schooling were over, Jefferson carried away from

Williamsburg the lessons and the high intellectual and moral standards of these cultivated gentlemen as models for his decisions and behavior. And they would help him to account to himself for his most valuable possession: his reputation.

> I had the good fortune to become acquainted very early with some characters of very high standing and to feel the incessant wish that I could even become what they were. Under temptations and difficulties, I could ask myself, "What would Dr. Small, Mr. Wythe . . . do in this situation? What course in it will ensure me their approbation?" I am certain that this mode of deciding on my conduct tended more to its correctness than any reasoning powers I possessed.[11]

As President, Jefferson looked back through the tunnel of turbulent events of nearly half a century to sum up his happy college years, his fortuitous friendships in Williamsburg, and he realized how lucky he had been: "I am indebted for first impressions which have had the most salutary influence on the course of my life."[12]

George Wythe was a born teacher. Even if the formal instruction of law was in its infancy and the Vinerian chair of English law at Oxford had only been established three years earlier, Wythe was astute and up-to-the-minute enough to adopt the three volumes of law books of its first incumbent, Sir William Blackstone, as fast as they came off the printing presses in London. For ten months out of that first year of studying law, Jefferson labored away on the voluminous marginal notes of *Commentaries upon Littleton.* Simpler texts were available, but Wythe insisted his students start with Coke. There were no shortcuts; it all had to be studied painstakingly if one were to understand the fundamentals of the law. There were plenty of dull moments, even for someone as studious as young Jefferson.

At first, Jefferson thought he could get through Coke in a single year. Before he stuffed Coke into his trunk and headed home to Shadwell for Christmas break, he was becoming thoroughly jaded by the hours of poring over case law, wishing not only that "the Devil had old Coke" but sure that even Job would begin "to whine a little under his afflictions" if he had such a course of study. "But the old fellows," he wrote his friend Page, "say we must read to gain knowledge, and gain knowledge to make us happy and admired. *Mere jargon.*"[13]

Grousing but proud, the nineteen-year-old student persevered and learned a lifelong admiration for the written constitution of laws that Coke had given the English-speaking world. And, one day, he would decry as monarchist and too pro-English "the substitution of Blackstone for my Lord Coke."[14] It was Coke who had challenged royal prerogative under the Stuarts in seventeenth-century England. Jefferson developed an early affinity for his magisterial if tedious defense of the rights of free-born Englishmen. He came to worry that Blackstone and the Scottish philosopher-historian David Hume were making "Tories of all England and are making tories of those young Americans whose native feelings of independence do not place them above the wily sophistries of a Hume or a Blackstone."[15]

Ten years later, Jefferson spelled out a merciless daily regimen of legal readings in a letter to an aspiring young lawyer, Bernard Moore, who had asked Jefferson's help in deciding which lawbooks to purchase. Jefferson told Moore much more than he had asked. He probably assumed it would not be the last time he was asked such a question and he labored over his answer, laying out a scientific progression of law readings. He "required," for common law, Coke and Bacon, Baron Gilbert, Fonblanque and Blackstone—the last a "perfect digest of both branches of law."

But the scientific regimen Jefferson outlined for Moore was not the half of his own daily labors over the books. By his own account, it only took him from eight in the morning to noontime. He believed there was "a great inequality" in the "vigor of the mind at different times of day," and he divided his day into five periods. Even before he tackled his legal studies each morning at eight, he had read "ethics, religion and natural law" for three hours "from five to eight," including Cicero, Locke's essays, Condorcet, Francis Hutcheson's *Introduction to Moral Philosophy,* and Lord Kames's *Natural Religion.* For his matinal reading, under the rubric of religion he read commentaries on the New Testament, the sermons of Sterne, Massillon, and Bourdaloue. For natural law, he read Vattel and Rayneval.

After he had survived this unimaginable seven-hour morning of philosophy and law—used by law professors ever since to quiet the complaints of generations of law students—Jefferson embarked on the third period of his day. After lunch, he read politics: more Locke, Sidney, and Priestley on government, Montesquieu's *On the Spirit of Laws,* Say's *Political Economy,* Malthus's *The Principles of Population.* In the afternoons, he allowed himself latitude for a meal, a break, perhaps conversation. But still he held himself to the reading of Greek and Roman history in the original as well as histories of Europe: Millot's *Modern History,* Robertson's *Charles V,* plus histories of England, America, and Virginia. The evenings were reserved for improving "*belles lettres,* criticism, rhetoric and oratory." He later wrote that "this portion of time [borrowing some of the afternoon when the days are long and the nights short] is to be applied also to acquiring the art of writing and speaking correctly." He urged other would-be lawyers to "criticize the style of any book whatsoever, committing the criticism to writing."[16] Between 1762 and 1765 in his *Literary Commonplace Book,* he abstracted some seventy entries from Euripides, his current favorite, eight from Homer, three from Herodotus, one from Livy, and many from the popular work of Thomson's *The Seasons* in his mature handwriting, a clear, strong, slanting italic. He also urged Moore to read Demosthenes and Cicero. He himself had paraphrased their *Orations* and practiced them on a long-suffering friend, probably Dabney Carr. At this time, he also bought a copy of Sheridan's *Elocution* to help him practice public speaking before his bar examination. Small wonder Jefferson emerged from his thirteen unparalleled years of studies as one of the widest-read, most cultivated, and most philosophical of colonial Americans!

He seems to have followed his sixteen-hour-a-day marathon routine of study consistently only at Shadwell; in Devilsburgh he was tempted away from his books more and more as time went on. According to his account books, he went to the races, bought music, drank punch with his friends, visited their homes, bought books—and pined for the gay social whirl whenever he visited Shadwell.

Jefferson knew that his thirst for knowledge, underwritten by a generous inheritance and inspired by a fortuitous group of teachers, could only be indulged for a limited time. And, indeed, each year he was in Williamsburg he seems to have studied less. His two years as a law clerk and the three years of independent study he pursued at Shadwell were in his own words, "a time of life when I was bold in the pursuit of knowledge, never fearing to follow truth and reason to whatever results they led and bearding every authority which stood in their way."[17]

Between the ages of nineteen and twenty-four, the determined young Jefferson studied to become a legal philosopher while most of his friends spent much of their time hunting, gambling, cockfighting, staking out western lands or wenching among their slaves. He read the kinds of books most of his friends shunned as too difficult. His more serious cronies rarely studied law more than a few months, a year at most, before standing for the necessary bar examinations. Did this mean that Jefferson was insecure about his preparations for the law? Was he simply unwilling to abandon the role of student and surrogate son to George Wythe until he absolutely had to? It is more likely that he knew how uncommon was his good fortune to have the opportunities for study and tutelage in Williamsburg and he postponed as long as possible taking up the life of a country squire and county lawyer at Shadwell.

He knew just how poorly prepared the rural practitioners could be and he had no reason to fear matching his wits and training with them. Whenever Patrick Henry came to Williamsburg to court, he stayed with Jefferson. Henry once looked over Jefferson's library—Jefferson had begun the lifelong habit of buying books in large numbers—and said, "Mr. J., I will take two volumes of Hume's *Essays,* and try to read them this winter." But when Henry finally returned them, Jefferson remarked to a friend later, "he had not been able to get half way into one of them."[18]

The young Jefferson who emerged from this chrysalis of "self-catechizing, this return into himself," was an impressive young lawyer who would be written up by a contemporary legal historian, Edmund Randolph, a distant cousin, in his *Essay on the Revolutionary History of Virginia.*

> Indefatigable and methodical, Jefferson spoke with ease, perspicuity and elegance. His style in writing was more impassioned and, although often incorrect, was too glowing not to be acquitted as venial departures to rigid rules. Without being an overwhelming orator, he was an impressive speaker who fixed the attention. On two signal arguments before the General Court, [between] Mr. Henry and himself . . . each characterized himself. Mr. Jefferson drew copiously from the depths of the law, Mr. Henry from the recesses of the human heart.[19]

By the standards of the time, Thomas Jefferson was delaying getting married, and yet he seems to have spent a good deal of time—late at night and on holidays when his friends embarrassed him into putting aside his books—with his friends John Page and William Fleming, and they seem to have insisted on ushering him into the company of young women, in particular one Rebecca Burwell.

They met during Jefferson's first year as a law student in 1762 when Jefferson was nineteen and she was sixteen and considered a mature—and eligible—young lady. Rebecca and her brother, Lewis, a college classmate of Jefferson's, were the orphans of a distinguished lawyer and plantation owner. They lived with their uncle, the lordly William Nelson, a longtime judge at York in Gloucester County. She had grown up with her cousin, William Nelson, Jr., who was one day to sign the Declaration of Independence. Rebecca Burwell's father had been on the Governor's Council when Jefferson's father had reported on the expedition to extend the colony's boundary and, as acting governor, had commissioned Peter Jefferson's map. If Lewis Burwell had not actually introduced Rebecca to him, Jefferson may have been matched up by his friend, Governor Fauquier, who had recruited Jefferson to join a few other good amateur musicians—a cellist, a harpsichordist, a flutist—to play in the weekly concerts at the Palace. From this vantage point, Jefferson could spy the latest arrivals among the belles coming to the capital.

For more than a year, Jefferson, stunned by Rebecca's attractions, could not summon up the courage to approach her. He talked about her and he wrote about her—nine letters survive. They are not *to* her, but to his friends about her in cryptic or coded language. His reticence was, as Jefferson had feared, "the subject of a great deal of mirth and raillery."[20] His friend Page had written his own confessions of love for Nancy Wilson, the object of his desires, in Latin, but too many men read Latin in those days for Jefferson's security. "We must fall on some scheme of communicating our thoughts to each other," he told his friend, Page, "which shall be totally unintelligible to everyone but to ourselves."[21] Jefferson concocted various code names for Rebecca—among them Adnelab, Becca, finally Belinda, an anagram from the Latin *compana in die*, "bell in day." Then he unleashed a torrent of Latin and Greek puns in letters that thinly masked his boyish sheepishness at what more to do about this first infatuation.

The first of his letters, which his friends seem to have treasured and kept not from any prescience of Jefferson's greatness but to remind them of the beautiful Rebecca, came from his pen as soon as he left Williamsburg for the Christmas holidays. Thus, he could not see her for several weeks. On Christmas day, 1762, he visited Fairfields, the home of his friend Will Fleming, as he made his reluctant slow march home. Even at Christmas, he was in no hurry to confront his mother. His visits home may have helped produce, by age nineteen, the migraine headaches that plagued him all his adult life.[22]

Jefferson's anguished letter to John Page, who was courting the sister of Rebecca Burwell, was written Christmas day:

> This very day, to others the day of greatest mirth and jollity, sees me overwhelmed with more and greater misfortunes than have befallen a descendant of Adam for these thousand years . . . excepting Job, since the creation of the world. . . . You must know, dear Page, that I am now in a house surrounded with enemies, who take counsel together against my soul and, when I lay down to rest, they say among themselves, "Come, let us destroy him." I am sure that, if there is such a thing as a devil in this world, he must have been here last night and have had some hand in contriving what happened to me.[23]

Apparently, sometime probably just before he had thrown Coke into his portmanteau and headed home, Jefferson had gotten up enough courage to talk to Rebecca and with, for him, amazing trepidation, had asked her to a cut a silhouette of herself for him to carry inside the back of his watch on his travels. On Christmas Eve, as Jefferson lay asleep,

> Do you think the cursed rats [at the Devil's instigation, I suppose], did not eat up my pocketbook, which was in my pocket within a foot of my head?[24]

The rats had destroyed a pair of silk garters hand-worked by Jefferson's beloved sister Jemmy, "and a half dozen new minuets" he was taking home to play with her. But "something worse" had also happened:

> You know it rained last night. . . . When I went to bed, I laid my watch in the usual place, and going to take her up after I arose this morning I found her in the same place, it's true, but . . . all afloat in water let in at a leak in the roof of the house. . . . I should not have cared much for this but something worse attended it: the subtle particles of the water with which the case was filled had by their penetration so overcome the cohesion of the particles of the paper of which my dear picture . . . [was] composed that, in attempting to take [it] out to dry . . . Good God! My cursed fingers gave [it] such a rent as I fear I never shall get over. . . . And now, although the picture be defaced, there is so lively an image of her imprinted in my mind that I shall think of her too often, I fear, for my peace of mind and too often I am sure to get through Old Coke this winter, for God knows I have not seen him since I packed him up in my trunk in Williamsburg.[25]

Then he pleaded for Page to write him back "everything which happened" at a recent wedding: "Was SHE there? Because if she was I ought to have been at the devil for not being there, too."[26] Then, to keep up the pretense of pining after some other girl, he asked Page to "remember me affectionately to all the young ladies of my acquaintance, particularly the Miss Burwells and Miss Potters":

> I would fain ask the favor of Miss Becca Burwell to give me another watch paper of her own cutting which I should esteem much more though it were a plain round one than the nicest in the world cut by other hands. . . . However, I am afraid she would think this

> presumption after my suffering the other to get spoiled. If you think you can excuse me to her for this, I should be glad if you would ask her.[27]

Precocious in the world of books, Jefferson at nineteen was painfully shy and forced by his own awkwardness to use an interlocutor in courting Rebecca Burwell. He obviously trusted his "dear Page" with his most embarrassing secrets and missed his friend's company acutely. That first letter was long, taking up much of Christmas day to compose in a guest room at Fairfields, up under the eaves with a leaky roof. There, as was his habit, Jefferson shut himself away from noisy celebration. Whenever he was away from Williamsburg and friends during those years, he unburdened himself to his best friend, to whom he gave his "power of attorney" in affairs of the heart.

Four weeks later, evidently, he wrote from Shadwell after he heard back from Page that a rival had entered the scene. He reported that his days there were so dull that "I have not a syllable to write about." His motive for writing quickly became evident:

> How does R. B. do? What do you think of my affair, or what would you advise me to do? Had I better stay here and do nothing, or go down and do less? Inclination tells me to go, receive my sentence and be no longer in suspense. But reason says, "If you go and your attempt proves unsuccessful, you will be ten times more wretched than ever."[28]

Indecisive, Jefferson stayed at Shadwell, superintending the construction of a flat-bottomed boat that he wanted to sail downriver to Williamsburg in the spring. He would name it *Rebecca* at a safe distance from its namesake, but he would not launch into any love affair hastily. "Why cannot you and I be married," he asked Page, "when and to whom we would choose?"[29]

So bored was he with life at Shadwell that he was thinking of building a house of his own, not near his mother, but in Williamsburg:

> I think to build. No castle, though, I assure you, only a small house which shall contain a room for myself and another for you, and no more, unless Belinda should think proper to favor us with her company, in which case I will enlarge the plan as much as she pleases."[30]

When, shortly after Jefferson's twentieth birthday, Page urged him to come down off his mountaintop and "go immediately and lay siege in form," Jefferson protested that his heart was set on going to Europe to make a grand tour and he was reluctant "to begin an affair of that sort now." His heart was set, he wrote Page. "I intend to hoist sail and away. I shall visit England, Holland, France, Spain, Italy (where I would buy me a good fiddle) and Egypt" and return via Scotland. "Have you an inclination to travel?" he asked Page. "Because if you have, I shall be glad of your company."[31] He was not ready to seek any public engagement to Rebecca Burwell or anyone else:

> No, no, Page, whatever assurances I may give her in private of my esteem for her or whatever assurances I may ask in return from her, depend on it, they must be kept in private.[32]

There is no reason to believe that Rebecca had given him any assurances:

> That she may, I pray most sincerely, but that she will, she never gave me reason to hope. . . . I should be scared to death at making her so unreasonable a proposal as that of waiting until I returned from Britain, unless she could be prepared for it.[33]

Delegating his friend to prepare her and to have a "confab" with Rebecca and making a lawyer's pleading that he could not talk to her, a ward, until he talked to her guardians first, he was, obviously, as he had put it, "scared to death." He was prepared for a rebuff, hoping only that it would be quick, not lingering:

> But the event at last must be this, that if she consents, I shall be happy; if she does not, I must endeavor to be as much so as possible.[34]

Attempting to apply his self-taught stoicism at Shadwell, he nevertheless braced for his return to Williamsburg. There, on the evening of October 7, 1763, after more than a year of letters and negotiations, he finally danced with Rebecca Burwell. The next day, he wrote Page about it. He had rehearsed what he would do and say as if preparing for a jury trial—with predictable results:

> Last night, as merry as agreeable company and dancing with Belinda in the Apollo [Room at the Raleigh Tavern] could make me, I

> never could have thought the succeeding sun would have seen me so wretched as I now am! I was prepared to say a great deal. I had dressed up in my own mind such thoughts as occurred to me, in as moving language as I knew how, and expected to have performed in a tolerably creditable manner.
>
> But, good God! When I had an opportunity of venting them, a few broken sentences, uttered in great disorder and interrupted with pauses of uncommon length were the too visible marks of my strange confusion!"[35]

Red-faced, Jefferson refused to say more until he saw Page in person. He was glad to escape to his lawbooks. "The court is now at hand, which I must attend constantly." But he was desperate to talk to his friend and seek his advice. "For God's sake, come,"[36] he pleaded with Page, rushing off in the wake of George Wythe.

Sometime later that month, Jefferson, still daydreaming of marrying Rebecca, was off to Richmond with his mentor, George Wythe. He was beginning to recover from his mortification, writing to Will Fleming as if surprised that he had survived a charming evening with another girl, Jenny Taliaferro, who resembled John Page's belle, but was "prettier": "I was vastly pleased with her playing on the spinet and singing." It had only taken him a few days to become disgusted this time with country people, he had so thoroughly come to enjoy his friends and his work in Williamsburg. His traveling companions on the legal circuit in the rough backcountry town of Richmond were a "disagreeable" crowd.

> I do not like the ups and downs of a country life: today you are frolicking with a fine girl and tomorrow you are moping by yourself. Thank God! I shall shortly be where my happiness will be less interrupted.[37]

In the autumn of 1763 or early winter of 1764, Jefferson talked once more with Rebecca Burwell. On January 19, 1764, he wrote again from "Devilsburgh" to Page, maintaining a conceit in the letter that it had been a legal conference with a male business counterpart to disguise a conversation about marriage:

> I then opened my mind more freely and fully. I mentioned the necessity of my going to England and the delays which would consequently be occasioned. . . . I said in what manner I should conduct

> myself till then and explained my reasons. which appeared to give that satisfaction I could have wished. . . .
>
> After the proofs I have given of my sincerity, he [Rebecca] can be under no apprehensions of a change in my sentiments. . . . He is satisfied that I shall make him [her] an offer, and if he intends to accept of it, he will disregard those made by others.[38]

In other words, Jefferson had offered to consider himself engaged to Rebecca Burwell before and during his grand tour of Europe and she had agreed to wait for him to return! He refused to take Page's advice and go see Rebecca again, to press her further for an answer. His fate, he argued, depended on her "present resolutions."

> By them I must stand or fall. If they are not favorable to me, it is out of my power to say anything to make them so which I have not said already. So that a visit could not possibly be of the least weight and, it is, I am sure, what she does not in the least expect.[39]

Refusing to discuss the subject any further with his best friend, Jefferson changed it, asking to borrow a lawbook.

But he could hardly have been surprised to learn from Page that Rebecca had been incredulous at his idea of a lingering courtship while he traveled around Europe, and his stubborn indolence, his refusal to make another effort, indicates that he was beginning to believe all hope was already lost. Four days later, he made one more feeble effort in a letter to Page: "Put *campana in die* [Rebecca] in mind of me; tell [Rebecca] I think as I always did."[40]

Apparently, no word of encouragement came back. Late the night of March 20, 1764, Jefferson, suffering from a violent headache "with which I have been afflicted these two days," wrote from Williamsburg, where he had returned to Wythe's office, to Will Fleming that he had learned the worst. Even now, he insisted on being "circumspect" about his intended courtship of Rebecca:

> With regard to the scheme which I proposed to you some time since, I am sorry to tell you it is totally frustrated by Miss R. B.'s marriage with Jacqueline Ambler which the people here tell me they expect daily. I say, the people here tell me so, for [can you believe it?] I have been so abominably indolent as not to have seen her since last October. . . . Well, the lord bless her, I say![41]

Jefferson was to suffer from migraine headaches at times of great stress all his life, but he tried to make light of his disappointment, to wax the stoic to his friend:

> You say you are determined to be married as soon as possible and advise me to do the same. No thank ye; I will consider of it first. Many and great are the comforts of a single state. . . . For St. Paul only says that it is better to be married than to burn. . . .

If St. Paul had known of "means of extinguishing their fire than those of matrimony, he would have earnestly recommended them."[42]

A month later, Jefferson narrowly missed running into Rebecca, now Mrs. Ambler, at the home of the Burwells, to which he had been invited but declined to go. "What a high figure I should have cut had I gone!" he joked to his friend Page, who was about to marry Frances Burwell. "When I heard who visited you there I thought I had met with the narrowest escape in the world." But he could not help intellectualizing the meeting in retrospect. "I wonder how I should have behaved? I am sure I should have been at a great loss." Then he chided his friend Page for lingering with the Burwell sisters: "If your mistress can spare you a little time, your friends here [Williamsburg] would be very glad to see you, particularly Small and myself. . . . Adieu, dear Page."[43]

Young Jefferson had resisted the undertow of a tide of young friends marrying immediately after college. With a sigh of relief, he went back to his studies. He would miss the excitement of courting a sought-after belle, but, still only twenty-one and aware that he had seen little of the world outside Virginia, he was not ready to marry. For her part, no record has been saved of what Rebecca thought of all this. She had two daughters. One later wrote a letter mocking Jefferson in the *Atlantic Monthly*,[44] the other married John Marshall, who became Jefferson's mortal enemy. He also undoubtedly felt he had enough responsibilities for his brother, sisters, and his spendthrift mother. His account books show he was constantly having to transfer money to her from his share of his father's estate; sides of pork and quarters of beef for the table, allowances for his older sisters, first, then the younger children, then expenditures so that his retarded younger sister, Elizabeth, could be "well-dressed,"[45] expenses he paid until she died at the age of thirty. He could well have thought that to take on the added responsibilities of a wife and family of his own would be more than he could manage from his estate until he had established his law practice, despite the example of so many young friends who had just come into their inheritances and promptly married.

For nine months after he had left Williamsburg to go home to Shadwell at Christmas 1762, Jefferson remained at his mother's house. Reluctant to return to the capital because of an outbreak of smallpox, he continued his first legal study assignment from Wythe—wading through Coke's encyclopedic four-volume *Institutes,* beginning with *Commentaries upon Littleton,* what legal historian Frank L. Dewey has called "a cruelly difficult book to read, even for an adult with some familiarity with the law. . . . It must have been a severe test for a novice not yet twenty."[46] By October 1763, Jefferson was back in the capital, attending the fall General Court session "constantly" with Wythe. He went to Williamsburg intending to stay, and he remained there for two years working in Wythe's office, attending court and legislative sessions and preparing to take his bar examinations.

During these two years, the emphasis of his studies shifted from the theoretical to the practical. He attended the twenty-four working-day General Court sessions beginning each October 10 and April 10. Sometimes called the Supreme Court of Judicature, the General Court of Virginia was a court of equity as well as a common law court, where a majority of the colony's land tenure cases were heard along with felony cases and many other legal matters. He was also obliged to attend the "rule days" between sessions when procedural matters were disposed of by the court clerk. He no doubt also accompanied Wythe to the hustings Williamsburg and James City County courts which met monthly in the capital.

A good deal of Jefferson's law practice—slightly over one half—related to land titles. A person who had obtained a patent for Virginia land from the colony had to "seat and plant" the land within three years and to pay an annual quitrent fee of one shilling for each fifty acres. Anyone who failed to seat and plant in three years or went three years or more into arrears on quitrents could be brought before the General Court by informers to show cause why the land should not be forfeited. Since there was virtually no other agency for enforcing the land laws in colonial Virginia, the statutes permitted the use of informers to sue violators and to keep any penalties collected. The first informer who could establish the patent holder's delinquency was rewarded, after filing a petition for lapsed lands, with a certificate entitling him to a patent to the land once he paid the appropriate fees. Jefferson's first and second cases as a lawyer were just such land patent cases. He spent long hours researching ownership in the land office, where all the patent records were kept, and he attended the

June sessions of the Governor's Privy Council, where the cases were heard and decided.

When Jefferson was not in constant attendance on his legal master and his cases, he enjoyed the run of Wythe's extensive law practice as he crammed for his bar exams. Still, he became a steady customer in the bookstore of the *Virginia Gazette,* buying the lawbooks he needed for his studies and that he would take home to set up his rural practice far from any law library. Buying books was one of Jefferson's lifelong addictions, a passion he early learned to justify: "a proper collection of books" must be provided for the young lawyer, he wrote a would-be apprentice of his own a few years later. "A lawyer without books would be like a workman without tools." According to the daily ledger of the *Gazette*'s sales, on February 15, 1764, he bought six volumes, quite an armload: Robert Richardson's and Joseph Harrison's two-volume guides to practice in the courts of King Bench and Common Pleas and Joseph Harrison's two-volume guide to the Chancery Court, each with examples of legal pleadings in the second volume. He came back a few days later to buy the *Attorney's Pocket Companion,* a "guide to the practice of law." Jefferson was making a characteristically thorough study of pleading, the heart of common-law jurisprudence. Then he faced the central part of a law student's hard and lonely training —long hours of copying out the legal instruments to help him learn how to draft his own pleadings one day soon.

He was also cramming the statutes of Virginia and of England. On October 3, 1764, he bought William Rastell's *Collection in English of the Statutes Now in Force,* which included all British statutes from the Magna Carta in the early thirteenth century through the reign of James I in the early seventeenth century, which, with the seven volumes on procedures, he took only ten weeks to plow through. He was back at the *Gazette* by April 30, 1765, bought *Virginia Laws Since the Revisal,* and, just before he sat for his bar exams in October 1765, he purchased, no doubt to help with last-minute cramming, *Grounds and Rudiments of Law.*

If Wythe followed the same method of precepting for each student, then Wythe told Jefferson to read legal theory on his own and then to learn the practical side of the law in the time he spent with his mentor. No doubt following the same course as Wythe's later student, John Marshall, young Jefferson spent the fall and spring terms with Wythe, "reading law and taking notes"[47] and attending court. Especially during his first year with Wythe, Jefferson attended his mentor's court appearances, even though it pinched his time for studying. At Shadwell, Jefferson could keep to his fifteen-hour sunrise-to-midnight routine of studying law and reading literature: in Williamsburg this was impossible, with all his court and office duties. By the time his years with Wythe were over, he was to make a rare complaint about Wythe's encroachment on his time in a letter to his cousin, Thomas Turpin, who asked him to serve as his son's preceptor:

> I was always of opinion that the placing [of] a youth to study with an attorney was rather a prejudice than a help. . . . We [lawyers] are all too apt [to shift] onto them our business, to encroach on that time which should be devoted to their studies.[48]

Jefferson could eke out only the odd hour to read poetry and philosophy —he was reading Euripides, the most modern of the Greek tragedians. He was trying to keep his private readings alive, and he bought books, many books, against his future leisure and enjoyment. He bought for his own library the books he had once read at Maury's: Milton's *Works,* Hume's six-volume *History of England,* Robertson's *History of Scotland,* Stith's *History of Virginia,* Sali's *Koran,* Bacon's *Philosophy,* volumes of Cicero as well as modern Italian poetry, a scientific dictionary. He bought and read everything by Laurence Sterne, the only novels he ever appreciated, later describing *Tristram Shandy* as "the best course of morality that was ever written."[49] As his expense records and the ledger of sales at the *Virginia Gazette* bookstore attest, Jefferson spent freely to build up his first library, and he spent time and money with his friends, rarely missing a play, a concert, or a horse race. He was teaching himself Italian, first attracted by the notations on the music he bought for his violin playing, then studying the scores of operas, the latest rage to arrive in Williamsburg from Europe. He seems to have tried his tongue at German, too, but that language stumped him.

In the spring of 1765, Jefferson's personal plans and problems, like those of many Americans, were suddenly subsumed in a crisis that gripped the American colonies for nearly a year and was the first major development in a ten-year series of confrontations with the mother country, England, that eventually mushroomed into the American Revolution. England's assertion of authority over its American colonies in the form of new taxes came at a time of British ascendancy and as-yet-unparalleled global influence that flowed from her spectacular victory in the Seven Years' War, which actually began when Jefferson was but a boy of thirteen and lasted until he was twenty. To be sure, England had not won alone, but her allies, colonial and Continental, had lost almost as much as her enemies. Virtually all of Western Europe lay ruined and impoverished. Vast armies of peasant soldiery had been marshaled all over Europe to decide which ruler should tax the most men. A hundred towns and cities had been burned to the ground. In many villages, only women and children were left to till the fields. One in nine Prussians, half a million, had been killed in seven years. The war had opened the doors of Europe to sudden and permanent changes. Sweden, once a European power, receded from importance. Russia lost a staggering 120,000 men but, marching into the west for the first time, opened a new era in European history. For France, the losses were catastrophic: France lost her American empire.

But the Seven Years' War and Britain's place as the preeminent world power were won not on the fields of Europe, but on the Atlantic and in the forests of North America. Colonial trade equaled imperial power. A century's expansion of overseas trade had financed the expansion of both the British and French fleets until, by 1758, England boasted one hundred fifty-six ships-of-the-line, while France only had seventy-seven. British naval squadrons destroyed two French fleets in 1759, off Lagos, Portugal, and in Quiberon Bay. French overseas trade had plummeted from thirty million livres in 1755 to four million in 1760.

As young Thomas Jefferson rode east in 1760 to go to college, much of North America, from the St. Lawrence basin to the Gulf of Mexico, was still in French hands. French fortresses dominated Pittsburgh, Chicago, Detroit. French possessions blocked English expansion westward. The doorway to New France was a great fortress on the bluffs at Louisbourg, on Cape Breton Island at the mouth of the St. Lawrence: all the supplies and troops for the French *habitants* of the American interior had to come up the St. Lawrence. An English flotilla of forty-two vessels and eighteen thousand troops besieged Louisbourg in June 1758. The British interrupted French reinforcements at sea; British artillery onshore finished the job.

That same summer of 1760, after the fall of Quebec and Montreal, all French possessions from the Arctic Circle to the Falkland Islands fell under British rule. France was bankrupt, her colonies gone. To gain its great victories, England squeezed its purses empty. Landowners had paid up to twenty-five percent of the assessed value of their acres each year in war taxes; poor men paid high new taxes on their beer and tobacco, and the middle class paid twenty-five percent of value on houses, deeds, offices, brandy, and spirits. If a man owned a house, he paid a tax not only on it but on every window in it. No wonder they were resisting postwar proposals for further taxes to pay the bill for their great victory. The funded British national debt by January 1763, according to Exchequer accounts, had reached £122,603,336—a staggering sum that increased by £7 million in interest the next year and by that much again in the next six months. Instead of bringing a harvest of mercantile profits, the end of the costly imperial war brought a severe depression in British trade. Facing a large postwar debt, heavy and unpopular taxes at home, and the need to support an army of at least ten thousand Redcoats along the frontier, in Canada and on Caribbean islands, the Ministry sought revenue from the American colonies. Banning the import into the colonies of foreign rum and French wines and doubling the duties already on the books on goods affecting every facet of American life, the Ministry set out to enforce and efficiently administer the new imperial taxes.

A second blow from London was aimed directly at Virginia. The new Currency Act affected plantation as well as town. During the war, Virginia had issued £250,000 of legal-tender money. Now, the British banned all colonial paper money. Only British sterling or gold coins or credit for raw materials shipped to England by colonists could be accepted to pay debts owed to British citizens. By passing the new Sugar and Currency acts, Parliament radically extended British home government control over her colonial governments regardless of their ancient charter privileges. To Americans already struggling with a sharp postwar business decline, the deflationary Currency Act, shrinking the cash supply, and the strict enforcement of higher import duties threatened to ruin the colonial economy. The Boston Town Meeting in May 1764, with the effects of collapsing American trade visible along its dying waterfront, denounced taxation by Parliament without American representation and urged united opposition by the colonies.

The first polite Virginia protest came in the fall of 1764. In the Assembly, Jefferson watched and listened as George Wythe was appointed to a committee to draw up addresses to the king protesting the proposed internal tax on both practical and constitutional grounds. Wythe's draft was considerably stronger, he told Jefferson, than the toned-down version finally passed by his more conservative colleagues, who probably heeded Governor Fauquier. The governor advised dignified language: the dutiful colonists were supposed to be "praying to be permitted to tax themselves."[50]

In the Virginia Assembly, Wythe joined Pendleton and Richard Henry Lee in drafting a polite yet adamant refusal to give English creditors a more favorable position than American creditors. While America depended on Britain, they pointed out that

> . . . This is not a dependence of a people subjugated by the arms of a conqueror but of sons sent out to explore and [settle] a New World for the mutual benefit of themselves and their common parent.[51]

Peaceful boycotts and humble prayers of petition were answered in February 1765, when the House of Commons passed a stamp tax on the North American colonies which, together with import duties, was supposed to raise one-third of the cost of the upkeep of Britain's colonial military establishment. The Currency Act of 1764 had asserted parliamentary control over colonial legislatures; the Stamp Act, the first direct internal tax ever levied by Parliament on the American colonies, taxed newspapers, almanacs, pamphlets, broadsides, all diplomas and legal documents, insurance policies, all licenses, even dice and playing cards. Until now, only the colonial assemblies had been empowered to set and collect taxes within the colonies, while the British collected external customs duties on imports at the water's edge and had to ask the colonies for special internal levies for money and supplies in wartime.

Word of this new tax was on shipboard bound for Williamsburg and other colonial capitals as Jefferson's flirtation with Rebecca Burwell came to its frustrating end. Americans instantly reacted, fearing the new tax was to be the first of many. The stamp tax fell heaviest on the most influential citizens—lawyers, printers, tavern owners, land speculators and settlers, merchants, shipowners and students—arousing broad-based opposition. The act also granted jurisdiction in cases of alleged violations to the admiralty courts, where British-appointed judges, not American juries made up of their neighbors, tried maritime law cases, including accusations of smuggling. This shift of jurisdiction raised fears of an assault upon the jealously guarded English right to trial by jury. The new tax stamps had to be purchased with scarce gold or silver.

Tensions had been building for years in Virginia, never very subservient to England. One contemporary British traveler had described the Virginians as "haughty and jealous of their liberties, impatient of restraint. . . . [They] can scarcely bear the thought of being controlled by any superior power."[52] Yet Virginians had been loyal and generous in supporting the British during the recent war, never refusing supplies or troops.

On the long Virginia frontier, trouble with the Indian neighbors had also been brewing for years. No Americans, as historian Robert Middlekauff has put it, were "more aggressive or greedy than the Virginians."[53] Invoking its seventeenth-century charter, Virginia was still claiming the entire region north and west of the Ohio River, most of what today is called the Midwest. Small groups and individual settlers had been slipping into the region for a quarter century, their numbers increasing after the defeat of the French and Indians. Earlier, while Peter Jefferson was still exploring and settling lands on the Virginia frontier, ambitious Virginia land speculators—including George Washington and several Lees

—had banded together to form the Ohio Company, receiving a royal charter for two hundred thousand acres south of present-day Pittsburgh. Washington had been on his way to claim these lands in 1754 when his party had clashed with the French, inadvertently touching off the last of the French and Indian Wars. This charter not only whetted Virginia appetites for more land but opened up competition with other colonies. Fur-trading companies staked out conflicting claims to the vast wilderness vacuum created by the collapse of French hegemony over the Indians, and thousands of squatters moved west to lands the Indians had long considered their own.

Many British officers who had led the attacks on the French and their Indian confederates shared the English frontiersman's hatred for the Indians but they tried to curtail white expansion into Indian country, sensing there would be trouble. In 1761, the home government, acting on information from Indian superintendents in America that there was widespread encroachment all along the frontier, had taken control of Indian land affairs out of the hands of colonial governors and forbade them to grant land even within the colonies if these grants would interfere with Indian rights. All applications henceforth were to be sent to the Board of Trade and Plantations in London, three thousand miles away. This failed, however, to stem the tide of settlement in the west and, on October 7, 1763, the British Ministry issued a royal proclamation closing the west between the Appalachian Mountains and the Mississippi River to white settlers and protecting it as an Indian reserve. Peter Jefferson would never have been able to build an estate in Virginia or beyond its western mountains had he been born a generation later. The century-and-a-half westward expansion of Virginia was stopped dead by this massive extension of British authority. From his hilltop at Monticello, Jefferson could see the western limit of white settlement. No wonder he decided to specialize in the form of land law made more than ever necessary by the new British rules.

As the colonial crisis worsened, the House of Burgesses passed statutes that named a London agent and a Committee of Correspondence to direct his activities before the Board of Trade and other British government agencies. George Wythe was one of eight burgesses appointed to the committee. His young clerk Jefferson now had duties that took him into the orbit of politics for the first time. He also was expected to follow the affairs of four other committees Wythe served on, including the key Committee of Trade.

It was at this moment, too, that the Parson's Cause came to the center of the controversy. Wythe had been retained in two of the four appeals by

Anglican clergy for back pay after the Privy Council in London had overturned the Two-Penny Act of 1758. During Jefferson's first year with Reverend Maury, the Virginia House of Burgesses had passed the Two-Penny Act, which provided that each clergyman was to be paid not the annual stipend of sixteen thousand pounds of tobacco (which had long been the custom) but in hard money at the fixed rate of two pennies for each pound of tobacco wages due instead of the prevailing fourpence per pound. It was wartime and, with marauding ships at sea, what tobacco reached England fetched a high price. Reverend Maury was one of the organizers and the spokesman for what became famous as "the Parson's Cause," one of the first courtroom struggles between colonial taxpayers and the British on the road to revolution.

When cash became scarce, the Virginia House of Burgesses had reinstated the old method of payment. Reverend Maury then drafted a successful province-wide clergy petition to the King's Privy Council in London that overruled the House of Burgesses and disallowed the Two-Penny Act. Then Reverend Maury and the other pastors sued for three years' back pay. Wythe was the presiding justice in his home county court, undoubtedly taking his clerk Jefferson to Elizabeth City with him when the first clergy lawsuit against the Anglican lay vestry went to trial. But the most celebrated confrontation between the clergy and the Anglican gentry, who had come to despise them, was instituted by Reverend Maury in Louisa County. The lay vestry of that county had, in March 1761, explicitly rejected Maury's plea for the £400 in back salary due him for the three years the Two-Penny Act was in effect. Maury chose to have his suit tried in neighboring Hanover County, where he assumed the Louisa vestry was less influential. But he inadvertently brought his case into a county where dissenting anti-Anglican sects were strong. Worse luck for him, this brought his case into the growing sphere of influence of young Patrick Henry.

When the case came before the court in November 1763, shortly after Jefferson entered Wythe's law practice, the presiding judge was Col. John Henry, a devout Anglican and Patrick Henry's father. Judge Henry departed from the previous county decisions in the Parson's Cause and ruled that the Two-Penny Act was null and void from its inception. Therefore the only question for the jury was how much to pay Maury. Presumably this was the difference between the market price for the tobacco and the twopence on the pound already paid him. Until Judge Henry's ruling, all of the parsons' cases had been in favor of the Virginia taxpayers, and the parsons had been unable to collect back pay. In a modern court of law, Judge Henry would have had to disqualify himself; he was not only an

Anglican but a vestryman of St. Paul's Church in Hanover County, where his brother, the Rev. Mr. Patrick Henry, was the rector of St. Paul's and stood to benefit handsomely from the ruling. As if to cure such blatant nepotism by administering a further dose, at this point the county collector of taxes retained young Patrick Henry to defend the taxpayers' interest against the clerics' raid on their wallets. There was a certain logic to the appointment: it seemed wise to hire someone who might have influence before the court if there was to be any hope of reversing Judge Henry's devastating ruling.

Maury two weeks later wrote a long letter to a fellow parson describing the scene as Patrick Henry entered public life by way of his family's court. After excusing two men from jury duty whose involvement in the case was apparent, the sheriff selected jurors from "the vulgar herd," as Maury put it, including at least three members of dissenting sects, all of whom had a vested interest in undermining the power and authority of the Anglican establishment. One was a relative of Patrick Henry. Several jurors were holders of considerable acreage who would be taxed heavily if the parson won his suit; the others, less wealthy yeomen farmers, could afford the outcome even less. When Maury protested these selections, Henry "insisted they were honest men and, therefore, unexceptionable . . . They were immediately called to the book and sworn."[54]

After a tobacco dealer established the value of the tobacco in the three years in question and Maury produced a receipt showing he had been paid roughly one-third of this amount, it should have taken only a little arithmetic to calculate his payment. But it was at precisely this moment that Patrick Henry made his political debut. The only contemporary account of what he said was by the obviously biased Reverend Maury, who called it a "harangue," but he nevertheless took the only notes. Henry opened by asserting that the Two-Penny Act, which had been overturned by the Privy Council,

> had every characteristic of a good law; that it was a law of general utility and could not, consistently with what he called the original compact between King and people, stipulating protection on the one hand and obedience on the other, be annulled. . . . A King, by disallowing Acts of this salutary nature, from being the father of his people, degenerates into a tyrant, and forfeits all rights to his subjects' obedience.[55]

With these words, Henry was making one of the first public appeals in America to John Locke's social-compact theory of government as a contract between the king and his subjects, and he evoked an immediate uproar. Maury's lawyer, Peter Lyons, "called out aloud and with an honest warmth," wrote Maury, "that the gentleman had spoken treason."[56] As others echoed the cry of "treason," Patrick Henry, uninterrupted by his father the judge, went on. Virtually unschooled in the law, young Henry had gone to the heart of the matter, drawing on a doctrine of tyrannicide first uttered by John Knox, the Scottish founder of Presbyterianism, two centuries earlier: the king's subjects were freed from obeying their sovereign when that sovereign did not fulfill his part of the compact. Scotch-Irish Presbyterian jurors in Hanover County Court must have known his words already, and they sat there nodding as Henry went on. Not only had the king broken the social contract but the Anglican clergy, he said,

> had most notoriously counteracted those great ends of their institution, that therefore, instead of useful members of the state, they ought to be considered as enemies of the community.[57]

Maury was amazed that Henry was allowed to proceed "in the same treasonable and licentious strain without interruption from the Bench, nay, even without receiving the least exterior notice of disapprobation." The parson considered Patrick Henry a "little petty-fogging attorney" whose "sole view in engaging the cause" was to "render himself popular."[58] And popular Patrick Henry became. After deliberating less than five minutes, the jury brought in a verdict ordering the tax collectors of Louisa County to pay James Maury one penny in damages. Henry received fifteen shillings, and in the next year his business doubled. Patrick Henry had made his first step toward building a reputation as an orator and defender of colonial rights against the claims of the clergy and Crown of England that carried him into the House of Burgesses as the representative of Louisa County little more than a year later.

When Henry took his seat in the House, not every burgess was impressed by his powers. George Wythe, one of his bar examiners, had been horrified at Henry's brief, slipshod legal training—something on the order of six weeks' reading—and had refused to sign his law license. Jefferson would later write that Henry was "totally unqualified for anything but mere jury causes"[59] before the county courts, where he could prevail with his oratory.

News that the Stamp Act had been enacted to help support British troops on the Virginia frontier reached Williamsburg early in April 1765. For six weeks, almost nothing about the act appeared in colonial newspapers, and no public body in America, including the Virginia House of Burgesses, seemed eager to be the first to criticize the home government.

Part of the formation of a Virginia attorney, as Jefferson wrote, was to "pay attention to what was passing in the legislature,"[60] especially since his mentor was so active a member of the House. Only thirty-nine of the one hundred sixteen burgesses remained in Williamsburg; the others either did not expect any discussion of the Stamp Act or had no reason to oppose it and, having concluded all business that interested them, had gone home. So long as Wythe remained, however, Jefferson had to stay in attendance. But he hardly had to be prodded. He probably had heard rumors, maybe even from Henry himself, that his friend Patrick Henry was preparing to make a protest speech against the Stamp Act. Henry was in the habit of staying with Jefferson whenever he came to Williamsburg and only a week earlier had been sworn in as the new representative from Louisa County on the frontier, where anti-English sentiment was running high. Undoubtedly Jefferson had heard that Henry had presented to the committee of the whole on May 28, 1765, resolutions to be debated the next day.

There was still no visitors' gallery in the Capitol, so Jefferson and his friend and fellow law student, John Tyler, stood that Wednesday, May 29, 1765, in a crowded hallway between the lobby and the burgesses' chamber, craning their necks to see and hear the debate. Patrick Henry, turned twenty-nine that day, had urged the House to go on record opposed to the Stamp Act. His own account of what followed was terse:

> I had been for the first time elected a Burgess a few days before, was young, inexperienced, unacquainted with the forms of the House and the members that composed it. Finding the men of weight averse to opposition and the commencement of the tax at hand and that no person was likely to set forth, I determined to venture and alone, unadvised and unassisted, on a blank leaf of an old law book, wrote the [Stamp Act Resolves].[61]

He wanted, first, a House resolution stating that Virginians had "all the privileges, franchises and immunities" of "the people of Great Britain." They were entitled to these rights "as if they had been abiding and born" in England. Furthermore, only persons chosen by colonists should decide "what taxes the people are able to bear." That, he argued, was "the distinguishing characteristic of British freedom." The people of Virginia had never forfeited their right to have "their own assembly,"[62] tax, and police themselves. There was really nothing new here: Jefferson's mentor, Wythe, in the chamber that day, had written much the same resolutions the year before in an attempt to avert a British levy on molasses imported into the colonies. But then Henry added new language, raising the temperature of the colonial protest significantly:

> Resolved, therefore, that the General Assembly of this colony have the *only* and sole *exclusive* right and power to lay taxes and impositions upon the inhabitants of this colony and that every attempt to vest such power in any person or persons whatsoever other than the General Assembly . . . has a manifest tendency to destroy *British* as well as *American* freedom.[63]

Patrick Henry was making more than an abstract assertion of colonial rights. This was nothing less than an explicit denial of British authority to tax the American colonies. Even the use of the word *American* was a relatively bold new idea. Most of Henry's listeners considered themselves English first and Virginian a close second. "American" implied a union of interests and rights set apart from England's, and Henry's comments, which he must have known would be quickly disseminated by the colonial press, called for Virginia to lead the way for an implicit and vigorous denunciation of the Stamp Act. According to Henry's own recollection,

> upon offering them [the resolutions] to the House, violent debates ensued. . . . Many threats were uttered and much abuse cast on me by the party for submission.[64]

But Henry was not prepared to submit. In a speech supporting his resolutions, he supposedly exclaimed, "Tarquin and Caesar had each his Brutus, Charles the First his Cromwell, and George the Third . . ." Before he could finish the phrase, red-robed Speaker of the House John Robinson cried, "Treason! Treason," as other burgesses took up the cry. But Henry stared the Speaker in the eye and finished his sentence: ". . . may profit by their example! If this be treason, make the most of it!"[65]

In the hallway, Jefferson, who had the habit of closing his eyes when he heard Patrick Henry speak, was stunned by the speech, its raw power, its effect on the veteran lawmakers. Fifty years later, he still recalled its impact:

> I well remember the cry of treason, the pause of Mr. Henry at the name of George III and the presence of mind with which he closed the sentence and baffled the charge vociferated.[66]

In his autobiography, Jefferson honored Henry's "splendid display" that day, calling his talents as a popular orator "such as I have never heard from any other man."[67] Jefferson admired Henry, but he was much too discreet ever to allow himself to emulate him.

The House of Burgesses voted the next day, May 30, passing the first four resolutions by small majorities. The fifth caused a heated debate and passed by a single vote, causing Peyton Randolph, an opponent, to exclaim to Jefferson as he shouldered his way out of the chamber, "By God, I would have given 500 guineas for a single vote!"[68] Carefully Jefferson tallied Henry's opponents—nearly every influential Virginian, including George Wythe. The next morning, May 31, he stood at the end of the clerk's table, waiting to see what was up next, and watched Col. Peter Randolph, his former guardian and a member of the Privy Council, for an hour or more thumbing through the volumes of the House journal, looking for a precedent for expunging a vote. Soon afterward there came a motion to expunge Henry's fifth resolve. Henry was not there to oppose it. He had pulled on his buckskins and ridden home to the hills. His absence was enough to split the vote, leaving the way open for Speaker Robinson to nullify it. There was a "small alteration,"[69] the governor reported to London. The "timid members,"[70] remembered Jefferson, had had a change of heart. The "young, hot and giddy members,"[71] as Governor Fauquier more accurately described them, indeed had had a change of heart. No trace of the May 31 debate was entered in the *Burgesses Journal,* and before there could be any further trouble, the governor dissolved the Assembly.

But Patrick Henry had released the genie of colonial resistance from the bottle of apparent imperial order. Before the summer of 1765 was over, despite the fact that Virginia officially defeated Henry's formal protest and the editor of the *Virginia Gazette* refused to print his resolves, other papers, in Maryland, Philadelphia, and even as far away as Newport, Rhode Island, carried them, sometimes in altered form. By September, the Rhode Island legislature passed Henry's resolves virtually intact. While Virginia had stopped short of official disapproval, nine of the thirteen mainland American colonies passed declaratory resolutions defining their rights. In all of these, as Henry had wished, the limits set on parliamentary authority excluded the right to tax the colonies. As for Jefferson, years later he admitted he had been stunned by Patrick Henry's "Homeric" oratory, remembering that his views at this stage were in "exact conformity" with Patrick Henry's, and noting that the votes backing Henry had been cast by delegates from the middle and upland country, his home country. Jefferson's view of politics was forever changed by Henry's Stamp

Act speech, and he always credited Patrick Henry with initiating the revolutionary movement in Virginia.

As soon as the June court term was over, Jefferson rode home to Shadwell, where he soon began a deep and systematic study of political philosophy. But first he had to pass his bar examinations. There was no reason to hurry back to Williamsburg: as the Stamp Act protest spread, lawyers all over America boycotted the courts, following the pattern set in New Jersey, where eight hundred lawyers surrounded a courthouse at New Brunswick and agreed to boycott the use of the stamps rather than affix them to documents as required by the new law. Jefferson remained at Shadwell until late October, where reports of the protest movement reached him.

He was still in the mountains when the stamps actually arrived in Williamsburg, and he must have heard of Governor Fauquier's heroic stand in the face of an angry crowd when Colonel Mercer, the stamp collector for Virginia, arrived with a strongbox of the detested stamps. Governor Fauquier had gone to the coffeehouse near the Capitol and was sitting on the porch with Speaker Robinson and several members of the Governor's Council when Mercer arrived, followed closely by a crowd that would have been considered a mob if its members had not been so well dressed. In a report to the Board of Trade in London, Fauquier wrote

> I immediately heard a cry, "See the Governor. Take care of him." Those who were pushing up the steps immediately fell back and left a small space between me and them.

Fauquier had been a popular governor. The face-off lasted only a few seconds. Then the royal governor stepped off the porch with the stamp collector "side by side through the thickest of the people," ignoring "some little murmurs."[72] They were unmolested, but Mercer soon resigned. Before the year was out, not a single stamp collector dared remain at his post in colonial America.

Amid the political turmoil brought on by the Stamp Act crisis, Thomas Jefferson made an auspicious political debut not as a fiery orator like his friend Patrick Henry but as a pragmatist, a young man of the Enlightenment analyzing a problem, proposing a solution, and using the existing political system to bring about change.

With the county courts of Virginia closed by the lawyers' boycott, Jefferson passed the first two years after he came of age in 1764 largely at home in Albemarle County. While he considered himself "yet a student of law" when Henry proposed his Resolutions in May 1765, according to the terms of his father's will, he had come into his estate in April 1764, when he turned twenty-one. He immediately declared his financial independence from his mother and his five guardians. Now the overseers worked for him and, while he allowed the management of his plantations to continue in some ways as it had before, in 1765 he began renting the four hundred acres at Shadwell and five slaves from his mother at a specified rate. He kept careful track of loins of pork and sacks of grain he provided her, bartering them against the rent. He took over operation of his father's profitable, water-powered grist mill on the Rivanna. While the executors remained legally responsible for the younger children, Thomas assumed his place as head of the family and manager of its business affairs.

As he studied the affairs of his estate, one of the first matters to which the new Albemarle planter turned his attention was an attempt to ease the shipment of his harvest to market. Thomas learned that a canoe could be used to navigate the lower Rivanna. Climbing into his own canoe, he explored the river. He already knew that no hogsheads of tobacco had ever been transported along the north branch of the James to its confluence with the larger river. As a result, the cash crops of the Jeffersons and other Rivanna planters had to be hauled tediously and expensively for miles to the James, where they were loaded on ships. Exploring the only seriously obstructed section of the stream, he observed that the Rivanna could be made navigable below Milton Falls, within a mile or so of his property, merely by removing loose rock.

Organizing a subscription drive among Albemarle neighbors, he raised £200 and brought the project to the attention of his neighbor, Dr. Thomas Walker, a member of the House of Burgesses. Together they worked to expand the project and line up support. In October 1765, an act of Assembly was passed "for clearing the great falls of James River, the river Chickahominy and the north branch of James River."[73] The act authorized as "laudable and useful" young Jefferson's first public project and allowed private subscriptions to carry out the work without any expense to taxpayers. Three committees of trustees were formed to raise subscriptions, sign contracts, and oversee the work. Jefferson was the organizer and most active member of the group assigned to the Rivanna clearance, and he worked closely with other prominent Albemarle gentlemen. The project took some five years to complete; moreover, Jefferson was to have long-lasting dealings with almost every person involved in the river project either in private or public life. For fully a generation, the "completely and fully cleared"[74] Rivanna carried Jefferson's and his neighbors' harvests toward market: Jefferson remained proud of his maiden political endeavor even when he was elected president more than thirty years later. He asked himself if his country was any better for his political efforts. When he drew up a list, in answer, of constructive public services in which he had been instrumental, he listed first his youthful Rivanna project and second his authorship of the Declaration of Independence.

Thomas Jefferson's decision to be a lawyer as well as a planter was not unusual: prominent Virginia planters often had adjunct careers as merchants, politicians, surveyors, or lawyers. Once the crops were in the ground or in hogsheads en route to London, they had abundant leisure, and their status as gentlemen required them to be reasonably well educated. But they were more noted for their ability to acquire land than to manage it. Stewards and overseers managed plantations, leaving planters free to enhance their incomes by avocation. Among enlightened younger planters such as Jefferson and Washington, there was a trend toward diversification: when tobacco sales or prices were down, planters were strapped. Even a sideline in law was no guarantee, however; when tobacco sales were down, lawyers went unpaid. No doubt knowing this, young Jefferson applied himself to learning ways to manage his plantation more efficiently. While it has been thought that Jefferson did not keep systematic accounts before August 1767, an expense account book beginning in 1764 records Jefferson's earliest known land transactions and shows that he was actively rearranging his inherited holdings by the age of twenty-two. A journal of land, stock, and slave transactions beginning in 1764 and running to 1778 shows young Jefferson buying to augment his family's holdings on Fine Creek in Cumberland County, spending one hundred pounds for four hundred acres on the Tye River in Amherst County, and picking up three hundred acres on Piney River, also in Amherst County, for ninety pounds. Even while he was studying law, young Jefferson was being infected by the land fever that consumed the cash of so many Virginia planters.

To keep track of his increasingly complicated financial affairs, Jefferson opened a series of account books. As a youth, he had studied "bookkeeping or merchants-accompts, after the Italian manner, by way of double-entry" in *The American Instructor: or Young Man's Companion;* the ninth edition, published in 1748, was in his library.[75] Three books were to be used: a waste-book, a journal, and a ledger. Out-of-pocket expenditures and collections were to be posted daily in the waste-book, then transferred with the particulars of the transaction to the journal. Finally, all transactions were to be transferred to the ledger, debits on the left page, credits on the right-hand facing page, where they could not be altered. Jefferson never followed these practices, preferring to experiment. He used a single-entry system: a transaction, he reasoned, was either a debit or a credit, but not both. He maintained special account books for his fees as a lawyer, but he did not separate the expenses of his practice from his personal expenses. At first he kept only a record of his cash and of his sale of stock, land, and slaves, but he gradually began to keep more complete records. Jotting down his out-of-pocket expenses on the blank pages of the thin, vest-pocket-size *Virginia Almanac,* he jotted down other personal observations in his memoranda books, which became not only a bookkeeping record but a formal personal journal that he was to maintain for more than sixty years.

Jefferson's account fell short of the state of accounting practiced in England at the time, but he kept business records all his adult life that, according to Steven Hochman, a business historian who has studied Jefferson's accounts, "are probably the fullest that survive of an American planter or government official of his era."

> This is because, in addition to keeping account books, he retained notes on business transactions and copies of his correspondence. . . . The question is not why Jefferson did not keep more sophisticated records, but why Americans in general did not.[76]

One answer, Jefferson might have told him, was the perpetual shortage of cash in America before the Revolution. Jefferson's ledgers reveal he often had to borrow cash from his servants: on November 29, 1768, he "borrowed of Jupiter"[77] one shilling threepence, repaying him two days later. He also had to borrow cash from relatives and friends, and he lent out small amounts when they ran short. To overcome the cash shortage, he bartered for goods and services, as did other Virginians. But even though the economy ran on barter, when exchanges of crops and goods were involved, it could be years, even decades before accounts could be settled, and much of the voluminous record-keeping was to carry over old accounts for many years. When Jefferson died in 1826, he still owed money for some debts he had incurred in the 1760s.

Jefferson had no reason to keep cumbersome double-entry books: all he wanted was as clear a picture as possible of what he owed, who owed him, how much income he was making, and how much cash he had on hand. While Jefferson was a meticulous record-keeper, recording much raw data such as disbursements, receipts, loans, and contracts, he was by no means an analytical accountant and could not tell from day to day or year to year exactly where he stood financially. He could, for example, have chosen to keep a ledger in which he preserved a comprehensive record of his assets and liabilities, but he did not. For him, the efficacy of his financial records was subservient to his need to draw conclusions, make decisions, and plot his future course. But his records enable us to study his thought processes as well as the ring of his contacts and his role in events.

While Jefferson continued his systematic study of the law, in October 1765, probably at the same time the House of Burgesses was endorsing his navigation project, he presented himself for bar examinations by Robert Carter Nicholas, John Randolph, and his preceptor George Wythe to qualify to practice in the county courts. Not that he ever intended to practice at the county level; he merely wanted access there. He passed his bars, then left Wythe and Williamsburg behind. The county courts remained closed by the lawyers' stamp boycott anyway, and Jefferson had decided a course unique among Virginia lawyers of his times. He would practice only in the higher General Court of Virginia—without an apprenticeship at the county-court level. It was only at the high-court level that Jefferson believed, as he wrote years later, that law could be pursued scientifically, as a "science may be encouraged and may live."[78] At the county-court level, where poorly trained lawyers swarmed after clients, there were no enlightened men of science, only "insects."

In July of 1765, before his bar examinations, Jefferson had gone home for a celebration, the marriage of his youngest sister, Martha, to his horse-racing schoolmate from Maury's and William and Mary, Dabney Carr, one of his most intimate friends. Martha married at nineteen, as had his older sister, Mary. But Jane, his favorite sister and the firstborn child, had never married. Jefferson seems to have been closer to her than he ever was to his mother. Jane had helped to raise Thomas and devotedly cared for their retarded younger sister, Elizabeth, four years her junior. She had made Jefferson's home visits to Shadwell bearable with her singing—Jefferson remembered her as "a singer of uncommon skill—and sweetness."[79] He considered her, of them all, his "equal in understanding," as his granddaughter later put it. He would always remember their evenings together. On "many a soft summer twilight on the wooded banks of the Rivanna," his granddaughter said, the family "heard their voices, accompanied by the notes of his violin, thus ascending together."[80] Recently they had been planting flowers together at Shadwell. At the time of Martha's wedding on July 20, 1765, Jane was twenty-five and considered a spinster in Virginia society.

Then, on October 1, 1765, ten weeks after Martha's wedding, Jane died. There is no record of the cause of her death, although it may have been smallpox, which raged in Virginia at this time; shortly after Jane died,

Jefferson journeyed all the way to Philadelphia to be inoculated, and for years, he fought a legal battle without fee against the opponents of legalizing smallpox inoculation in Virginia. Jane's death was the first death in his family since his father's. He could never write directly about her loss and he probably never accepted it. When he was an old man, sitting in church, he was moved to tears when the congregation sang a psalm he had first heard sung by Jane. His granddaughter averred that, as an old man, he still spoke of Jane "in terms of warm admiration and love as if the grave had but just closed over her."[81] And, six years later, when he began to build on his nearby mountain, he could not stand the idea of leaving Jane behind at Shadwell. His plans included a family cemetery, and he arranged to have her body exhumed and carried to his hilltop. There is no hint of any sense that his mother had lost her firstborn child as Jefferson took her daughter's body away from her. Over her plot he designed a temple of meditation. In an *Account Book,* he neatly penned his intention to

> choose out for a burying place some unfrequented vale in the park where is "no sound to break the stillness but a brook that, bubbling, winds among the weeds; no mark of human shape that had been there, unless the skeleton of some poor wretch, Who sought that place out to despair and die in" . . . Let it be among ancient and venerable oaks; interspersed some gloomy evergreens. The area circular, about sixty feet in diameter, encircled with an untrimmed hedge of cedar, in the center of it, erect a small Gothic temple of antique appearance. . . . Let the exit look on a small and distant part of the Blue Mountains. . . . In the middle of the temple an altar, the sides of turf, the top of plain stone, very little light, perhaps none at all, save only the feeble ray of an half-extinguished lamp.[82]

In the pages of the *Account Book,* he also wrote her melancholy epitaph, in Latin, rich in puns and inspired by the philosopher-gardener William Shenstone.

Jane Jefferson
Ah, Jane, best of girls!
Flower snatched away in its bloom!
May the earth weigh lightly upon you!
Farewell for a long, long time.[83]

The next spring he planted flowers in her memory, and he began to take notes that he would eventually transfer into his famous *Garden Book*, making his first notation, "Purple hyacinth begins to bloom," on March 26, 1766. On April 6, he noted "Narcissus and Puckoon open"; one week later, on his twenty-third birthday, "Puckoon flowers fallen."[84] He carefully wrote down the births and deaths of flowers and eventually of crops, of wheat and tobacco and vegetables, for the rest of his life. That spring until he left for a trip north for smallpox inoculation, he noted the appearance and rapid disappearance of spring flowers. His first notations were of fragile flowers that were beautiful and gave the first pleasure in the springtime and then quickly died.

With the flowers of spring came renewal. Still numb from Jane's death, Jefferson, now twenty-two, returned to Williamsburg in April 1766 to apply to practice before the General Court. He began to read intensively about politics, his new passion. There was a one-year waiting period before he could begin to practice law, most likely designed to prevent students from stealing the clients of their preceptors. Jefferson, probably prompted by questions raised by the Stamp Act crisis, used the time to plunge into Bolingbroke's five-volume *Philosophical Transactions.* It was becoming a lifelong trait, a cycle of deep, analytical scholarship following a great loss. He had immersed himself in Greek and Latin after his father's death. Now, with Jane gone, there was Bolingbroke. One day, he would come as close as he ever did to a personal comment when he wrote to his grandson, Thomas Jefferson Randolph, "These little returns into ourselves, this self-catechizing habit leads to the prudent selection and steady pursuit of what is right."[85]

Jefferson had copied out moralistic lines from Bolingbroke after his father's death; now he returned to copy out fifty-eight pages in his *Literary Commonplace Book* in his mature, vertical, tight handwriting. In all, he commonplaced some ten thousand words of Bolingbroke, six times more than those of any other author. He studied especially four long essays, posthumously published only ten years earlier. Viscount Bolingbroke had, in the words of Dr. Samuel Johnson, fired a "blunderbuss against religion."[86] The essays made a powerful impression on Jefferson as he worked over them during the next two years while the colonial crisis continued. He was, to a remarkable extent, to adopt Bolingbroke's philosophical views: his thoroughgoing materialism, his rationalistic rejection of metaphysics and of all speculation that went beyond the human mind, his uncompromising belief in reason as the sole and final arbiter of knowledge and worth. He also absorbed Bolingbroke's opinion of churchmen as corrupters of Christianity, and he came away with a strong skepticism about the historical accuracy of the Bible.

After Jane's death, he could not stand to stay at Shadwell until he was allowed to begin practicing law. For years he had longed to travel around Europe and as far as Egypt; now, he settled for a trip to the north as far as New York. Even now, he had to persuade himself that there was some practical benefit. Smallpox raged intermittently in the Tidewater, yet inoculation against it was considered too dangerous. The technique in use, "variolation," was generally believed to spread rather than curb the dread disease. Dr. George Gilmer, a nephew of Jefferson's neighbor Dr. Thomas Walker, had been trained as a physician in the elite medical school in Edinburgh, but he was forbidden to carry out the inoculation procedure in Virginia. He wrote his "particular friend" Jefferson a letter of recommendation to Dr. John Morgan in Philadelphia.

It was Jefferson's first time outside his native colony, his first three-month trip, from May to August 1766. He had counted on the lively company of his fun-loving college roommate Frank Willis, but Willis had to start early and left Jefferson to travel alone.

As he forded the Pamunkey River in a one-horse carriage, he almost turned it over "in water so deep as to run over the cushion as I sat on it," he wrote at the beginning of a series of playful letters to John Page from stops along the way. In Annapolis, he visited the lower house of the Maryland Assembly in session:

> I went into the lower house, sitting in an old courthouse which, judging from its form and appearance, was built in the year one. I was surprised, on approaching it, to hear as great a noise and hubbub as you will usually observe at a public meeting of the planters in Virginia. The first object which struck me . . . was the figure of a little old man dressed but indifferently with a yellow queue wig on . . . [He was] the speaker, a man of . . . very little [of] the air of a speaker.

The clerk of the Assembly read "with a schoolboy tone and an abrupt pause at every half dozen words." Accustomed to the dazzling opulence of Williamsburg, Jefferson poked fun at the Maryland lawmakers. "The mob (for such was their appearance)" sat with their hats on and "were divided into little clubs amusing themselves in the common chit chat ways." Jefferson was "surprised to see them address the speaker without rising from their seats, and three, four and five at a time without being checked."

If Jefferson was unimpressed by Maryland's government and government buildings, he found its houses "in general are better than those in Williamsburg" and the harbor "extremely beautiful." He arrived in time to take part in the "rejoicings here on the repeal by Parliament of the Stamp Act."[87] He drove on to Philadelphia, where there were fireworks and salutes of cannon, but Jefferson was more excited by the apple orchards he found along the road; he made a mental note to introduce apples into the garden he was planning on his mountain. In New York, he visited Trinity Church. He had recently purchased his first book on architecture from a cabinetmaker in Williamsburg. His taste for architecture and travel only whetted by the journey, Jefferson was nonetheless out of time. He sailed south to begin his important if short-lived career as a Virginia attorney.

5

"An Untiring Spirit of Investigation"

God works wonders now and then,
Behold, a lawyer, an honest man.

—*Poor Richard's Almanac*, 1752

ONE CHILLY MOUNTAIN morning in March 1767, Thomas Jefferson, five years of uncommonly rigorous study of the law behind him, tugged up the pair of overalls he always wore when he went riding and nudged his horse west from Shadwell toward Rockfish Gap and the great valley of the Shenandoah. His talkative brother-in-law, Dabney Carr, was at his side as they headed for the opening of the spring court term at Staunton, seat of the vast new frontier county of Augusta. In the distance, the pale diaphanous green of hardwoods veiled hillsides in range after range of the Shenandoahs. Along the banks of the Rivanna as it cut through South Mountain, brick-red tobacco fields lay freshly turned and puddled as field slaves labored to prepare another harvest. The two young lawyers urged their horses around muddy ruts in the road, their mounted slaves keeping up behind them.

Ever since Jefferson's weekend visits home from Reverend Maury's classical academy with his high-spirited schoolmate, their outings together had been a source of unmitigated enjoyment. Jefferson admired Carr's ability to relish the ordinary. As he would write some years later to their mutual friend, John Page, Carr possessed "the art of extracting comfort from things the most trivial. Every incident in life he takes [so] as to render it a source of pleasure."[1] Since Carr's marriage to Jefferson's younger sister Martha two years earlier, Dabney Carr had become more Jefferson's friend than his brother-in-law. Dabney Carr was everything that Thomas Jefferson was not: dashing, quick, easygoing, voluble, and light-hearted. They also had interests in common: in the law because, after college, Dabney had concluded his legal studies in the standard two years and then began county-court practice and already had nearly five years' practical experience from which to advise his friend Jefferson; in horses, for both men were excellent riders. Horses had, in fact, become a passion for Jefferson. He liked few things better than to gallop off on a good mount or go for a breakneck drive in his two-seat phaeton with "a pair of keen horses."[2] A lifelong servant testified to Jefferson's habit of taking the reins, even when he'd brought along a driver. "Whenever he wanted to travel fast *he'd* drive; he would drive powerful hard himself."[3] And while he loved to watch horse races, he gambled little on them. He studied horses, and when a fine racehorse was retired, he often bought it after its racing days were done.

Of former schoolmates who remained his friends, Jefferson had come to consider Carr the closest. When Carr died, Jefferson praised his friend's character, his "handsome imagination enriched by education and reading." Patrick Henry's only rival in spontaneous oratory on the Virginia county-court circuit, Carr, according to Jefferson, displayed "correct and ready elocution," "impressing every hearer with the sincerity of the heart from which it flowed."[4]

For five years Carr had crisscrossed the colony, trying cases in court, meeting clients in taverns, paying calls on friends. Now, Jefferson was getting the benefit of his fund of experience. The two men attended the 1767 spring term of the Albemarle County Court, where Jefferson had recently assumed his father's former seat as a magistrate. Court days in Charlottesville lasted a full week ending on Friday. Jefferson and Carr then had part of the weekend together before the forty-mile ride over muddy roads to Staunton, where the Augusta County Court session opened the next Monday morning. There was not even time to pick up fresh laundry at Shadwell. Jefferson's *Expense Book* shows he paid a washerwoman a shilling to rinse out his linens as soon as he arrived in Staunton. He also had to pay a blacksmith five shillings to reshoe his horse, since he had pushed it hard over the rough mountain roads.

The farms they passed were little more than stump-filled clearings in the wilderness. The overmountain region did not then produce tobacco; only occasional stands of hemp, grown to make rope and reap a small royal bounty, were in evidence. With no field crops there were few slaves. West of the mountains only six estates could afford (or needed) slaves. While half the families owned a few books, especially a Bible, most of the frontiersmen, who had come only a few years earlier, brought all their possessions in a wagon or on a horse. Few homes boasted a bookcase, a wingback chair, or a bureau. Fewer than half the homes had kitchen utensils; only about a third had beds. On cold nights livestock often lived inside the cabins. One family in six had a table or chairs. Buckskin or coarse homespun cloth was the clothing of necessity, and almost every family had a spinning wheel.

That Sunday night, from a rise in Valley Pike where it intersected Middle Trail, Jefferson and Carr could make out the scattered roof lines of Staunton shrouded in heavy woodsmoke. In town, on Augusta Street, they passed rows of rough-timbered taverns that flanked the elegant red-brick Trinity Church. At Mill Place, set back on a deep lot with stocks on one side and gallows on the other, they could probably make out the tidy stone Augusta County courthouse.

Court days jammed this frontier crossroads. Rooms were scarce, and the two young gentlemen had to settle for accommodations in one of the grimy, dark, low-ceilinged taverns Jefferson had so loathed since first going on the hustings with George Wythe. Jefferson arrived in Staunton, according to his memorandum book, with the best of all possible recommendations in his pocket. He was George Wythe's protégé. He was also the newest member of the elite General Court bar. And he was the son of Peter Jefferson. He also already had in his pocket, even before he arrived, his first Augusta County client. And that first client was Gabriel Jones, the king's attorney, Augusta's foremost lawyer, and the father-in-law of Jefferson's former guardian.

Aristocracy in any age is about social ease and connections. Jefferson had been the schoolmate, at Reverend Maury's, of the son of the Augusta County Court clerk, John Madison. That first day of his frontier law practice, Jefferson went to the trim little courthouse and was ushered into the presence of the redoubtable Mr. Madison, who had managed court affairs in the county since its founding almost twenty years earlier. For four decades in all, affable John Madison prided himself on initiating green young members of the bar into Augusta's legal fraternity. Taking his newest charge in tow, the jovial old clerk led the courtly young Jefferson around Staunton, a noisome aggregation of crudely fashioned log houses, stores, and taverns crowded with equally rough-hewn Scots-Irish mountain men, most of them dressed in hides. Nowhere in colonial America was rapid growth more apparent than on the slopes of Virginia's Blue Ridge, where, as its greatest landowner, William Byrd III, put it, the settlers swarmed south from Pennsylvania "like the Goths and Vandals of old."[5] Staunton was one of a string of valley towns running north-south from York, Pennsylvania, to Camden, South Carolina. Land for new arrivals was a fought-over commodity. The county seat of America's largest county, Augusta attracted frontiersmen who had traveled hundreds of miles over muddy horse tracks to record their claims to land, pay off their fees, have contracts drawn, signed, and witnessed.

At the highest level, the Province of Virginia was governed by a two-house Assembly, its upper house, the appointive Governor's Council and its lower house, the elective House of Burgesses. Some forty interrelated families, including Jefferson's mother's family, the Randolphs, with great wealth and tireless effort, dominated Virginia society and government, controlling all levels from the Governor's Council and General Court on down to the Church of England parish vestries and all county courts. With the provincial capital at Williamsburg days or even weeks away from the homesteads of many settlers, the county courts on the southwest frontier governed the overmountain region. The county court not only settled legal disputes but wielded, as historian Gordon S. Wood puts it, "a general paternalistic authority" over the county by handling "a wide variety of what we today would call 'administrative' tasks, drawing on the community for help." As a result, the Augusta County courts were as much the "instruments of government as they were judicial bodies."[6] They assessed taxes, granted licenses, oversaw poor relief, supervised road repairs, set prices, and upheld community moral standards.

Virginia's court system closely resembled England's. Slavish in imitation, the county courts were ruled by what Jefferson later bitterly criticized as a self-perpetuating oligarchy that handled the bulk of Virginia's legal business. Each court was composed of four or more justices of the peace who exercised civil and criminal jurisdiction in all cases but admiralty. Appeals could be taken to a superior court—the General Court or the chancery court in the capital—in civil matters which involved at least ten pounds sterling or which concerned land titles or boundaries. In criminal cases, the county court tried all misdemeanors and bound over defendants for the grand jury of the General Court. If the accused was a slave, the county court had sole jurisdiction. Appeals could be taken from the superior courts to a supreme Court of Appeals, made up of judges of the superior courts who were not only superior-court lawyers but invariably were drawn from Virginia's aristocratic families.

On court days, the Augusta courthouse took on the look of a county fair where cattle, horses, hogs, and slaves were auctioned while peddlers and prostitutes threaded through crowds of frontiersmen. Every building nearby was crammed with goods, with entertainers and professional gamblers and unwary wagerers.

By the end of Jefferson's first day, according to local legend, John Madison led a tired Jefferson to a rude tavern with a loft where a torchlight card game was in progress. Madison had hidden a powder horn under his greatcoat. Surreptitiously he worked loose its stopper, leaving a trail of black grains on the plank floor around the table as he introduced Jefferson. Then, still trailing a trickle behind him, Madison left the room. Just as an unlucky gambler began to curse the Devil, Madison ignited the powder train. Seconds later, the blast sent Jefferson and the other gamblers shouting and surging out of the smoke-filled place.

Rough frontier humor was a hallmark of a lawyer's life in the strapping new Augusta territory. Only a year earlier, Indian raids were within a day's ride. In 1766, eight Indians had attacked a wilderness cabin on the south fork of the Shenandoah and killed a Mennonite minister, his wife, and their three children. Jefferson did not shrink from the assorted hazards of frontier practice. After five years hunched over his lawbooks, he cheerfully embraced his new way of life. Admitted to the bar of the Virginia General Court at the Capitol in Williamsburg in 1766, for the next eight years, until the deepening revolutionary crisis closed down the provincial courts forever, Jefferson made the law his principal occupation. His decision to bypass county-court practice and seek admission to the General Court automatically put him in a small legal elite. When he was admitted in 1767, only eight lawyers in the largest and most populous colony in America practiced before the General Court bar. When twenty-four-year-old Jefferson came to the bar, his rivals included the scholarly George Wythe; the brilliant orator Edmund Pendleton; kinsman Peyton Randolph, educated at William and Mary and the Inns of Court; provincial treasurer Robert Carter Nicholas; constitutional lawyer Richard Bland; Thomson Mason; John Blair; and John Mercer.

Under the General Court rules, Jefferson could expect ample business once he became known. The court was inundated with cases, some taking as long as ten years to come to trial. In the April term of 1771, for instance, when Jefferson's practice was at its peak, there were 1,912 cases on the court's *ready* docket for five or more years, with hundreds more "at the rules," in the preliminary stages of pleading and preparing written evidence before they were docketed. It was not unique: *Hite v. Fairfax* had been at the rules or on the docket since 1749 and was decided in 1771.

Then it was appealed and was not finally disposed of until 1786, some thirty-seven years after suit was first brought. One case Jefferson became involved in, *Dunbar v. Washington,* had languished "at the rules" for twenty years before Jefferson was hired in December 1770. To handle this staggering load were seven or eight General Court judges, who acted as appellate court for county and provincial court cases, as well as trial court, court of equity, and court of law. The General Court met only twice a year for exactly twenty-four working days each session. An appeal to extend the court sessions had been vetoed by the king in 1748. There were no more court days in 1767 than had been allowed in 1705, but the population had increased more than five-fold. While a few General Court lawyers were well trained at the Inns of Court in London and a few more under the demanding tutelage of George Wythe, in general both the judges and lawyers (Patrick Henry was but one example) fitted Jefferson's description, in his posthumously published book, *Reports of Cases Determined in the General Court of Virginia.* They were drawn "from among the gentlemen of the country, for their wealth and standing, without any regard to legal knowledge."[7]

As his caseload increased (and it was bound to, as the number of lawyers diminished), young Jefferson could be assured plenty of business. One estimate is that only one hundred cases were actually tried each year, but, under common General Court practice, each side of a court case was represented by two General Court lawyers. Jefferson was destined to move from being George Wythe's student to his courtroom partner. Both of the plaintiff's and defendant's two lawyers were entitled to speak as long as they wished, no matter how repetitiously, further slowing the process. To increase even further Jefferson's chances for success in this elite log jam was the fact that more lawyers were leaving General Court practice than entering. Robert Carter Nicholas, recently appointed treasurer of the colony, was not taking any new cases and was sharing much of his work with John Blair. In 1770, Blair dropped out to become clerk of the Governor's Council. When Blair, in his turn, decided to shut down his practice, Nicholas did the same.

Unlike most gentlemen-planters who dabbled at the law but were forbidden by their code of honor to let on that they relied on the law for income, Jefferson pursued it vigorously, making only modest and conservative changes in his inherited estate and leaving his farms in the hands of overseers much of the year to allow him time to pursue his law business. During this little-known but pivotal period of Jefferson's life, in a good year up to half his cash income came from practicing law.

While many of his records and much of his correspondence from the early years of his practice appear to have been destroyed in a fire, enough can be reconstructed from scattered account books and court records to determine that he practiced law aggressively. He attempted to fuse his scientific Enlightenment impulse to experiment with a sustained effort to reduce the day-to-day practice of law to a new and systematic order.

Within five years, before he turned thirty, he managed to emerge from the swarm of hundreds of pettifogging Virginia county court lawyers—"a disagreeable crowd" and "tedious,"[8] he confided at different times in letters to friends—to become one of a handful of the most prominent barristers, distinguishing himself as the leading land-tenure lawyer in the fluid upland region. Gaining a reputation for fearlessly confronting the status quo, Jefferson bearded the provincial squirearchy for attempting to corner the frontier land market illegally. He opposed the superstitious gentry for preventing smallpox inoculation. And he dared to raise in court the delicate questions that were troubling him personally and philosophically, especially slavery and adultery.

The law was no gentlemanly sideline for Jefferson as it was for several of his planter kinsmen. He vividly pursued legal business and had up to five hundred cases on his books in one year. He surprised himself in his interest in legal practice. He later said he had thought he would have preferred to be a legal philosopher, studying the theory of the law rather than practicing it, but he quickly distinguished himself for his elegantly researched, brilliantly argued cases, drawing particular enjoyment from applying his scholarly knowledge of land-tenure law.

One legal historian has called Jefferson "cool, unemotional, prepared, more lawyer than advocate, stating the issues as he stated everything he touched with a clarity born both of industry and genius." Those who wished "entertainment," said J. W. Davis, "took as their lawyer Patrick Henry, but those who wanted to win their cases [hired] Jefferson."[9] James Madison heard Jefferson in the courtroom. Jefferson, he said, spoke "fluently and well."[10] In court, Jefferson applied the maxim that he later said had derived from his years of abstracting Coke's *Littleton,* "the most valuable of all talents, that of never using two words where one will do."[11]

By the time he decided to abandon his legal practice and plunge full-time into revolutionary politics, he represented many of the colony's wealthiest and most prominent citizens, including many of the leading lawyers. For the rest of his life, he would remain eager to take up his lawbooks at critical moments to pursue, quite personally, what he considered to be justice. In his quiet, determined way, he locked horns with colonial, state, and federal judges, ultimately using his legal knowledge and experience to reform and reshape the law, first while revolutionary governor of Virginia, later as president of the United States.

Passing from a bookish, lonesome boyhood to a busy, responsible dual career as lawyer and planter with no real pause for the pleasures of youth, Jefferson in his early twenties became paterfamilias to a large family that included his mother, at forty-seven an old woman by the Virginia standards of the time, and her four children still living at home: his unmarried younger sister, Elizabeth, who was mentally retarded; his fifteen-year-old sister, Lucy; and twelve-year-old twins. With responsibility for this sizable group, no wonder he saw no immediate need to take on the added responsibility of a wife and children of his own.

The mature Jefferson cut a striking figure, according to one of his longtime house slaves, Isaac, who, when he was interviewed after Jefferson's death, remembered that "nary man" in Virginia "walked so straight as my old master." Jefferson was "neat a built man as ever was seen in Virginia," a "straight-up man; long face, high nose."[12] One of Jefferson's overseers described him as a "well-proportioned" six feet, two and a half inches high, "straight as a gun barrel." He was "like a fine horse; he had no extra flesh."[13] As he took over the full burden of being Squire Jefferson, he had to consolidate his patrimony, look for new lands to augment his fifty-three-hundred-acre cash-poor estate and sell off acres too farflung to be profitable. Britain's currency laws left him, like other Virginia landlords, with little cash unless he pursued it as a lawyer or as a merchant, which was considered by many to be beneath his gentlemanly status. As historian Frank L. Dewey puts it, "Jefferson went after it."[14]

Though Thomas Jefferson was reserved and appeared reticent among strangers, he was rarely as passive as he sometimes seemed. When John Adams finished his law studies and hung out his shingle, he waited for business to come to him. Not so Jefferson. Legal historian Dewey estimates a typical chancery case at the trial level consumed more than eight years from commencement to trial. The typical common-law case, except for felonies, took just as long. Appeals consumed another two to three years. The leisurely pace of the judges, leaders of Virginia's plantation society, was slowed further by the fact that they wore many appointive hats bestowed by the Crown.

An ambitious young lawyer, on the other hand, could not hope to build a practice in the genteel environment of the General Court in Williamsburg; he had to go out onto the hustings where most law was practiced. The great bulk of Jefferson's practice was related to land patents, most of them on the frontier, and Jefferson aggressively staked out his practice.

Colonial Virginia law forbade him to practice in both county and General Courts, but he could go where county-court lawyers practiced and lingered and then obtain referrals in difficult cases and have their jurisdiction removed to the General Court in Williamsburg. The county courts met periodically, usually quarterly, for several days. Jefferson decided to concentrate on the two large counties west of the mountains, Albemarle and Augusta, where most of the turnover in land patents was taking place.

A day's ride from Shadwell in the vast Shenandoah Valley lay one of the most fertile fields in colonial America for land-tenure law. In the autumn of 1766, once the mandatory one-year waiting period between bar examinations and admission to the General Court expired, Jefferson went in search of clients. Not only was his first client a lawyer but many more were destined to be; with his deeply researched knowledge of land law and his excellent recommendations, he was almost immediately considered a lawyer's lawyer.

Over the next seven years, Jefferson took on clients from nearly every county in Virginia, even though, except in Williamsburg, Charlottesville, and Staunton, he did not attend court days to solicit business. Since he sat on Albemarle's county court and owned so much land in the county, he had to attend court in Charlottesville, but he devoted his primary efforts to the raw Augusta County where, as he had known since his clerkship with Wythe, many claims for land ownership were in dispute. On paper (based on the Virginia Charter), Augusta County sprawled from the western slopes of the Blue Ridge hundreds of miles west to the Mississippi, from present-day Rockingham County on the north to the North Carolina border. Where his father had once dragged surveyor's chains through the wilderness, young Jefferson now settled disputed claims to land titles.

Between Peter Jefferson's surveying expedition of 1745 and the beginning of the French and Indian Wars in 1756, the governor and Council had granted more than one million acres of the trans-Allegheny lands to a handful of speculating companies. Three major groups—the Robinson-Nelson faction from the lower Tidewater; the Washington-Lee-Mason faction from the Potomac; and the Loyal Company, including Peter Jefferson and his neighbor, Dr. John Walker—all jockeyed for official favor, larger land grants and higher land prices.

The fastest-growing region of Virginia, Augusta was filling up not only with immigrants from eastern Virginia but also with recent German and Scottish immigrants trudging south from overcrowded Pennsylvania. Thousands of land-hungry settlers were trekking south to the Virginia valleys. Immigration, high birth rates, and low mortality rates gave North

American colonies the fastest-rising population in the Western world. Between 1750 and 1770, the numbers doubled from one million to over two million. Immigrants were pouring in, on average thousands each year; from the British Isles, Englishmen, Scots, Protestant Irish; from Europe, Swiss and Germans. These new immigrants joined uprooted colonists who had been confined along a hundred-mile-wide strip of Atlantic coast until the French were driven out and the Indians were subdued. Eastern soil, overcultivated, had already become depleted. Older towns were becoming overcrowded. Younger sons were no longer willing to wait for their fathers' lands. In addition, newly arrived Scots-Irish, Swiss, and Germans trudged south along roads on both sides of the Blue Ridge, leaving new country market towns in their wake from York, Pennsylvania, to Camden, South Carolina. The prize they all sought was a choice parcel of the half billion acres of interior lands between the Appalachians and the Mississippi vacated in 1763 by the French.

In *Notes on Virginia,* Jefferson himself explained the cumbersome process of patenting wilderness lands:

> An individual wishing to appropriate to himself lands still unappropriated by any other, pays to the public treasurer a sum of money proportioned to the quantity [of land] he wants. He carries the treasurer's receipts to the auditors of public accounts, who thereupon debit the treasurer . . . and order the register of the [Virginia] land-office to give the party a warrant for his land. With this warrant from the register, he goes to the surveyor of the county where the land lies on which he has cast his eye. The surveyor lays it off for him, gives him its exact description, in the form of a certificate, which . . . he returns to the land-office, where a grant is made out and is signed by the governor. This vests in him a perfect dominion in his lands, transmissible to whom he pleases by deed or will, or by descent to his heirs if he dies intestate.[15]

Jefferson visited the two frontier county courts for several days, usually three, at the opening of their quarterly sessions, which continued up to ten days. It was an efficient and convenient way to see several clients at once, to discuss cases with them, to collect money owed him. He could also, after consultation with clients, carry their appeals from the county-court cases to superior courts in Williamsburg, and he could serve as a party to cases where he was executor, guardian, property owner, or witness. Claims to land that was virtually free could be surveyed and filed for a modest fee, but they could be challenged if the land were not settled and cultivated within the required time.

By 1767, nearly one million acres of Virginia land were improperly patented. While Jefferson hobnobbed with his friends Carr and Patrick Henry and college roommate Will Fleming, he was busily making it clear that he was the *only* General Court lawyer who lived and practiced in western Virginia. On the opening days of court sessions in Augusta and Charlottesville, he conferred with lawyers at the courthouses. Then, not waiting for the session to end, he headed east to Williamsburg or sent ahead a courier to file papers in the superior courts and pay fees as he continued on his rounds. The pace was hectic the four times each year when both county courts met only a weekend and forty miles apart. Jefferson's years as a lawyer got him out of his books and into the world.

While Jefferson never traveled farther west than present-day West Virginia, he saw more of the interior than most eighteenth-century Americans, and he came to know frontier thinking, the ceaseless quest for more land, the "new" men and women beyond the rooted society of the Atlantic seacoast and the Chesapeake Tidewater. He would build his political fame on an understanding of the American continent that linked him with other expansionists such as Washington and Benjamin Franklin, both of whom sought at this very time vast personal claims to western lands so recently wrested from French domination.

As he rode from his clients in western Albemarle and Augusta counties to the provincial law courts in Williamsburg to file his lawsuits, he paid calls along the way on old school friends, many of them newly married, lively, and full of fun, some of them showing off their firstborn babies. There were hours of gossip and music and endless rounds of chess, one of Jefferson's passions, especially with old friends John Page and Will Fleming. Sprinkled among his letters, which he still wrote in Latin to fellow former members of the Flat Hat Club, were kindly reminders to bring along a chessboard to gatherings of Tidewater society (Jefferson would bring the chessmen). But more and more he looked forward to long discussions of politics and literature. "I reflect often with pleasure on the philosophical evenings I passed at Rosewell on my last visits there," he wrote to Page. He was especially enchanted by Page's wife:

> I was always fond of philosophy, even in its dryer forms, but from a ruby it comes with charms irresistible. Such a feast of sentiment must exhilarate and lengthen life at least as much as the feast of the sensualist shortens it. In a word, I prize it so highly that, if you will at any

> time collect the same *belle assemblée,* on giving me three days previous notice, I shall certainly repair to my place as a member of it.[16]

Jefferson's peregrinations around Virginia made him more and more popular. He was remembered for his unfailing cheerfulness and his dapper dress. Always "very neat," he was fond of short breeches and white stockings that showed his good strong legs and "bright shoe buckles."[17] More and more he wore light, Virginia-made cloth coats and a red waistcoat.

One of his colleagues at the bar of the General Court, Edmund Randolph, describes Jefferson at this time:

> He pursued the law with an eager industry. . . . Reserved toward the world at large, to his intimate friends he shewed a peculiar sweetness of temper and by them was admired and beloved. . . . He panted after the fine arts and discovered a taste in them not easily satisfied with such scanty means as existed in a colony. . . . It constituted a part of Mr. Jefferson's pride to run before the times in which he lived. [He was] an admirer of elegance and convenience."[18]

With the freedom of movement during his earlier bachelor days, he came and went from Shadwell, spending more and more time in Williamsburg, where he lodged at Charlton's on Duke of Gloucester Street and took his meals at the Raleigh Tavern across the street. Striding up the wide avenue to the coffeehouse or to Mrs. Vole's or Ayscough's Tavern for a drink or entertainment, he haunted the playhouse during the spring of 1768. It was the Virginia Company of Comedians that trod the boards, and Jefferson rarely missed a performance: Shakespeare's *The Merchant of Venice,* Addison's *The Drummer,* Otway's *Venice Preserved* and *The Orphan,* and Gay's *The Beggar's Opera* were among the high-spirited fare. When he turned twenty-five, he was with his family at Shadwell studying German and reading Bolingbroke's political philosophy. The brief birthday visit provided a respite from a hectic schedule. In recent weeks he had traveled to Goochland Courthouse on the frontier and back to Williamsburg. There, according to his *Account Book,* he visited his favorite coffeehouse, attended a play, and was on the road again. Visiting the Lewises at Warner Hall for three days, he tipped the servants two shillings sixpence on his departure, and wrote it down. He stopped the next day at White Hall to visit Frank Willis, his prankster-friend from Reverend Maury's and William and Mary, but he left in time to be ferried across the Ware River before dark. Back in Williamsburg, he paid seven and a half shillings, three days' wages for many a laborer, to see a live elk, and he attended another play. In a few days he left Williamsburg again to visit the Pages at their splendid mansion at Rosewell for another three days, then returned to Williamsburg for a round of coffeehouse and playhouse visits.

His account books show that, in his increasingly frequent visits to the capital, Jefferson attended the races and placed small bets on the horses, gambled modestly at cards and dice, paid to see puppet shows, a magician, an elk, a tiger, a "great hog."[19] He paid to hear music of any kind played by a chamber orchestra or by an organ grinder or on a glass armonica, a contraption of water-filled tubes of various sizes pumped by a foot-pedal and tuned by moistened fingers, while a woman sang to its accompaniment. In the afternoons, in the evenings, he searched out music or played it on his fine fiddle, and according to one of his servants, when he rode or walked, he loved to hum minuets or sing. An overseer recalled, "When he was not talking, he was nearly always humming some tune or singing in a low tone to himself." Added his manservant, Isaac, "Hardly saw him anywhere but what he was a-singing."[20] He rode out again to visit the Ludwells at Greenspring on April 24. On this breathless round, he made no note in his *Account Book* of the beauties of a Tidewater spring.

At the beginning of May, Jefferson once again stayed with the Burwells and the Pages on his way to Hanover Court House and, after picking up a little business, he went on home to Shadwell. Seeing all his domesticated friends seems to have had its effect, and he gave the first solid indication that he was thinking of settling down and building his own house. He signed an agreement with a Mr. Moore "that he shall level 250f square on the top of the Mountain [the future Monticello] at the N.E. end by Christmas."[21] By May 20, 1768, when the mountain laurel whitened the dark trees along his route, he rode back to Staunton for the Augusta County Court spring term. Here, in the mayhem of court days, his horse ran away from him and he had to pay "a negro for finding my horse, 5 shillings."[22]

Apparently the horse was to get little rest: four days later, a fresh batch of caveats in his saddlebag, Jefferson was on his way back in Williamsburg. There, with a fat bonus just paid him by a client, he bought a fine new Italian violin for five pounds. (It may have been a Stradivarius.) He celebrated the purchase at the coffeehouse, where he no doubt regaled his friends with its sweet sounds. In a little less than two months, the restless young Jefferson had ridden five hundred miles and had met with clients in five counties as well as the provincial capital. On paper, at least, his law practice was thriving. His legal *Account Book* shows that, in 1767, his first year of practice, he handled sixty-eight cases. He was paid forty-three pounds, four and three-quarter shillings and had "legal fees due" of two hundred fifty pounds, fivepence. In his ledger, he lumped together the payments and accounts receivable for the year, showing "total profits" of two hundred ninety pounds, four shillings, for the first year of practice.[23] In actuality, after expenses and fees that he advanced for clients and fees that he carried on the books unpaid, he had cleared only ten pounds, half of which he plunked down on a good fiddle. A modern accounting historian, Steven Hochman, takes issue with Jefferson's including in "legal fees due" many cases that were still in process "and might not be finished for years." Jefferson, he points out, "ignored his office expenses, most notably his very considerable investment in law books, but also the cost of such items as paper and ink. . . . To some extent, his living expenses in Williamsburg should have been deducted as well as his travel expenses when they were connected with his practice."[24]

Thomas Jefferson's legal practice depended on the fact that he was the only General Court lawyer who was accessible to westerners, especially to county-court lawyers not qualified to appear before the high court in Williamsburg, where land patents were decided and could be revoked. Since his private practice emphasized caveats and petitions for lapsed lands—proceedings dealing with the patenting of land—he correctly analyzed that his greatest usefulness was on the frontier. "When he entered upon the practice of the law," reported General Court colleague Edmund Randolph, "he chose a residence and travelled to a distance which enabled him to display his great literary endowments and to establish advantageous connections among that class of men who were daily rising in weight."[25] Forty-one of the first sixty-eight items he listed in his casebook for 1767 involved Augusta clients, twelve were from Albemarle County. His decision to concentrate on cultivating clients in Augusta County, where large tracts of undeveloped land were held by speculators but where many of their land titles were imperfect, proved to be shrewd. But his rapid rise as a methodical specialist whose lawsuits threatened to break up the monopoly of absentee landlords on western lands inflamed many of the Virginia gentry and led to his first bruising confrontation with the landowning oligarchy.

Jefferson attended all four sessions in Staunton that first year—Gabriel Jones hired him to defend two slander cases in Williamsburg, where Jefferson also defended his mentor, George Wythe. "My friend," Jefferson noted in his *Fee Book*. "Take no fee."[26] When the case finally came to court four years later, Jefferson won it.

Jefferson's specialty of challenging applications for land patents was a timely success because, at the very moment he hung out his shingle, a loophole in the Virginia land law encouraged an element of collusion. The system was supposed to rake in hard-money emoluments for royal government officials. It was also supposed to spur settlement along the frontier by allowing legal informants to sniff out dilatory developers: there was still no bureaucracy to ferret them out. By encouraging informants to pay up the delinquent filing fees and quitrents and take over the lands themselves, the system was supposed to open up wilderness lands for settlement. But a few Virginians had discovered the gaping loophole in this land-tenure law, as John Blair, deputy auditor for the colony, explained to Governor Fauquier in May 1767:

> A custom prevails that, to save their lands without paying quitrents, they get a friend to petition for them as lapsed. . . . These petitions, which keep off all other petitioners from entering, are, by collusion, kept many years on the docket without paying. The Court has lately, on a motion, resolved that, if not decided in three [court sessions], they shall be struck off the docket and so left open to any other petitioner. But that [petitioner], by contrivance, may be another friend and, at last, take out perhaps a new grant in a friend's name and then all arrears are lost. Some great tracts are now held in this manner and many smaller, to the great prejudice of His Majesty's Revenue.[27]

Thomas Jefferson's second law case was just such a "friendly" caveat for Augusta County lawyer Gabriel Jones. Their nominal client was Nicholas Meriwether, of Albemarle County, "friendly" petitioner for the real client, Francis Meriwether, of Amherst County, who held a patent to 650 acres on Tuckahoe Creek in Bedford County. Francis Meriwether brought the case to Jones and Jefferson; Nicholas Meriwether allowed his name to be used as petitioner of record. The case was still on the General Court docket more than seven years later, which made it a success for Jefferson and his client, since Meriwether still held the land without parting with a penny of the patent fees. What he paid Jefferson was far less than the patent fees and quitrents would have cost him.

Jefferson frequently used delay as a legal weapon. He not only engaged successfully in caveats for others—they made up fully one-half of his practice—but he wheeled-and-dealed in them himself. In October 1767, he arranged to have "friendly" caveats filed against him "for my three tracts of land in Bedford" by Francis Meriwether, his erstwhile client and a colleague on the Albemarle County magistrate's bench, as the nominal petitioner. Again, in May 1771, after patenting Bedford lands he had just purchased for one hundred pounds, he arranged "friendly" caveats against himself, this time using his mother as nominal petitioner, keeping his lands on the court docket for another two and a half years before the caveats were dismissed by the court. In all, he astutely bought six years' time to speculate in Bedford County land without parting with his precious cash for recording fees or quitrents. Meanwhile, he successfully staved off other would-be speculators.

Young Jefferson's legal machinations brought him more and more clients. In November 1767, Patrick Henry and his father-in-law, John Shelton, of Augusta County, hired Jefferson to file six "friendlies" for tracts of one hundred fifty to fourteen hundred acres for Henry's half brother. Shelton had already tied up the lands in southwesternmost Virginia nearly twenty years before. He had no money. He had mortgaged the lands to Patrick Henry, who in turn hired Jefferson, who then in turn only charged Patrick Henry half his usual fee after four years of footdragging.

Thomas Jefferson's caveating skill soon led to a get-rich-quick scheme concocted by sharp Augusta County land dealers. According to legal historian Frank Dewey, caveating had been "essentially a retail operation" until Jefferson's former Latin-school classmate John Madison, Jr., the son of the Augusta County clerk, and several Shenandoah land speculators came to Jefferson to enlist his aid "on a wholesale basis."[28] In March 1769, Jefferson filed fourteen petitions in the name of James Greenlee against John Buchanan, a dealer in western lands. In his *Account Book* for March 24, 1769, Jefferson wrote, "Greenlee's name is used, but J. Madison junr. and James McDowell are partners."[29] The Greenlee caveat was only a prelude to a much-more-ambitious scheme: young Madison was the prime mover for a secret company that included William Waterson, Hugh Donaghue, and Andrew Johnston, all of Augusta. Through Madison, the group hired Jefferson to identify *all* delinquent lands in Augusta County and to bring caveats against all the owners. Jefferson noted in his *Account Book* for July 29, 1769,

> Send 20 [shillings] which H. Donaghue left with me to [Virginia] Auditor's clerk to send him a list of all the lands in Augusta for which rents are 3 years behind, and to set down the particular years, by whom and when paid.[30]

A little more than a month later, Jefferson wrote surreptitiously in Latin to warn his friend, John Walker,

> Are your father's grazing lands in Augusta safe? That is, may they not be subject to lapse through lack of cultivation or quitrents? A word to the wise: a great loss will soon come to those who are otherwise.[31]

According to modern standards of legal ethics binding conflicts of interest, Jefferson was committing a serious breach of ethics by warning his personal friend to defend himself against caveats for imperfect title when the action was paid for by Jefferson's client. Already, Jefferson was putting personal loyalties above the law. Writes legal scholar Frank Dewey of the affair,

> Confidentiality of client information is at the heart of modern legal ethics. That Jefferson wrote the warning in Latin and in language that was cryptic even in translation suggests that he was uncomfortable about what he was doing and indicates that confidentiality was understood even then to be part of the lawyer's compact with his client.[32]

When Jefferson's friends ran into trouble, he dug down into his green lawyer's bag and used every trick he had learned to help them. William Nelson, a member of the Governor's Council, sued Jefferson's college friend, Frank Willis, for debt. Jefferson boldly defended Willis without fee in a series of five separate actions. As a modern legal historian puts it, "The sluggishness of the General Court system favored defendants. Because prompt enforcement was not available, there was little incentive to pay debts promptly."[33] In his *Account Book* for April 9, 1768, Jefferson wrote, "Appear for Willis and use every dilatory."[34] Jefferson removed the Gloucester County case to the General Court in Williamsburg, where he could represent Willis, using a writ of habeas corpus. After eighteen months of "in the rules" preliminaries, the Willis case appeared far down on the twenty-second-day list of a twenty-four-day calendar: it dragged on from October 1769 until April 1774, when it was finally dismissed. There is no evidence that Willis ever paid a farthing on his debts. Jefferson finally arranged an out-of-court settlement and actually got the plaintiff to pay *him* a hefty ten pounds (about a thousand dollars today) "for attending his business in Willis' affairs."[35]

The activities of the Madison-Waterson land-caveating group had begun in April 1769, when Waterson hired Jefferson to file five petitions and fifteen caveats. The target of five of the caveats was Col. William Preston, a partner with Jefferson's father in the Loyal Company who had been twice elected a burgess from Augusta County. An important pioneer in southwest Virginia, he was one of the few slaveowners in the region and he dealt in western lands on a large scale. Even before a summons could officially be served on Colonel Preston to defend his challenged land patents, he learned of Jefferson's caveats. In his *Account Book,* Jefferson wrote what happened next:

> Caveats were entered . . . April 19. Col. Preston, suspecting this, borrows money of Col. Syme, and Francis Smith, his father-in-law, posts down to Williamsburg, where he arrives May 18. He hires one of the boys [law clerks in the land office] to receive his works [patent filings] which were antedated. . . . After this, he comes to Waterson's attorney [myself], tells him he [Jefferson] had entered caveats for lands which he [Colonel Preston] had patented. The attorney, on seeing this, scratches out the caveats. It seems there happened to be some blank patents signed in the [land] office. Col. Preston confessed all this to William Crow . . .[36]

Only a few days after this face-off, an undeterred Waterson brought forty-four more caveats to young Jefferson, who passed the list on to John May, his agent in Williamsburg, who in turn submitted it to the secretary of the colony. There, as the list was culled, the slow process of filing and summoning began.

Meanwhile, on July 29, 1769, Jefferson forwarded another 270 Waterson caveats to John May for screening. That same day, according to his *Account Book,* Jefferson forwarded to John May some 127 caveats and 12 petitions in the name of Andrew Johnston, another of the Waterson group. By April 1770, the Waterson group had filed 450 caveats and petitions. In addition to the cases Jefferson entered in his casebook, the April 1770 General Court docket lists 70 petitions.

Taking the cases on a speculative contingent-fee basis, Jefferson stood to earn the standard £2.10 rate per case if he won them. In all, he stood to gain as much as £1,300, more than $130,000 in 1993 terms, more than half his earnings for his entire eight years of law practice. But he never collected a farthing. His clients' activities outraged the Virginia gentry. At first Colonel Preston's friends had trouble identifying the members of Waterson's "combination." Thomas Lewis, who had accompanied Peter Jefferson on the 1748 boundary-line expedition, called the participants in "the business of caveating here" little more than "villains" from "the highest to the lowest" against whom the original land patentees were "defenseless." Lewis had "attempted to form a petition" to the governor and his council "to oppose the practice of caveating." Lewis urged Colonel Preston to join a group of frontier land speculators mobilizing to ride on Williamsburg. He expected the Council "will be moved with compassion." Lewis urged Preston, "Be there. If we can do good, it will be a pleasure to us that never can be felt by any that are in this cursed combination!"[37]

Evidently Preston did go to Williamsburg with Colonel Lewis and his land-speculating friends. When the General Assembly reconvened on May 21, 1770, Colonel Preston's lawyer, Edmund Pendleton, submitted a bill "to compel certain persons to find security in certain cases."[38] The bill was referred to a special committee made up of John Blair, president of the Governor's Council, Edmund Pendleton (Colonel Preston's lawyer), Jefferson (now a member of the House of Burgesses from Albemarle County), and his land speculator neighbor, Dr. Thomas Walker, the other burgess from Albemarle. A change in Virginia land law resulted from the speculator-dominated committee's finding that

> divers litigious persons have, of late, preferred petitions . . . for grants of land under pretence that they were forfeited for non-payment of His Majesty's quitrents, or for want of seating or planting. . . . Others have entered caveats . . . pretending that the rules of government have not been complied with. . . .[39]

The new statute required petitioner or caveator to post cash security for all court and legal costs including witness fees and lawyers' fees and costs, effectively killing all but "friendly" caveats. From 1770 on, only the rich land speculator could stay in the business of frontier land development in Virginia. The Assembly, prodded to alarmed activity by Colonel Preston and his cashless land-cornering friends, shrugged at the institutional collusion of "friendly" caveats and winked at the widespread evasion of payments of land rents to the Crown.

Thomas Jefferson had learned a costly lesson in the influence-peddling of patrician Virginia. "There can be little doubt," holds legal historian Dewey, "that the statute was aimed at Jefferson's clients."[40] In its clever way, the Assembly had rebuffed a massive legal threat by Jefferson and the "new men" of the west. Pretending that it was concerned that plaintiffs would lose caveat cases and could not afford to pay the winner's costs, it was concealing its fear that the plaintiffs, new money men such as Madison and Waterson, would win. The Augusta group, probably on Jefferson's advice, made one more attempt to keep the cases alive by referring them to arbitration, but no one would serve on the arbitration panel.

A gleeful Thomas Lewis reported to Colonel Preston on August 28, 1770, that Jefferson, Walker, and John Buchanan of Augusta had tried to get the Assembly's arbitrators to force the issue, but the four arbitrators feared "ill consequences."[41] Of Jefferson's original caveats on behalf of

Waterson and Company, he wrote in his copy of the April 1771 General Court docket, "all these [were] dismissed for want of security."[42] In June 1772, all of Waterson's remaining 305 caveats were dismissed, 125 cases for failure to give security, and 180 for unstated reasons. In an unprecedented action, in order to cut off any further possibility of new caveats with security, the Governor's Council allowed Preston and the other defendants in the Waterson caveats three more months to perfect their patents. Clearly the Governor's Council did not want to put its own land-speculating friends out of business. A massive legal assault on the privileged first settlers of the frontier counties by Jefferson and Madison threatened the established gentry and the very fabric of frontier society. For all of his troubles, Jefferson received two horses and a harsh lesson.

Madison brought him one more group of cases in 1770, this time against James Patton, one of the heirs and executors of a Shenandoah founding father. Jefferson, who had been informed that Colonel Preston was one of the heirs, insisted on a written agreement that he be paid full fees, win or lose. He actually collected two-thirds of them. Jefferson apparently took on the case because he was assured his client could afford to post security for all costs. But when the cases finally went to trial two years later in April 1773, they were dismissed with all court costs assessed against Jefferson's client. At every turn, Jefferson and his clients had been blocked and punished. Rebellious land lawyer Thomas Jefferson had been soundly thrashed by the gentry. On April 23, 1773, Edmund Pendleton exulted to his client, Colonel Preston, that Jefferson and his client had been given "a total defeat."[43]

Although the bulk of Jefferson's legal business had come from his Augusta clients from 1767 to 1770, his first three years of practice, his crushing as a caveat lawyer must have seriously weakened his practice on the frontier. In the next three years, he neglected Augusta and the new frontier counties and quietly consolidated his affairs at the capital in Williamsburg and in his native Albemarle County. After his third year of practice, when he showed £370 in fees on the books, he considered himself a success and obviously thought he could begin to pick and choose his cases. In 1772, he rode to Staunton only twice, only once in 1773, not at all in 1774, his last year in law practice.

But while his account books paint a grim picture of financial failure on the Augusta frontier, his public stand had only enhanced his reputation in Williamsburg. On June 14, 1768, little more than a year after starting to practice, Jefferson noted that William Byrd III, the largest landholder in Virginia, "retains me generally."[44] In addition to his retainer, he received additional fees for actual cases and obviously regarded Byrd's clientage as a mark of success. Even after the legislature killed his clients' wholesale caveating, Byrd continued to retain Jefferson. Burwell and Thomas Nelson, two members of the Governor's Council, which had rebuffed his clients, gave him business shortly afterward. When the Council president, John Blair, Jr., gave up his General Court practice to become clerk of the Council in 1770, he turned over part of his practice to Jefferson, as did Robert Carter Nicholas, the acknowledged leader of the Virginia bar, when he retired in 1771. Even Colonel Preston's lawyer, Edmund Pendleton, hired Jefferson in one case. In the Waterson-Madison case, a tenacious Thomas Jefferson, while he had lost considerable money at first, had demonstrated his mastery of land-tenure law even as he had conspicuously paid his dues, in only three years of practice moving from the backcountry county court to prominent practice in the capital by the age of twenty-seven.

In all, according to his precisely written and perfectly preserved leatherbound casebook, he took on some 939 cases between February 12, 1767, and November 9, 1774. But, in cashless colonial Virginia, each year he was finding it more and more difficult to collect his fees. While, on the books, the fees owed him had increased from £293.4.5 in 1767 to £521.5.10 by 1770, his fourth year in practice and the peak year of his earnings, he collected a meager £213 before expenses, little more than twenty thousand dollars in 1993 money. Legal historian Frank Dewey, who has studied Jefferson's law practice very carefully, contends that "it was not financially rewarding."[45]

In his old age, Jefferson took a dim retrospective view of his years as a lawyer. He wrote that the "venality" of most lawyers "makes me ashamed that I was ever a member" of the bar. Lawyers, he wrote another friend, "question everything, yield nothing and talk by the hour." His answer would have been to let only "men of science"—educated men—practice law. And when legal reform did come from England in the form of *Blackstone's Commentaries,* Jefferson assailed

> Blackstone, whose book, although the most elegant and best digested in our catalogue, has been perverted more than all others, to the degeneracy of legal science.[46]

He condemned the "great herd" of county court lawyers for imbibing Blackstone's "smattering of everything." Jefferson railed at the typical lawyer's "indolence" that persuaded him that he was a legal expert after mastering three volumes of Blackstone's lawbooks. He called them "Blackstone lawyers," deriding the majority of Virginia lawyers as "ephemeral insects of the law."[47]

Jefferson also worried about the casual training of judges. When the Revolution made reforms possible in Virginia, then-Governor Jefferson wrote George Wythe a scathingly honest assessment of his erstwhile colleagues:

> I think the bar of the General Court a proper and an excellent nursery for future judges if it be so regulated that [scientific education] may be encouraged and may live there. But this can never be if an inundation of insects be permitted to come from the county courts and consume the harvest.[48]

His personal response was a sustained attempt to systematize Virginia law by bringing scientific order to it. Increasingly he considered Virginia law paramount over English law, but he found it inaccessible. In 1768, Jefferson began to gather manuscript reports of hundreds of Virginia cases. There were few lawbooks in America at this early date; most of them came from England: only thirty-three lawbooks were published in America before 1776. The only Virginia lawbook published in a century and a half of court practice was Webb's *The Office and Authority of a Justice of the Peace;* the only collection of Virginia statutes passed in the colony's first century was not even in printed form but had been gathered in three manuscript volumes some thirty-five years earlier by his kinsman, Sir John Randolph. Jefferson borrowed these volumes from Randolph's son, Attorney General John Randolph, Jr. By 1795, when he turned his own resulting volume over to Virginia's Committee on Publications, Jefferson's collection, largely the result of his efforts between 1768 and 1774, was the most complete collection of Virginia law reports in America. Jefferson later wrote that he had spared neither time, trouble, nor expense in gathering additional precious manuscripts and what few printed copies he could glean. The later case reports were based on notes of trials Jefferson had attended or in which he had participated; of eleven cases he reported, he had taken part in three. The earlier reports he based on notes made by his uncle of thirty-one cases heard a generation earlier. Ultimately his law reports were published, but not until 1829, three years after Jefferson's death, under the title *Reports of Cases Determined in the General Court of Virginia from 1730 to 1740 and from 1768 to 1772.* One contemporary historian called the work "a monument of his early labors and useful talents."[49]

Early and late in his life, Jefferson realized that most lawyers were neither as well trained nor as painstaking about their research as he was, yet he continued to write bitterly on the subject. To his successor as president, James Madison, on the eve of the War of 1812, he wrote:

> I have much doubted whether, in the case of a war, Congress would find it practicable to do their part of the business. . . . That a body containing one hundred lawyers in it should direct the measures of a war is, I fear, impossible . . .[50]

Jefferson's dealings in western Virginia lands on behalf of clients were interspersed with his own meanderings in search of choice lands, which he bought in his own right, and with his own awestruck study of the natural beauties of the Blue Ridge region, which he would describe copiously some fifteen years later in *Notes on the State of Virginia.* As soon as he turned twenty-one, his *Account Book* in the Huntington Library shows, he unsentimentally sold off one hundred acres of land on Fine Creek in Cumberland County to raise one hundred pounds in ready cash. The land had been part of his inheritance from his father, but Jefferson decided it was too far from Shadwell to be efficiently managed. In all that first year of his stewardship over Shadwell, young Jefferson had sold off seven hundred acres in Amherst County for a much-needed one hundred ninety pounds. He wanted the money to buy a larger tract, some four hundred acres closer to home on the Tye River in Amherst County, which he purchased the same year for one hundred pounds. That same year, he also located, purchased for ninety pounds, staked out, and surveyed, as his father had taught him, some three hundred acres on the Piney River in Amherst County. In that first year of youthful land speculation, Jefferson had increased his holdings by six hundred acres.[51]

In 1767, Jefferson began acquiring lands in Bedford County. By October he had purchased three tracts there, still steadily increasing his holdings, which he cryptically noted in his *Expense Book,* until he had purchased, by the beginning of the Revolution in 1775, some 4,847.5 acres, virtually doubling his landed estate and increasing his harvests of tobacco proportionately. He bought and sold land, ending up with fewer but more valuable acres. He purchased 298 acres on Tomahawk Creek for one hundred pounds on September 3, 1767. Two weeks later, he sold 560 acres there for the same amount. It was here that he would eventually build Poplar Forest, his secluded summer retreat.

His peripatetic travels included his happy discovery of the Natural Bridge. Between 1771 and 1773, he was able to buy the limestone bridge and 755 surrounding acres. In *Notes on the State of Virginia,* written at a time when he had traveled little outside Virginia, he called it "the most sublime of Nature's works." Betraying the long hours he had spent riding and hiking and gazing up at its great "semi-elliptical" limestone arch, he wrote about its dramatic effect on the first-time visitor and detailed his scientific study of what most have considered but a natural oddity. His notes give a rare glimpse into his scientific turn of mind as he climbed and measured the width and depth of the Cedar Creek gorge and the high arch across it:

> It is on the ascent of a hill which seems to have been cloven . . . by some great convulsion. The fissure, just at the bridge, is by some measurements 270 feet deep. . . . It is about 45 feet wide at the bottom and 90 feet at the top. . . . This, of course, determines the length of the bridge and its height from the water. . . . A coat of earth . . . gives growth to many large trees.

Jefferson's excitement as he gazed down from the height of limestone into the canyon beneath comes through:

> Though the sides of this bridge are provided in some parts with a parapet of fixed rocks, yet few men have resolution to walk to them and look over into the abyss. . . . You involuntarily fall on your hands and feet, creep to the parapet and peep over it. . . . Looking down from this height about a minute gave me a violent headache. . . . [This painful sensation is relieved by a short but pleasing view of the Blue Ridge along the fissure downwards].

When Jefferson climbed back down from the natural archway, he was transported by the beauty around him. Perhaps no other passage of his writing is so euphoric, so romantic years before there was a Romantic Age:

> Descending, then, to the valley below, the sensation becomes delightful in the extreme. . . . It is impossible for the emotions, arising from the sublime, to be felt beyond what they are here: so beautiful an arch, so elevated, so light and springing, as it were, up to heaven, the rapture of the spectator is really indescribable.[52]

Jefferson's writings about his travels, so clear, so deep, are rarely so elegiac.

When he visited "the only remarkable cascade" in Virginia, Falling Spring in Augusta County, he catalogued it scientifically, matter-of-factly: it could not compare to Niagara "as to the quantity of water composing it," but it is "half as high again." He had powerful curiosity and a special fascination for mountains and caves, most notably Madison's Cave on the north side of the Blue Ridge. To get into the cave, "which is in a hill 200 feet perpendicular high," he had to climb almost straight up the face of the hill. The hill was "so steep that you may pitch a biscuit from its summit into the [Shenandoah] river which washes its base." Descending into the earth some three hundred feet, Jefferson speculated about the source of basins of water he found. When he reached the main chamber of the cave, he craned his neck to study the mud-colored stalactites above him, the thrusting rock formations around him.

> The vault of this cave is of solid limestone from twenty to fifty feet high through which water is continually percolating. . . . This, trickling down the sides of the cave, has encrusted them over in the form of elegant drapery. . . . Dripping from the top of the vault generates stalactites of a conical form, some of which have met and formed massive columns.[53]

Jefferson came away from his early travels firmly convinced that his native country, Virginia, was one of the most spectacularly beautiful on earth. Before he saw the Alps, he raved about the Blue Ridge, "The passage of the Potomac [River] through the Blue Ridge is perhaps one of the most stupendous scenes in nature." Where the Shenandoah and Potomac meet, "You cross the Potomac . . . pass along its side through the base of the mountain for three miles, its terrible precipices hanging in fragments over you . . . This scene is worth a voyage across the Atlantic." As noteworthy to Jefferson as the natural splendor he attempted to describe was the fact that there were people who had passed their lives within half a dozen miles "of the Natural Bridge on the Potomac Gap and have never been to survey these monuments of a war between rivers and mountains."[54]

Able to buy what he considered most beautiful, Jefferson employed his surveyor's eye and tools and legal skills to set aside for himself breathtaking tracts of Virginia. He had been spending less and less time at Shadwell, not two months in the year, as he confided to Thomas Turpin early in 1769. He was restless, his wanderings reflecting his readings. In his *Commonplace Book,* he copied out, from Edward Young's popular meditative poem, "The Complaint, or Night-Thoughts on Life, Death and Immortality," a verse he quoted time and again in later years:

For what live ever here? With laboring step
To heed our former footsteps? Pace the round

Eternal? To climb life's worn, heavy wheel,
Which draws up nothing new? To beat, and beat
The beaten track?[55]

While Jefferson's peregrinations seem random, his purchases were not. In addition to selling off faraway acres and buying lands closer to home, in 1773 he bought a town lot in Richmond, where he had considerable legal business. Even during the Revolution, he was busily buying and patenting land: in June 1778, he purchased 1,850 acres of land on the Fluvanna River in Albemarle County.

In part because he had the good luck to have been left an unencumbered estate by his father, in part because Jefferson and his family lived well within their means at this period, spending little on luxuries or speculations, Jefferson could continue taking on the cases he wanted, many of them without fee. Successful in his own gentlemanly terms, he could afford to charge George Wythe no fee when his preceptor was sued by a ship's captain. He charged Frank Willis no fee in four cases involving his debts. At his own expense, he brought a "friendly" caveat against his kinsman Thomas Mann Randolph for four hundred acres on Southwest Mountain, near Edgehill, and he defended Randolph against another claimant for the same land, at least eight times representing this kinsman without any fee.

Jefferson declined fees whenever cases came his way that would have established the freedom of persons their masters claimed were slaves. He had problems getting paid, but his collection problems were no more acute than those of other leading Virginia practitioners. Fees were set by law and had not been increased for half a century, and with little cash in the colony, many clients never did pay up. A notice appeared in the May 20, 1773, *Virginia Gazette,* signed if not actually written by Jefferson on behalf of five out of seven of the other members of the General Court bar, complaining that "the fees allowed by law, if regularly paid, would barely compensate our incessant labors, reimburse our expenses and the losses incurred by neglect of our private affairs." An "unworthy part of our clients"[56] withheld any payment at all. As a result, Jefferson and his associates announced that henceforth they were requiring full payment in advance for opinions and half-fee before taking on any case. The next year, Jefferson, while not always following this announced policy, managed to collect nearly half his fees (he made numerous exceptions). But, after six years of practice, he had collected only £797.10.5 and was still showing £1,324.16.6 in uncollected fees, roughly two-thirds of his gross earnings. His net earnings were even worse. Ominously, Jefferson's records showed him having "total profits" of £2,119.7, the sum of the two figures, a grossly optimistic overstatement.[57]

Even as he attracted clients from among Virginia's landed aristocracy and distinguished himself as a rising young member of its ranks, Jefferson was becoming disenchanted with the all-controlling country gentry. As he turned his back on scheming frontier land speculators and shifted his law practice away from the "insects" of the county courts and gravitated toward the capital at Williamsburg, he began to read about, think about, and argue in open court questions of conscience and law—slavery, adultery, divorce, religion, and power—which now vied for his interest with the mere making of money.

One of the earliest cases Jefferson took on as a young lawyer of twenty-four involved adultery. He wrote with graphic, down-to-earth precision this casebook notation:

> David Frame [Augusta] directs me to issue writ in Scandal against James Burnside [Augusta]. . . . Burnside said he caught Frame [who is a married man] in bed with Eliza Burkin. . . . [He] put his [hand] on Frame's as he lay in bed with the girl and felt it wet, and then put his hand on his [erased and inked over] and felt it wet also."[58]

It was the only time Jefferson wrote explicitly about sex. On the later occasions when he wrote about adultery, it was always indirectly, and with compassion. After living in France for a few years, he wrote of "good" and "bad" passions and he became absolutely playful on the subject of violations of the marriage vows. At age sixty-six, ex-President Jefferson wrote to a close friend about the relationship between morality and religion, that

> reading, reflection and time have convinced me that the interests of society require the observation of those moral precepts only in which all religions agree, for all forbid us to steal, murder, plunder or bear false witness.[59]

Significantly, he excluded adultery from the list. Under the heading of "morality" in *The Jefferson Cyclopedia,* John Foley listed everything Jefferson wrote on the subject (at least what was known in 1900) and it amounted to nearly three thousand words. Under "virtue," Jefferson wrote about five hundred words. There is no heading for chastity.

Between his first year of law practice at twenty-four and his presidency, Jefferson's views on sexual matters evolved from listening to tall stories among young male friends to taking testimony in numerous marital cases resulting from the exaggerated eighteenth-century code of honor to participation in a society where sexual contact was perhaps as easy as in the twentieth century. There is no evidence that he was ever promiscuous in his eighty-three-year-long life, during which he was married for only eleven years. But, at least from the age of twenty-five, Jefferson was fond of married women and, on at least two occasions, he fell intensely and passionately in love with one.

During the summer of 1768, Jefferson's friend and neighbor, Jack Walker, went off with his father, Dr. Thomas Walker, as part of Virginia's delegation to the Indian treaty talks at Fort Stanwyx on the New York frontier. They were gone four months. Jefferson and Jack Walker had been to school together at Reverend Maury's and at William and Mary, where they were cronies in the Flat Hat Club. Their fathers had been closest friends and partners in the Loyal Company of land speculators. Dr. Walker had been one of Peter Jefferson's executors. The sons remained close for many years, socializing together, riding together, eventually serving together in the House of Burgesses as delegates from Albemarle County.

When Jack Walker became engaged, Jefferson had spread the word to their mutual friend, John Page: "Jack Walker is engaged to Betsy Moore, and desired all his brethren might be made acquainted with his happiness."[60] Jefferson had known Betsy's family for years: two of her brothers were at William and Mary during his time there. He had been a bridesman, part of the groom's party, when Walker married Betsy Moore in June 1764.

When Jack Walker rode north to Fort Stanwyx as clerk to the Virginia delegation in July 1768, he entrusted his wife and their baby daughter, and their home at Belvoir, five miles from Shadwell, to Jefferson's particular care. Jefferson was considered a friend by Betsy's family, too, and she trusted him. He gave elaborate advice to her brother, Bernard, on the lawbooks he should purchase for his library and on how to prepare himself for a legal career, and, when her brother's business affairs foundered, Jefferson backed him financially and represented him without fee. He often visited Betsy's father's home in King William County and they exchanged specimens for their gardens. When Jack Walker drew up his will before riding off to the Indian conference, he named Jefferson his chief executor.

It was evidently sixteen years later before Betsy Walker told her husband what had happened that summer of 1768 while he was away so long and his virile, handsome friend, Thomas Jefferson, had paid frequent calls for one reason or another after long rides over from Shadwell. In 1784, the outraged husband, by now an avowed political enemy of Jefferson's, kept Betsy's account to himself except to write Jefferson about it. For nearly twenty more years, Walker remained silent.

6

"All Men Are Born Free"

Those who labor in the earth are the chosen people of God . . . his peculiar deposit for substantial and genuine virtue.

—THOMAS JEFFERSON, *Notes on the State of Virginia*

IN FEBRUARY 1769, Thomas Jefferson, nearly twenty-six, wrote from his mother's house at Shadwell to inform his uncle, Thomas Turpin, that he could not take on the law tutoring of his cousin Philip as the family had expected. In a rare letter surviving from his early years, Jefferson explained to one of his former guardians, "My situation, both present and future, render it utterly impossible" because "I do not expect to be here more than two months in the whole [year] between this [month] and November next."

There had been an obvious and major change in Jefferson's plans. A few months earlier, in December 1768, in the first province-wide election since Jefferson had completed his own legal studies, he had successfully stood for the seat that had once been his father's in the Virginia House of Burgesses. More of his time would now be consumed by legislative affairs in the provincial capital at Williamsburg. And, while Jefferson had recently broken ground for a small house of his own on a hilltop two miles away from his mother's house, he clearly intended it only to be a bachelor's quarters for his rare visits home. Besides, the work was lagging. A severe shortage of limestone needed to make mortar would slow bricklaying for a year. As confident as anyone naïvely setting out to build his first house, Jefferson wrote, "I propose to remove to another habitation, which I am about to erect." The house was to be "on a plan so contracted" that there would be only one guest room, and southern hospitality combined with the far-flung nature of settlement to dictate that young Jefferson reserve the "one spare bedchamber" for weary travelers. There was no room for a live-in law student.[1]

The letter to Turpin is the earliest surviving record that Jefferson was about to break ground for his own mountaintop home, although there is an entry in his *Garden Book* as early as August 3, 1767, that shows he had already begun planting trees on the little mountain's slopes.[2] Once again, he could not conceal a little envy of his brother-in-law, Dabney Carr, on this score. "This friend of ours," he wrote John Page, "in a very small house, with a table, half a dozen chairs, and one or two servants, is the happiest man in the universe." In his modest new house, Jefferson hoped to emulate Carr, who lived a Spartan existence "with as much benevolence as the heart of man will hold but with an utter neglect of the costly apparatus of life."[3]

As his letter to Turpin makes plain, Jefferson had no interest at this stage in becoming a mentor to his cousin or anyone else. He was disenchanted with the traditional apprenticing system for training lawyers. In a letter at once remarkably candid and critical of the English legal system, he hinted that he resented his long-drawn-out clerkship under George Wythe:

> Placing a youth to study with an attorney was rather a prejudice than a help. . . . We are all too apt, by shifting on them our business, to encroach on that time which should be devoted to studies.

All a law student needed "is to be directed what books to read and in what order to read them." He "strongly recommended" that young Turpin "put himself into apprenticeship with no one." As busy as he was, Jefferson made time to formulate "a plan of study which will afford him all the assistance a tutor could" without wasting young Turpin's time "for the emolument of another." If Philip Turpin was willing to take Jefferson's advice, Jefferson pledged "as often as possible to take your house in my way to and from Williamsburg" to observe his young cousin's "progress in science" and give him pointers.[4]

In fact, Jefferson could have taken on young Turpin, dragging him about the countryside to haul his books, to do his research in court records, to catch his every word, but by declining to take this unpaid and unpaying assistant, Jefferson was once again refusing one of the reciprocal obligations that came with his status as a gentleman. He did not, at least consistently, believe in the network of money and social connections in the way he was supposed to. He had refused to capitalize on his Randolph family connections when he had chosen his own law teacher. Now he refused his family obligations by sloughing off young Turpin, who would eventually study law in Edinburgh, Scotland, quite possibly on Jefferson's advice.

The real and deeper reason for Jefferson's reticence to assume the traditional chore of taking on one's kinsman as an apprentice was that, more and more, Jefferson wanted to be in Williamsburg, not looking over the shoulder of a clerk in Charlottesville, even if he had a proper office there. The colonial crisis was becoming more complicated and Jefferson had eagerly plunged into provincial politics at this exciting moment. Jefferson evidently had little hope that the problems with England would be solved quickly. He intimated to the elder Turpin that he saw "no reason to expect at any future day to pass a greater proportion of my time at home."[5] For some time he had maintained rented rooms in Williamsburg. His increasingly prestigious practice before the General Court in the capital and his recent election were pulling him away from the west just at the moment he was beginning serious work on a home of his own.

His youthful political debut of five years earlier, the Rivanna River project, had less to do with his plunge into politics than a steadily deepening interest in colonial affairs that dated from the Stamp Act crisis. After repeal of this hated measure and the reopening of the county courts, there had been a brief lull in the face-off with the British Parliament but, testing its assertion that it was the ultimate power in all imperial affairs, Parliament had passed a new set of revenue-producing colonial duties, the Townshend Acts of 1767, which had loomed on the Virginia political horizon as Jefferson had ridden west to launch his legal career. These new parliamentary customs duties had quickly and radically altered the nature of the quarrel between colonial legislatures and England by establishing a system of customs commissioners who were to operate onshore, on American soil for the first time instead of from ships offshore. Now the taxes were internal and intended to raise revenue for the British. Parliament was challenging the exclusive authority of colonial governments to raise their own taxes. The steep new duties, levied on tea, glass, lead, painter's colors, and paper imported into the colonies, were intended to wring £40,000 a year from the colonies to help the British defray military expenses in America and to help pay fixed salaries for royal officials in the colonies. This took a cherished charter right, the power of the purse, out of the hands of colonial assemblies that had long clutched the purse strings as a way of controlling corrupt or incompetent royal officials. So far, the assemblies had been able to frustrate repeated Crown attempts to establish a fixed Civil List of royal officials who would be appointed and paid directly from England. It was a thoroughly alarming development. The British were seeking to take tax money from American colonists by means of parliamentary taxation instead of the time-honored levying of taxes by the colonists' own representatives elected to provincial assemblies which, in turn, only in emergencies voted voluntary subsidies to the home government but otherwise routinely had set and paid all official salaries and public costs. But what really riled many colonists was that Townshend intended to use the money thus raised to keep order in America—that is, to bar further protests such as the Stamp Act riots—at the same time trimming back on colonial liberties by making colonial governors and judges financially independent of the colonial assemblies.

To enforce the odious new measures efficiently, the British Parliament legalized writs of assistance, which gave customs officers authority to call on provincial officers for assistance and authorized them to search private warehouses and homes for smugglers' contraband. This controversial step had been authorized since the widely flouted Molasses Act of 1733, but never had it been systematically enforced before the 1760s, when customs officers in Boston and Philadelphia had begun wide-scale seizures of illicit cargoes. In Boston, James Otis had unsuccessfully argued before the Massachusetts Supreme Court that the writs were illegal. Now, in 1769, instead of eliminating the writs, Parliament was extending their use. At the same time, the Townshend Acts created vice admiralty courts in colonial seaports. On the books since the fourteenth century and rivaling in importance the common-law courts at Westminster, the High Court of Admiralty could create regional branches in districts headed by vice admiralties. There had long been attempts to establish vice admiralty courts, without trial by jury, in the colonies, but they had not been successful. As early as 1678, royal governors had received, as part of their commissions, appointment as vice admirals. Gradually vice admiralty courts had been organized in America and their jurisdictions widened beyond that of their English counterparts to enforce the unpopular Acts of Trade against smuggling. Repeatedly the British had extended the admiralty courts to include prize cases (1708) and timber (1722). Convicted smugglers could appeal only to London to the High Court of Admiralty, a ruinously expensive procedure. Since 1748, the king's Privy Council had also been taking appeals. In 1766, only one year before passage of the Townshend Acts, the Privy Council had assumed sole appellate jurisdiction.

As Jefferson had begun his law practice in the Virginia backcountry, the Townshend Acts further extended the vice admiralty system, establishing courts in Halifax, Boston, Philadelphia, and Charleston. Turning over the appellate jurisdiction in America to these new courts, the Townshend Acts forbade "further right of appeal to England." No longer could accused smugglers expect leniency at the hands of juries of their neighbors; no longer could they petition over the heads of detested royal appointees to London for justice.

Thomas Jefferson had been in Williamsburg for the General Court session in October 1768, when a new royal governor had arrived to replace Jefferson's friend, Francis Fauquier, who had died six months earlier after a long illness. One night in late October, Jefferson was in the crowd as fireworks illuminated the capital to welcome the Right Honorable Norborne Berkeley, Baron de Botetourt, an impecunious lord of the bedchamber to George III. Young Jefferson was still too obscure to be invited to the sumptuous welcoming dinner at Raleigh Tavern, but soon enough he was to come to the attention of this fiftyish Tory resident governor-general, described by one anonymous revolutionary as "a cringing, bowing, fawning, sword-bearing courtier." It was the arrival of Lord Botetourt, a viceroy of the Crown, that prompted Jefferson to run for the House of Burgesses. Dissolving the old Assembly, as was his royal prerogative, Lord Botetourt had issued writs for election of a new one.

By mid-December 1768, the freeholders of Albemarle County—any man owning fifty unsettled acres or a plantation of at least twenty-five acres with a house on it—crowded into the courthouse in Charlottesville, for rum, cider and cakes, and a little handshaking. There is no record of a heated contest. Jefferson's father's closest friend, Dr. Thomas Walker, just back from treaty-signing at Fort Stanwyx with *his* son, was reelected without opposition. Jefferson challenged land-rich Edward Carter, whose Albemarle holdings as well as his estate on Carter's Mountain towered over the young Jefferson's, but Carter had failed to attend the last legislative session and Jefferson no doubt capitalized on this fact. It was a friendly contest, though, with Carter and Jefferson sharing the cost of the rum and Jefferson, according to his account, laying out a modest amount for cakes and other drinks.[6] It was more a political initiation for Jefferson than a brilliant upset. He ran for office as soon as he could, in the first election after he finished his law studies and in the first Assembly election since the Stamp Act crisis. As historian Dumas Malone put it, "The freeholders of his County elected him as soon as they very well could."[7] After a bit of "burgessing," as the pouring of up to one and a half quarts of liquor per man and the handshaking and back-slapping was called, after what one of Jefferson's friends called "swilling the planters with bumbo," the freeholders ushered him into what had been, on the founding of the county, his father's seat. This was no secret ballot: every freeholder, as well as he was still able, called out the name of his candidate.[8] But it was not just his father's former position that gave young Jefferson a claim on a seat in the House of Burgesses. He had established himself as a lawyer and as a member of the Albemarle gentry. More than that, it was the land, his deep involvement in it, and his attitude toward it, that gave him his influence in land-hungry Virginia at so young an age.

Thomas Jefferson's public life began on April 6, 1769, when the new burgess arrived in Williamsburg with his trusted servant, Jupiter. In the intervening month before the Assembly convened, he caught up on a backlog of legal casework. On Monday, May 8, he took the oath of office in the crowded Council Chamber at the Capitol with all the other burgesses. The ritual of initiation was impressive. After being sworn in the chamber of the upper house, the burgesses retired to their own chamber across the hall. Soon they were summoned again to attend the governor, resplendent in a scarlet coat, in the Council Chamber in a ceremony that mimicked the opening of Parliament in London. Jefferson had watched six matched white horses draw His Excellency from the palace along Duke of Gloucester Street to the Capitol in his ornate state coach emblazoned with the Virginia coat of arms. Governor Botetourt then sent them back to elect a speaker: they unanimously elected Jefferson's cousin, Peyton Randolph, the man who, only four years earlier, had worked so hard to obliterate Patrick Henry's Stamp Act resolutions. How far toward the radical position this Virginia grandee had come! The Speaker's mace laid on the table, Lord Botetourt was informed of Randolph's election. Then the burgesses once again paraded to the Council Chamber to learn that the governor approved their choice. Once formally recognized, Speaker Randolph claimed the traditional rights and privileges of the House, now sitting for the one hundred fiftieth year, the oldest legislative assembly in the New World. These ancient rights included freedom of speech and debate, exemption of members from arrest while the House was sitting, the protection of their estates. Promising to uphold their privileges, Lord Botetourt made his first speech.

When the burgesses returned this time to their own chamber, they resolved, as was the custom, on presenting His Lordship with an address in thanks for his "very affectionate" comments. The Speaker, Peyton Randolph, had known his young kinsman since his clerkship, when Wythe had introduced them, and he was happy to launch the young lawmaker. He chose the new burgess from Albemarle County to draft a series of resolutions answering the governor. There was little hint of originality or revolutionary thinking in Jefferson's first brief official expressions in this "most humble and dutiful address," punctuated with genuflections to "his Majesty's sacred person and government." Yet, in his last resolution, he combined obsequious loyalty with the firm self-confidence of this largest and wealthiest constituent assembly in colonial America: England's "interests, and ours, are inseparably the same." He concluded with a cautious wish for Governor Botetourt's cooperation, that "Providence and the royal pleasure may long continue his Lordship the happy ruler of a free and happy people."[9] Jefferson's resolutions were accepted by the House, and the Speaker named him to a committee to turn them into a remonstrance to the royal governor. In committee, he fared less well. He was chagrined that his resolves were considered too lean and terse and were thoroughly rewritten by Robert Carter Nicholas, the stooped-over treasurer of the colony, a leader of the powerful Carter clan who was, at first, Jefferson's chief critic in the committee charged with drafting the resolutions. The sensitive Jefferson long remembered and later commented that Nicholas insisted on drawing up "an address more at large," which he did with "amplification enough," which was accepted by the whole House. Still smarting forty-six years later at the fledgling's humiliating first flight, he wrote a biographer, "Being a young man as well as a young member, it made on me an impression proportioned to the sensibility of that time of life."[10]

In all, Jefferson was appointed to three committees in his first session, where he wrote much and said little as one of the more obscure of the one hundred eight members. Jefferson was at first content in this company of men, most of whom he already knew. The courtly yet affable Speaker Randolph was kind to his young kinsman; Patrick Henry, still rough-cut but clearly more prosperous, regarding the proceedings askance from the sidelines; Edmund Pendleton, the brilliant nemesis of Jefferson's mentor, George Wythe, adroitly steering Virginia's reactions to Parliament and king. Jefferson made new friends, including George Washington, a dignified, usually silent thirty-seven-year-old former assemblyman up from Fairfax. There was little dissent among them in their closed committee sessions. A little over a week into their proceedings, on May 16, 1769, they unanimously voted a series of resolves to become known as the Virginia Resolves, which delineated their rights as colonists but could not help but be regarded as provocative by the new royal governor.

Since Jefferson had heard Patrick Henry orate against the Stamp Act in this chamber four years earlier, he had closely followed the reactions of his fellow Virginians to parliamentary measures. His kinsman, Richard Bland, had written what has been acknowledged as the most intricate literary conceit of the revolutionary period, *The Colonel Dismounted or the Rector Vindicated,* in 1764, ridiculing the clergy in the Parson's Cause controversy, and he was a first-rate scholar of English constitutional history and law. Jefferson had read Bland's *An Inquiry into the Rights of the British Colonies,* in which he denied the assertion of some British lawmakers that Americans enjoyed virtual, if not real, representation in Parliament and could therefore be justly taxed by the faraway Parliament. Young Jefferson agreed with Bland that British taxation of American colonists was, as he later put it, part of a "connected chain of Parliamentary usurpation." "Right and power," Bland had argued, "have very different meanings and convey very different ideas." Power, "abstracted from right," cannot give "a just title to dominion." And it is not possible to "build right upon power": the result is "brutal power" which becomes "an irresistible argument of boundless right."[11] He immediately aligned himself with the subtle and literate Bland's radical faction, even though this meant risking the wrath of more conservative followers of the late Governor Fauquier, who had dubbed the tax protesters "young hotheads."[12] But as the young lawmaker took his seat in a colonial assembly, it was the suspension of the legislature of the Province of New York that troubled him most.

In 1765, within months of passage of the Stamp Act, Parliament passed the Mutiny Act, designed to improve discipline in the British Army throughout the empire. Including a provision for quartering troops in private houses, the act alarmed Americans, who evaded the requirement by refusing to recognize any clause of the new law that did not specifically refer to overseas British possessions. At the specific request of the British commander in America, Gen. Thomas Gage, Parliament closed the loophole by passing the Quartering Act, requiring that the colonies provide barracks and supplies for British troops in America, eliminating the odious billeting of troops in private homes. Colonial legislatures were reluctant to vote funds for barracks, realizing that their compliance could be taken as acknowledgment of Parliament's right to tax them without their consent. Some assemblies refused to vote supplies of food, firewood, or straw for soldiers' bunks.

American resistance infuriated many Britons. In 1766, a new and tougher Quartering Act authorized the use of taverns and empty houses for billeting troops. When the royal governor of New York asked the New York Assembly to vote provisions for British troops, the Assembly refused full compliance on the grounds that the colony, as British Army headquarters for North America, would be unfairly burdened. When the Assembly rejected a second and third request, pleading insufficient funds, soldiers and citizens clashed in the streets and the royal governor prorogued the Assembly. On June 15, 1767, the king gave his assent to a Townshend Ministry act that suspended all of New York's legislative powers until the Assembly complied with the Quartering Act. The colonial struggle had taken a grave turn. When New Yorkers elected a new Assembly, it, too, was dissolved; a third Assembly was elected in January 1769, one month after Jefferson's election in Virginia, but it had still not accepted the Quartering Act. Assemblies all over America had tensely watched the impasse.

The Virginia House of Burgesses had once again spoken out against parliamentary obduracy. In its 1768 session, the burgesses had argued that Virginians could never admit they were at the mercy of the will of the English Parliament, a position Jefferson firmly espoused. Now the 1769 session once again turned its attention to the Townshend duties, still on the books after two years of noisy colonial trade boycotts against goods carrying its imposts. Jefferson believed that Parliament had neither right nor authority to impose its will over the colonies in collecting either internal or external taxes, a position more radical than most burgesses were willing to take on the record. They were opposed to laws passed by Parliament to raise revenue by regulating external trade, but were willing to allow Parliament to assert measures for "preserving a necessary dependency."[13] Jefferson refused to make such a nice distinction; he was coming to view the controversy over internal versus external taxation a sham.

Another controversy involving colonial legislatures had erupted in the previous session in Massachusetts, and it had Virginia lawmakers worried. In the Bay Colony, where there was a longer and more violent history of resistance to British taxes than in Virginia, a board of customs commissioners was ensconced and enforcing the new duties on glass, paper, dyes, and tea, four commodities in popular demand. In February 1768, the Massachusetts House of Representatives had sent other American legislatures a circular letter calling for a united front against the Townshend duties; the British colonial secretary, in turn, suspended the Massachusetts House until it should repeal the circular letter. The Virginia Assembly responded to this latest threat by petitioning king, Lords, and Commons to repeal the duties and directed Speaker Randolph to write to the Massachusetts Speaker to applaud Massachusetts's "attention to American liberty."[14] This was new language; until this time, the word *American* had been applied by Englishmen to describe native American Indians. As late as 1771, when Benjamin Franklin presented his commission as the new agent for the Massachusetts Assembly to the British colonial secretary, the word was still a pejorative term. Franklin wrote that he was considered by British officials "too much of an American"[15] to hold high imperial office. The British assault on colonial assemblies convinced many Americans that Parliament and Ministry would go to any length, even to the extent of destroying representative government in the colonies, to establish Parliament's supremacy. The Virginia Assembly had informed the Massachusetts Assembly in 1768 that it intended to act in concert "with other colonies in their application for redress."[16] As Virginia and other colonies supported the Massachusetts circular letter, the imperial government angrily ordered all royal governors to treat the letter as subversive and dissolve any assembly joining the protest against Townshend duties. Lord Botetourt carried such an order to Virginia with him, yet the slow, smooth conciliatory speeches of the new Virginia governor had done nothing to separate the northern and southern colonies.

When the British Colonial Office ordered the Massachusetts House to rescind its circular letter, Massachusetts refused. The colonial secretary was Lord Hillsborough, described by Franklin in a letter to Massachusetts Speaker Thomas Cushing as "proud, supercilious, extremely conceited (moderate as they are) of his political knowledge and abilities, and inimical to all who dare to tell him disagreeable truths."[17] To put teeth in Hillsborough's order, Parliament, a few months before Jefferson took his seat, voted to approve Hillsborough's heavy-handed policy and revive an ancient law that permitted the British government to arrest and transport to Westminster for trial any person accused of treason outside the kingdom, a clear reference to the supposed traitors in American assemblies from Massachusetts to Virginia, including Peyton Randolph and Patrick Henry. The word *treason* was not an empty one for subjects of England of the time. Jefferson had been a little boy when Scottish lords had been executed after the Rising of 1745, and their skulls still graced spikes at the gates of London.

On Jefferson's ninth day as a lawmaker, the Virginia House of Burgesses accepted Parliament's challenge, declaring that Virginia burgesses alone, not members of England's Parliament, had the right to lay taxes on Virginians, also asserting their right to petition the king over the head of Parliament and their right to seek the concurrence of other colonial legislatures. Where Governor Botetourt had hoped for silence, the burgesses complained loudly about the "dangers and miseries"[18] that they would suffer if they were transported to England for trial.

Summoning the Assembly into the Council Chamber, Lord Botetourt, once again dressed in official scarlet, lectured them: "I have heard of your resolves and augur ill of the effect. You have made it my duty to dissolve you, and you are dissolved accordingly."[19] That Tuesday noon, May 16, 1769, Jefferson joined his now-former Assembly colleagues, the "late representatives of the people,"[20] as they marched down Duke of Gloucester Street and crowded into the Apollo Room of the Raleigh Tavern. Electing Peyton Randolph moderator, the now-extralegal assembly voted to adopt an "Association," an agreement that barred the importation or consumption of British goods. Tall, powerful Colonel Washington, an ex-burgess from Fairfax County, took command of the discussion. He presented a set of resolutions adopted in March by Philadelphia merchants. The Virginia resolves were drawn up by Washington, Mason, and Richard Henry Lee of Chantilly in Westmoreland County, who had lost his awe of the English as a student in London. Washington, one of the wealthiest planters as a result of his marriage to the richest widow in Virginia, was typical of an increasing number of Virginia planters who had become disillusioned with the endless, useless process of sending humble addresses, memorials, and remonstrances to London. Earlier that year, Washington had come to the conclusion that a boycott of English goods—the American colonies accounted for one-third of all British exports—was the only peaceable solution: "How far their attention to our rights and privileges is to be awakened or alarmed by starving their trade and manufacturers remains to be tried."[21] Washington was already the most militant of the burgesses, although he gave no hint of it that day. In a private letter to George Mason of April 5, 1769, he had derided "our lordly masters in Great Britain," who would "be satisfied with nothing less than the deprecation of American freedom." He was prepared to take up arms to defend his liberty, but "arms, I would beg to leave to add, should be the last recourse, the *dernier resort.*" The first, he suggested, should be "starving their trade and manufacturing."[22]

The list of banned imports was extensive and was to make life more difficult for the small number, including Jefferson, who heeded it conscientiously. It included slaves, food, liquor, wine, fruit, meat and cheese, oil and candles, tables and chairs, watches and clocks, sugar, pickles and candy, mirrors, carriages, cabinets and upholstery, jewelry, gold and silver, "ribbon and millinery of all sorts," lace, "India goods of all sorts [this included tea]," silken robes, woolens and linens [except for sewing at home], shoes, boots, saddles and all manufactured leather goods. The sweeping ban on British imports was to take effect September 1, and life in the farming colony of Virginia was to become much more austere.[23]

Washington and his radical faction favored a ban on exports as well, but the more moderate majority voted him down. The Association agreement that Washington took out of his pocket was signed that night by 94 of 115 of the "principal gentlemen" of Virginia, most of whom had so recently served in its Assembly. And then they all drank toasts to the king, the queen (whose birthday they would all celebrate the next day), various members of the royal family and, significantly, several leading pro-Americans in Parliament who had spoken out against the Ministry's colonial policies. On May 18, 1769, at age twenty-six, Thomas Jefferson was the sixteenth to sign, marking him as a troublemaker in British eyes from that moment.[24]

Nearly half a century after his active practice as a Virginia lawyer, Thomas Jefferson finally told someone what he thought of county justices who had blocked him in a variety of his attempts to systematize and modernize the legal profession, efforts that began with his wholesale land caveating in western counties and led to a complete overhauling of Virginia law in the Revolution. In an exchange of letters on the nature of constitutional government, by-then ex-President Thomas Jefferson revealed that he continued to think that the real threat to any lasting legal reform in Virginia lay in self-perpetuating family oligarchies that controlled the county benches. Jefferson lashed out at

> the vicious constitution of our county court . . . self-appointed, self-contained, holding their authorities for life and with an impossibility of breaking in on the perpetual succession of any faction once possessed of the bench. They are, in truth, the executive, the judiciary and the military of their respective counties, and the sum of the counties makes the state.[25]

Jefferson was summarizing old fears of country politicians like himself who had railed at the corruption of the Court in England until they had banded together to bring down the Stuart kings twice—in the English Revolution of the 1640s and in the so-called Glorious Revolution of 1688. His country ideology became central to his opposition to corruption in the British imperial system, but his first encounters with it took place at the county level. There were few more frustrating incidents in his short-lived legal career than his entanglement in the aftermath of the anti-inoculation riots in Norfolk County that began in the summer of 1769.

The summertime prevalence of smallpox in Williamsburg certainly contributed to Jefferson's grudging long stays at home at Shadwell and it propelled him all the way to Philadelphia for an inoculation that he could not find closer to home in Virginia, although inoculations had been carried out secretly there for twenty years. Until 1796, when a vaccination was developed, those who wished protection against the dread disease had to run the risks of inoculation. This process consisted of transmitting infected matter from a person with a mild case of smallpox to the person being inoculated in an attempt to cause a similar mild case that would produce lifelong immunity. Inoculated patients had to be quarantined until the disease ran its course. The practice was rarely fatal, but neighbors often objected, for fear of contracting the disease unintentionally. When inoculation had first been introduced in America in 1721, Bostonians rioted despite the fact that prominent clergymen and physicians had promoted it.

In 1768, two people died near Yorktown, in Norfolk County, Virginia, after inoculated patients supposedly were released prematurely from quarantine. That June and again in May 1769, the confrontation in Norfolk between pro- and anti-inoculation factions led to riots. The public inoculations were probably the first ever performed in Norfolk County and had the support of the mayor of Norfolk, Cornelius Calvert. The two doctors who carried out the inoculations, Dr. Archibald Campbell and John Dalgleish (who had practiced the year before on his apprentice) were indicted by the county court.

The first riot came after Dr. Campbell and some of his friends tried to have Dr. Dalgleish inoculate their families. They chose a quarantine site in Norfolk, but anti-inoculation leaders objected. So Dr. Campbell scheduled the inoculations for June 25, 1768, at his own plantation three miles out of town, and agreed to limit the number he invited to the doctors' immediate families. After the inoculations, popular outrage grew. Dr. Campbell, under pressure, agreed to move his patients to the Norfolk pesthouse, the final shelter for the terminally ill. Dr. Campbell insisted that the pesthouse be cleaned up first. But Joseph Calvert, the brother of the mayor, would not wait, and he organized a mob that attacked Dr. Campbell's house and chased the inoculated patients the three miles to the pesthouse in a storm. Two days later, Dr. Campbell's plantation house burned.

Reward offers by colonial officials brought no response, but an unsigned account of the riot in the *Virginia Gazette* stated that when Dr. Campbell had met with anti-inoculation leaders to limit the inoculation, one opponent had said that, if he had not limited it, his house would have been destroyed that night. When no arrests took place, eventually Dr. Campbell came to Thomas Jefferson, on December 20, 1769, and asked Jefferson's opinion whether there was enough evidence to bring a civil suit against the man who had since identified himself in a subsequent *Gazette* article as having threatened to level the doctor's house. Jefferson thought there was insufficient evidence.

Yet Jefferson was keenly interested in the controversy, and he was ready to support publicly his belief in the need for inoculation. He had already received a letter from another lawyer retaining him "on behalf of the sufferers by the riot in Norfolk."[26] In the October 1768 General Court term, Jefferson filed three civil suits on behalf of the Campbells and another victim, Norfolk merchant James Parker, against four leaders of the anti-inoculation faction. Six years later, when the onset of revolution closed the courts, none of the three cases had come to trial. Other victims had hired another lawyer, who turned his Norfolk clients over to Jefferson when he retired. Those cases never came to trial, either. Criminal charges, brought by both factions, moved faster. In the April 1769 session, *The King v. Calvert et al.*, an indictment for riot was entered on the docket with *The King v. Dalgleish,* an indictment for nuisance.

In May 1769, before either case could come to trial, another riot broke out. Mayor Calvert, fearing that some of his slaves might be exposed to smallpox while unloading ships in the harbor, had Dr. Dalgleish inoculate them at the pesthouse. The next day, the borough magistrate, Mayor Calvert's brother, Maximillian, ordered Dr. Dalgleish arrested and jailed by another brother, Borough Sergeant Joseph Calvert. That night, a mob damaged Mayor Calvert's house and Dr. Campbell's town house. When they surrounded Parker's house, they were repulsed by the merchant and his friends with a show of guns.

More suits followed, more civil cases came by inheritance to Jefferson, and three more indictments were issued by the General Court. On October 8, 1769, with Jefferson in the courtroom, the first case came before the amateur judges of the General Court, all members of the Governor's Council with the royal governor, Lord Botetourt, presiding. Botetourt thought the mayor and physician had acted prudently by carrying out the slave inoculations in the pesthouse, but several members of the General Court were opposed to inoculation. Judge John Page, father of Jefferson's close friend, voiced his opinion "that Dr. Dalgleish and Calvert intended to spread the infection." He and another Council member wanted the doctor kept under bail, but Governor Botetourt, ordering that there be no further inoculations "except in cases of necessity," ordered the doctor released.

After the hearing, the governor was accosted by an angry William Nelson, a judge of the General Court and member of the Governor's Council who had refused to disqualify himself even though he was the most powerful man in Norfolk County. James Parker, the Norfolk merchant who had armed his servants and friends and successfully defended his home, wrote a friend in London that Nelson, "extending his right arm, his face as red as fire," shouted at the governor, " 'If I had the power I would hang up every man that would inoculate even in his own house.' "[27]

On Monday, April 16, 1770, Thomas Jefferson was retained to defend Dr. Campbell and Dr. Dalgleish before the Virginia General Court against the charge of nuisance, which was to come before the supreme Virginia court only two days later.

Their clientage in this facet of the complicated smallpox controversy could not have come at a more inopportune time. Scarcely two months earlier, on February 1, 1770, while Jefferson had been in Charlottesville on business, fire whipped by winter winds raged through his family home at Shadwell, utterly destroying it in a matter of minutes. Jefferson had dined early, at about two o'clock, with his mother and his family, and then ridden to Charlottesville, three miles away. Several hours later, a slave brought him the news and told him that no one had been hurt. According to family lore, Jefferson's first question was whether his books had been saved. "No, Master, all lost, but we save your fiddle."[28] Presumably it was the fine Italian violin his family knew he treasured. A few books had been saved and two or three beds. They were moved to outbuildings, where his mother and the children crowded in, away from the mountain winter. Jefferson's small house at Monticello was far enough along for him to spend the next nine months, whenever he was home, in a single finished room.

If he wrote or said a word about his mother's terrible loss of everything she owned, the home Jefferson's father had built for her, all its handsome furnishings and her family's heirlooms, there is no hint of it in the letters that survive. Perhaps Jefferson was too deep in shock, but he was obviously angry that no one had saved his prized possessions, and his coldness toward her once again may mean he blamed his mother for the loss of his father's house as well as his precious library and law records.

Not only most of his books, but all of his legal notes and records apparently were destroyed. He poured out his agony to John Page in a letter he wrote three weeks after the fire. He attempted to be lighthearted, but Jefferson was never good at breeziness—or grief. "My late loss may perhaps have reached you by this time." Indeed, it must have. The *Virginia Gazette* probably had just reached Page with its grim announcement of February 22, 1770:

> We hear from Albemarle that, about a fortnight ago, the house of Thomas Jefferson, Esq., in that county, was burnt to the ground, together with all his furniture, books, records, etc., by which that

> gentleman sustains a very great loss. He was from home when the accident occurred.

Jefferson seemed to have no sense that Page could have heard of the disaster from anyone other than Jefferson himself, or that he was important enough as a burgess and General Court attorney to be written up in the Williamsburg newspaper. To Page, he reported

> . . . the loss of my mother's house by fire, and in it, of every paper I had in the world, and almost every book. On a reasonable estimate I calculate the cost of the books burned to have been £200 sterling [about $20,000 in today's money]. Would to god it had been the money; then had it never cost me a sigh!

Never mind the great emotional loss of his birthplace or its small fortune in furnishings. But the effect on his legal practice and his private studies was devastating:

> To make the loss more sensible, it fell principally on my books of common law, of which I have but one left, at that time lent out. Of papers, too, of every kind I am utterly destitute. All of these, whether public or private, of business or of amusement, have perished in the flames.

Jefferson had made "some progress" before the fire preparing his cases for the spring term, including the smallpox case,

> having, as was my custom, thrown my thoughts into the form of notes, I troubled my head no more with them. These are gone, and "like the baseless fabric of a vision, leave not a trace behind."

His records and legal documents and notes "furnished me with [the status] of the several cases" and they "have shared the same fate." Jefferson's law practice was in ruins. He had not yet fathomed the depth of the disaster, and he was still too stunned to work out a detailed plan. He had been working feverishly "to recollect some of them," but there were so many holes in his memory "I can make nothing of them." The General Court had a firm rule against continuances of cases, despite its chronic backlog of several years or, perhaps, because of it. "What am I to do then in April?" Yet he was preparing to argue that his case was "singular."

All that Jefferson could salvage from Shadwell's ashes was some comfort that, while "I am burned out of a home," he was "advanced so far in preparing another." Had he not already begun work on Monticello, "I might have cherished some treasonable thoughts of leaving these my native hills."[29]

From the ashes, Jefferson apparently salvaged his fee and casebooks and his commonplace books. (Did a slave save them after all or did Jefferson have them with him away from Shadwell or up at his new mountain house at the time of the fire?) As he scrambled to reconstruct his law notes, he also tried to remember as much as possible of his financial dealings and recast new ledgers. He seemed concerned only with replacing his books. None of his other belongings seemed to matter that much to him. He began to buy lawbooks at a rapid clip and he borrowed what he could not buy until replacements could come from England. To his surprise, he was able to win a few vital continuances from the General Court while he interviewed clients all over again and prepared to go reopen their cases. Before the Norfolk anti-inoculation case came to trial, Jefferson was able to reconstruct the lost brief for *Godwin v. Lunan,* one of the most important cases to be tried in Virginia before the Revolution. Second only to his authorship of the Declaration of Independence, Jefferson was proud of his fight for religious toleration in Virginia. From his research, his courtroom practice, and from five years as an Anglican vestryman in Fredericksburg, Jefferson was developing a lifelong opposition to the powerful connection between church and state. He once quipped that the law of New England was composed of the law of God—except where there were local statutes to the contrary. When he compiled his book of law reports, Jefferson added an appendix, "a disquisition of my own," on "whether Christianity is a part of the Common Law." In it, he dealt with an area of the law that he found deeply troubling:

> The most remarkable instance of judicial legislation that has ever occurred in English jurisprudence, or perhaps in any other . . . the adoption in mass of the whole code of another nation and its incorporation into the legitimate system by usurpation of the judges alone, without a particle of legislative will having ever been called on. . . .

In his book, Jefferson went on to excoriate the unholy alliance between church and state in England, which had made judges the accomplices of the clergy:

> And thus, they incorporated into the English code, laws made for the Jews alone, and the precepts of the [Christian] gospel . . . and then arm[ed] the whole with the coercions of municipal law.[30]

In the months before fire had destroyed his law library at Shadwell, Jefferson had been deeply engaged in studying the ecclesiastical laws of Virginia. He already considered the laws of Virginia and England separate, and Virginia's superior, and he began a lifelong campaign to separate church and state, first by attacking a series of ancient English judicial decisions that held that Christianity was part and parcel of English Common Law.

As with so many of his public crusades, Jefferson's deepening interest in the connection of church and state sprang from a combination of private experience and the serendipity of legal business coming his way. He had become interested in the method of ordaining and selecting parish clergy almost concurrently. Scottish-born James Ogilvie, a young friend of Jefferson's neighbor, John Walker, had come to Williamsburg from Aberdeenshire and wanted to be ordained an Anglican priest. He had routinely applied for a letter of recommendation from the Reverend James Horrocks, the ordinary in Virginia for the bishop of London, who had control over Church of England ordinations in the colonies. At first refused, he applied again; this time Horrocks gave him a letter to the bishop. At his own expense, he had sailed to England with General Court lawyer James Mercer, a partner with the Walkers and Peter Jefferson in the Loyal Land Company. To his horror, he found that the ordinary had written the bishop that the candidate "goes home in opposition" to the ordinary's wishes. "Nothing could have been more artfully couched to do me harm,"[31] Ogilvie wrote back to Jefferson on March 28, 1770. Although Jefferson scarcely knew Ogilvie, he knew Walker and shared the view of many Virginia gentry that the unpopular and ambitious Horrocks, who had been pushing to be made the first Anglican bishop in America, should be brought up short. Indeed, Horrocks may have sniffed a rival in the popular Ogilvie. Jefferson rolled up his sleeves to line up support for Ogilvie's ordination. He and his friend John Walker "took proper measures for prevailing on the commissary to withdraw his opposition." Jefferson wrote an affidavit—"a certificate of your conduct in life"—and presented it to Horrocks. He arranged for Ogilvie to draw whatever funds he needed from his London agent. And he found two vacant parishes, one in the gift of his old teacher, Reverend Maury. He even busied himself with lining up a bride for Ogilvie: Jefferson's February 20, 1771, letter to

Ogilvie was the first hint that "I, too, am in that way";[32] he had found the woman he wanted to marry, but he did not identify her.

In *Godwin v. Lunan,* which Jefferson argued before the General Court in Williamsburg in October 1771, he was retained with his old mentor, George Wythe, as counsel for Anglican vestrymen and churchwardens in Nansemond County who were seeking the removal of a curate, Patrick Lunan, "of evil fame and profligate manners."[33] The cleric stood accused of profane swearing, drunkenness, adultery, and exposing himself to his congregation. Worse still, he had neglected his parish duties and had disavowed his belief in Christianity. It was an important case on the road to revolution, one of the series of cases brought against Church of England clergymen that revealed the growing disenchantment for the privileged position given Anglican clergymen by Virginia law. In it, Jefferson was calling for a Virginia decision to take precedence over that of an English court. He spoke for a large number of Virginia gentry who had little but scorn for the Anglican clergy who were kept by their tax money. Jefferson contended in his brief for *Godwin v. Lunan* that English judges had usurped from the English people and their representatives in Parliament the establishment of religion by law, thereby denying religious freedom to all who dissented from the state Church of England.

In *Notes on Virginia,* Jefferson gave a brief sketch of Virginia's religious laws that he found shocking, the "many vicious points" of a favored position the Church of England had been granted in the colony and what the established clergy had done with its monopoly over religion. "The first settlers were emigrants from England, of the English Church, just at a point of time when it was flushed with complete victory over the religions of all other persuasions." Given complete power over making and executing church laws in Virginia, the Anglican clergy "showed equal intolerance in this country with their Presbyterian brethren, who had emigrated to the northern government [Massachusetts]." The "poor Quakers," as Jefferson described them, "flying from persecution in England" in the 1660s, "cast their eyes on these new countries as asylums of civil and religious freedom." But Virginia was free only for those in "the reigning set."[34] A series of Virginia statutes made it a crime for parents not to baptize their children, thus subjecting the unbaptized Quakers to arrest. The harsh anti-Quaker laws prohibited their meetings, made it a crime for any master of a vessel to bring a Quaker into Virginia, and had ordered the arrest of all Quakers until they swore oaths to the colony and Crown: oath-taking of any kind was against their firmest belief. If expelled Quakers had returned, they were jailed the first and second times but executed the third. No Quaker was allowed to have a meeting in his house or near it; no one could entertain even a single Quaker in his house or distribute Quaker books.

But not only the radical Quakers were persecuted. Anyone accused of an undefined heresy against the Church of England could be burned at the stake. For a young deist who had ceased taking Anglicanism seriously in college, Jefferson was jolted by the 1705 Act of the Houses of Burgesses, still on the books, that stipulated that

> if a person brought up in the Christian religion denies the being of a God, of the Trinity, or asserts there are more gods than one, or denies the Christian religion to be true, or the Scriptures to be of divine authority, he could be barred on the first offense from any office or employment ecclesiastical, civil or military; on the second by disability to sue, to take any gift or legacy, to be guardian, executor or administrator[35]

and be jailed for three years without bail. Fathers could lose the custody of their children under this provision, Jefferson found, and the children put "into more orthodox hands." Jefferson came away considering Virginia's laws of religious establishment "religious slavery under which a people have been willing to remain who have lavished their lives and their fortunes for the establishment of their civil freedom."[36]

Jefferson seized the *Lunan* case as an opportunity to do massive research into the legal history of the church-state linkage. In his provocative study, the increasingly iconoclastic young lawyer discovered that numerous English judges over the past three centuries had merely proclaimed that Christianity was part and parcel of English Common Law. As a law student, Jefferson had read Coke's proposition that "the common law was grounded on the law of God." But Jefferson did not delve at length into the question of church and state until a Virginia parish vestry tried to dismiss the allegedly corrupt Reverend Lunan.

Jefferson, himself an Anglican vestryman since his coming of age, had followed the clash between gentry and clergy in the Parson's Cause controversy. He knew he spoke for many Virginia lay leaders of the church in the officially Anglican colony who had little but scorn for pastors who cared more for tobacco crops than congregations. Nevertheless he was going before the wealthiest and most conservative planters, who had been appointed to the Governor's Council and thus made up the colony's supreme court. He boldly challenged a century and a half of the acquiescence of Virginia civil judges to the church law of England.

An eminent federal judge and legal scholar of the twentieth century, Edward Dumbauld, wrote in 1978 that Jefferson insisted on the important distinction between ecclesiastical law and common law, between church government and civil government, a distinction that was "very clear and important." Of "equal"[37] importance, Judge Dumbauld noted, was the distinction Jefferson made between church law and religious doctrine. Further, Jefferson, in his study of laws involving churches, had found that many church matters, such as ownership of property, were derived from sources other than the Bible, such as proclamations by the General Assemblies of Virginia. This youthful legal research by Jefferson and his labor in the General Court of Virginia were eventually to lead to his political doctrine of a "wall of separation,"[38] as he put it in an 1802 letter, which should exclude government from intruding into the realm of religion, and vice versa.

Lunan gave Jefferson the opportunity to research the rights held by lay patrons of Anglican churches, such as himself. Jefferson wrote the brief and turned it over to George Wythe to argue before the General Court. Jefferson agreed with Blackstone that there were three kinds of churches. In a donative church, the patron, either the Crown or a subject, who has endowed the church with money or property has the right to invest the cleric without involving the bishop. Where the Crown was patron, the king could entrust visitation, which included the power to discipline or dismiss clergy, to a commission of parish visitors. In the second category, presentative churches, the candidate must be presented to the bishop for his examination and his admission to the vacant pulpit or rejection. If the bishop rejected the clerical candidate, the patron of the parish could then sue the bishop in a civil court for impeding his right. In a third category, a collative church, the bishop himself was the patron. In Virginia, where the only legally recognized churches were Anglican, and where there was no resident bishop, all were donative, with the Crown as their patron, Jefferson argued. Legislation approved by the king had given parish vestries, made up of laymen such as Jefferson, the right to nominate their own clergy without consulting the faraway bishop of London, this nominal bishop over all Anglicans in America. Since neither the king nor the bishop of London visited in person parishes in the colonies and had not deputized a separate colonial commission of visitation, it was the General Court which was to exercise the king's powers as his chancellor. Jefferson's and Wythe's argument, if it had been upheld in England, would have constituted a milestone in separating the powers of church and state in Virginia and, by precedent, throughout the American colonies. The General Court decided the case in favor of Jefferson and Wythe and their clients, the parish vestrymen. They upheld the jurisdiction of Virginia courts over the church for the first time. But the royal attorney general of Virginia, John Randolph, who was Jefferson's uncle, contended that Virginia had no general ecclesiastical jurisdiction, and he appealed the case to London. Three years later, when the Revolution closed the Virginia courts, the case was still pending.

More than the blackened boards and fused silverware of Jefferson's family home lay in the charred foundation hole of Shadwell. Around it, at first with his sisters Jane and Martha, and then alone on visits home from Williamsburg, Jefferson for five years had been laying out half a dozen gardens of flower and vegetable beds. In that first full year of gardening in 1767, in addition to those first hyacinth and narcissus, and those practical plantings of peas, Jefferson had festooned the grounds of Shadwell with numbered beds of carnations, marigolds, amaranth, coxcomb, lilac, Spanish broom, and laurel. And he had set out interplantings of native plants and exotic imports—almonds and cayenne pepper mixed with "twelve cuttings of gooseberries," "suckers of roses," lilies and wild honeysuckles amid strawberry roots, celery, lettuce, radishes. As he sowed lettuce, radishes, broccoli, and cauliflower one day, he noted the flourishing "flower-de-luces just opening." He was surprised when strawberries bore fruit after only one year and he began nearly sixty years of weighing, measuring, counting, writing it all down:

> Strawberries come to table. Note this is the first year of their bearing, having been planted in the spring of 1766. . . . On an average the plants bear 20 strawberries each. 100 fill half a pint.[39]

He had been careful to bracket notations he had not personally observed. He had been away when the larkspur and the poppies, the pinks and hollyhocks and the carnations had bloomed that first summer, so he noted, "by information of Mrs. Carr,"[40] his sister.

August 1, 1767, was the first time in all his writings that Jefferson wrote the name of his little mountain, Monticello, the new home to which he now had to transplant himself. That year, too, he also began to keep a wine cellar, laying in his first cases of Madeira.

He had been long torn between the bustling cultural life of Williamsburg and his upland home country. Much of his pleasure reading that year after the fire was from the Roman poet, Horace, whose lines had consoled him when his father had died, whose philosophy, expressed in *Ars Poetica,* of "sweetness and utility" so satisfied him. With Dabney Carr, he had first climbed Monticello as a boy and sat under tall trees on their weekends away from Reverend Maury's school. The schoolboy had written in a round hand,

> O rural home: when shall I behold you! When shall I be able
> Now with books of the ancients, now with sleep and idle hours
> To quaff sweet forgetfulness of life's cares![41]

Now, the mature man, the last remnants left by his father in ruins, turned to the indestructible words of Horace, his lifelong favorite and, in his small brick house on Monticello, copied out his longest extract from Horace, a careful condensation of the Roman poet's second Epode. Gilbert Chinard, an earlier biographer, pointed out that Jefferson excised the purely Roman elements of the poem, including its irony, turning it into a hymn of glorification of the southern plantation he now chose as he turned away from all thought of city life for the beauty and solitude of his mountaintop.

> Happy the man who, far away from business cares, works his ancestral acres with his steers, from all money-lending free; he avoids the Forum and proud thresholds of more powerful citizens.

From Horace, he gleaned a romanticized farm life. Editing out more than half of the second Epode, Jefferson excluded the soldier, the sailor, the fisherman, the hunter riding after his pack of hounds. His Sabine farm was Spartan, not Epicurean. Like Cincinnatus, he tilled his own acres, with only a few trusted slaves to drive his tired oxen home at night. Jefferson was already disgusted by his brushes with the opulent plantation-owning Virginia grandees of the Governor's Council and the General Court. He found it more

> . . . pleasant now to lie beneath some ancient ilex-tree, now on the matted turf. . . . Meanwhile the rills glide between their high banks; birds warble in the woods; the fountains splash with their flowing waters, a sound to invite soft slumbers.

In his copyings from Horace, Jefferson sounded a new and yearning note from his solitude. He wanted a hardworking, dutiful farm wife:

> But if a modest wife shall do her part in tending home and children dear, piling high the sacred hearth with seasoned firewood against the coming of her weary husband . . .

She was to raise his children and pour his home-grown wine and fix him "unbought" meals.[42]

Jefferson was thinking, now that the flames had flushed his family from the settled slave routines of Shadwell, of a new way of life on his mountain. He seems to have quickly dismissed any idea of rebuilding Shadwell. His mother and brother and sisters went to live at the plantation of her brother, Sir John Randolph. Jefferson concentrated on building an environment of his own that he could control. In his solitude at Monticello another relationship, that between master and slave, which Jefferson had known all his young life, now troubled him. He had begun to question whether he needed slave labor on his mountain. As a lawyer, he was at his most prosperous, and he began, in his legal work, in his readings and writings, to question whether any man had the right to subsist on the perpetual bondage of another.

Jefferson was not the first to confront this question. As early as 1736, Virginia's largest landowner had appealed to Parliament "to put an end to this unchristian traffick of making merchandise of our fellow creatures."[43] As soon as Jefferson had been elected to the House of Burgesses at twenty-five, he later wrote in his autobiography, "I made one effort in that body for the permission of the emancipation of slaves, which was rejected."[44] As a freshman burgess, he had been too junior in the legislature to put forward the bill, his first legislative act, which would have made it possible to free a slave simply by registering the emancipation in the county court, as was allowed in Georgia and North Carolina.

At the time, Virginia law allowed emancipation only by proving "meritorious services" in court. In this bill, put forward by his respected kinsman, Richard Bland, Jefferson asked the right of emancipation by free choice of the slaveholder be granted all Virginians. When Bland introduced the bill, Jefferson seconded it. It was shouted down by the burgesses. Bland "was denounced as an enemy of his country," Jefferson later reported.[45] At the time, under royal government, "nothing liberal could expect success." Jefferson blamed the unwillingness of his relatives and neighbors to ban slavery as well as other forms of bigotry and intolerance on the British:

> Our minds were circumscribed within narrow limits by an habitual belief that it was our duty to be subordinate to the mother country in all matters of government, to direct all our labors in subservience to her interests.[46]

The laws of England allowed slavery and the slave trade, and its economy thrived on the products of slave labor, allowed since Charles II had chartered the Royal African Company in 1660, which had grown into the world's leading slave trader.

Later that first year in the Assembly, Jefferson once again had to wrestle with the question of slavery, this time in a less abstract form. A mulatto shoemaker and sometime carpenter named Sandy, who was thirty-five and had been left to Jefferson by his father, stole one of Jefferson's workhorses and rode away. That the man had committed the felony of stealing a horse while seeking his freedom tempered Jefferson's public idealism with pragmatic concerns. Jefferson wanted to be able to free a slave, but he wanted to control his slaves' fate, to decide who would be emancipated and when. Sandy was one of the slave artisans that Jefferson and his father had trained and because his skills made him more valuable, he had been given privileges that liberated him from the harshest field labor. Well trained, well housed, well fed, Sandy was expected to be grateful and loyal. By running away to seek his freedom, he was an ingrate, a bad example to other slaves, and twice a felon for stealing himself and his master's horse.

In the September 7, 1769, *Virginia Gazette*, Jefferson placed a long, detailed advertisement that made Sandy's flight from Virginia all but impossible. Undoubtedly written by Jefferson himself, it affords a rare snapshot of the ambivalent attitude of the young man who had just publicly proclaimed, for the first time, his opposition to slavery: in it, Jefferson offered a scale of rewards ranging to a whopping ten pounds ($1,000 today) if the man was returned. It was an offer certain to attract the notice of bounty slave hunters:

> Run away from the subscriber in Albemarle, a mulatto slave called Sandy, about 35 years of age, his stature is rather low, inclining to corpulence, and his complexion light; he is a shoemaker by trade, in which he uses his left hand principally; can do some coarse carpentry work and is something of a horse jockey; he is greatly addicted to drink, and when drunk is insolent and disorderly. In his conversation, he swears much, and his behavior is artful and knavish. He took with him a white horse, much scarred with traces, of which it is

> expected he will endeavor to dispose; he also carried his shoemaker's tools, and will probably endeavor to get employment that way . . .[47]

Evidently the ad was effective. Sandy does not seem to have gotten very far before, with or without the horse, he was hauled back to Monticello where he worked for three more years. Could he expect harsh treatment, any punishment other than a return to perpetual servitude? It is difficult to say, but one longtime Jefferson slave, Isaac, an artisan whom Jefferson taught to be a tinsmith, testified: "Old master very kind to servants."[48] Just so they did not try to leave his possession without his permission. Jefferson never came to terms with this unruly slave-laborer Sandy. On January 21, 1773, Jefferson wrote in his *Account Book* that he had sold Sandy to his cousin Colonel Charles Lewis for one hundred pounds.[49] He was only the second slave Jefferson sold. Little is known of this first slave Jefferson ever sold, but the year after Sandy ran away and was recaptured, Jefferson sold a woman known as Myrtella. The only note of the transaction appears at the top of the new *Account Book* he opened after Shadwell burned down. He sold her for sixty-seven pounds in February 1770, immediately after the fire. Just below her name is Sandy's, with his price. The transactions are listed opposite Jefferson's western land transactions under the heading "personal stock to lands and slaves sold."[50]

All his adult life, Thomas Jefferson seems to have tossed and turned in an agony of ambivalence over the dilemma of slavery and freedom. Repeatedly he sought to have public institutions relieve him of the burden of his conscience while he tried to avoid giving offense to his close-knit family and the slaveowning society of his beloved Virginia. He knew slavery was evil, he called it evil and spoke out against it in a series of public forums, but he would only push so far—and then he would fall back on a way of life utterly dependent on slave labor.

In the mid-eighteenth century, historian Gordon Wood estimates, "one half of colonial society was legally unfree." Half a million African-Americans had been reduced to lifetime hereditary servitude. Slaveowners such as Henry Laurens of South Carolina, who would lead the war of independence from what they perceived to be enslavement by the British, had several hundred slaves he saw as "poor creatures who look up to their master as their father, their guardian and protector, and to whom there is a reciprocal obligation upon the master."[51] Slavery was not restricted to the south. By 1750, one out of every five families in Boston owned slaves; black slaves constituted twelve percent of the people of Rhode Island. In 1746, more than one-quarter of New York City's workforce was black slaves, with half the households in the city holding at least one slave. In the fields of southern New Jersey, slaves labored to make Quakers rich, and abolitionist John Woolman was left to harangue an empty annual meeting in Philadelphia while his coreligionists went to lunch. Slavery was pervasive in the rural south, so much so that, as Jefferson pointed out in his *Notes,* children, black and white, slave and free, were "nursed, educated and daily exercised in . . . the most boisterous passions, the most unremitting despotism on the one part, and degrading submissions on the other."[52]

Not only black slaves lacked freedom: white servitude was widespread. In 1759, according to Benjamin Franklin, most of the labor in the middle colonies was carried out by indentured servants brought from Britain, Ireland, and Germany. Somewhere between half and two-thirds of all immigrants to colonial America had come as indentured servants who worked for five to seven years to pay off their ships' passage. Among them were, between 1718 and 1775, an estimated fifty thousand British and Irish convicts and vagrants shipped to America and, in effect, sentenced to seven to fourteen years of servitude, sometimes even to life. As the decade of the American Revolution began, half of America was not free. If Thomas Jefferson ever harbored any hope of being able to abandon the system of plantation agriculture that had come to depend on slaves, it was during his legal career. He seemed to settle for being a humane slaveowner who abhorred the more violent aspects of the practice. In 1770, for example, he refused to defend Isaac Bates, who had whipped a black woman to death.

His first eloquent, ingenious pleading that all men are created equal came in the case of *Howell v. Netherland,* which pitted him against George Wythe. According to Jefferson's casebook, on October 18, 1769, one month after his ill-fated attempt in the House of Burgesses to legalize emancipation in the Virginia county courts, Samuel Howell, "a pauper," as Jefferson described him, came to him and asked him to help win his freedom. Jefferson knew that, in a case involving paupers, especially a servant suing for freedom, a trial could be expedited: no money was to be made by long, drawn-out procedures. "The plaintiff's great-grandmother," Jefferson noted, "was a white woman and had a daughter by a negro man, whose grandson the plaintiff is." Howell "sues for freedom," Jefferson wrote. "Charge no fee."[53] On December 15, 1769, Jefferson took out a writ at his own expense to bring suit in the General Court. At the very next court term, April 1770, the case came to trial.

Under a Virginia law passed in 1705, "if any woman servant shall have a bastard child by a negro or mulatto, or if a free Christian white woman shall have such bastard child by a negro or mulatto, in both the said cases, the churchwardens shall bind the said child to be a servant [*sic*] until it shall be of thirty-one years of age."[54] Howell's grandmother, therefore, a mulatto begotten of a white woman by a black man, was bound by the 1705 law to serve until the age of thirty-one. During her servitude, she gave birth to Howell's mother and, during *her* servitude, she gave birth to Samuel Howell, who was suing for freedom from the master to whom he had been sold and who was claiming his services until he was thirty-one.

Jefferson was well aware that the judges of the General Court were not trained lawyers steeped in legal precedent, but that they knew the law of slavery, the Bible, and Anglican church beliefs. He knew their prejudices, too: they abhorred the idea of a white woman bearing a black man's child, and they believed this could only mean she had been raped. They also believed that, had she consented, all of her descendants would be black. Samuel Howell was fair skinned, and Jefferson played off this prejudice by arguing, in biblical terms, that the sins of the father should not be visited upon the children to the third generation. In his posthumously published book of law reports, Jefferson ranked *Howell v. Netherland* as one of the fourteen most important cases from the decade before the Revolution and one of only three of his own cases included in that number. He incorporated his own handwritten notes into the *Reports.*

Jefferson rose before the General Court in April 1770 and stunned the jurists with the vehemence of his peroration. He pointed out that the 1705 law had been revised by an act of 1723 that bound any child of a female mulatto to serve as long as its mother had to serve. Howell had been born in 1742, during his mother's servitude. The family that had owned his family since his grandmother's enslavement had sold him to Netherland, who claimed his services until he was thirty-one (Howell was now twenty-seven or twenty-eight). Jefferson cleverly contended that the Act of 1723 imposed servitude on Howell's mother but not upon him, since the 1723 statute said that the "child," but not the grandchild of a female mulatto slave, should be subjected to servitude. Adroitly Jefferson was sidestepping the fact that the law of 1705 had been subsumed by the law of 1723. As Judge Dumbauld has pointed out,

> . . . The statute operates with equal effect upon plaintiff's [Howell's] mother in the same manner as it did upon his grandmother. . . . The mother was also a "female mulatto"; she too was "by law obliged to serve" till the age of thirty-one; she too, "during the time of her servitude," had a "child born of her body" (to wit, the plaintiff). Therefore, by virtue of the Act of 1723, "such child"

> [the plaintiff Howell] must serve his mother's master or mistress until he attains the same age [thirty-one] to which "the mother of such child" [that is, Howell's mother] "was obliged by law, to serve . . ."[55]

Even if Jefferson was on shaky legal ground, he felt he had a strong moral and philosophical position, and he pressed it boldly. As he scanned the row of a dozen of colonial America's wealthiest slaveholders, he became almost brilliantly sarcastic at first, and then he gave the judges a Sunday sermon they could recognize:

> I suppose it will not be pretended that the mother being a servant, the child would be a servant also under the law of nature, without any particular provision of the act. . . . Under the law of nature, all men are born free. Everyone comes into the world with a right to his own person, which includes the liberty of moving and using it at his own will. This is what is called personal liberty, and [it] is given him by the Author of nature, because [it is] necessary for his own sustenance. The reducing of the mother to servitude was a violation of the law of nature. Surely then the same law cannot prescribe a continuance of the violation to her issue, and that too without end, for if it extends to any, it must be to every degree of descendants. . . .
>
> The act of 1705 makes servants of the first mulatto; that of 1723, extends it to her children, but that it remains for some future legislature, if any shall be found wicked enough, to extend it to the grandchildren and other issue more remote. . . .[56]

Jefferson got no further; pounding his gavel, Lord Botetourt cut him off. The young Jefferson's impassioned eloquence proved to be in vain. "Wythe, for the defendant, was about to answer," Jefferson recorded in his *Reports,* "but the Court interrupted him, and gave judgment in favor of his client."[57]

Jefferson's summary of the verdict was more typically understated than his argument had been. Infuriated, the General Court justices did not even wait for Wythe to say more on the subject; he, too, was well known to be a liberal on the subject of slavery. Amid the turmoil, the usually genial Lord Botetourt pounded his gavel and the judges clamored for a verdict in favor of the slaveowner. It was the first time Jefferson had spoken the words in public—"all men are born free"—six years before he wrote the Declaration of Independence. An absolute statement of the natural-rights philosophy he had already adapted wholeheartedly, his heretical words were far too radical for most of his Virginia countrymen. But he had said them, and he would record them in his law reports, and he had said them first in the legal defense of a black slave.

On May 21, 1770, the General Assembly reconvened in Williamsburg, immediately facing the latest challenge from London. In 1769, Parliament quietly allowed the Quartering Act to expire and repealed all of the Townshend duties except threepence in the pound sterling on tea. But to except tea was to leave the Declaratory Act—that Parliament was supreme "in all cases whatsoever"—in force as if the Townshend Acts had not been repealed at all. Instead of easing the colonial crises, Parliament had provoked a new round. Radical Virginians began to talk of further boycotting English goods by tightening the nonimportation agreement. In the spread-out plantation country of the south, the strategy on nonimportation had not worked so well as in the northern and Middle Colonies, where the population was concentrated in towns and committees of association could bring pressure to bear on the recalcitrant merchant-importer. In the northern colonies, imports from Great Britain in 1769 had been cut by two-thirds. In Virginia, imports were actually increasing, despite the fact that the majority of planters were opposing parliamentary duties, probably because merchants and planters had hoarded British goods in anticipation of the boycott.

In late June 1770, the burgesses, acting unofficially, invited prominent merchants, mainly from port towns and the Tidewater to join them in Williamsburg to draw up and sign new articles of association. On June 22, 1770, Peyton Randolph was the first to sign; merchant Andrew Sprowle, "Chairman of the Trade," the elected representative of more than one thousand merchants, was second. Col. Thomas Jefferson—he had been commissioned only two weeks earlier at age twenty-seven as the lieutenant of Albemarle County, its highest office—was tenth in the entire colony to sign the nonimportation agreement. The new association went further than the previous year's boycott, setting up county committees of enforcement, and was intended to remain in force until Parliament totally repealed its offensive internal duties or one hundred associates agreed otherwise.

7

"The Pursuit of Happiness"

No partnership can oblige continuance in contradiction to its end and design.

—THOMAS JEFFERSON'S "NOTE ON DIVORCE," 1773

IN THE YEAR or two after the fire on the mountain at Shadwell destroyed so many of Thomas Jefferson's links with his past, he thought much about bonds and bondage of various kinds. He all but cut his remaining ties with his mother, leaving little more than the auditing of her financial affairs, remaining close only to his sister Martha, through his friend and brother-in-law, Dabney Carr. The flames both drove and liberated him belatedly, at age twenty-seven, from living under his mother's roof, making inevitable and more urgent the new construction atop Monticello. More than ever, his closest ties were of another kind, the connections between the life of his mind—his pleasure readings, his legal scholarship—and his resultant personal and political actions. As Jefferson the lawyer pondered the perpetual links of bondage between slave and master, he thought, wrote, and spoke of marriage and divorce, of union and separation, of personal liberties and what he called the "servility" of colonial Americans to the mother country, England. Even as he uprooted himself from home and family at Shadwell, he planned and planted copiously with all the paradoxical faith of the Age of Reason for his new and well-ordered universe at Monticello. As he thought about unions that should be made or dissolved, he began to hint, in his letters, that he was thinking of eventually making a family of his own. Yet, at the same time that he was considering the subject of a wife, he took on the role of counsel in a complicated courtroom drama resulting from a notorious divorce case and he prepared to introduce in the Assembly a bill allowing divorce for the first time in Virginia. By late 1771, he came to conclude that "no partnership can oblige continuance in contradiction to its end and design."[1]

In part at least from his readings of Horace, Jefferson had imbibed not only a strong sense of the spirit of a place and the urge to build Monticello but a belief that, if he were ever to amount to anything, it was because he had had a good father. He seems to have set himself twin goals after Shadwell burned: to build a house that reflected his classical tastes and to find himself a suitable wife to manage it and perpetuate the Jefferson line. Disdaining more and more the typical rambling Tidewater English-derived rural architecture around him, he studied the classical revival of Roman architecture that had taken place in Italy in its sixteenth-century Renaissance and he fell in love with the style and design of the Roman villa.

The main house at Shadwell destroyed, his mother, sisters and his brother at first crowded into an overseer's farmhouse. Jefferson chose to leave them behind and to rent rooms in Charlottesville until construction could be finished at Monticello by the autumn of 1770. He used the cold, wet months to plan for warm weather with all the thoroughness of a military campaign.

"This winter is employed in getting [lumber for] framing, limestone and [in] bringing up stone for the foundation," he wrote. It was a major logistical and financial operation and, according to Jefferson's notebooks, among other things it involved finding enough water (Jefferson figured that "a bed of mortar which makes 2,000 bricks take 6 hogsheads of water")[2] and limestone to make the mortar and enough timber to make the forms to pour some 310,000 bricks that would be needed for his house and its dependencies, including the one-room South Pavilion, where he was to live while the main house was built.

As strapped for cash as any American colonist, Jefferson bartered. He had already tried the technique: in 1769, while he had been away at Williamsburg representing his district, Jefferson had paid John Moore and his crew in wheat and corn to level the northeast end of his mountaintop. In his *Account Book,* he calculated how much it would cost. "I think a middling hand in twelve hours (including breakfast) could dig and haul away the earth of four cubical yards."[3] In July 1769, the construction work had begun in earnest as George Dudley arrived with his gang of black bricklayers on the mountain to make the first forty-five thousand bricks. Jefferson paid Dudley twelve pounds, in cash, cloth, molasses, and salt to feed and clothe his slave crew. By September 1769, as he was elected to his second legislature, Jefferson's hired crews had cleared a park a third of a mile in circumference in front of the homesite on the north side of the mountain and ordered eight thousand chestnut fence rails to be hauled. As early as the spring of 1769, a year before the fire, as he had waited for weather warm and dry enough to dig excavations, Jefferson had begun planting trees on the southeast slopes. The farmer in him knew there were tasks for every season. In his *Garden Book,* he recorded that his plantings followed a ridge. Beginning at its bottom, he set out rows of pear trees, cherries, New York apples, peaches, apricots, nectarines. In the hollow between ridge and hill, he planted pomegranate trees, figs, peaches "for inoculating apricots,"[4] and walnut trees.

After Shadwell burned, Jefferson pushed his foremen. He hired Will Beck and another crew to dig the cellar for the South Pavilion despite the cold, wet mountain weather. A skilled housewright and bricklayer, Stephen Willis, arrived on the mountain in March, the month after the fire, with his squad of skilled slave artisans to speed the work of erecting the small South Pavilion. Willis's team framed and built the elegant brick cottage in a single summer. Beck's crews finished digging the well in twenty-three days, working nonstop for thirty two days and nights to keep up with Willis. Then Beck's crew turned to digging the huge cellar hole for the main house. Willis found enough time that summer of 1770 to build the walls of a seventeen-by-thirty-four-foot stone cottage that eventually housed the overseer, the house servants, and a weaver's shop. After Jefferson went off to Williamsburg for the spring court session in April 1770, work lagged in his absence. A relative, Mary Walker Lewis, wrote, "Your Negroes have been sickly since you left us."[5] When he came back in May, the work was going well. He hired more workers, relieving some of the burden on his own slaves. He hired a slave, Phil, from Will Beck for ten pounds a year to drive a team and wagon hauling rock and lumber up the mountain. Beck's crews dug ten-foot-deep foundation holes and laid courses of stone to support the building's walls as another crew worked day and night burning lime to make the mortar.

The workmen followed plans produced by Jefferson himself. For the design of his mountaintop villa, Jefferson, a poor freehand draftsman himself, turned to pattern books of Palladian-style villas. "Palladio is the Bible," he told a friend, urging him to get Andrea Palladio's four-volume 1570 treatise on classical Roman architecture "and stick close to it." Poring over Palladio's engravings as well as a copy of Robert Morris's *Select*

Architecture, Jefferson found in Morris the exact prototype of the stylish country house "proposed for a single gentleman": a two-story center block with balanced one-story wings and flanking courtyards. Jefferson traced Plate Number 3 of Morris and scribbled in the dimensions. Center block, twenty-four by twenty-eight feet; wings, seventeen by seventeen; in all, twenty-five hundred square feet. A master bedroom would take up one wing, the dining room, the other: a generous entrance hall and parlor filled the center downstairs, and there would be a large library and two more bedrooms upstairs. For the obligatory classical portico, Jefferson sketched in four freestanding columns. While Jefferson experimented with other designs, he settled on Morris and turned over the drawings to his housewright while he finished designing the outbuildings—the kitchen, privy, laundry, storehouse, and workshops. He copied out, working with rule and compass in precise scale, a drawing or set of drawings for every room, frontal elevation, and fireplace, carrying out his measurements to four and five decimal places. As architectural historian Jack McLaughlin puts it, "His architectural drawings are symbolic of his kind of compulsive personality." Not only did he feel the "need to save and account for everything—money, books, records, facts," but his careful, slavishly perfect drawings were "an absurdity in the building trades where carpenters and bricklayers are often lucky if they can keep to the inch rather than the ten-thousandth of an inch." Jefferson's drawings, concludes McLaughlin, were more than instructions to artisans, however, "they were mathematical absolutes that recorded universal verities."[6] Attacking architecture as he did the law, Jefferson researched his house plans systematically and thoroughly. His search had led him to Palladio, and his first house was completely derivative. But he scrapped Palladio when it came to the dependent outbuildings: these he decided to place out of the view of the approaching traveler, below grade, set into his mountain in a pair of submerged L-shaped wings. The basement of the house was to be level with the outbuildings on the downward slope; a roofed corridor connected the dependencies to the main house. The outbuildings had aboveground doors and windows for light and air. It was an ingenious, elegant plan that Jefferson drew up at age twenty-six and, despite all the other changes at Monticello, over the years, it still survives. One thing further expressed his emerging philosophy. Monticello, like so much about Jefferson, faced west to the mountains and the future that drew him. Its back was turned upon Europe and the past.

By September 1770, as work went on at Monticello, Jefferson called off Will Beck and paid him five shillings to carry a wagonload of his lawbooks to Williamsburg, where a new session of the Assembly and the General Court had been convened by Lord Botetourt. Jefferson was rapidly replenishing his library, and his purchases reflected a change in his literary tastes. His storehouse of legal knowledge and classical lore he now considered amply stocked: he turned now to more modern works of philosophy, politics, poetry, and fiction.

There was a brief interlude in the colonial crisis in the fall of 1770 and on into early 1771 as the Association's boycott of English imports fizzled. Although the Declaratory Act and the last of the Townshend duties, the tea tax, remained on the books, Americans smuggled their tea from the Dutch West Indies. No new measures came from London. On the contrary, when Lord Botetourt reconvened the Assembly in the fall, his tone was most conciliatory: there was no ministerial design to ask Parliament to impose any further taxes on America. In the next session, indeed, the Ministry would ask Parliament to remove the unpopular taxes on glass, paper, and dyes because these duties were contrary to the principles of commerce. Further undercutting the protests of radical Virginians such as Jefferson and Washington was Lord Botetourt's announcement that the king had approved the westward extension of the boundary line between land-hungry Virginia and the homeland of the Cherokee Indians.

In the lull from politics, Jefferson was able to move into his handsome little brick cottage on the south portico at Monticello in October 1770. As soon as the court session ended that month, he had Jupiter and Will Beck load up the wagon with his trunks of books and clothes for the journey west from Williamsburg and up his mountain. Within a year of the fire, Jefferson was writing that his one-room house, "like the cobbler's, serves me for parlor, for kitchen and hall, I may add for bed chamber and study, too. . . . I have hope, however, of getting more elbow room this summer."[7]

The death of Lord Botetourt in October 1770 made any fresh confrontation between Virginians and the Crown impossible until a new governor was sent out. The General Court session had already begun and Jefferson was undoubtedly in a pew at Bruton Parish Church for His Lordship's funeral and followed the coffin to the college chapel, where it was interred beneath the floor. No doubt Jefferson voted aye to the handsome monument to Botetourt that graced the last peacetime royal governor's memory. As his part in the elaborate funeral ceremonies, Jefferson contributed two whole loins of venison, which he paid Will Beck to fetch in his wagon.

As Jefferson prepared to move in, he unpacked crates of books that he had bought in Williamsburg. He was replacing his lost library rapidly. As he turned them in his hands and shelved them, he turned them over in his mind. In a letter to his friend Robert Skipwith, he listed the books he recommended for starting a library that would be "improving as well as amusing." The list and his explanation of his choices reflected his widening interests in reading. It contained, unsurprisingly, histories of Virginia, England, Scotland, France, the Roman Empire, the Bible; there were the predictable books on law and philosophy, and as might be expected, Jethro Tull's *Horse-hoeing Husbandry,* but there was also an imposingly long list of books of drama, poetry, and fiction. His tastes were becoming more modern, more political—Dryden and Pope, Molière and Rousseau, Smollett, Richardson, Fielding, Goldsmith, Swift. He urged his friend Skipwith in a letter that was an essay on the virtues of reading to stock his library and his mind with imaginative writings, fiction and poetry, a scandalously radical idea in the eyes of many in eighteenth-century Virginia. He urged Skipwith to pay

> A little attention . . . to the nature of the human mind evinces that the entertainments of fiction are useful as well as pleasant. That they are pleasant when well written, every person feels who reads. But wherein is its utility, asks the reverend sage, big with the notion that nothing can be useful but the learned number of Greek and Roman reading with which his head is stored? I answer, everything is useful which contributes to fix us in the principles and practice of virtue. When any signal act of charity or of gratitude, for instance, is presented either to our sight or imagination, we are deeply impressed with its beauty and feel a strong desire in ourselves of doing charitable and grateful acts also. On the contrary, when we see or read of any atrocious deed, we are disgusted with its deformity and conceive an abhorrence of vice. Now, every emotion of this kind is an exercise of our virtuous dispositions. . . . Dispositions of the mind, like limbs of the body, acquire strength by exercise. But exercise produces habit . . . the exercise being of the moral feelings, produces a habit of thinking and acting virtuously. We never reflect whether the story we read be truth or fiction. If the painting be lively, and a tolerable picture of nature, we are thrown into a reverie. If we awaken, it is the fault of the writer.
>
> I appeal to every reader of feeling and sentiment whether the fictitious murder of Duncan by Macbeth in Shakespeare does not excite in him as great horror of villainy as the real one of Henry IV by Ravaillac as related by Davila? And whether the fidelity of Nelson

> and generosity of Blandford in Marmontel do not dilate his breast and elevate his sentiments as much as any similar incident which real history can furnish? Does he not in fact feel himself a better man while reading them, and privately covenant to copy the fair example? We neither know nor care whether Laurence Sterne really went to France, whether he was there accosted by the poor Franciscan, at first rebuked him unkindly, and then gave him a peace offering; or whether the whole be not a fiction. In either case, we are equally sorrowful at the rebuke, and secretly resolve we will never do so. We are pleased with the subsequent atonement, and view with emulation a soul candidly acknowledging its fault, and making a just reparation.
>
> Considering history as a moral exercise, her lessons would be too infrequent if confined to real life. Of those recorded by historians, few incidents have been attended with such circumstances as to excite in any high degree this sympathetic emotion of virtue. We are therefore wisely framed to be as warmly interested for a fictitious as for a real personage. The spacious field of imagination is thus laid open to our use, and lessons may be formed to illustrate and carry home to the mind every moral rule of life. . . . A lively and lasting sense of filial duty is more effectually impressed on the mind of a son or daughter by reading King Lear than by all the dry volumes of ethics and divinity that ever were written. This is my idea of well-written romance, of tragedy, comedy, and epic poetry.[8]

In the autumn of 1770, soon after Jefferson settled into his little house, when he wrote to send off the list of books to Skipwith, he invited him to bring his wife, Tibby, and spend some time at Monticello. Jefferson seems to have been lonelier than he expected after leaving Shadwell. In February, when he had written to John Page about Shadwell's burning, Jefferson had felt cut off from his old friends: "You may be all dead for anything we can tell here." "Currus," as the old school chums called Dabney Carr, "speaks, thinks and dreams of nothing but his young son."[9]

As he rearranged his affairs after the fire, Jefferson began to step out more than he had when he was in Williamsburg, and not alone. Once he had inveighed to this same John Page (who might now have remembered and chuckled) against dapper young friends: "I am sure that the man who powders most, perfumes most, embroiders most and talks most nonsense, is most admired."[10] They were "monkey-like animals," half made by God, half by tailors and barbers. Jefferson had succumbed to the bagwig only on court days until now, but now, in the spring of 1770, Jefferson began sending Jupiter around Williamsburg to buy hair powder and buckles and theater tickets. That spring, he began to pay court to a handsome young widow, Martha Wayles Skelton.

Jefferson had known Bathhurst Skelton at William and Mary, and no doubt he was aware that Skelton had married Martha Wayles at the time most of Jefferson's college friends were pairing off. According to Williamsburg tradition, Martha Wayles was beautiful—"a little above medium height, slightly but exquisitely formed."[11] Biographer Harry Randall, who interviewed Jefferson's friends and family, said "her complexion was brilliant—her large expressive eyes of the richest shade of hazel—her luxuriant hair of the finest tinge of auburn."[12] She was also a graceful rider and dancer and enjoyed long walks; she was well read, preferring fiction. As if these were not enough attributes of consequence to Jefferson, she loved talking and laughter and music. Playing harpsichord and spinet exceptionally well, she sang as sweetly as his sister Jane, and may have reminded him of this first woman he had loved. The boy who would not have cut in when she danced with Bathhurst Skelton at the Apollo Room now did what he had learned as a lawyer: he went after her.

A beautiful young widow whose father was one of Virginia's wealthier planters had many suitors. But Jefferson once again mounted a campaign in which he showed himself to advantage. His invitation to Skipwith to come to Monticello, with its gardens and deer park, its breathless prospect of the Blue Ridge, its fine house under construction, was part of his attempt to pry Martha away from her father's house, where she lived with her three-year-old son, and the daily teatime assault of rivals for her affection. Jefferson was capable of a stratagem or two. Martha Skelton was Tibby Skipwith's sister, and he invited her to accompany the Skipwiths to Monticello. Her sister would be her chaperone. He knew that Martha loved books, and he must have held their conversations on a literary plane, high ground that he could defend admirably against her other suitors, so much so that Skipwith had asked Jefferson to select a library for him. Did Jefferson aim to impress? Would not Skipwith undoubtedly put Jefferson's list, ostensibly intended for him, and his essay on the moral utility of imaginative readings before Martha to read? As he finished his long list, Jefferson had playfully alluded to Mount Olympus and its Arcadian pleasures: "Come to the new [Olympus], from which you may reach your hand to a library formed on a more extensive plan." As if the hundreds of volumes he had listed were not dazzling enough for a woman who loved to read! He would put his guests up in the stone farmhouse and they could feast upon literature out of the hot afternoon sun. Separated from each other by but a few paces, the possessions of each would be open to each other. With the Skipwiths decorously chaperoning, Jefferson could meet Martha at "a spring, centrally situated; it would be the scene of every evening's joy."

> There, we should talk over the lessons of the day or lose them in music, chess or the merriments of our family companions. . . . The heart thus lightened, our pillows would be soft . . .
>
> Come, then, and bring our dear Tibby with you, the first in your affections, and second in mine. Offer prayers for me, too, at that shrine to which, tho' absent, I pay continual devotion. In every scheme of [my] happiness, she is placed in the foreground of the picture, as the principal figure. Take that away, and it is no picture for me.[13]

For nearly two years, Jefferson's horse wore grooves in the road, spring and fall, from Williamsburg to John Wayles's estate at The Forest, with servant Jupiter along to bring the violin and books he brought as presents. To win the long siege of courtship, he faced Martha's father, standing astride the path of any applicant for his daughter's hand—and her major inherited share in his fortune. Wayles, born in Wales, had made a fortune in law and speculated in land and slaves. Who was this upstart Jefferson? As his heart raced at each visit, he raced the construction work atop Monticello. In a letter to James Ogilvie, he revealed as early as February 1771 that he feared the obstruction of "the unfeeling temper of a parent who delays, perhaps refuses to approve her daughter's choice." There was no question of a "want of feeling in the fair one,"[14] but the loss of Shadwell made him appear a poor man in the eyes of her father. John Wayles wanted a merger with a man who could manage more than twice Jefferson's acres, and the slaves to cultivate them. Jefferson had come to admire his fellow Welshman, who had resisted many of the usual temptations of the great planters, who lavishly indulged their fantasies as transplanted English aristocrats by recklessly pursuing idleness and luxurious gambling. Wayles could remember how Spartan existence in Virginia had been when he arrived in 1740, and how it had changed. "In 1740," he wrote, "I don't remember to have seen such a thing as a Turkey carpet in the country except a small thing in a bedchamber. Now nothing are so common as Turkey or Wilton carpets, the whole furniture, rooms, elegant and every appearance of opulence."[15] Exasperated at times, Jefferson wrote to his agent to procure him a coat of arms, to "search the Herald's office for the arms of my family."[16]

Enchanted by what Jefferson later described as Martha's "sweetness of temper,"[17] he brushed past Wayles's objections, visiting The Forest at least three times in 1770. She was all "spriteliness and sensibility."[18] He was transformed to find someone with his tastes who appreciated him. What joy he must have felt to discover their shared interest in the novels of Laurence Sterne, author of scandalously savage black humor that was all the rage in English drawing rooms. Had Jefferson begun to question his own heavy dose of book-learning from the pages of *Tristram Shandy,* with its lampooning of academics, doctors, and soldiers, before he wrote Robert Skipwith's somewhat-lighter list for a library? No doubt.

Whatever combination of literary allusions, shared interests, or pragmatic considerations were also involved, Jefferson and Martha Wayles Skelton fell in love, and soon Jefferson was singing both minuets as he rode to The Forest and Martha's praises in the drawing rooms of Williamsburg society. By the spring of 1771, a mutual friend, a Mrs. Drummond, was writing Jefferson,

> Let me recollect your description, which bars all the romantical poetical ones I ever read. . . . Thou wonderful young man, indeed I shall think spirits of a higher order inhabit your airy mountains—or rather mountain, which I may contemplate but can never aspire to.
>
> Persevere, thou good young man, persevere—she has good sense, and good nature, and I hope will not refuse . . . of your hand, if her heart's not engaged already.[19]

If this note caused Jefferson any more anxiety, even if his seductive invitation to Monticello was ignored, still, Jefferson rode ever more frequently to The Forest as 1771 progressed.

As early as February 20, 1771, Jefferson, usually frugal, was sending off to London for an extravagant wedding present for his beloved Patty, as her family called her in private. He ordered her a clavichord, to be custom made and shipped that summer. By September 1771, Robert Skipwith had entered into the lists in Jefferson's behalf, and he told Jefferson so. He wanted Jefferson as his brother-in-law and he wanted Jefferson to succeed with Martha so that he and his "dearest Tibby" could

> be neighbors to a couple so well calculated and disposed to communicate knowledge and pleasure. . . . My sister Skelton, Jefferson I

wish it were, has all the qualities which promise to assure you the greatest happiness mortals are capable of enjoying.[20]

According to family lore, one day in the fall of 1771, two other suitors for Martha's hand arrived at the same time at The Forest and, as they climbed to the veranda and were shown into the waiting room, they could make out the strains of music coming from the parlor: the notes of harpsichord and violin were perfectly meshed. They had no question who was playing the harpsichord and they quickly guessed who was playing the violin. And when they heard Jefferson and his Martha singing together, stanza after stanza, they saw no point in staying.

By November 11, 1771, if Jefferson's pocket account book is any indication, he had won over even Patty's father. As he rode home to Monticello to prepare for the wedding six weeks later, Jefferson tipped Wayles's servants more heavily than usual. There was much to be done, and Jefferson must have ridden at an even more breakneck pace than usual. He sent off Will Beck to Williamsburg with the wagon to fetch a long list of things, including draperies. He fired off a letter to his agent in London, cursing the trade boycott he had helped to enact and enforce. He had, he wrote, "seen a forte-piano and am charmed with it. . . . Send me this instrument." It would be his wedding present to Martha. "Add also 1/2 doz. pair India cotton stockings for myself" and "1/2 doz. best white silk and a large umbrella with brass ribs covered with green silk." Jefferson the bridegroom had become a man of fashion. And to finish his house for Martha, he added, in a note of desperation, "procure me an architect . . . as soon as you can."[21]

On December 23, Thomas Jefferson, now twenty-eight years old, returned to The Forest from Monticello to sign the wedding bond and marry twenty-three-year-old Martha Wayles Skelton; his cosigner was Martha's brother-in-law, Francis Eppes:

> Know all men by these presents that we, Thomas Jefferson and Francis Eppes, are held and firmly bound to our Sovereign Lord the King, his heirs and successors, in the sum of £50 current money of Virginia. . . . If there be no lawful cause to obstruct a marriage intended to be had and solemnized between the above bound Thomas Jefferson and Martha Skelton of the County of Charles City,

widow, for which a license is desired, then this obligation is to be null and void; otherwise, to remain in full force.[22]

Jefferson paid the forty shillings to the royal governor for the marriage license and waited the next eight days: no "lawful cause" came forward to bar his marriage. In the meantime, his mother, brother, four sisters, and scores of relatives and guests rolled up the driveway to The Forest after long journeys from all over the colony. On New Year's Day, 1772, two Anglican priests presided, each receiving five pounds (about $500 in today's money). There were fiddlers to stir the country dances as the guests devoured an enormous black wedding cake crammed with pounds of fruit, wine, brandy, and dozens of eggs.

For days the celebration went on: it was January 18 before the newlyweds drove off toward Monticello. They stopped over at Tuckahoe, the Randolph estate, where Jefferson had lived for seven years as a boy and where he now visited the man, "Tuckahoe Tom," who had been his childhood friend. Snow was falling heavily as they pushed on toward Monticello. The light phaeton was making little headway, the weary horses fighting their way through chest-high drifts by the time Jefferson turned into the lane at Blenheim, one of the Carter plantations, eight miles short of Monticello. The next day, ignoring the worst blizzard in Virginia in decades, they left their phaeton and plowed on toward Monticello on horseback, finding the barely visible road up Monticello's slope at midnight. There was no sound from the mountain wind howling over three feet of snow. The slaves and white workmen had long since gone to bed. Jefferson led Martha through the pitch-black gloom to the one-room cottage. There were no further notations in Jefferson's pocket account book for nineteen days.

On September 27, 1772, a month early, their first child, a small and sickly girl, was born. Jefferson named her after her mother. The infant, sick for six months, was finally nursed by a slave, Ursula, part of a family of slaves who had come to Monticello in payment of a debt owed Patty's first husband. The baby survived and thrived. Jefferson's first child was to become his robust, cheerful, beloved Patsy, his comfort through more than a half century of tumult. But her mother was not well. Fragile, weak, she had had a difficult pregnancy. Her frail condition justifiably worried her father. On October 20, 1772, nine months after she had left The Forest, her father wrote, "I have heard nothing about dear Patty since you left this place."[23] Jefferson hovered near her, skipping the spring 1772 term of the General Court, and he did not attend the House of Burgesses session. He was losing interest in all but a few law cases. He read much of the fall of

1772, superintending the construction at Monticello and supervising the harvest of wheat, corn, and tobacco. The basic outline of the house was now apparent. There were new problems here, too; a winter of unusually heavy snows in the mountains was followed by the worst floods of the eighteenth century. Jefferson's mill, a major source of income vital to his farms, was swept away when the Rivanna flooded. At a time when he was having increasing difficulty collecting his legal fees, he now had to pay to transport and grind his crops at a neighbor's mill. More and more, he thought about the cash shortage that exacerbated so many other colonial problems, the refusal of the British to allow Americans to manage their own affairs. He read deeply about other forms of government—republics, confederations. He took on a major case in the autumn of 1772 that required him to think and read for a full month about the most intimate of affairs and how Americans were utterly dependent upon England even on questions of the bonds of matrimony and divorce. To obtain a divorce in colonial Virginia was impossible. Virginia was governed by the law of England. Under English law, it literally took an Act of Parliament to obtain a limited divorce "from bed and board," what is today called a legal separation. Divorce was as impossible in the 1770s as it had been for Henry VIII. A marriage could be terminated only by annulment. Thomas Jefferson attempted to bring about reform by introducing Virginia's own divorce law in the House of Burgesses after he acted as counsel in a scandalous Williamsburg affair.

The case, which Jefferson later rated among his three most important, came his way on November 25, 1772, according to his *Account Book,* but he must have known of it previously, not only because it involved a close friend and a client but because it had long been the topic of winks and whispers in Williamsburg. On May 21, 1771, Dr. James Blair of Williamsburg, son of the longtime president of the Governor's Council and twice acting governor, married Kitty Eustace. The bridegroom's brother, John, was Jefferson's friend, a colleague at the bar of the General Court who, when he became clerk of the Council, had recently turned over his cases to Jefferson. The bride had arrived with her mother in Norfolk in 1769, after the death of her father, a physician. In August 1770, the wife of Norfolk merchant James Parker wrote back to a mutual friend in London that "Mrs. Eustace and her pretty daughter" were "really two very agreeable ladies."[24] Parker himself was less flattering:

> Kitty Eustace, the chip which you forwarded to this country, is to be married immediately after the General Court to Dr. Blair. I think the doctor will marry more than one. The mother and her must go

> together. She's a clever, managing sort of a lady and has played her cards exceeding well.[25]

The marriage immediately broke up. Mrs. Blair moved down the street with her mother "to another house not 100 paces off"[26] and sued her husband for separate maintenance and specific performance of a premarital agreement. Arbitration failed. The court denied Mrs. Blair's first claim and ordered reconciliation. Gossipy James Parker reported that Blair, his wife, his mother-in-law, and his sister were among the guests at a dinner party at the Governor's Palace in May 1772:

> Nothing was talked of but separation. Matters were painted blacker than they really were, and she is acquitted of everything but not allowing him to have a fair chance ever since they were married. By the prudent counsel of Dr. Campbell they are seemingly reconciled and she was to let him have a push the night we left the Court. Common report says she is not capacious enough.[27]

What Williamsburg gossips may not have known was that Dr. Blair had been quite ill in London a year before his marriage. He had been on his way home from medical school in Edinburgh when his classmate, Dr. Arthur Lee, wrote to Blair's father that the young doctor had been "seized" with "a violent nervous disorder that effects his brain and every part of his body."[28] His ailment may have made him impotent.

Blair's marriage to Kitty was further complicated by the arrival in Williamsburg of a new royal governor, John Murray, Earl of Dunmore. Now the scandal intensified. St. George Tucker, a William and Mary student who boarded at Mrs. Eustace's house, somehow obtained a letter from Kitty to Dr. Blair's sister in which she admitted she had not married the doctor, ten years her senior, for love. The unsigned letter to Kitty's mother, said, "experience will conform it, that Dr. Blair is incompetent [*sic*]." Mrs. Eustace showed the letter to her young lodger Tucker, who urged her to ignore it. Aware that public opinion condemned her for leaving her husband, Kitty decided to resume her marriage. But Dr. Blair became ill again. When he recovered, he heard that Mrs. Eustace had shown the anonymous letter around.

He also heard another rumor, that Kitty had had a secret liaison with Governor Dunmore, and wrote an accusatory letter to His Lordship. Governor Dunmore called in Dr. Blair's brother, John, and told him that, unless Dr. Blair retracted the charge, brother John could resign as clerk of the Governor's Council. Dr. Blair retracted. "And so the matter stands,"[29] wrote merchant Parker to their London friend. "The old lady persists her daughter is still a maid and that the Dr. never had and indeed cannot do as a man should do." In November 1772, Jefferson noted in his *Account Book* that he had been employed by Dr. Blair, defendant in a chancery suit "for a specific performance of the condition of a bond which was to give her half the estate of defendant (Dr. Blair)."[30] The Blair family had asked Jefferson to look into the possibility of obtaining a divorce.

The newlywed Jefferson did not attend the only meeting of the Virginia House of Burgesses between May 1771 and March 1773. He was evidently planning to propose a bill to allow Dr. Blair's divorce when the Assembly reconvened in the spring of 1773. At home at Monticello, he plunged into research on divorce laws from his reconstituted law library, working intensively from November 25 to December 26. His research went beyond divorce: it entailed the very relationship of an empire and its colonies and whose laws should govern. The Blair case was a test of his emerging theory of self-rule for Virginia.

In attempting to win a legislative divorce for Dr. Blair, he faced, as legal historian Frank Dewey puts it, "a formidable problem, given the severity of the rules governing divorce that prevailed throughout most of the English-speaking world two hundred years ago."[31] Virginia had no law of its own on divorce; its law had previously been held to be the same as the law of England, and Virginia, in 155 years of settlement, had never granted a divorce, even for adultery. As Jefferson reported, Parliament had only granted an absolute divorce by a special act for the first time in 1669. But Parliamentary divorces were rarely granted, only one to three times a year, and then only after the petitioner obtained a legal separation from the ecclesiastical courts and a verdict at law for "criminal conversation," in other words, adultery. Virginia (like all the other southern colonies) had adopted the ecclesiastical marriage and divorce laws of England. Jefferson had come to believe that English laws passed after America was colonized could not be applied to the colonies and that 1607, when the English had landed at Jamestown, was the cutoff date. His notes outline English divorce laws before that date. This made the parliamentary divorces after 1669 inapplicable in Virginia. Jefferson was not entirely alone in advancing this legal logic. In his papers in the Library of Congress are unpublished notes he took of another case, *King v. Dugard,* argued before the General Court in April 1773. Dugard had been indicted for bigamy under an English statute enacted in 1603. Attorney Thomson Mason argued that the law did not apply to the colonies because Parliament had no power to legislate for the colonies after they were "seated"; he placed that date in Elizabeth I's reign, *before* James I. In the *Dugard* case, Attorney General

John Randolph conceded for the Crown that the year 1607 "had ever been considered as the era at which the right of legislating within ourselves commenced and the obligation of English statutes ceased."[32]

In his comprehensive study of divorce, Jefferson was attempting to build a case that would persuade the General Assembly to follow Parliament's example in granting a divorce by special act, but to grant it on the grounds that Parliament's authority in the matter had never been recognized in Virginia. If his notes are any indication of the order of arguments he was preparing for the Assembly, he had laid the groundwork for the dissolution of the union between colonies and mother country which took place three and a half years later, basing his argument primarily on natural law. In his new library, Jefferson could consult Baron Samuel Pufendorf's treatise, *The Law of Nature and Nations,* written in Latin and first published in England in 1672 and including dissertations on marriage and divorce. Jefferson's 1815 inventory of his library included the French version of Pufendorf as well as the 1749 English edition. In *Howell v. Netherland,* Jefferson, in seeking freedom for a third-generation mulatto, had cited Pufendorf's treatise for his own proposition that "under the law of nature, all men are born free."[33] Indisputably, Jefferson was familiar with natural law theory by 1772: Jefferson cited Pufendorf more than twenty times. Jefferson was prepared to argue, like Pufendorf, that the natural law provided a rule of reason that transcended judge—or Parliament—a law, as Pufendorf had put it, that was "that most general and universal rule of human actions to which every man is obliged to conform, as he is a reasonable creature."[34] According to legal scholar Dewey, he also read Scottish philosopher David Hume and John Locke and French philosopher Montesquieu at this time, not only a Pufendorf translator's footnotes to Locke. But Locke had been a Jefferson favorite since Jefferson's days at William and Mary and his dinner-table debates with Dr. Small, Governor Fauquier, and George Wythe. He came to consider Locke, as he wrote to John Trumbull in 1789, one of the "three greatest men that have ever lived, without any exception."[35] (The others were Isaac Newton and Francis Bacon.) Jefferson had been introduced to all three Enlightenment heroes by Dr. Small fully ten years earlier. In 1771, Jefferson had included "Locke on government," obviously his *Second Treatise on Government,* on the short list of books on politics that he recommended for Skipwith's library. He prefaced this choice list with the words, "Of politics and trade, I have given you a few only of the best books,"[36] clearly indicating that he thought himself conversant with all of them by this time.

It is very unlikely that Jefferson had time to reread, at this time, all the books he cited in his notes on divorce, but the main outlines of his study are clear. He was thoroughly conversant with Enlightenment literature and he was applying natural rights to the laws of marriage as he had to the relationships between slave and master and church and state. He was already a confirmed skeptic on matters of religion, using the lower case when he wrote the words *god* and *bible.* He had studied the writings of English statesmen-philosopher Henry Saint-John, Viscount Bolingbroke ever since the Stamp Act crisis of 1765. Bolingbroke had a strong influence on Jefferson's growing skepticism toward a variety of forms of authority. As Douglas Wilson points out, Jefferson commonplaced an amazing amount of Bolingbroke's *Philosophical Works,* in excess of ten thousand words, six times as many as he copied out from any other author and, in all, fifty-four entries comprising forty percent of his *Literary Commonplace Book.* More important, he adopted wholesale most of Bolingbroke's political tenets: his "thorough-going materialism," his "rejection of metaphysics and all speculation that ventures beyond the reach of human apprehension," his uncompromising belief in reason as the "final arbiter of knowledge and validity."[37] It was from his embracing of reason as well as his experiences and studies as a lawyer that Jefferson was coming to regard churchmen and theologians as corrupters of Christianity, and he considered the Bible as only a book of stories unless he could scrutinize and document them independently. He was eventually to write a nephew advising him to submit religious tenets to rational scrutiny and to treat the Bible as any other history book: "Fix reason firmly in her seat, and call to her tribunal every fact, every opinion. . . . Read the bible then, as you would read Livy or Tacitus." This view was straight-out Bolingbroke. But as a shrewd lawyer who had to persuade amateur judges and Anglican lawyers, Jefferson had to cite Scripture to his purpose. He included a section of biblical arguments, as well as some oblique references, quoted from others of his favorite writers, including John Milton, who had written four essays urging relaxation of the English law of divorce and who favored a return to Mosaic law, which permitted a husband to discard his wife almost at will. "Scripture commands to put away all obstacles to piety."[38]

His arguments for divorce began with his conclusion that it was "cruel to continue by violence a union made at first by mutual love but now dissolved by hatred. Citing Hume's *Essay Concerning Human Understanding,* he found this was "to chain a man to misery till death."[39] Echoing Hume, he wrote that "liberty of divorce prevents and cures domestic quarrels" and "preserves liberty of affection (which is [a] natural right)." The object of marriage "is propagation and happiness . . . where can be neither, should be dissolved. Should add education," he noted to himself.

Jefferson saw marriage as a contractual obligation that, once fulfilled, could be dissolved: "procreation, education and inheritance [had been] taken care of, [there was] no necessity from [the] nature of [the] thing to continue the society longer, but it may be made at the time of contracting dissoluble by consent, at a certain time or on certain conditions, as other compacts." Here, he was quoting almost verbatim from Locke's *Second Treatise on Government.* Jefferson's own reason for finally giving up his cherished bachelor existence atop tranquil Monticello peeks through his next argument for permitting divorce—citing Pufendorf, he noted the "moral obligation on man to propagate," following immediately with "Nature has excited the appetite, God had forbidden a meretricial gratification—he therefore enjoins a new marriage." Then, Jefferson notes the "private desires of an heir."[40]

Had he wandered into introspective reverie? If so, he returned to the Blair case, arguing that "constant residence and admission of copulation is implied in [Pufendorf's] covenant." For page after page, Jefferson listed the divorce "practices of several nations." He underscored one he found in the forward-looking codification of Prussian laws: a divorce could be granted if one partner was "not *fit* for marriage and the other ignorant of it." Jefferson concluded his draft arguments for what would have been Virginia's first divorce with some "miscellaneous observations."[41] Among his earliest recorded philosophical jottings, they reflected the writings of Pufendorf, Hume, Montesquieu, Locke. Bending over his pen in a quiet corner at Monticello, he wrote in note form the arguments he intended to press on the House of Burgesses in April 1773.

> In begetting children, parties [are] bound by nature to provide till they can provide for themselves. This done, no being can reproach him with injury in dissolving marriage. . . . [Divorce] restores to women their natural right of equality. . . . [It is] cruel to confine divorce or repudiation to [the] husband, who has so many ways of rendering his domestic affairs agreeable. . . . There should be no such thing as a divorce *a mensa et thoro* [a legal separation without a divorce] because of injury done to the innocent party.

In his final observation, he noted that marriage had never been "declared indissoluble till by the Popes, who made marriage a sacrament and so took cognizance of it to themselves."[42]

Whether Jefferson ever finished writing his notes on divorce is unknown. On December 26, 1772, after nineteen months of an unconsummated marriage, Dr. James Blair died. There is no record that Jefferson ever made public his notes, and now there was no need for his legislative divorce bill. Eventually, in November, 1773, when Kitty Eustace Blair sued the Blair family for her sizable dower right to Blair's slaves, the case came before the General Court. Jefferson took copious notes for his book of law reports. "The evidence," he wrote, "was voluminous and indecent." There was a long discussion of Dr. Blair's impotence. Kitty Blair, Jefferson noted, had departed from Blair's house "not without some hint of adultery."[43] But no one pressed Dr. Blair's initial charge of infidelity between his bride and her kinsman, the new royal governor. "The suspicions of adultery were with Lord Dunmore," wrote Jefferson, "who, presiding at the court at the hearing of the [case], might be the reason why those suspicions were not urged." But Kitty Blair won. Writing home to England, her mother heaped praise on one of her lawyers, Patrick Henry. "He shined in the cause of justice backed by the law."[44]

It was the kind of performance by Henry that left Jefferson sputtering. While their political views "conformed exactly," Jefferson said that Henry could not draw up a bill or a brief "on the most simple subject which would bear legal criticism or even the ordinary criticism which looks to correctness of style and ideas, for indeed there was no accuracy of idea in his head. His imagination was copious, poetical, sublime, but vague. . . . He said the strongest things in the finest language, but without logic, without arrangement."[45] If Thomas Jefferson ever ranted, it was on the subject of his friend Patrick Henry.

His criticism of Henry was only an indication of his own growing disillusion with the lawyers, the judges, and the English laws that blocked his attempts at system and logic at every turn. While his years of research and argument had prepared him for far more important forums, in not one of his most major cases had he to succeed. His legal career was also a financial failure. On paper he was earning a considerable income. But, as the currency crisis in the colonies continued, few clients were paying him. Each year from 1767 to 1771, his earnings on paper had risen sharply, from £293 in 1767–68 to £521 in 1770–71, the equivalent of roughly $50,000 a year today. As he turned away from caveat cases and devoted his attention more to briefing and pleading important cases before the General Court, his gross earnings dropped nearly fifty percent in 1772 and rose slightly to £295 in 1773, his last full year of practicing law. But his collections rarely exceeded fifty percent of his legal fees due, in some years only fifteen or twenty percent. Although he seems to have made a real effort to be paid after his marriage, his income was modest. After expenses and advances on behalf of his friends and relatives—he represented his cousin, "Tuckahoe Tom" Randolph thirteen times in eight years, charged no fees, and advanced £62 for him, a dead loss—he averaged only £200 a year (a mere $20,000 a year today), a paltry sum for all his education and efforts. Without the income from his farms and his occasional land sales, Jefferson could not have afforded the gentlemanly luxury of practicing law in colonial Virginia. Finally, a series of personal and political crises in 1773 and 1774 led him to abandon his practice, sacrificing two-thirds of the roughly £1,400 of accumulated debts owed him by his clients. By August 1774, when he turned his remaining 132 active cases over to his twenty-one-year-old cousin, Edmund Randolph, Jefferson was thoroughly disgusted with the British colonial court system and was masterminding an aggressive attack to shut it down.

8

"God Gave Us Liberty"

Public service and private misery are inseparably linked together.

—Thomas Jefferson to James Monroe, May 20, 1782

In early March 1773, Jefferson rode east with Dabney Carr toward Williamsburg for an earlier-than-usual session of the House of Burgesses. It had been a year since the Assembly had met and nearly two years since Jefferson had been in his seat from Albemarle. Carr was riding with him this time as a newly elected delegate from Albemarle. Before the Assembly was convened by the royal governor, Lord Dunmore, however, the two men huddled with three other members of the young radical faction—Patrick Henry, Richard Henry Lee, and Francis Lightfoot Lee—at Raleigh Tavern "to consult on the state of things." From their standpoint, they agreed, the state of things in Virginia was all "lethargy." Most Virginians were choosing to ignore "passing events,"[1] as Jefferson put it, in other colonies and in England. Without the Assembly as the hub of political activity in the colony, the protest movement had gone from weak to ineffectual. Lord Dunmore no doubt knew that the radicals would make little headway so long as he left them scattered on their farflung plantations. But he needed tax money to renovate the College of William and Mary and he wanted to redecorate the Governor's Palace, so he had made the terrible mistake of calling the Assembly.

Jefferson and Carr had temporarily left behind one set of common interests for another. Dabney Carr never tired of talking about his little boy, and now, Carr's wife was pregnant again for the sixth time in eight years. Jefferson's infant daughter was just beginning to flourish at six months: both men's wives were bedridden. But their personal worries were subsumed by a grave new turn in the colonial crisis. The British government, unable to gain evidence, witnesses, or convictions against smugglers in colonial courts, had just decreed that American prisoners were to be transported to England to stand trial before predictably hostile judges. This step thoroughly alarmed Jefferson. It came after an armed British revenue schooner chasing a smuggling ship in Narragansett Bay off Rhode Island had run aground and had been boarded and burned to the waterline by more than sixty attackers organized and led by prominent local merchants. When no witnesses had come forward, the British had set up a special investigating commission made up of royal chief justices and other high British colonial officials, but they, too, could find no witnesses willing to testify. In New England, to inform on smugglers was considered a far worse crime than smuggling.

Once at Williamsburg, Jefferson and his radical friends, in a series of private evening meetings at the Raleigh Tavern, reviewed the nonimportation movement in the colony and the reasons for its failure. As long ago as 1770, it had become obvious that the nonimportation associations were doomed. It had proven impossible to hold antiministry colonial public opinion at fever pitch year in and year out, especially after partial repeal of the Townshend duties had left only a trifling tax on tea. Boycotts in Boston and Philadelphia had collapsed. When Virginia's associators, including Jefferson, had attempted to hold a second boycott meeting in Williamsburg on December 14, 1770, so few delegates attended that William Nelson, president of the Governor's Council, had reported to Lord Hillsborough, the colonial secretary in London, "such lukewarmness as convinces me that this [association] will soon die away and come to nought."[2] Jefferson still held out hope that Parliament would relent on its tea tax and, himself violating the association, sent off to London two hogsheads of tobacco to pay for a long list of articles he wanted for his new house, including the custom-made clavichord for Patty:

> You will observe that part of these articles [such as are licensed by the Association] are to be sent at any event. . . . Another part [being prohibited] are only to be sent if the tea act should be repealed before you get home. . . . I am not without expectation that the

> repeal may take place . . . I believe the Parliament want nothing but a colorable motive to adopt this measure.[3]

But Jefferson underestimated the Ministry's determination to maintain a token tax to assert its supremacy over colonial legislatures. Jefferson champed at the bit to confirm his order from London for Patty (he had changed his order from clavichord to piano) even as he conceded the defeat of the boycotting Association:

> The day appointed for the meeting of the associates is not yet arrived. . . . However, from the universal sense of those who are likely to attend, it seems reduced to a certainty that the restrictions will be taken off.[4]

Jefferson told his London agent to send him the part of the February list that included "some shoes and other prohibited articles":

> . . . If, contrary to our expectations, the restrictions should be continued, I can store or otherwise dispose of them as our committees please.

Tossing radicalism aside and indulging himself anyway, he wrote that he had "seen a forte-piano and am charmed with it . . . let the case be of fine mahogany, solid, not veneered . . . the workmanship of the whole very fine, and worthy of the acceptance of a lady for whom I intend it."[5]

A major factor in the collapse of the Virginia boycott had been the conciliatory tactics of Lord Botetourt. Jefferson later observed that, had Botetourt lived, there might have been no revolution in Virginia. His Lordship had so anesthetized His Majesty's subjects in Virginia that, as Jefferson put it, "Our countrymen seemed to fall into a state of insensibility to our situation."[6] When Lord Botetourt died, the General Assembly had moved to erect a "very elegant statue"[7] to him. But the diplomatic old baron had been replaced by the youthful, militant Earl of Dunmore, a soldier and former royal governor of New York who had arrived with his wife, seven children, and "a set of his drunken companions [who promptly] sallied about midnight from his Palace" and attacked the coach of the provincial chief justice. The "coach was destroyed and the poor horses lost their tails."[8] When the short, muscular, prematurely graying Lord Dunmore was sworn in, no one knew he was the last royal governor of Virginia.

Convening his first Assembly, Dunmore faced a stiff debate, unrecorded in the House journals, on the subject of reviving customs duties on the purchase of slaves. The debate came nearly two years after Jefferson's futile attempt to win freedom for the mulatto slave Samuel Howell before the General Court. Jefferson had been on his honeymoon at Monticello when Burgess Benjamin Harrison successfully moved on March 20, 1772, "that an humble address be prepared to his Majesty" asking the king's assistance in averting "a calamity of a most alarming nature"—continued entrapment in the slave system. Speaker Peyton Randolph appointed Harrison chairman of a committee including Archibald Cary, Edmund Pendleton, Richard Henry Lee, Robert Carter Nicholas, and Richard Bland to write an address to the king. On April 1, 1772, two and a half years after Jefferson's first attempt at a ban on importing slaves had been shouted down by many of these same assemblymen, the burgesses unanimously adopted the resolution. All slaveowning planters, they condemned "the importation of slaves into the colonies from the coast of Africa" as a trade of great inhumanity. If the trade was allowed to continue in direct competition with free labor, "we have too much reason to fear [it] will endanger the very existence of your Majesty's American dominions." The Virginia resolution warned that the powerful influence of English slave traders was retarding "the settlement of the colonies with more useful inhabitants."[9] Lord Dunmore's written instructions from the Colonial Office precluded his supporting the resolution with any vigor, but he passed it on to London.

No direct answer came from king or Parliament but, coinciding with the arrival of the Virginia petition in England, Baron Mansfield, lord chief justice of the King's Bench, on June 22, 1772, published his famous Somerset decision. James Somerset, an African allegedly sold in Virginia, was helped to sue for his freedom after he set foot in England. Wrote Lord Mansfield:

> The state of slavery is so odious that nothing can be suffered to support it but positive law. Whatever inconveniences, therefore, may follow from a decision, I cannot say this case is allowed or approved by the law of England. . . . Therefore, the black must be discharged.[10]

The institution of slavery began to die that day in England but, by Jefferson's own logic, Justice Mansfield's decision had no effect in Virginia or any other British colony.

As they conferred at the Raleigh Tavern in Williamsburg in mid-March 1773, Jefferson and his radical young friends concluded that many of their associates in the Assembly had grown too conservative, lacking the determination that the times demanded. The extension of parliamentary power and judicial authority over Americans alarmed them. Jefferson had few misgivings that he was in the company of the other most radical burgesses at these secret evening meetings. Among his compatriots were the Lees, who had been in the forefront of the network of American colonists determined to expand and unite colonial protests against spreading British encroachments into colonial affairs and the resultant suppression of Americans' civil liberties. Arthur and William Lee had been active in London radical circles for five years. Leaving Virginia for London, Arthur Lee pledged to send "speedy and accurate information of the real designs of the British Ministry . . . to leading men in the colonies," so that they could "harmonize in one system of opposition."[11] The Lees corresponded with George Mason and George Washington in Virginia and with Samuel Adams and Joseph Warren in Massachusetts, bringing their correspondents into direct contact with each other. Arthur Lee also put John Dickinson of Philadelphia and Samuel Adams in touch with the foremost radical in Virginia, Richard Henry Lee: it was this particular Lee who had smuggled the rough draft of the first Association from Philadelphia to Mason and Washington. Samuel Adams had urged this network to set as its top priority the alerting of all colonists to the latest British inroads.

Now, in March 1773, Jefferson and the handful of the Virginia radicals pledged to draw up a set of resolutions to protest in general terms in the House of Burgesses all Parliamentary proceedings that tended "to deprive them of their ancient legal and constitutional rights." Jefferson wrote the resolution and vowed to fight in the Assembly to set up, in his words, a standing "committee of correspondence and inquiry" to "obtain the most early and authentic intelligence" of acts of Parliament and the British Ministry and to "maintain a correspondence and communication with our sister colonies."[12] In the House, Jefferson gave the task of introducing the resolve to Dabney Carr, who, on March 12, 1773, delivered a fiery maiden speech promoting intracolonial cooperation. In the resolution, Jefferson warned especially of the consequences if the Rhode Island commission had been empowered to transport prisoners to England. Jefferson's resolution passed the House without a single dissenting vote. Jefferson was proud of Carr's debut; his brother-in-law had already earned a reputation for oratory on the county-court circuit. Jefferson wrote that Carr had given his maiden speech "with great ability" praising his "reasonings" and "the temper and moderation with which they were developed."[13] The burgesses unanimously authorized Virginia to take the lead in communicating the "most early and authentic intelligence" of British innovations affecting all Americans to her "sister"[14] colonies.

Amazingly the royal governor did not see the danger contained in the resolution, which opened the door for the first time for unified colonial action. "There are some resolves which show a little ill humour in the House of Burgesses, but I thought them so insignificant that I took no notice of them," he wrote to the colonial secretary in London.[15] More concerned about a mild rebuke from the burgesses for his heavy-handed roundup of counterfeiters in the colony, Lord Dunmore prorogued the Assembly. But he was too late: the next day the eleven-member standing Committee of Correspondence went on, quite legally, with its meetings at Raleigh Tavern. Jefferson and his radical young friends, though they were outvoted for the moment by more moderate burgesses on the committee, had succeeded in creating the first colony-wide standing revolutionary committee in America. And now Jefferson's writings would circulate to other colonies, who would follow his example and Virginia's lead. The new Committee of Correspondence would be the only Assembly committee functioning in Virginia for the next year as the revolutionary crisis accelerated, and its creation was widely hailed and emulated. With a stroke, the precedent of committees of correspondence led to creation of protest groups in every American colony except Pennsylvania in the next few months.

Jefferson enjoyed nearly two idyllic months at Monticello in the spring of 1773 with Patty. The two read to each other and Jefferson dandled little Patsy, now at last growing plump and strong. He worked long hours in the garden on the south slope of Monticello, right outside the South Pavilion. He had put in a vegetable garden on the flattened plane he had dug at the top of the slope where his orchard started to cascade downhill, just beyond the row of mulberry trees that lined the road up his mountain. On this wide terrace, he planted twenty-four square beds of vegetables, and he took the most complete notes yet in his *Garden Book.* He "distinguished" his plantings "by sticking numbered sticks in the beds," assigning them a code in the *Garden Book.* For the first time, some of the names were in Italian: number 15, radishes, was "radicchio di Pistoia," number 26, Spanish onions, was "cipolle bianche di Tuckahoe," number 45, carrots, "Carote di Pisa."[16] A new influence had come into his life. Jefferson had renewed his study of Italian: while he had failed to master German, he could now amuse himself by studying in Latin, Greek, French, Anglo-Saxon, and Italian, which he found handy as he adapted Italian Renaissance motifs for Monticello, whose construction was now far advanced.

In early May 1773, Jefferson had to break away to ride to Williamsburg to assist as counsel in the Blair case. This time he went without Dabney Carr, who had court business in Cumberland County. In an undated letter to his wife, Carr wrote that he had been ill, but that "I have been as well today as any day for a week past." He complained of "a small chilliness not to be called an ague." He had "so little fever that I have done my business in court" but he had decided not to go on to the county court in Amelia "unless I continue as well as I am." His "worst complaint" was that "there is a very good dinner before me" and "I have not the stomach to eat a mouthful."[17] Carr pushed on toward his home in Goochland County but apparently decided to stop off to recover at Monticello. He made it as far as Charlottesville where, before continuing the last few miles to Monticello, he stopped to see Dr. George Gilmer. No doubt weakened by the savage spring snowstorm that had lashed through the Blue Ridge on May 4, Carr had plodded on alone over the frozen, cracked roads day after day: now he could go no farther. He sent a slave to bring his wife, but she was still recovering from childbirth and before she could reach his side, Dabney Carr, not quite thirty, died. Dr. Gilmer did not know why: he called it what they always did in the eighteenth century when they weren't sure, a "bilious fever."

Carr's widow, Jefferson's sister Martha, was still in her twenties. Carr's death stunned her. She had been too weak to leave her bed when Carr had ridden off for the last time. She had raised herself on her bed, as Harry Randall wrote, "to catch a parting glimpse of him as he rode past her window, but she saw merely his moving hat. For months the moving shadowy hat was ever passing before her eyes."[18]

By the time Jefferson could hurry back from Williamsburg, they had buried Carr at Shadwell. Unable to express his grief in any other way, Jefferson took up his pen and, in the *Garden Book,* after the last notation he had made for fragile spring plantings, wrote that "the Blue Ridge of mountains" was "covered with snow," that "a frost . . . destroyed almost everything."[19] Of all men Jefferson knew, he had loved Dabney Carr the most. Once they had stared out over the blue mountain ridges and sworn that they would lie together here when they died.

Together, as adolescent boys home from Reverend Maury's doing homework under the great oak on Monticello, Jefferson and Carr had read aloud from Horace's odes. The lines they chanted were among the first Jefferson had copied out into his commonplace book so soon after his father's death. Together they had read of Jupiter's delight in driving "his thundering steeds and flying car through a sky serene."[20] As he searched now for the words for his best friend's gravestone, they lay there, round and innocent on the page:

> *A dreadful storm has narrowed heaven's expanse,*
> *And rain and snow are bring Jove to earth . . .*
> *Pale Death with foot impartial knocks*
> *At the poor man's cottage and at princes' palaces.*[21]

Now Jefferson brought Carr to lie near him. He ordered two men to dig a grave out of the harsh clay. As he always did, he channeled his grief into his *Garden Book,* recording on May 22 that "two hands grubbed the graveyard eighty feet square equals one-seventh of an acre in $3^1/_2$ hours so that one would have done it in seven hours and would grub an acre in 49 hours equal four days." Then he had Dabney's body disinterred and brought over in a bare, rattling wagon. He arranged to place a marble slab over his friend under their oak tree and he composed its inscription:

> To his virtue, good sense, learning and friendship this stone is dedicated by Thomas Jefferson who, of all men living, loved him most.[22]

Then he brought to Monticello the living: his sister and Dabney's children. Carr had honored him by naming him one of his executors. Holding seven-year-old Peter close to him, he decided to educate the boy as his son.

Less than two weeks after Dabney Carr died, a rider dismounted at Monticello with more grim news: John Wayles, Patty's father, had died at the age of fifty-eight. The death of the blustery lawyer, whom Jefferson had come to like, forever changed Jefferson's life. Wayles was land-rich (Patty's share of her father's estate was eleven thousand acres) and had 135 slaves.

The terrible irony of Martha Wayles Jefferson's inheritance may have hit Jefferson even before John Wayles's death. For five years, Jefferson had publicly opposed slavery; now, as master of his wife's affairs, he managed one of the largest slave-holdings in America. He did not record his views on his father-in-law's miscegenation with his slave concubine Betty Hemings, matriarch of many of his slaves, but it was well known at the time that, second only to slave trading, Jefferson detested the idea of miscegenation.

At the time of John Wayles's death, the Jeffersons had thirty-four slaves, none of them children, including those Jefferson had inherited ten years earlier and the body servants Patty had brought with her from The Forest as a bride. Jefferson had sold only two slaves in ten years, and had bought slaves only once: in January 1773, in payment of a debt owed Patty, he had accepted a slave family, George and Ursula and their two children, from his friends the Flemings. Suddenly the number of his slave dependents was 169, many of them newly arrived from Africa. An ad appeared in three successive weekly issues of the *Virginia Gazette* in October 1772:

> Just arrived from Africa, the ship Prince of Wales, James Bivins commander, with about four hundred five healthy slaves, the sale of which will begin at Bermuda Hundred on Thursday the 8th of October and continue until all are sold.
>
> John Wayles
> Richard Randolph[23]

Wayles had written to his son-in-law later that month that "the sale of slaves goes on slowly"[24] and that, as a consequence, he would have to put off visiting Monticello as planned to see his new granddaughter. Wayles and his partner managed to sell few if any slaves in the following months because of the cloud on slave trading that had resulted from the House of Burgesses antislavery petition to the Crown a year earlier.

When Wayles died, the terrible task of selling off the newly arrived slaves devolved on Jefferson and his two brothers-in-law: Wayles had divided his vast estate equally among his three daughters. Jefferson apparently refused to sell the slaves at first, preferring to sell off some of Wayles's lands to pay his vast debts, which had to be divided three ways, too. Much of Wayles's estate was not liquid; his debts had to be paid in sterling. Worse, most of them were in sterling to British merchants and factors. Wayles had borrowed money to make money. He had not wasted money on luxuries like so many Virginia planters. In the autumn of 1772, Jefferson had learned from a Glasgow merchant who had just toured England that there had been a "sad revolution" among many British merchants brought about by a crackdown on credit by the Bank of England. The result, wrote Alexander McCaul in July 1772, had been that "many houses in London esteemed eminent and wealthy failed," bringing down many in Edinburgh and Glasgow. "It has thrown a damp on public credit and it will be sometime before it is perfectly restored."[25]

As Jefferson tackled the task of paying off his father-in-law's bills, he found that Wayles owed £10,200 sterling to three British bankers in addition to £6,770 sterling for the shipload consignment of slaves, a debt that technically devolved on Richard Randolph, Wayles's surviving business partner. Jefferson felt that to shift this debt to Randolph was immoral, however. Wayles's assets, his farms, undeveloped lands, and slaves, if sold, potentially outweighed his liabilities two-to-one. Partly because of British refusal to allow a Virginia currency or to approve reforms to inheritance laws, Wayles had been unable to sell off any of the newly imported slaves. Rather than be forced into the business of selling slaves, Jefferson decided to sell off lands, accepting notes against future payment.

Jefferson did not castigate his late father-in-law for extravagant speculations. He, like so many Virginians, blamed the British merchants and British mercantilist policies, especially the Crown's refusal to allow Americans their own currency even as they insisted on payment in scarce gold and silver. Since the mercantilist system had been fortified by the Navigation Acts in the mid-seventeenth century, the planters had been content to send all their tobacco to the mother country. There, ninety percent of it was reexported to the Continent and other British possessions, often yielding upwards of £200,000 a year to the British Treasury and lining the pockets of merchants, warehouse owners, and thousands of workers with the profits of the tobacco trade. But the planters had no control over prices, which had been declining for years even as the prices of everything they could not manufacture in America—an ever-growing list—was climbing. Until the 1770s, Virginians had not objected to the trade laws, which included guaranteed markets for all colonial harvests. But Jefferson and Wayles were part of a new breed of increasingly self-sufficient Virginians who believed that credit, not trade, was the real problem. Over a thousand Scottish resident factors had come to Virginia and Maryland at midcentury, buying and selling and encouraging debt by extending easy credit. As the planters settled deeper and deeper into a quagmire of debts, their anger and frustration with England grew. As Jefferson explained the system,

> The advantages made by the British merchants on the tobacco consigned to them were so enormous that they spared no means of increasing those consignments. A powerful ending for this purpose was the giving of good prices and credit to the planter, till they got him more immersed in debt than he could pay without selling his lands and slaves. They then reduced the price given for the tobacco, so that let his shipments be ever so great, and his demand of necessaries ever so economical, they never permitted him to clear off his

> debt. These debts [became] hereditary from father to son for many generations so that the planters were a species of property annexed to certain mercantile houses in London.[26]

Jefferson estimated the overall planter debt at a staggering £2 million to £3 million by the 1770s, equivalent to one hundred times that amount today.

As Jefferson was undoubtedly aware, the first political crisis brought about by planter debts had surfaced in 1766, immediately after the Stamp Act crisis. The death of John Robinson, for nearly thirty years the treasurer of the colony and Speaker of the House of Burgesses, revealed that, in exchange for political support, he had handed out £100,000 from the Virginia treasury to his debt-ridden clique of planters. Paper money issued during the French and Indian Wars was to be retired when collected as taxes. Instead of burning it, as the law required, Robinson had embezzled it, lending it to his strapped friends, expecting them to cover the loans before they could become known. Among the 240 planters who had accepted Robinson largesse were Jefferson's father, £52, and John Wayles, a considerable £445. The money technically was owed to Robinson's estate, which had to repay all of it to the colony. Many planters had been humiliated by the scandal; quite a few were ruined. Yet Jefferson remained always the optimist: on July 15, 1773, only six weeks after Wayles's death, Jefferson and the coexecutors, the husbands of Patty's sisters, printed another ad in the *Gazette*, offering in all fifty-four hundred choice acres scattered around Virginia. One of Wayles's largest debts was for the shipload of slaves, and land had to be sold off quickly to satisfy the English slave trader. Unfortunately Jefferson and his coexecutors accepted commercial paper for the lands: the notes were later paid back with badly depreciated money during the Revolution and Jefferson was forced to pay the Wayles debt all over again. It was a mistake of far-reaching consequences, but Jefferson had no hint of it at the time.

When Jefferson had married Martha Wayles, he did not have any reason to believe that his financial condition would change in the near future. He had visited Wayles shortly before his father-in-law's death to help systematize the mass of papers accumulated in his land transactions. In his autobiography, he remembered Wayles as "a most agreeable companion, full of pleasantry and good humor." When Wayles died, according to Jefferson, the portion of his estate coming to the Jeffersons after payment of all debts—"which were very considerable"—was about equal to Jefferson's patrimony. Jefferson's impression at the time was, he remembered, that the inheritance "doubled the ease of our circumstances."[27] Once again, Jefferson was consolidating his holdings closer to Charlottesville. The one major exception was Wayles's lands in the southwest, in Bedford County, where Jefferson had so recently bought the Natural Bridge site. By mid-1773, Jefferson decided he was now a wealthy man, about £6,000 richer (roughly $600,000 today) than he had been before John Wayles's death. Until now financially conservative, Jefferson could now think seriously about abandoning the unprofitable part of his law practice and retaining only his specialty of caveating so that he could divide his time between administering his now doubled estates of some seventeen thousand acres, looking after the disposition of the 405 newly arrived slaves and the accommodations for the one-third of them to be brought to Monticello. If there was any time left over, he would devote it to his building projects at Monticello, his gardens and orchards and horse breeding and, most of all, to a more prominent stance in the growing political crisis facing the American colonies. Ironically it was slavery that suddenly made Thomas Jefferson free of his tedious, unremunerative law practice and enabled him to devote his energies to Americans' freedoms.

In January 1774, Jefferson received the rolls of 135 slaves he had selected from all of Wayles's plantations. He evidently paid particular concern to keeping families together. He listed them in family groups except for the artisans; there were fifty-four field hands, nine unskilled workmen, two carpenters, two blacksmiths, three watermen, one shoemaker, fifty-nine children, and five aged or infirm.

The wagonloads of slaves who were transported up the hill to Monticello included Betty Hemings's mulatto family. Isaac, a slave born the next year, later described the Hemings family, brought to Monticello in fulfillment of one of John Wayles's debts, who became the Jefferson family's favored house slaves. There were three darker Hemingses, Bett, Martin, and Mary, and two light mulatto boys, James, nine, and Bob, twelve, the first of the Wayles slaves brought to Monticello a year after his death. Martin was to become Jefferson's butler; James, his body servant; Bob, a groom. Their mother, Betty, Wayles's concubine, came to Monticello in 1775 and was given her own cabin, along with considerable independence. She raised "pullets" and "fowls" which she sold to Jefferson, according to his account books. Betty, described by her grandson Isaac as "a bright mulatto woman," bore a son, James, who had probably been fathered by John Wayles, and a beautiful daughter named Sally, who, Isaac said, was "mighty near white." Sally was the same age as Jefferson's oldest daughter, Patsy, and was to be her body servant. According to Isaac Jefferson, "Sally was very handsome: long straight hair down her back." Her father probably was Nelson Jones, a white carpenter at Monticello. According to Madison Hemings, interviewed after the Civil War, his grandmother Betty Hemings "had seven children by white men and seven by colored—fourteen in all."[28]

Still another personal loss befell Jefferson shortly after the deaths of Dabney Carr and John Wayles: his mentally retarded sister, Elizabeth, died at twenty-nine. Since his twenty-first birthday, Jefferson had been responsible for her. He had watched out for her financially and had kept separate her accounts, though, largely because of his mother's interference, he was unable to control her, and to prevent her, among other excesses, from shopping sprees at stores in Charlottesville and Richmond. For a man so discreet, the absence of discretion in his sister was vexing. And, when a series of earthquakes rocked the buildings at Monticello on February 21 and 22, 1774, Elizabeth had run outdoors. In the raw mountain winter weather, the confused Elizabeth wandered away. In his *Account Book*, Jefferson noted on

> March 1. My sister Elizabeth was found last Thursday being Feb. 24.[29]

The exact cause of her death remains unknown.

Strapped as usual for cash, the land-rich Jefferson had to sell off two old bookcases to pay the Reverend Charles Clay to perform the funeral service. Three weeks later, Jefferson's second daughter was born; he named her Jane, after his mother and his sister.

But if Jefferson was keeping a balance sheet between all the personal losses and gains of this period of his life, there were some offsetting elements to be taken into account. As if by magic, there had appeared at Monticello that winter a talkative Italian exile turned wine merchant named Philip Mazzei, who had come to Virginia, all uninvited and unannounced, accompanied by Jefferson's London merchant-agent, Thomas Adams. It was Adams who had interceded for Jefferson in trying to win holy orders for his friend Ogilvie, and Jefferson had sent Adams long lists of goods he wanted for Monticello, including the wedding-present piano for Patty. Mazzei, forty-three, had been trained as a surgeon in Florence and practiced in the Middle East before settling in London, where he had been a wine merchant for many years. A well-known horticulturalist, he had sailed to Virginia to introduce the culture of grapes, olives, and whatever fruit trees would flourish there, and he had brought his own crew of Italian vineyard workers with him. Thomas Adams was leading Mazzei on a tour of Virginia en route to Augusta County, where he hoped to sell him land. When Mazzei had arrived in Virginia with Adams and his Italian laborers, they had visited briefly at The Forest with Patty Jefferson's sister Elizabeth and her husband, Francis Eppes. They had arrived at Monticello one November evening on their way to the Shenandoah Valley and Adams's lands. Rising the next morning before the others, Mazzei and Jefferson went for a long walk through his gardens and orchard and construction site and then down the roundabout pathways with which Jefferson had girdled his mountain. They talked about botany and vines and classical antiquity. Jefferson almost offhandedly showed Mazzei a handsome tract of four hundred acres of lands adjoining Monticello that he was willing to sell; if the Italian wanted, Jefferson would throw in a gift of two thousand acres more. When they had returned to Monticello, a defeated Thomas Adams said to Jefferson, "I see by your expression that you've taken him away from me. I knew you would do that."[30] Unwilling to let Mazzei leave until, like a fledgling plant, he was established, he put up Mazzei as his houseguest for months while the Italian summoned his Tuscan workers and built a house he called Colle, two miles from Monticello, and planted his first hillside vineyards, all with the help of a £2,000 subscription drive Jefferson arranged with his friends.

The spring of 1774 promised to be a fertile period for Jefferson as he "laid off ground to be levelled for a future garden." According to his *Garden Book* entry for March 31, it was enormous: 668 feet long, 80 feet wide. Planting apple and cherry trees sent him by his Albemarle neighbor, Michael Woods, he then set out almonds, apricots, "198 cherries of different kinds from Italy, about 1500 olive stones." As he had awaited the birth of his second child, he was spending more time with his new friend Mazzei, and he tried out the romantic-sounding Italian names of the specimens he planted: "lamponi . . . Raspberries . . . 3 rows; fragole Alpine . . . Alpine strawberries . . . 3 rows; fragolini di giardino, large garden strawberries . . . 1 row." He used prosaic English for "a bed of parsley, 62 red cabbage, radishes." There were two day's pauses in his notations for the day before Jane Jefferson was born and the day of delivery itself, but the very next day, Jefferson was back in the garden: "The peas of March 24 come up."[31]On April 6, Jefferson made elaborate notes on the first planting of grape vines he had ever attempted. "They were planted by some Tuscan vignerons who came over with Mr. Mazzei."[32] The next day he planted hills of Italian watermelon seeds from Pistoia and Naples, cantelopes from Melun, muskmelons. Then, in a high meadow, he supervised the planting of rice. For five weeks the planting went on. As the root vegetables were dropped into trenches and hoed over, Jefferson watched and timed the laying of a stone wall. But a killing frost on May 4 destroyed almost every plant that had come into leaf as well as all of Mazzei's. Jefferson wrote it all down and hurried off to Williamsburg, where the royal governor, Lord Dunmore, had finally convened the Assembly.

Seeking to rule Virginia without the interference of Virginians, Dunmore had prorogued the General Assembly of March 1773, when Jefferson and his friends had won creation of the standing Committee of Correspondence, until June 1773, then to August, November, and finally to May 1774. In setting the May date, Governor Dunmore either forgot that a Virginia statute that fixed the fees of various officials expired on April 12 or he simply assumed the General Assembly would renew it retroactively or routinely enact a new one. In the meantime, officials including court clerks and sheriffs continued to perform the duties of royal government. Coroners continued to conduct inquests, surveyors made surveys, constables whipped slaves and servants, sheriffs chased runaways and kept jails under their watchful eyes, in every case receiving the fees that had been prescribed for thirty years. When the General Court opened its semiannual session in April, neither judges nor lawyers seemed concerned: criminal and civil trials were held normally. The lack of a fee bill became an issue on May 4, 1774, when a court clerk began to tot up the court charges for a creditor to slap on a debtor who was expected to remove his property from Virginia. The clerk, Benjamin Waller, a lawyer, asked the General Court how much to charge the winning party for the period after the fee bill had expired on April 12. The court ruled on May 4 that the fees would be the same as the fees in the expired bill.

The next day, May 5, the General Assembly convened and the day after that its Committee for the Courts of Justice, chaired by Richard Henry Lee, was assigned the task of recommending which fees "are fit to be revived and continued."[33] On May 10, the committee recommended not reviving the fee bill. Giving no explanation, the committee created widespread speculation. Holding up the fee bill would pressure the secretary of the colony, one of its major beneficiaries, and bring to a standstill many government operations in Virginia, but it also could close the colony's courts, making it impossible for British creditors to collect money in Virginia and thereby inducing them to put pressure on the British government to be more conciliatory toward her colonies. Such pressure had worked before in the Stamp Act crisis, which had closed the county courts for a year. Jefferson was not involved with the Lee committee's recommendation (he did not reach Williamsburg until May 9) but he was present on May 11 when the House of Burgesses overruled Lee's committee, ordered that the fee bill be revived, and referred its drafting to the Committee on

Propositions and Grievances, of which Jefferson was a key member. The committee included five members of the General Court bar—Jefferson, Henry, John Randolph, Pendleton, Mercer—and the entire spectrum of Virginia politics, from radical Richard Henry Lee to conservative John Randolph. But before the committee could meet, news of the Boston Port Bill reached Virginia.

The slight tax on tea left over from the Townshend Acts of 1773 finally pushed the long impasse over colonial taxation into an open and more violent stage that eventually wrecked the First British Empire. To save the East India Company from bankruptcy, by helping it get rid of a £17 million surplus of tea in the colonies, Parliament passed a new Tea Act in April 1773. All duties were remitted except threepence on the pound. The new act changed the method of sale from public auction—which allowed Americans to profit as factors and retailers—to a complete company monopoly on importation and sales, which further angered American merchants, who now faced financial ruin. Now rich Boston merchants united with political radicals such as Sam Adams in the protest movement. On November 26, 1773, three British East India Company tea ships reached Boston. Colonists prevented their unloading. Royal Governor Hutchinson refused to issue permits for the ships to leave the harbor. There was a twenty-day waiting period after the first ship reached the port for the tea consignors ashore to pay the customs duty on the tea or face seizure. Fearing that the cheap East India Company tea—a fraction of the cost of smuggled Dutch tea—would be an "invincible temptation"[34] for Bostonians, radical leader Sam Adams, who was backed by some of the wealthiest tea smugglers, organized a "tea party." On December 16, the night before the customs deadline, thinly disguised "Indians" boarded the ships and jettisoned 342 lacquered chests of tea worth more than $100,000 today.

While less radical Bostonians decried this violent turn, radical patriots had taken command of the colonial protest movement. Even though one hundred Boston merchants offered to pay for the tea and Benjamin Franklin, London agent for the Massachusetts Assembly, urged them to do so, the radicals refused. England retaliated by passing the Boston Port Acts, with Lord North, the first minister, personally shepherding them through Parliament. One of these so-called Intolerable Acts closed the port of Boston. That the trade and the livelihoods of an entire seaport could be sacrificed to British demands for revenge of tea-dumping by a handful of radicals appalled Jefferson: "This is administering justice with a heavy hand indeed!"[35] His words struck a responsive chord among other radicals.

Although electrifying rumors of the British blockade of Boston port had been circulating in Williamsburg for days, on Thursday, May 19, 1774, the *Virginia Gazette* printed "An Epitome of the Boston Bill" two weeks into the General Assembly session. Over the weekend, Jefferson, Patrick Henry, Richard Henry Lee, and a handful of the younger burgesses became convinced that the majority of Virginians, like their conservative Assembly colleagues, were napping while dangerous events were moving quickly closer, and they determined to act. On Monday, May 23, they called a secret meeting in the vacated Council chamber at the Capitol ostensibly "for the benefit of the library in that room."[36] The group decided to introduce a resolution calling for "a day of fasting, humiliation and prayer"[37] on June 1, the day the port of Boston was to be closed, to dramatize the danger and bring their message directly to the public; not since the outbreak of the French and Indian Wars on the Virginia frontier in 1755 had there been such a convocation.

Although the author's identity was kept secret, Jefferson undoubtedly drafted the resolution after rummaging through the Puritan John Rushworth's familiar *Historical Collections,* a set of seventeenth-century bound documents that could supply precedents for all occasions. The resolve warned Virginians of "the hostile invasion" of Boston in "our sister colony" whose commerce was "to be stopped by an armed force" which "threatens destruction to our civil rights and the evils of civil war." Jefferson and his faction called for "one heart and one mind firmly to oppose, by all just and proper means, every injury to *American* rights."[38] The next morning, using flawless psychology, Jefferson and his co-conspirators paid a visit on the colonial treasurer, the widely respected "grave and religious" Robert Carter Nicholas, requesting that he introduce the resolution in the House. How could he decline? That same day, Nicholas introduced the resolution. Jefferson wrote in his autobiography that it "passed without opposition;"[39] Nicholas wrote there was one objector out of one hundred members. George Washington later wrote that there were far more radical protests already drawn up, especially one by Richard Henry Lee, but the House had expected another full month to complete its business before getting around to resolutions.

On May 26, the fast-day proclamation appeared in the *Virginia Gazette,* where the red-faced royal governor read it. At 3 P.M. that same day, Lord Dunmore angrily summoned the House of Burgesses to the Council chamber. He particularly objected to the phrase *sister colonies:* this smacked of forming an American government without the approval of Parliament, connecting colony to colony by cutting across the lines that connected colony and home government. This time he did not stall for time by proroguing the Assembly, he dissolved it. "I have in my hand a paper published by order of your House, conceived in such terms as reflect highly upon His Majesty and the Parliament of Great Britain, which makes it necessary for me to dissolve you, and you are dissolved accordingly."[40]

As the more conservative members stormed out in consternation to gather up their papers from the House chamber, Jefferson and his radical friends exulted, his glow still radiating forty years later when he penned his autobiography. "The lead in the House on these subjects being no longer left to the old members," Jefferson's radical young-turk faction, "agreeing that we must boldly take an unequivocal stand in the line with Massachusetts," had determined to act "under the conviction of arousing our people from the lethargy into which they had fallen." He added simply that "we cooked up a resolution,"[41] but the resolution was no simple concoction that allowed the crisis to simmer down. When Lord Dunmore dissolved the Assembly this time, he did so before a new fee bill or a militia bill could be passed. Virginia's courts now could indeed be closed down indefinitely and the colony left defenseless.

The next morning, May 27, 1774, a rump session of one hundred or more of the dissolved House of Burgesses, with former Speaker of the House Peyton Randolph acting as its moderator, crowded into the Apollo Room of the Raleigh Tavern. By reconvening without royal approbation at the tavern that day, the Virginia burgesses were taking the longest and boldest step yet toward independence. Taking the lead in the American protests, Virginia's self-confident aristocrats emphasized a new doctrine: an attack on one colony was considered an attack on all. Radical leaders proposed further protests against the Boston Port Bill, including a halt to all trade with Britain. Merchant James Parker of Norfolk wrote to his friend Charles Steuart in London: "There was some violent debate here about the Association. . . . George Mason, Patrick Henry, R.H. Lee, the Treasurer [Nicholas] were for paying no debts to Britain, no exportation or importation and no courts."[42] But other Assembly leaders, including Edmund Pendleton and House Speaker Peyton Randolph, opposed them. And some still sat on the fence. While George Washington objected to British taxation without American consent—"They have no right to put their hands in my pockets"—and called the Boston Port Acts "an invasion of our rights and privileges,"[43] that night he attended a ball given in honor of Lady Dunmore and that day dined with the governor twice: the previous morning they had gone for a ride together. Washington still needed to be in the governor's good graces to obtain bounty lands in the west. The final item of heated debate at the Raleigh Tavern on May 27 was over the courts and official business depending on fees. Pendleton argued that the Assembly had expected to renew the fee bill. Jefferson, in an argument he expanded and later published in his *Reports*, contended that, under English law going back many centuries, fees of officials could be established only by the legislature or by ancient custom. A fee was not a matter of custom unless its origins antedated written authority. If it were based on written authority, no matter how old, it did not meet this standard. The only logical outcome was a complete shutdown of the courts. Aggressive, brilliant, Jefferson's attack brought about the county-court closure he had strived for, a step that would involve a maximum number of Virginians in the debate over Britain's anti-American measures. The session ended after the extralegal burgesses voted to call a colony-wide convention on August 1, fully three months later, each county to send elected delegates. What Parker evidently did not know was that the final item of business at Raleigh Tavern, proposed by Richard Henry Lee, was a perfectly illegal annual meeting of the recently created Committee of Correspondence with delegates from other colonies. The committee decided to send letters to sister committees set up in other colonies calling for a continental congress that was not the sort of ad hoc gathering they had convened during the Stamp Act crisis, but a permanent American assembly. The word *annually* was a red flag to many burgesses more moderate than the Jefferson-Lee faction, and the meeting broke up in pandemonium.

The next morning, a worried Jefferson joined the Committee of Correspondence as it drew up an urgent request for the other colonies to hold an "extremely important"[44] Congress.

On the morning of May 30, after three-fourths of the burgesses had gone home, dispatches arrived in Williamsburg for the Committee of Correspondence from Boston, Philadelphia, and Annapolis. Committee Chairman Peyton Randolph called an emergency meeting of the twenty-five burgesses left in town. As the crisis worsened, Boston radicals called for a general boycott of colonial trade with Britain and the British West Indies until Parliament repealed the Port Act. Philadelphia patriots echoed the call for economic countermeasures. Maryland seconded Boston's proposals and proposed cancellation of all debts owed British merchants and, further, asked for a trade boycott of any colony that refused to join in the resistance movement. While many Virginians present called the cancellation of debts dishonorable, they agreed with Samuel Adams's appeal from Boston that it would be "dishonorable" to leave Boston "to struggle alone."[45] But Jefferson and his compatriots did not presume to represent the wishes of all Virginians. They voted unanimously to invite former burgesses to sound out their constituents and return to Williamsburg. In the circular letter probably drafted by Jefferson and definitely signed by him (his was the ninth signature), the committeemen concluded,

> Things seem to be hurrying to an alarming crisis and demand the speedy, united counsel of all those who have a regard for the common cause.[46]

If Governor Dunmore would have objected once again to the term *sister colonies,* he would certainly react hotly to the term, boldly written and signed, *common cause.* It smacked of conspiracy and treason. But Lord Dunmore had no police force to break up the crowd of Virginians who gathered that afternoon to cheer the new resolves. As worried Williamsburgers filed into Bruton Parish Church on June 1, Jefferson and other organizers of the spreading protest movement fanned out across Virginia to seek support for the convention they had called in Williamsburg for August 1. With his neighbor, John Walker, who had recently replaced his father in the Assembly, Jefferson hurried back to Charlottesville. As Jefferson recalled in his autobiography,

> We returned home and in our several counties invited the clergy to meet assemblies of the people on the 1st of June to perform the ceremonies of the day, and to address to them discourses suited to the occasion. The people met generally with anxiety and alarm in their countenances, and the effect of the day thro' the colony was like a shock of electricity, arousing every man and placing him erect and solidly on his center.[47]

But not all of the Church of England clergy were so eager to lead what were, though cleverly disguised, nonetheless protest meetings against the enforcement of laws passed by Parliament in response to crimes committed by radical Bostonians. The Rev. Thomas Gwatkin, principal of the grammar school at William and Mary, who had been "appointed" by the disbanded burgesses' resolution "to preach a sermon suitable to the occasion,"[48] asked to be excused, as Lord Dunmore reported to London. The Rev. Thomas Price, chaplain of the House of Burgesses, agreed to deliver the sermon. On the morning of June 1, the streets of Williamsburg were thronged and the governor and his wife stayed inside the Palace. According to the *Virginia Gazette,*

> Every inhabitant of this city and numbers from country, testified . . . in the most expressive manner by attending the worthy and patriotic Speaker [Peyton Randolph] at the courthouse and, proceeding from thence, with the utmost decency and decorum, to the church.[49]

Reverend Price led the packed church in reading the 103rd Psalm, intoning the somber promise of "righteousness and judgment for all who are oppressed," but warning, "neither will he keep his anger forever." Then, building on the words of Genesis 18:32, he preached the theme, "Oh, let not the Lord be angry."

What may have been Thomas Jefferson's last lawsuit as a practicing attorney involved, of course, land. Petitioning on behalf of George Mason, who was seeking warrants for lands in Fincastle County, the pleading filed with the General Court and Governor's Council was a closely reasoned argument on the cutting edge of land tenure law. Lord Dunmore, on behalf of the Crown, was insisting that all unoccupied lands in Virginia belonged to the king, who could sell them and then receive feudal quitrents in perpetuity. Jefferson argued that King James I had granted all land rights to the Virginia Company in 1609, giving up any royal claim. Further, King Charles II in 1676, when the Crown had absorbed the defunct Virginia Company, had ordered that the colony was to issue all settlers fifty acres of free land. For 150 years, the Crown had made no further attempt to claim vacant or deserted lands. The land rights of the original settlers had become "interwoven" with the colony's constitution. Now, Lord Dunmore, carrying instructions from the Crown, was threatening to auction off all empty or legally unprotected lands at a much higher than usual minimum price for the benefit of the Crown treasury and to collect annual quitrents in gold or silver. This new British policy greatly added to the burden of anxiety felt by Jefferson and other land-hungry planters, whose crops ruined the thin topsoil and whose future depended on cheap and abundant lands in the west. By order of the King's Privy Council of April 7, 1773, the Crown had stopped further grants of land by colonial governors. By order of February 3, 1774, the Privy Council set forth a plan to dispose of lands by auction. For a century and a half, the British government had "never in the least interfered" in Virginia land tenure. Now, wrote Jefferson, the king was "as much bound by the Act of his Royal Predecessors as any Private Subject." Virginians could not be bound, he argued, by "any instruction or late regulations respecting the ungranted lands in this colony."[50] The Mason case never came to court, but its filing demonstrates the concern of leading Virginian landowners in the Crown's latest "innovation."

But the spring and summer of 1774 were seasons of anger and anxiety all across Virginia, all up and down the Atlantic seaboard and the Allegheny frontier. British policy, bad weather and crops, bad government and restive Indians seemed to conspire in contributing to a generalized anxiety. As printing presses poured out a stream of pamphlets debating the issues of the great crisis, Indians on the southern frontier chased mountain settlers from their roughest farm clearings in what became known as Lord Dunmore's War. It seemed as if Lord Dunmore had planned the Indian war as a diversion from the tax revolt. Tensions had been building in the Shawnee hunting grounds called Kentucky, south of the Ohio River, since 1750, when Dr. Walker and a small party prospecting for land had discovered the Cumberland Gap, a natural pass through the Appalachian Mountains. In 1752, a Pennsylvania trader, captured as he floated down the Ohio River in a canoe, escaped and brought back reports of enormous herds of buffalo, wild turkey, and deer in the heart of Kentucky's bluegrass country. His glowing reports fueled westward expansion only interrupted by the French and Indians Wars. In the spring of 1769 as young lawyer Thomas Jefferson had plied his caveat law in the Virginia backcountry, backwoodsman Daniel Boone of North Carolina and a friend had secured financial backing and supplies from a land speculator, and headed through Cumberland Gap, for two years roaming Kentucky, living off the land like the Indians they imitated. Captured by Shawnees, they escaped by hiding in thirty-foot-high vegetation. Trespassing in the forbidden, British-protected forest lands again in 1771, trapping and hunting until they emerged with a fortune in furs, Boone and his companion were ambushed near Cumberland Gap. All their furs, horses, and supplies seized, the whites were sent home on foot, empty-handed and bitter after two years' work.

White faces in Kentucky profoundly disturbed the Shawnees, who knew that the British had closed the frontier at the crest of the Appalachians by their 1763 proclamation. In 1773, a company of land speculators armed with land bounties issued to Virginia soldiers during the French and Indian Wars claimed that they were not bound by the proclamation of 1763. The Shawnees, rejecting their importunities, decided to resist and began to organize. They were refused aid by the Iroquois Six Nations and the Cherokees, who were kept loyal to the English by skilled agents, but the Mingos and Ottawas joined the Shawnees.

As the Boston crisis loomed, Lord Dunmore made war inevitable by claiming Kentucky under Virginia's 1609 sea-to-sea charter. He sent Dr. John Connolly to Pittsburgh to enforce Virginia's title. En route, Dr. Connolly, an astute land hunter, aroused Virginia frontiersmen against the Indians. Shawnee chief Cornstalk called for peace and proposed a conference. Working through Pennsylvania Quakers who respected Indian rights, he sent peace messengers to Connolly at Pittsburgh. But Lord Dunmore had decided that his best policy was simply to drive the Indians out of Kentucky. In an initial attack by Connolly's adherents, thirteen Shawnees were killed: the Shawnees killed an equal number of whites in revenge. Learning of the Indian counterattack, Lord Dunmore, without summoning a new Assembly, declared war on the Indians, illegally calling up the colony's militia more than a month after the Militia Act had expired and riding west at the head of fifteen hundred frontiersmen as hundreds of Cornstalk's warriors crossed the Ohio to face them.

White and Indian atrocities ensued. Near the junction of Yellow Creek and the Ohio, at Logan's Camp, thirty-five miles west of Pittsburgh, Capt. Michael Cresap and his party killed one Indian and captured another on April 27, 1774. Three days later, Daniel Greathouse lured some Indians to an "entertainment" at Logan's Camp and murdered six of them. The half-breed chieftain, Logan, known as a great friend of the whites, lost a brother and a sister in the slaughter. He took thirteen scalps in retaliation as the Shawnees went to war. On June 10, Dunmore called out the militia of southwest Virginia and, riding west, personally established headquarters at Pittsburgh. Early in August, one of his commanders, Maj. Angus McDonald, raided Shawnee villages on the Muskingum River, one hundred miles west of Pittsburgh. Dunmore led a force of nearly two thousand Virginia militia down the Ohio and sent Col. Andrew Lewis with another thousand men down the Kanawha to rendezvous deep inside Indian country. The Shawnees, led by Cornstalk, mustered a thousand Shawnees, Miamis, Hurons, and Ottawas to attack Lewis before Dunmore could send reinforcements. In the largest battle with the Indians before the Revolution, the Virginia militia smashed the Indians in a major battle on October 10 near Point Pleasant, Ohio. Indian resistance collapsed.

The war Lord Dunmore precipitated against the Indians prompted Jefferson to turn his mind to the plight of America's Indians, a subject that would fascinate him and draw him to write about it again and again. Since the visits of Indians to Shadwell to meet with his father on their way to the provincial capital, Jefferson had been fascinated by their powerful oratory. As lieutenant of Albemarle County, he had to follow the frontier war closely, and he interviewed officers and other visitors coming from the west, collecting his own version of Lord Dunmore's War, which he would eventually publish in his *Notes on Virginia.* He reserved his greatest admiration for Chief Logan, and he would not be swayed from it even after Logan killed many Virginians. Jefferson's version of the cause of the frontier war was quite different from Lord Dunmore's. The war had been prompted, he wrote, when "a robbery was committed by some Indians on certain land-adventurers on the river Ohio." According to Jefferson, "the whites in that neighborhood undertook to punish this outrage in a summary way." Captain Cresap, who had been involved in land schemes on the Ohio for at least twenty years, proved to be the most ruthlessly opportunistic, deliberately provoking a war on the Indians on orders from Lord Dunmore. The actual fighting was carried out by one thousand Shenandoah Valley militia led by Col. Andrew Lewis, who had fled to Augusta County after murdering his landlord in Ireland. Lewis's militia marched 160 miles in nineteen days over the mountains, through forests, and across rivers and swamps to attack the Indians at Point Pleasant. (Seven of these riflemen became Revolutionary War generals, six others commanded regiments.) "Leading on these parties, [he] surprised, at different times, travelling and hunting parties of the Indians, having their women and children with them, and murdered many." Among the victims of these ambushes was "unfortunately the family of Logan, a chief celebrated in peace and war, and long distinguished as the friend of the whites." After this "unworthy" attack, wrote Jefferson, Logan had "signalized himself in the war which ensued."[51]

After Lord Dunmore's Virginia militias crushed the Shawnees, Mingos and Delawares on the Kanawha, the Indians had "sued for peace" but Logan "disdained to be seen among the suppliants," instead sending a speech by messenger to Lord Dunmore. Jefferson somehow secured the text and copied it whole into his *Notes on Virginia:*

> I appeal to any white man to say if he ever entered Logan's cabin hungry, and he gave him not meat, if he ever came cold and naked, and he clothed him not. During the course of the last long and bloody war, Logan remained idle in his cabin, an advocate for peace. Such was my love for the whites that my countrymen pointed as they passed and said, "Logan is the friend of white men!" I had even thought to have lived with you, but for the injuries of one man . . . Cresap, [who] the last spring, in cold blood, and unprovoked, murdered all the relations of Logan, not sparing even my women and children. This called on me for revenge. I have sought it. I have killed many. I have fully glutted my vengeance. For my country, I rejoice at the beams of peace. But do not harbor a thought that mine

> is the joy of fear. Logan never felt fear. He will not turn on his heel to save his life. Who is there to mourn for Logan? Not one.[52]

Thomas Jefferson, himself facing an unusually uncertain future, casting off the law practice that had so attracted and absorbed him and preparing to leave his wife and babies and the sanctuary of Monticello, identified thoroughly with Chief Logan. After years of listening to the flatulent pleadings of planter-lawyers and watching the indecisive behavior of Virginians in the colonial crisis, he found the conduct and the rhetoric of Logan, his "eloquence in council, bravery and address in war," inspiriting. "I may challenge the whole orations of Demosthenes and Cicero and of any more eminent orator if Europe has furnished [anyone] more eminent, to produce a single passage superior to the speech of Logan."[53] As if to fling aside the mantle of his classically steeped past, he turned to embrace the romantic rhetoric and politic courage of this quintessential American.

9

"Let Those Flatter Who Fear"

Kings are the servants, not the proprietors, of the people.

—Thomas Jefferson, "A Summary View of the Rights of British America," 1774

In July 1774, Virginia was shaken repeatedly by news of the British crackdown on Boston radicals. Virginians reacted strongly to the successive shocks of Lord Dunmore's dissolution of the House of Burgesses and his illegal call-up of the Virginia militia for his unauthorized war on the western Indians. At the same time the law setting court fees had expired, as had the Militia Act. In every county, committees of former burgesses formed ranks with other leading citizens protesting British measures to constitute themselves Committees of Safety and, in the absence of any other local authority, enforce the trade boycott with England and keep the courts closed. Each week's *Virginia Gazette* threw more logs on the fires of controversy that now began to make evident the divisions of public opinion over what response was appropriate. With the crisis clearly coming to a climax, old friends and families began to split apart. Thomas Jefferson's cousins, Peyton and John Randolph, were among those who parted in this time of trouble. They were descended from Jefferson's maternal great-grandfather William. The sons of Sir John Randolph, who had been king's attorney general of Virginia and Speaker of the House of Burgesses, the brothers had divided and rotated these high offices. As fifty-three-year-old

Peyton Randolph left his chair as Speaker of the House in May 1774, he shifted easily down Duke of Gloucester Street to Raleigh Tavern to preside as moderator of the rump assembly, then was chosen to preside over the August colony-wide convention. His younger brother, John, had studied law at the Middle Temple in London and had been admitted to the bar at Westminster. John was clerk of the House of Burgesses from 1752 to 1765, and his Williamsburg home was the cultural hub of the capital, a magnet for young, literate Virginians such as Jefferson.

John Randolph had succeeded his older brother as attorney general in 1765 at the time of the Stamp Act crisis. Jefferson and the new attorney general were close enough for Randolph to entrust Jefferson with his precious collection of manuscripts of his father's law cases for Jefferson to incorporate in his volume of law reports, and he lent Jefferson many books after the fire at Shadwell. Jefferson loved to go to the home of stately, witty, elegant Cousin Randolph to play his fiddle with this accomplished amateur musician, who owned a fine brilliant amber Cremona violin, probably made by Nicolò Amati in 1660, with ebony fingerboard and tailpiece and ivory string pegs. Three years earlier, Thomas Jefferson and John Randolph had entered into an odd agreement. At something of a tongue-in-cheek ceremony before all the lawyers of the General Court, they had agreed that, "in the case the said John shall survive the said Thomas," Randolph was to have the choice of one hundred pounds' worth of Jefferson's books. If Randolph died first, his executors were to deliver to Jefferson "the violin which the said John brought with him into Virginia together with all his music composed for the violin."[1] Jefferson had also acted as an informal mentor for Randolph's son, Edmund, who was ten years Jefferson's junior. When Jefferson decided to give up his law practice in the spring of 1774, he offered it first to John Randolph, who decided to turn it over to his son.

But by mid-July 1774, Attorney General John Randolph had become a leading spokesman of the royalist position and stood diametrically opposed not only to Peyton Randolph's moderate friends but even more so to Jefferson and other "hotheaded" young radicals. Randolph and Jefferson had only one point of agreement in their writings that summer: that there was "no law sufficient" to warrant Lord Dunmore's war against the Shawnees.[2] Early in July, a pamphlet titled *Considerations on the Present State of Virginia,* undoubtedly authored by Randolph, was printed by Clementina Rind's printing office. While increasingly considered a Tory who panted after royal favor and condescended to those less well born than himself, Randolph's opinions still carried considerable weight among conservative Virginians and were tolerated out of consideration for his family. Randolph wrote that, while America might "be capable, some time or other, to establish an independence," all that was needed for the present was a little tinkering here and there with the grand machine of the British imperial system to replace or repair a few worn-out parts and America could remain sheltered and coddled indefinitely by benevolent Britons. But, first of all,

> We ought, in the first place, to declare, in the most public manner, that the act of the Bostonians in destroying the property of the East India Company was illegal and ought not to be countenanced. This will evince our uprightness.[3]

Jefferson and his adherents agreed with only one thing John Randolph had to say: Lord Dunmore had no right to raise troops for any reason without the assent of the House of Burgesses to a new militia act. When and against whom might he again raise and lead an army?

Randolph was not alone in expressing the views of Americans loyal to king and Parliament. The growing split, deepened by Parliament's intransigent measures, raised serious questions for the established Church of England clergy, upholder of established English authority, including that of Parliament. The Rev. Jonathan Boucher, a distinguished Maryland Anglican and tutor to George Washington's stepson, Jackie Custis, followed his convictions when he published *A Letter from a Virginian to Members of Congress*, which powerfully defended parliamentary supremacy and insisted on colonial subordination. Boucher, who placed a pair of loaded pistols on the pulpit seat as he preached, argued that the American colonies were only a small part of a "British community."[4] The majority of the British were represented in Parliament and governed the empire through it. The colonies owed the obedience to that majority and thus to Parliament. In his sermons, soon published and widely read, Boucher contradicted Jefferson and the radical view "that the whole human race is born equal and that no man is naturally inferior." Men obviously were not equal in ability, capacity, or virtue. The Creator had ordained "some relative inferiority and superiority." Boucher, pistols cocked, boomed a closely reasoned divine right of kings as an antidote to Jefferson's egalitarianism:

> As soon as there were some to be governed, there were also some to govern. . . . The first man, by virtue of his paternal claim, was first invested with the power of government.

All subsequent governments were derived from Adam's authority. "Kings and princes" derived their authority "from God, the source and origin of all power," not from "any supposed consent or suffrage of men." While Boucher's logic was sound enough, it ignored ten years of political change, and his tone was insulting to many American protesters. He called them "knaves," "quacks in politics," and "imposters in patriotism" who were manipulating the "credulity of the well-meaning, deluded multitude" of "forward children who refuse to eat when they are hungry" and "vex their indulgent mother."[5]

The rising tide of resentment against loyal supporters of the British was already leading some Virginians to flee to England. Boucher would leave, as did eventually every Anglican clergyman who insisted on offering public prayers for the king and queen. As George Washington had prepared to travel north to the Continental Congress, he paused to sell off the home and elegant furnishings of his friends George and Sally Fairfax, of Belvoir. Where Washington had, as one recent biographer put it, "talked and laughed, danced and walked" with young Sally and had once "said so many goodbyes to her,"[6] now he conducted the eighteenth-century equivalent of a yard sale. Would-be buyers pored over furniture where young Virginia gentry had once been so merry, and Washington himself bought curtains, a sideboard, dining-room chairs, a chest of drawers, carpet, mirror, and Sally's bed coverlets, pillows, and bolsters, shipping them to Mount Vernon as he headed north to Congress.

In the summer of 1774, while Jefferson was entering a period of intense philosophical speculation, he was also extremely busy winding up his law practice and plunging into his new role as a spokesman for colonial resistance to oppressive British policies. He returned to Williamsburg for the mid-June session of the General Court. Even though he was preparing a devastating report that would deny the court its right to continue to conduct business, he personally attended the last session to clear up a record number of caveat cases. Of 159 caveats considered at the 1774 meeting, 69 were Jefferson's. He finally disposed of some forty cases, according to his *Fee Book.* He had planned to abandon most of his practice for two years now: the Wayles inheritance provided the wherewithal, but his *Fee Book* reveals that he stopped keeping complete records of the money owed him at the end of 1772. He did not even bother to tally the "profits" of his law practice in 1773 or 1774. As legal historian Frank Dewey argues convincingly, "Economic considerations played an important role in his withdrawal from a legal career."[7] It was only a few days after Jefferson had run an ad in the *Virginia Gazette* chastising "the unworthy part" of his clients who did not pay their bills, that he learned of the death of John Wayles, which left him, or so he long thought, a wealthy man. Unhappy with long absences from Monticello, he put his thirteen years of drudgework and caviling as law clerk and counselor behind him. In June 1774, at age thirty-one, Jefferson entered a courtroom as a lawyer for the last time, even though he could not know it at the time (he still hoped to hold on to his land caveating work after the courts reopened). He could not foresee, of course, that, as a result of the fast-moving colonial crisis, the General Court would never hear another civil case.

Jefferson left his law practice with mixed feelings. He continued criticizing and hoping to reform the law all his life, twice helping to establish chairs for its study in Virginia colleges. In 1810, he wrote an uncharacteristically cynical letter to dissuade a federal judge he had appointed from putting his son to the law. Instead, he urged the boy's second choice, medicine:

> Law is quite overdone. It is fallen to the ground, and a man must have great powers to raise himself in it to either honor or profit. The mob of the profession get as little money and less respect than they would by digging the earth. The physician is happy in the attachment of the families in which he practices. If, to the consciousness of having saved some lives, he can add that of having, at no time, from want of caution, destroyed the boon he was called to save, he will enjoy, in age, the happy reflection of not having lived in vain, while the lawyer has only to recollect how many, by his dexterity, have been cheated of their right and reduced to beggary.[8]

Yet from his years at the bar, he had learned much. He had gained great skill as a speaker and writer, gained a wide acquaintance with all sorts and classes of Virginians. His confrontations with the elite of the General Court and of Tidewater aristocracy had convinced him that he had a mind at least as good as any of theirs. He had now the self-assurance to promote the views that his years of reading, thought, and argument as a lawyer had helped him to hone. Already, views he had voiced in the General Court were echoing in the pamphlet prose of leading radicals. Another General Court practitioner, Thomson Mason, wrote a series of essays under the pseudonym, the "British American." In the seventh of the series, published in the *Virginia Gazette,* Mason held that British law could not bind Virginians beyond the first day of their settling in America in 1607, as Jefferson had argued in the *Lunan* case.

It was a short step for Jefferson from being one of Virginia's best-known lawyers to becoming one of its most effective politicians.

Trouble with the Indians on the frontier in the summer of 1774 helps to explain why it was not until July 23 that Thomas Jefferson and John Walker were able to summon the farmers of Albemarle County to days of fasting and prayer all over the hill country. In broadsides publicizing the meeting at St. Anne's Church in Charlottesville, Jefferson cited the peril to "sister colonies" and warned of "the dangers which threaten our civil rights and all the evils of civil war."[9] Jefferson timed Albemarle's call to a day of fasting, humiliation, and prayer for the weekend when the voters of Albemarle County would be crowding into Charlottesville for the county elections the next Tuesday. Lord Dunmore had signed warrants for a new Assembly election on July 8, but when the same burgesses had been elected, he prorogued it again, causing consternation. For the second time in two weeks, the freeholders went to the trouble of assembling from their remote farms. After jamming St. Anne's on Saturday, July 23, they lingered to drink punch and talk politics at local taverns, reconvening at the county courthouse on Tuesday, July 26. That summer, in addition to the causes of irritation within Virginia and along its frontiers, the Albemarle freeholders had received a steady series of shocks from outside their dominion that had only begun with the news of the Boston Port Acts. On May 20, Parliament passed the Administration of Justice Act, known locally as the "Murdering Act," which protected Crown officials from having to stand trial for capital offenses in Massachusetts by allowing the transfer of such cases elsewhere. On the same day Parliament had enacted the Massachusetts Government Act, which all but annulled the charter of the Bay Colony, bestowing on its royal governor and his appointive Council practically dictatorial powers. This act eviscerated popular local control by banning town meetings without prior written consent and permission to act upon a set agenda from the royal governor. The Massachusetts Act also moved the offices of the Board of American Customs Commissioners to Salem, out of Boston and away from the reach of Boston's crowds, officially diverting all trade away from the closed port city. A few weeks later, on June 2, Parliament revived the Quartering Act after a five-year lapse, further broadening royal authority to house Redcoats in occupied homes. At this very moment, the Virginia freeholders' suspicions that the British Ministry was following a definite plan to wipe out colonial liberties was confirmed by the *Virginia Gazette* for July 21, which carried a proclamation by Lt. Gen. Thomas Gage, the new military governor of Massachusetts and commander of all British land forces in North America. Gage had ordered the Boston Committee of Correspondence to disband and he threatened criminal action against anyone who sought to disrupt trade with Great Britain.

A torrent of political tracts by Jefferson, Robert Carter Nicholas, Arthur Lee, Thomson Mason, and John Randolph mingled that summer in the columns of the *Virginia Gazette* with resolutions from at least thirty-one of sixty-one counties. "The sense of the counties is taking,"[10] Richard Henry Lee wrote to Samuel Adams on June 23, 1774. There were so many resolutions in support of unified protest against the British that on August 4 the *Gazette* finally pleaded lack of space for the resolves of a dozen counties. Of the tract writers, four—Jefferson, Mason, Nicholas and John Randolph—were General Court lawyers writing anonymously. Randolph was the royal attorney general: he had a double reason for anonymity. Jefferson's uncle had become the spokesman for the Loyalist viewpoint in Virginia.

Jefferson was typical of all of the tractarians, a man educated at a time when men still could agree on what comprised an education. All of them —and, it was assumed, their readers—were versed in the writings of the political philosophers Harrington, Hume, Locke, Halifax, Montesquieu, Sidney, and Bolingbroke, as well as moral philosophers, such as Grotius, Francis Hutcheson, Pufendorf, and Vattel. Moreover, all could recite the views of historians from Thucydides and Tacitus to the Earl of Clarendon. All were familiar with the system of checks and balances, of laws and traditions that made up the English Constitution. "They have been so repeatedly wrung in our ears," bemoaned Nicholas, that "the veriest smatterer in politics must long since have had them all by rote."[11] Every literate freeholder was expected to understand what made up the constitutional "rights of Englishmen," what were the "natural laws"—what had been considered right and wrong since it had been established by "nature,"—and what were their "natural rights," their belief that, from a state of nature, man had entered into society by a contract and therefore possessed rights that neither he himself nor his posterity could lose or dispose of; that, as a result, property must be represented in any government empowered to tax that property. Sovereignty, they had come to believe, resided in the people, although it was not always clear who the "people" were. While the tract writers and freeholders had this same ideological base, only Jefferson among the tractarians affirmed in this summer-long debate over liberties brought up the question of slavery, maintaining that no person had a right to hold another person in bondage.

Freeholders had been meeting since June 8, when the citizens of Fredericksburg had led the way. A constant stream of resolutions, newspapers, and letters from other colonies was brought by messenger to Jefferson as a member of the Committee of Correspondence, at Monticello, where he had spent most of the summer reading and writing. That summer, as the temperature outside soared, so did Jefferson's. For five weeks, in one of his characteristic bursts of research and writing, he immersed himself in colonial rights and grievances and the British response to American protests. He brought his first effusion as a political philosopher down from his mountain to the Albemarle freeholders on July 26.

The Albemarle farmers sat, stunned, not only at Jefferson's resounding phrases, but at his angry tone. The people of British America, he asserted, were subject only to English laws they had adopted at their first settlement or to laws made by their own legislatures. "No other legislature whatever may rightfully exercise authority over them," Jefferson declared. In his first paragraph, he was boldly asserting a new doctrine, which went beyond all the other Virginia protesters. Jefferson was denying the authority of king, Parliament, and English law over America!

The resolution Jefferson proposed stated that the natural and legal rights of the American colonists had "in frequent instances been invaded by the Parliament of Great Britain." That foreign parliament had illegally deprived the people of Boston of their trade. "Such assumptions of unlawful power" were "dangerous to the rights of the British empire in general." He suggested that all liberty-loving Englishmen should make "common cause"[12] with the Americans. Jefferson proposed that his neighbors immediately boycott all imports from England. If Parliament did not reopen the port of Boston, repealing all customs duties on American imports and bans on American manufacturing, Americans should stop all exports to all parts of the British Empire on October 1, 1775. The resolutions finally called for Jefferson and John Walker to carry them to the provincial convention on August 1. The resolutions passed unanimously, thunderously, and Jefferson rode back to Monticello to put the finishing touches on a second version, a Declaration of Rights, which he had prepared for the impending Virginia Convention. It was an abbreviated, more concentrated statement of American rights, but it went even further by stating a willingness "to join with our fellow subjects in every part" of the British Empire in a movement to restore all "rightful powers which god has given us."[13] Further, Jefferson called for an immediate discontinuance of all commerce not only with England but with any other British colony that did not boycott British goods.

Beginning in 1770, shortly after British troops shot down Boston protestors in the Boston Massacre, Jefferson had begun studying systems of government, following Diderot's injunction in that Bible of the Enlightenment, the *Encyclopédie:* "Everything must be examined, everything must be shaken up, without exception and without circumspection."[14] He was not seeking a philosophical system to adopt whole. As Merrill Peterson has pointed out, Jefferson "was distrustful of philosophical systems generally," considering them "prisms of the mind."[15] He regarded thought as a tool for reshaping life, not for absorbing some grand design. His thinking was pragmatic, always as unfinished as his house at Monticello would be. But that was the whole point with both his thinking and his constructions, the doing of them. The delight was to finish neither, but to revise, constantly. He borrowed fully to assemble an eclectic set of principles which, he believed, provided the greatest flexibility, dynamism, durability. To prepare for the future, he reached back. He brushed aside whole systems. Years later, asked to be a godfather, he refused: "I had never sense enough to comprehend the articles of faith of the Church,"[16] he replied. Already a confirmed deist who believed in natural religion and morality, he regarded the clergy of the established Church of England as part of the problems of the British Empire, not as a solution. In concluding his brief in the *Lunan* case in 1774, he had written, "In truth, the alliance between church and state in England has ever made their judges accomplices in the frauds of the clergy, and even bolder than they are."[17] It was at about this time, this fecund summer of 1774, that, questioning the legal foundations of the established church, he penned a little essay in his commonplace book under the title, "Whether Christianity is a Part of the Common Law."[18]

For nearly fifteen years, Jefferson had followed the developments and writers of the Enlightenment, which had its roots in early eighteenth-century England. His three personal patron saints were Bacon, Newton, and Locke. While remaining a nominal Anglican and serving as a parish vestryman, Jefferson had drifted away from the Church of England as a student about the time he had begun to study "moral sense" Enlightenment philosophy under the tutelage of Dr. Small at William and Mary. As an old man, he wrote to John Adams in 1823:

> I can never join Calvin in addressing *his* God. He was indeed an atheist, which I can never be; or rather his religion was demonism. If ever man worshipped a false God, he did . . . not the God whom you and I acknowledge and adore, the Creator and benevolent Governor of the world, but a demon of malignant spirit.[19]

His commonplace books contain numerous excerpts from the religious thoughts of Locke and Shaftesbury and his disciple Francis Hutcheson. A third-generation Presbyterian minister, Hutcheson gave enormously popular lectures at the University of Edinburgh, included James Boswell and David Hume among his students, had rejected Calvinist orthodoxy, and was once tried by the Presbytery of Glasgow for teaching "false and dangerous"[20] doctrines. Hutcheson's "moral sense" philosophy asserted that moral goodness could be measured by the extent to which one's actions promoted the happiness of others. He also agreed that it was possible to experience a God-given knowledge of good and evil without resorting to the studying of God. Moral-sense philosophy weighed virtue in social terms: "That action is best which accomplishes the greatest happiness for the greatest numbers."[21] One of Hutcheson's disciples, Thomas Reid, held that "moral truths" could be divided into truths "self-evident to every man whose understanding and moral faculty are ripe" and truths that had to be "deduced by reasoning from those that are self-evident."[22] Another Scottish exponent of the moral-sense school was Henry Home, Lord Kames, whose thoughts Jefferson commonplaced copiously and who was listed under three headings in Jefferson's book-buying recommendation to Skipwith in 1771. Jefferson's study of Kames as early as 1767 led to his conviction that primogeniture in Virginia, the law requiring the leaving of all property to the firstborn son, had been unjustly transported from England and become early entrenched there. Jefferson had studied Kames's *Essays on the Principles of Morals and Natural Religion* during his student days, his boyish marginal notations surviving in one of the few books to escape the flames at Shadwell. From Kames, young Jefferson learned that "there is a principle of benevolence in man which prompts him to an equal pursuit of the happiness of all."[23] There were echoes of Kames in contemporary Scot Adam Smith's philosophical writings. "All constitutions of government," Smith wrote, "are valued only in proportion as they tend to promote the happiness of those who live under them."[24] And there were echoes of all these Scottish moral philosophies in Jefferson's political writings between 1774 and 1776.

If Jefferson had any religious credo, it was a utilitarian faith in progress. With Bacon, he believed that mysteries beyond human understanding should be set aside so that the mind was freed to attack real obstacles to happiness in life. Like the philosopher Baron de Holbach, who wrote that "man is unhappy only because he does not know nature,"[25] he believed that enlightenment provided a route to happiness. If man studied nature, he could bring himself into harmony with the natural order of his environment and use its laws to set himself free. He saw this as the pursuit of happiness that was his right as well as his deepest desire. Because there were individual definitions of happiness, societies needed the freedom that would allow pluralism and tolerance. Jefferson believed that limitless progress was possible, that man had all the "necessities" for progress, if not perfection:

> Although I do not, with some enthusiasts, believe that the human condition will ever advance to such a state of perfection as that there shall no longer be pain or vice in the world, yet I believe it susceptible to much improvement and, most of all, in matters of government and religion, and that the diffusion of knowledge among people is to be the instrument by which it is effected.[26]

It is not from the Scottish religious reformers but from English and European writers of the seventeenth- and eighteenth-century Age of Reason that Jefferson drew his evolving notions of government. From Bacon, the grandfather of the English Enlightenment, Jefferson had learned to use his powers of observation and question any opinion, regardless of its source. He adhered to Bacon's admonition to apply reason and learning to the functions of government to improve society. Jefferson was influenced by Newton's *Principia,* which held that the universe was a great clock invented, made, and set in motion by a deity, but he had adapted Newton's view to his own quest for a world of order and harmony. Like Newton, Jefferson did not believe in miracles. Jefferson's third hero from the time of his boyhood studies was Locke, who had joined the empiricism of Bacon and Newton to the realm of politics. Locke's *An Essay Concerning Human Understanding* for the first time fed his natural optimism and gave him hope that mankind could be improved by education. From Locke and his Scottish adherents, Jefferson had adopted the theory of the *Second Treatise of Government* that legitimate authority to govern was derived from the consent of the governed, which had first been granted while mankind had still been in a "state of nature" when all human beings were by right free and equal. Locke underpinned all of Jefferson's political thought.

As early as 1765, shortly after the jolt of the Stamp Act crisis, Jefferson had begun to study various forms of government. Encouraged by George Wythe, he had dug back through books of law and history to the times before the Norman conquerors had imposed a yoke on his independent Anglo-Saxon forebears. He admired Alfred the Great, the greatest king of the Anglo-Saxons. Ironically, both Thomas Jefferson and the young prince who became King George III (they were born within a year of each other) were studying Alfred as their youthful ideal at the same time. Jefferson, who first came to admire the Germanic tribes from reading Tacitus, honored the Anglo-Saxons for instituting trial by jury and the basic institutions of English representative government, and he considered the Norman French to be corrupters of the purer Anglo-Saxon culture. Jefferson had also studied the rules of descent of English Common Law, which led him to the conviction that the early peoples of Europe and Asia Minor had been free and independent, and had elected their own executive officers.

Jefferson's novitiate as a burgess in the spring of 1769 brought about a fundamental shift in his interests. He turned abruptly from his law practice to study history and government, extensively commonplacing his readings. In the summer of 1769, immediately after the closing of the Assembly, he had ordered from London a list of fourteen books, every one of them dealing with theories of government. By December they arrived at Williamsburg. The bill for the books detailed a "very elegant" copy in "gilt marble of the *Petits Jus Parliamentum,* Gordon's two-volume *History of Parliaments,* and the "very scarce" *Modus Tenendi Parliamentum* and *Determinations of the House of Commons.* But the books that were to be at the core of Jefferson's studies of government were John Locke's *Two Treatises on Government,* Burlamaquis's *Natural Law,* Ellis's *Tracts on Liberty,* Ferdinand Warner's *History of Ireland* and *History of Civil Wars,* Petty's *Survey of Ireland,* Ferguson's *Civil Society,* Stewart's *Political Economy,* and Montesquieu's *Spirit of Laws,* a comprehensive collection of political philosophy. After 693 entries on the law, Jefferson turned his attention to the histories of the early peoples of Europe. Flipping through Ferdinand Warner's *Ireland,* he noted that "the ancient Irish" gave "the inheritance to the strongest." From Locke, the Scottish pediatrician-turned-philosopher, Jefferson abstracted a single paragraph, including words that were to become central to his philosophy of empire:

> . . . A king, *elected by the people,* is one of the branches to whom the people have deputed the power of making laws. . . . They have never bound themselves to submit to any laws but such as have received the approbation of the Commons, the Lords and the king so elected.[27]

From Simon Pelloutier's *History of the Celts,* published in France in 1751, and from Abraham Stanyan's *Grecian History,* published in London in 1739, he had absorbed a view that the ancient states had treated their colonists liberally, their power based on "the free consent of the people."[28] Into his commonplace book, Jefferson poured his notes on the structure of Dutch and Swiss confederate republics and he wrote detailed, precise notes on the exact origins of the Angles, Saxons, and Jutes who had settled the British Isles. If Jefferson loathed the Norman conquerors of England, he admired greatly the more recent French political philosophers such as his favorite, Montesquieu. In all, he copied out twenty-eight pages of excerpts from *Spirit of Laws.* No doubt he imbibed from Montesquieu, as did others of the Founding Fathers, his firm belief in the theory of separation of powers into legislative, executive, and judicial branches of government as a bulwark of liberties. Between 1774 and 1776, Jefferson made a systematic study of Montesquieu, whose encyclopedic *Spirit of Laws* more immediately influenced his writing during this important period of his life than had his more-remote readings of Locke and the Scottish philosophers. One recent historian called *Spirit of Laws* "the most comprehensive treatise on politics produced during the Eighteenth Century."[29] Montesquieu scientifically and tersely laid down the structure, underlying principles, necessary laws, and social policies of four distinct forms of government: the democratic republic, the aristocratic republic, monarchy, and despotism. He described the English government as "a republic disguised as a monarchy."[30] His treatise was attacked by Voltaire and other contemporaries for its factual errors and oversights. An older, more experienced Jefferson eventually would reject Montesquieu as too fond of the English and erroneous for insisting that a republic could only exist in a small country. But in 1774, Jefferson recognized that Montesquieu offered him a vast amount of political information available nowhere else, especially his careful discussion of the democratic republic, based on a mosaic, its pieces easily recognized by Jefferson as classical sources on political practices and customs from ancient Athens, Carthage, and Rome. Most important, since 1770, Jefferson had been dipping into Montesquieu to study the nature of republican confederacies. As early as August 1771, Jefferson listed Montesquieu twice on the list of books recommended to Skipwith, one entry for his "Spirit of laws" the other under the heading "rise and fall of the Roman government."[31]

As a young law clerk, Jefferson had been copying into his legal commonplace book notes from the works of William Robertson and Francis Sullivan when he had stumbled onto Montesquieu. Like so many of his generation, the same generation that made the catchy aphorisms of Benjamin Franklin's *Poor Richard* the bestselling American writings, Jefferson was captivated by Montesquieu's ironic prose, even on the weightiest of subjects. "On the slavery of Negroes," Montesquieu wrote such comments as

> It is impossible for us to assume that these people are men because if we assumed they were men one would begin to believe that we ourselves are not Christians. . . . Sugar would be too expensive if the plant producing it were not cultivated by slaves. Those concerned are black from head to toe, and they have such flat noses that it is almost impossible to feel sorry for them.[32]

Montesquieu's mocking, memorable axioms had a way of sticking and echoing. "On every despotic government, it is very easy to sell oneself; there, political slavery more or less annihilates civil liberty."[33]

Once he began to read Montesquieu, Jefferson went on to abstract thousands of words, what he considered the heart of his work. He read the French and copied it out in French, then wrote his own observations in English, abridging Montesquieu, making him clearer, always with an eye to the practical. He gathered evidence, as if for a law court, to support his theory of elective kingship, citing examples in Greece, early England, Denmark, and Sweden. His republicanism emerged from the selections he wrote down. Ignoring Montesquieu's sections on monarchy, aristocracy, and despotism, he wrote down the ways to construct a democratic republic. He copied down *verbatim* Montesquieu's admonition to keep track of those possessing voting rights so that, on any given vote, it can be known whether all citizens had voted or what portion. Failure to keep accurate voting records, he quoted Montesquieu, had been "one of the principal causes of [Rome's] ruin."[34] Jefferson studied carefully Montesquieu's answers to the question of how great a role the people collectively should play in a democratic republic, in a pure democracy when the people themselves met in an assembly to pass laws. But how much power should the people have, aside from attending meetings of a popular assembly where laws are passed? Jefferson copied down verbatim this paragraph:

> A people having sovereign power should do for itself all it can do well, and what it cannot do well, it must do through its ministers. The people . . . need to be guided by a council or a senate. But in order for the people to trust it, they must elect its members. . . . The people are admirable for choosing those to whom they should entrust some part of their authority.[35]

Here, in Montesquieu, brushing aside kings and lords, Jefferson found the ideas that were to form the embryo of an American democratic republic with its Congress and Senate. Through Montesquieu, from ancient Greece and Rome, Jefferson was studying and writing down how to think about, then how to construct, a new form of government. He passed over, without commonplacing a word, Montesquieu's caveats about the political capacity of the people and his warning that their influence should extend no further than passing laws in the assembly and choosing their natural superiors as their magistrates. Such an idea was abhorrent to Jefferson's growing faith in democracy.

The promising spring of 1774 turned into a season of crisis for Jefferson. Frail Patty Jefferson's recovery from childbirth was slow. Her pregnancies were emotional crises as well as physical ordeals: her mother had died in childbirth. Jefferson stayed close to her, nursing her, playing with his toddling eighteen-month-old, Patsy, and Dabney Carr's little boy, Peter. As Patty recovered, they worked together in the garden, in the evenings reading poetry and fiction to each other in the newly finished parlor of Monticello. But the days were punctuated by the pelting of hoofs up the roundabout road as messengers brought the news of a dozen nearly weekly, ever-bolder resolutions and days of prayer in other counties strung together by the drumbeat of ominous news from the north. As the day for the Virginia Convention approached, Jefferson pored over his books, by July 26 turning from writing the resolutions of the Albemarle County freeholders to drafting a declaration of rights for the Virginia Convention. He wrote quickly and surely. He did not stop to check every detail. He did not intend publication. He completed a draft to present in Williamsburg. He left blanks for dates. He knew full well what a fateful step he was taking: the last time an English lawyer had prepared such a document was in 1628, when Puritans petitioned King Charles I to agree to their rights in writing on the eve of a long civil war. Jefferson's declaration proclaimed his boldest assertion based on his years at the Virginia bar. He dared to speak for all Americans when he wrote that Parliament had no authority whatsoever over Americans, who were only "subject to the laws which they had adopted at their first settlement" and any laws accepted by their own legislatures. For the first time, however, he did not petition the king of England as his subject or even as an Englishman. Americans held their privileges "as the common rights of mankind." Their rights had been confirmed by their own political institutions and by charters of rights confirmed by earlier kings. Boldly Jefferson wrote on. The "natural and legal rights" of Americans "have in frequent instances been invaded by the parliament of Great Britain." He did not capitalize the word *parliament* and he spoke of Great Britain as a foreign country. "The closing of Boston was an assumption of unlawful power." He called for Americans to make resistance to the British "its common cause" and exert their "rightful powers which god has given us" to reestablish their constitutional rights "when, where and by whomsoever invaded." As a first step, he called for a boycott of English goods; as a second, he called for a general congress of "the several American states." He no longer considered them colonies. He called them states. Once Parliament rescinded its illegal invasion of Boston, the American Congress would consider granting England its trading privileges with the Americans. His nerves in tatters, his usually superb digestion soured by a spastic colon, which always beset him at times of great stress, he was forced to submit to frequent debilitating interruptions. Jefferson had been shocked when the British Parliament in the Act for the Suppression of Riots and Tumults in the Town of Boston, had ordered that Americans be transported to England for trial, and he was well aware what the penalty could be for the documents he was writing. The usual punishment for treason was hanging until the prisoner was about to lose consciousness, at which time he was lowered, disemboweled, hanged again until dead, then beheaded and quartered.

But only a few days remained before he had to leave for Williamsburg. Probably the same day, July 26, Jefferson plunged on with his writing, this time penning a position paper on the crisis under the working title, "Instructions to the Virginia Delegates in the Continental Congress." He had no doubt that a majority of Virginians would send delegates from their county conventions to Williamsburg with instructions to approve the intercolonial congress of protest. For years, Jefferson had been arguing with George Wythe, and anyone else who would listen in or out of court, about the origins of American customs, statutes, institutions, as he formulated his own doctrine about the political relation between us and England. He believed that the relationship between Great Britain and her colonies was exactly the same as that of England and Scotland before the Act of Union of 1707 and the same as her present relation with the House of Hanover in Germany: they had the same executive, but no other political connection. Only Wythe, the dean of Virginia lawyers, had agreed with him. His passionate research had led them back nearly a thousand years in English history and beyond to pre-Norman times. He had come to believe that feudalism in England had been achieved by Norman imposition of a yoke of self-serving fictions on the ruins of Anglo-Saxon natural rights. Only by going back to the early eleventh century before the Norman conquest of England in 1066 could Americans find the proper relationship between a king and former subjects living beyond the bounds of his legal jurisdiction but willing to acknowledge him as their executive head. Stripped of his ministers, Parliament, and royal lockup, this kind of king would be only the first among equal public servants who should be judged on his record, which, in the case of recent English kings, had been, Jefferson adjudged, wretched. Other American writers had spared the king from their excoriations, placing all blame on his ministers and Parliament. Jefferson pulled no punches. "His majesty has made the civil subordinate to the military," he wrote. The king himself intended to enforce "arbitrary measures" that British monarchs had begun in England after the Norman Conquest and extended to the colonies, including, worst of all, claiming all the lands in England *and* America and introducing all their feudal tenures and taxes from France. But "America was not conquered by William the Norman, nor its lands surrendered to him or any of his successors."

Jefferson began his first major political statement, rooted in law and history, by recalling that the Saxon ancestors of the English had emigrated from their native lands in Germany and no authority over them had ever been imposed by the country from which they departed. Why, then, should the British government exercise any authority over American colonists? Although Great Britain had from time to time assisted and protected the American settlements, Parliament was not thereby entitled to take unto themselves any legal supremacy over the colonies and had "no right to land a single armed man on our shores." As soon as the American settlements had been established, the people had adopted English Common Law "to promote public happiness" and had continued their union with the mother country by submitting to the "same common sovereign who was thereby made the central link connecting the several parts of the empire."

Jefferson next protested parliamentary restraints on colonists' natural right to trade with any place in the world. He pointed out that, more than a century ago, under the so-called Navigation Acts passed by Parliament in the reign of Charles II, Virginians had learned "what hopes they might form from the justice of a British Parliament, were its uncontrolled power admitted over these states." Here, again, he called the American colonies "states," implying sovereignty and independence from England. Jefferson warned that there had been in more recent times repeated instances of royal tyranny under George II and George III. He lashed out at the "family of princes" whose "treasonable crimes against their people" had brought on the "sacred and sovereign rights of punishment": Charles I's beheading. Usurpation of power had not been confined to external trade and commerce but was systematically being extended to steadily encroaching British regulation of America's internal affairs: the Sugar Act of 1764, Stamp Act of 1765, Declaratory Act, Townshend Acts, the act suspending the New York Assembly from 1767 to 1769, and the Boston Port Act of 1774 all usurped the rights of colonists to regulate their own internal affairs. He took dead aim at English land tenure laws, of which he was an acknowledged expert, calling attention to "aggressions and encroachments now of King and now of Parliament in granting lands." "The true ground on which we declare these acts void is that the British Parliament has no right to exercise authority over us." Recent usurpations by Parliament were "acts of power, assumed by a body of men, foreign to our constitutions and unacknowledged by our laws." Under natural law, he asserted, the "British Parliament had no right to exercise its authority over us," because "one free and independent legislature" could not "take upon itself to suspend the powers of another, free and independent itself." Jefferson's words time after time rang with the defiance that would resound two years later in his Declaration of Independence:

> Single acts of tyranny may be ascribed to the accidental opinion of a day; but a series of oppressions, begun at a distinguished period, and pursued unalterably through every change of ministers too plainly prove a deliberate, systematic plan of reducing us to slavery.

Jefferson had thus turned angrily to the king: "He is no more than the chief executive of the people, appointed by the laws and circumscribed with definite powers to assist in working the great machine of government." It was now King George III's duty to resume a long-discontinued practice and veto the "passage of laws by one legislature of the empire which might bear injudiciously on the rights and interests" of another part of the same empire. But he warned the king not to abuse his veto powers over colonial legislatures, "more particularly, in the matter of dissolutions," as Lord Dunmore had done only three months earlier. "For the most trifling reasons and sometimes for no conceivable reasons at all, His Majesty has rejected laws of the most salutary tendency." When royal governors exercised the king's veto by dissolving legislatures, the sovereignty reverted to the people as a matter of right. "From the nature of things," he exclaimed in his best natural-rights philosophy terms, "every society must at all times possess within itself the sovereign power of legislation." The king's authority over America was limited. He was bound by laws. He was a party to a social contract and he governed according to limits and regulations established by that contract. The empire was made up of virtually independent parts connected loosely according to rules agreed to by its members. And while he knew he had the floor in the colonial debate, Jefferson unleashed another lonely, highly risky attack on the institution that was coming to beset all his days if not, too, his nights. These phrases could undercut his whole appeal in the ideas of his friends and neighbors:

> The abolition of domestic slavery is the great object of desire in these colonies, where it was, unhappily, introduced in their infant state. But, previous to the enfranchisement of the slaves we have, it is necessary to exclude all further importations from Africa. Yet our repeated attempts to effect this by prohibitions and by imposing duties which might amount to a prohibition, have been hitherto defeated by the King's negative. Thus, [the king prefers] the immediate advantages of a few British corsairs to the lasting interests of the American States, and to the rights of human nature, deeply wounded by this infamous practice. . . . This is so shameful an abuse of a power trusted with his Majesty. . . .

Jefferson concluded by depicting the American colonists as a free people who were now claiming their rights. It would be unwise for the king and his ministers to persist in allowing one part of the empire to sacrifice the rights of another part. The colonies did not wish to separate themselves. But let no power on earth except their own elected legislatures attempt to tax or regulate their properties! His tone was honest, angry, unparalleled in the decade-long American colonial struggle for its vehemence. And, as if he had not already stretched out his neck for the English hangman's noose, Jefferson read a concluding lecture to the king:

> No longer persevere in sacrificing the rights of one part of the empire to the inordinate desires of another, but deal out to all equal and impartial right. . . . Let no act be passed by any one legislature which may infringe on the rights and liberties of another. . . . These are our grievances, which we have thus laid before his Majesty, with that freedom of language and sentiment which becomes a free people, claiming their rights as derived from the laws of nature, and not as the gift of their chief magistrate. Let those flatter, who fear: it is not an American art. . . . Kings are the servants, not the proprietors of the people. Open your breast, Sire, to liberal and expanded thought. Let not the name of George the Third be a blot on the page of history. . . . It behooves you to think and act for yourself and your people. . . . The whole art of government consists in the

> art of being honest. . . . It is neither our wish nor our interest to separate from her. We are willing, on our part, to sacrifice everything which reason can ask, to [restore] tranquility. . . . On their part, let them be ready to establish union on a generous plan. . . . The God who gave us life, gave us liberty . . . the hand of force may destroy, but cannot disjoin them. This, Sire, is our last, our determined resolution.

At the heart of Jefferson's *Summary View,* hastily written in ill health after five years of reading, study, and political experience, was an uncompromising statement—not brokered and edited like his later, more famous documents—of what he considered a desirable framework for a constitutional union. Devoid of elaborate arguments and proofs of the nature of that union, it emphasized a workable system of government. He did not try to locate sovereignty in any one American region, New England or Virginia: he spoke for all British America. His demands were no doubt too radical for the rulers of the existing empire. He called for a fundamental change to a union that rested on a foundation of the rights of the individual to give full expression to life in the New World. Only in his first inaugural address more than a quarter century later did he paint his political principles with such bold, broad strokes. God and "the laws of nature" gave certain rights "equally and independently to all." Sovereignty belonged to the whole people, the "state." There could be no compromise with absolutism in any form, whether it was usurped by minority or majority: "Bodies of men as well as individuals are susceptible of the spirit of tyranny." The people and their rulers had to heed a higher law: "Force cannot give right."

The British Empire, made up of "the several states" that took direction from the king as "chief magistrate," was a dominion whose power was to be distributed among the agencies of government. The king was to be most carefully "circumscribed with definite powers." Subject to the "superintendence" of the people, the king was only to assist in working "the great machine of government." The king's duty was to receive protests and exert "the only mediatory power" between states. The real power was legislative and remained "at all times" within each community. When the king or his appointees dissolved legislatures, "the power reverts to the people." If the king violated this social contract, the people could exert "sacred and sovereign rights of punishment, reserved in the hands of the people for cases of extreme necessity."

Even the most radical opposition Whigs in England, such as Edmund

Burke and Charles James Fox, believed that the imperial bond stretched between inferior colonies and the superior British realm, with Parliament holding supreme power. In America, all but a few radicals shared this view in 1765 at the time of the Stamp Act crisis. By the time civil war began in 1774, many conservative spokesmen, such as Thomas Hutchinson of Massachusetts, William Franklin of New Jersey, and Joseph Galloway of Pennsylvania, clung to this view of the empire, which had once been revolutionary in the "Glorious" Revolution of 1688, when Parliament superseded the Crown in sovereignty over overseas possessions. Ironically, Jefferson's view was, in English eyes, high Tory, preferring king to a representative Parliament.

By 1774, even the most conservative Americans were thinking of colonial reforms that would have granted America greater home rule. Galloway's Anglo-American legislature would have possessed all the powers "necessary for regulating all the general police and affairs of the colonies." In England, Dissenter spokesman and economist Dr. Richard Price, the close friend of Benjamin Franklin, urged Parliament to confine itself to regulating external trade. But even radical American reformers who recommended creating an imperial legislature with American representation did not deny Parliament's sovereignty. Few Americans questioned that the colonists were subjects of the king. Only the most extremist, including Jefferson, argued that the settlers had quit their allegiance on emigrating to America. By 1770, Benjamin Franklin also subscribed to this view: "The right of migration is common to all men, a natural right. The colonists used that right and seated themselves out of the jurisdiction of Parliament." Americans, Franklin had argued in private, were within the king's dominions because they *voluntarily agreed* to be his subjects when they took his charters. Jefferson had studied Virginia's charter carefully and found that the colonial charters had lapsed with the beheading of Charles I and, when a new charter was granted in the Restoration, it was, he noted, "a voluntary act not forced by a conquest."[36] In late 1774, no one, not even Jefferson, was ready to sever allegiance to the king or to separate the American colonies from the British Empire.

Making a hasty copy, taking no time to correct a few minor points he could have checked easily in his library if he'd had enough time, Jefferson started toward Williamsburg. But he had to stop along the road, too sick, drained, and dehydrated from dysentery to go on in the scorching July heat. Sending off one copy by messenger to Peyton Randolph, moderator of the convention, he sent another copy to Patrick Henry, hoping he would introduce the resolution. To John Walker, his fellow delegate from Albemarle, Jefferson sent his proxy to sign any documents, including his own anonymous "instructions," approved by a majority of delegates. Then he returned to Charlottesville, surely aware that his sudden illness had deprived him of a place in Virginia's delegation to the very Continental Congress he had just proposed.

Patrick Henry apparently never read Jefferson's brilliant summation of American grievances. "Whether Mr. Henry disapproved the ground taken," wrote Jefferson in his autobiography, "or was too lazy to read it (for he was the laziest man in reading I ever knew) I never learned, but he communicated it to nobody."[37] Jefferson penned these words sixteen years after Henry was in his grave. The copy Jefferson dispatched to Peyton Randolph had a better fate. The moderator, presiding at his own house on Duke of Gloucester Street, "informed the convention he had received" Jefferson's paper "and he laid it on the table for perusal."[38] Randolph's nephew, Edmund, who had recently taken over Jefferson's law practice, recollected years later hearing the resolutions read aloud and "distinctly" remembered "the applause bestowed on most of them."[39] But not every delegate had reason to applaud, and the resolutions were not adopted. Old and young applauded Jefferson's heretical prose but voted a tamer resolution that took Virginia back to the 1767 arguments over "internal" versus external taxes, which Jefferson had shredded. The Virginians were not ready to declare that Parliament had no authority over them, "the leap I proposed being too long as yet for the mass of our citizens," as Jefferson wrote later. Some, such as constitutional lawyer Richard Bland, recoiled in alarm at the logical prospect of an armed revolt implied in Jefferson's ultimatum. The prudent majority favored keeping "front and rear together."[40] Set aside, Jefferson's proposed resolutions were not sent off to be printed with the official convention resolutions. But some of Jefferson's admirers gave his work the title, *A Summary View of the Rights of British America* and put two hundred copies through the press of the *Virginia Gazette.* Before the end of 1774, *A Summary View* was reprinted in Philadelphia, where the First Continental Congress, Peyton Randolph presiding, began to meet on September 4, and was twice reprinted in London. But Jefferson had gone much beyond the position of the radical Whig opposition to the Ministry, including that of Edmund Burke, who still called for reconciliation. The name missing from the title page of the *Summary View* had become internationally known as that of the most radical writer in America. Thirty-five years later, in 1809, Jefferson looked back to assess the *Summary View.* He said it "was not written for publication. It was a draft I had prepared for a petition to the king. If it had any merit, it was that of first taking our true ground, and that which was afterwards assumed and maintained."[41]

Too ill to travel to Williamsburg, passed over for the Virginia delegation to Philadelphia, Jefferson sharply criticized the pallid protest that resulted from the Virginia Convention of 1774. At Monticello, as he read over a copy of the instructions given to Virginia's delegates, he noted the defects he found. Once again, merchants had been appeased. The newest Association agreement provided that the sanctions would not take effect until November 1, fully three months later. In the meantime, "we are permitted to buy any goods imported." Once again, at this time of badly needed sanctions against the British, imports from England would actually *rise* as merchants crammed their stores and warehouses in anticipation of a long boycott. At the same time, Americans would be barred from building up their manufacturing, so vital to end dependence on British manufacturers. "We are not allowed to import the implements of manufacturing." Equally insane, he found, was a ban on importing books. Even more annoying to him was the convention's decision to allow imports of wine, coffee, and other luxury goods, even though they carried the obnoxious Townshend duties.

That a watered-down trade boycott had replaced his strongly worded instructions to delegates left Jefferson still sputtering sarcastically many years later when he recalled in his autobiography these "instructions very temperately and properly expressed, both as to style and matter."[42] Not willing to give offense to king and Parliament, Virginia's grandees fumed at the vague document drawn up at Peyton Randolph's. "The American grievances are not defined," he wrote. "We are allowed to continue commerce with the other parts of the British empire, though they should refuse to join us." And Virginians would be able to pick and choose the protests they preferred to avoid being susceptible to the more extreme thinking of such radicals as Jefferson and the Massachusetts Adamses. This clause "totally destroys" the unity of the American colonies, "which was the very purpose of calling a Congress." At the end of the printed Association of 1774, Jefferson scratched angrily with his pen: "Upon the whole, we may truly say we have left undone those things which we ought to have done. And we have done those things which we ought not to have done."[43]

Jefferson, after making his chief prerevolutionary contribution to the patriotic cause in the summer of 1774, was now in the vanguard of the revolutionary movement. Recovering from severe dysentery at Monticello, he waited tensely for reactions—from the Continental Congress convened in Philadelphia, from a newly elected and unpredictable Parliament in London. Many Americans shared Jefferson's unease. "The first act of violence," wrote John Dickinson, Pennsylvania delegate to the Congress, "will put the whole continent in arms from Nova Scotia to Georgia."[44] Poised at the abyss of revolution and civil war, Congress backed away, despite the fact that more disquieting news arrived from London like the ominous distant rumbling of a summer thunderstorm. Neither Parliament nor the king's privy councilors directly answer the numerous humble petitions, genuflections and threats of trade boycotts piling up at the office of Lord Dartmouth, Secretary of State for the Northern Department, but an indirect answer came in the form of the Quebec Act.

As frustrated as Jefferson was by the conservatism of his fellow burgesses in August 1774, had he been well enough to be elected to represent Virginia at the First Continental Congress in September, he might have boiled over at its outcome. When fifty-six delegates from twelve colonies—Georgia feared an Indian war without British arms for defense—gathered in Philadelphia on September 5, Peyton Randolph was elected president and radical Charles Thomson, head of Philadelphia's Sons of Liberty, its secretary. Each delegation had one vote. At first, radicals seemed in control of the Congress. The choice of the site of the Congress at Carpenter's Hall had been a symbolic victory, since the Pennsylvania State House was still in the control of conservatives loyal to the Penn family and the British Crown. "To my surprise," wrote Penn party leader Joseph Galloway, the move from the statehouse, expected location of such a major gathering, to the small, carpenters' union hall, "was privately settled by an interest made out-of-doors"[45]—the radicals of Philadelphia, Massachusetts, and Virginia. Massachusetts radicals had adopted the Suffolk County Resolves a week before Congress convened. The resolves, rushed to Philadelphia by post rider Paul Revere, declared the Boston Port Acts unconstitutional, urged Massachusetts to form a revolutionary government, raise militia, withhold taxes from the Crown until the coercive Boston acts were repealed, and meanwhile enforce economic sanctions against Britain. In a radical coup, the Continental Congress endorsed the Suffolk Resolves.

Second only to the closing of Boston in Congress's list of crises was the passage of the Quebec Act, a radically conservative set of British laws to govern territories conquered from the French. A path-breaking piece of legislation, the act replaced nearly fifteen years of martial law and seven years of secret meetings and bureaucratic in-fighting to emerge at exactly the worst possible moment. The king's hard-line advisers, influenced strongly by Guy Carleton, royal governor of Canada, decided that, since they were rewriting American policy anyway, they would issue a startlingly new charter for a vastly enlarged Province of Quebec, which was to stretch all the way south to the Ohio River, north to James Bay, and east to the Bay of Fundy. Intended as a model of colonial reform, the Quebec Act created a highly centralized government ruled by a royal governor with military powers and his councilors, appointed by the king and serving at his pleasure, with no elected lower house representing the people and holding the power of the purse strings over royal favorites sent out from London. All taxes were to be voted by Parliament in faraway London, which the Quebec Act openly proclaimed supreme, and were to be subject to royal veto. The religious and civil rights of the French Catholic majority were to be guaranteed and, certainly alarming to Jefferson, the law courts were to follow French law, thus dispensing with trial by jury. The extension of Quebec's borders to include the Indian country west of the Proclamation Line of 1763 cut off Virginia from much of her claimed western territory and thoroughly provoked stockholders of major land companies, including Jefferson and his neighbors, the Walkers, George Washington, and several of the Lees, Benjamin Franklin of Philadelphia, and other investors from Massachusetts to South Carolina who had been pressing claims to western lands for up to a quarter of a century. The Quebec Act, as Franklin put it, was "an arbitrary government on the back of our settlements dangerous to us all."[46] To radicals such as Franklin and Jefferson, the Quebec Act made it obvious that England intended to tighten control over other American colonies.

With the Massachusetts House of Representatives banned, the Bay Colony under martial law as part of the Boston Port Acts, and an illegally commissioned royal militia at war in the Ohio Valley, such a view was not too wild-eyed. As the Continental Congress debated in Philadelphia, the loss of the vast backcountry beyond the Appalachians provoked frontiersmen, merchants, and planters, aghast at the destruction of their chances for more land in the west, into an angry chorus of protest against the latest British repressions. An attempt at compromise, a call for formation of a commonwealth of colonies under the British Crown, was drafted by Benjamin Franklin's Loyalist son, William, the royal governor of New Jersey and by Joseph Galloway, Speaker of the Pennsylvania Assembly and the leading force in that colony's more conservative politics. The Franklin-Galloway Plan of Union, the logical outcome of Jefferson's call for an empire of colonies under the rule of king but not Parliament, was defeated by a single vote in Congress, then tabled. But the time for compromise had passed. Galloway was hounded out of Congress. The Philadelphia Sons of Liberty, led by radical Charles Thomson, delivered a box to his mansion containing a noose and a torn-up life-insurance policy. Galloway retired to his farm, and the British Commonwealth of Nations was not formed for another century and a half. Its chief conservatives cowed into silence, the First Continental Congress voted to support Boston's request for an immediate boycott of imports from England. The Congress denounced the Boston Port Acts and the Quebec Act, in all denouncing thirteen revenue measures passed by Parliament since 1763 as unconstitutional, especially the extension of Admiralty Courts, the dissolution of colonial assemblies, and the peacetime stationing of British regular troops in colonial towns. The delegates pledged to support economic sanctions until Parliament repealed its intolerable acts. Further, the Congress set forth the rights of the colonists in addresses to the king and to the British and American people. Agreeing to reconvene in May 1775 if their grievances were not remedied, the six-week Congress adjourned.

10

"I Speak the Sentiments of America"

Times like this call up genius which slept before and stimulate it in action.

—David Ramsey

Regaining his strength at Monticello in the autumn of 1774 as other members of the Virginia Committee of Correspondence kept him abreast of the developing opposition to British measures, Jefferson, noting in his *Garden Book* that the winter of 1774–75 was "the most favorable winter ever known in the memory of man,"[1] nonetheless remained in anxious expectation of news from England, where parliamentary elections were taking place. Learning that the First Continental Congress had passed a new association barring all imports from Great Britain after December 1, 1774, Jefferson, who had been pushing hard to finish construction of the main house atop Monticello, wrote to the Norfolk Committee of Safety, offering to destroy a shipment of sash frames he had ordered from England before the association was formed. Just what Jefferson planned to do for windows cannot be known, but the revolutionary committee allowed the import. Under Jefferson's watchful eyes, another fifty thousand bricks were fired and laid in late 1774 as the "middle building," containing the library, parlor, drawing room, and master bedroom, was completed.

Unable to see into the future, Virginians lived in uneasy peace. At plantations all along the James River, the annual round of dress balls proceeded. A young tutor visiting a Lee plantation wrote in his diary of a dinner "as elegant as could be expected when so great an assembly were to be kept for so long a time. . . . For drink, there were several sorts of wine, good lemon punch, toddy, cider, porter, etc." The young Presbyterian, fresh from the dour academy at Princeton, took note of the dashing gentlemen and gay ladies whose "silks and brocades rustled and trailed behind them." In their card rooms, planters "toasted the sons of America, some singing liberty songs, as they called them, in which six, eight, ten or more would put their heads near together and roar."[2]

Lacking clairvoyance, Jefferson indulged some of his favorite activities: building, gardening, buying and selling land. He drew up the charter of a joint stock company for his new friend and neighbor, Philip Mazzei, buying a fifty-pound-sterling share in a scheme to cultivate silk, grow wine grapes, and raise olive trees on the Mazzeis' slopes near Monticello, all without slave labor and relying on Italian vignobles imported from Tuscany. From April 1774, his notebooks were crammed with plans and expenditures to produce wine in the first large-scale viticulture experiment in North America. That he was not alone in his inability to foresee events is testified by the other shareholders, who included both George Washington and Lord Dunmore. According to local legend, Jefferson was able to greet the thirty vignobles in their own Tuscan accent. The men, who had heard only English for many months, wept.

Virginians each week anxiously scanned the latest political news from London and other American colonies in Clementina Rind's *Gazette*. There, in early February, Jefferson read the ominous report of the king's annual speech from the throne in which George III declared to a belligerently anti-American majority of Parliament that he was determined to uphold its "supreme authority." From the *Gazette* later that month, Jefferson also read that the king had received without comment the conciliatory petition of the First Continental Congress.

As each ship from London seemed to bring more ominous news, many Americans shared Jefferson's foreboding. Poised at the abyss of revolution and civil war, Jefferson and all of the Continental Congress had backed away in 1774, unable to believe that neither Parliament nor king nor privy councilors would ignore the numerous humble petitions and threats of trade boycotts.

In mid-March 1775, after noting that the peaches were already in blossom, Jefferson climbed into his phaeton and lashed his horse north toward Richmond. He had called moderator Peyton Randolph to hold the second Virginia Convention there, so as to be farther from the wrath of the royal governor and the cannon of the king's ships at Williamsburg. He had been elected first in a long slate of candidates in February, shortly after Lord Dunmore had, for the fifth time, postponed a General Assembly. On March 20, 1775, Jefferson was among ninety-five delegates to the second convention who assembled in the Town Church. Within a week, another twenty-five delegates had crowded in. Of sixty-one counties and four town corporations, only the predictably loyalist College of William and Mary was not represented. Jefferson's elder cousin, Attorney General John Randolph, the college's burgess, boycotted the convention. Of the 125 delegates, 105 had been elected to the House of Burgesses at least once: most had served as long as Jefferson had or longer. But no session of the burgesses had ever drawn such excited attendance. Especially noticeable was a last-minute infusion of new members from the frontier counties, able to participate now that the fighting in Lord Dunmore's war against the Indians was over. Recent scholarship on the subject of Virginia's counties "suggests the possibility of political upheaval in several constituencies" which had sent more uncompromising delegates than those who had brushed aside Jefferson's radical instructions in August 1774. Indeed, thirty-seven delegates, some thirty-one percent of the representatives sent to the second convention, had not attended the first. Twenty-one of the delegates were lawyers, fifteen were merchants, three were doctors, three ministers; the remaining eighty-four were primarily planters. Virtually all were experienced leaders of civil, church, and militia affairs, practically all had served as justices of the peace and parish vestrymen, and almost all were militia officers of field or company grade. Thomas Jefferson, at age thirty-one, was nine years younger than the average delegate, but was a justice of the peace, parish vestryman, and militia colonel. He had gone to school with many of these men, from his boyhood days in the log schoolhouse at Tuckahoe with Thomas Mann Randolph, now the delegate from Goochland County, to his college days with his old neighbors, John Walker and John Harvie, and from the new district of Augusta County West, his old William and Mary roommate, Will Fleming of Cumberland County. His father's best friend, Dr. John Walker, the first Virginian to explore Kentucky and, at age sixty-one, one of the oldest delegates, represented a county in which he owned land. Many, like George Washington, had served in other assemblies. All had one more thing in common: they were risking, as very public rebels, a great deal.

As Virginia's rebellious gentry crowded into plain white St. John's Church high on the banks of the James River in Richmond, and crowds pressed around the open doors and windows, all up and down the east coast of America, the leaders of Great Britain's colonies were assembling in emergency meetings, deserting their plantations, courtrooms, and countinghouses to take part in one of the most extraordinary and creative political movements in American history. Spurred to action after ten years of political and economic crisis by England's "hostile invasion of a sister,"[3] as Jefferson put it, they were embroiled in a worsening struggle with the mother country that required as well as created their talents. Their strong opinions had been crystallized and polarized by readings and writings that included some four hundred identifiable political pamphlets on the Anglo-American crisis over a twenty-year period, many of the shrillest and most important printed on the thirty-eight overtaxed American printing presses in the months between the arrival of the news of the British crackdown and the First Continental Congress. Many of the men who had participated in this unprecedented pamphlet war were converging to take part in the various assemblies, conventions, congresses, and committees responding to England's Intolerable Acts, as the patriots now called the parliamentary measures against Massachusetts. In the main they were men who "spoke, wrote, and acted with an energy far surpassing all expectations."[4]

For two days the delegates crowded into the tiny church overlooking the James, listened to their delegates to Philadelphia and approved the Continental Congress's resolutions. On March 23, 1775, the fourth day of the week-long Richmond convention, the delegates heard a long resolution from the then-important colony of Jamaica calling for nonresistance to British measures, either commercial boycott or military defense. From the provincial committee of Maryland had come another resolution, this one calling for "a well regulated militia."[5] The Fairfax County, Virginia, committee, concurring with Maryland, had issued a "call to arm for defense."[6] From Augusta County had come delegates' instructions calling on "the officers and men of each county of Virginia to make themselves masters of the military exercise."[7] When Patrick Henry introduced these resolutions, older, richer, more respected men who had, only a few months before, been in the forefront of the protest movement—including Richard Bland, Robert Nicholas, Benjamin Harrison, Edmund Pendleton—drew back. Nicholas, old, stooped, profound, declared that his only intention in coming to this convention had been "to have errors rectified and not to alter or destroy the English constitution."[8] Others, including George Washington of Fairfax County, argued that, while proposing defense, the formation of extralegal militias was a thinly disguised provocation, a threat of war that would place Virginia in the false position of appearing not to resist armed conflict while actually inviting it. Independent militia companies were already being formed quietly: why publicize them? And did they not have to wait to hear an answer to their olive-branch petition to the king?

Henry, in the third pew, rose again. He was galled by the idea of nonresistance. He spoke at first in a calm, reasoning tone, moving that Virginia "be immediately put into a state of defense."[9] Henry, who had refused to shake off his Celtic country accent, spoke of "the yearth" and of "men's nartural parts being improved by larnin'." Backcountry delegates thrilled as the man who was the voice of their countrymen rose. So did Jefferson, who later wrote of Henry:

> In private conversation, he was agreeable and facetious while in general society appeared to understand all the decencies and proprieties of it; but in his *heart,* he preferred low society, and sought it as often as possible.

Henry spoke without note or manuscript and Jefferson left no account of what Henry said next, but he had often opposed Henry in open court, and he later helped to explain why he could not remember:

> When he had spoken in opposition to *my* opinion, had produced a great effect, and I myself [had] been highly delighted and moved, I have asked myself when it ceased, "What the Devil has he said," and could never answer the enquiry.[10]

Part classically trained orator (he had read Virgil and Livy as a schoolboy), part rustic preacher, Henry was no stranger to the men in this assembly—but they were utterly unprepared for what he said next.

Tall, slightly stooped, his face pale and waxen, his forehead deep-furrowed, Henry glowered under heavy brows: "This is no time for ceremony." Everyone knew of Silas Deane's letter from Connecticut that thirty thousand men were poised around Boston and could be called up on two days' notice to fight. "The question before the house is one of awful moment to this country; for my own part, I consider it as nothing less than a question of freedom or slavery."[11] An eyewitness later said that the tendons of Henry's neck now stood out "like whipcords"[12] as his voice grew louder and louder. As he said the words "freedom or slavery," Henry stooped more, crossed his wrists:

> Should I keep back my opinions at such a time through fear of giving offense, I should consider myself as guilty of treason towards

> my country, and of an act of disloyalty towards the majesty of Heaven, which I revere above all earthly kings.

He paused, waiting for his meaning to sink in. The church was still, the crowds outside the open windows hushed. Jefferson sat quietly. "For my part," Henry continued, "I am willing to know the whole truth, to know the worst and to provide for it." He saw no reason to trust the British Ministry. "Suffer not yourselves to be betrayed with a kiss." Were fleets and armies being prepared in England "necessary to a work of love?" His unanswerable questions poured out as his voice rose to its climax: "Let us not deceive ourselves longer,"[13] he argued.

"I repeat it, sir, we must fight. An appeal to arms and to the God of Hosts is all that is left to us." His sentences swept on. "They tell us, sir, that we are weak, unable to cope with so formidable an adversary. But when shall we be stronger? Will it be the next week or the next year?" The words he said seemed inevitable now. He went down on one knee, his wrists locked as if in chains, his head down.

> There is no retreat but in submission and slavery. Our chains are forged. Their clanking may be heard on the plains of Boston. The war is inevitable. And let it come! I repeat it, sir, let it come!
>
> It is in vain, sir, to extenuate the matter. Gentlemen may cry, peace, peace—but there is no peace. The war is actually begun! The next gale that sweeps from the north will bring to our ears the clash of resounding arms! Our brethren are already in the field! Why stand we idle here?

What was it, he demanded, that the gentlemen wanted?

> Is life so dear, or peace so sweet, as to be purchased at the price of chains and slavery? Forbid it, Almighty God!

Rising up, yanking apart his imaginary manacles, Henry then roared out,

> I know not what course others may take, but as for me, give me liberty

Henry had raised aloft an imaginary dagger; now, he intoned, "or give me death,"[14] plunging it into his heart on the final word and sinking back into his seat. The convention sat transfixed. Outside the church, a man who had faced Henry through a window, at length found his voice: "Let me be buried at this spot." (Many years later, he was.) An old Baptist minister recorded that he was "*sick* with excitement."[15]

Ten years earlier, Jefferson had stood in the doorway of the House of Burgesses as Patrick Henry had electrified an earlier assembly, with his Virginia Resolves, only to be shouted down with cries of "Treason, treason!" Now, there was only stunned silence. Later, Jefferson would write down his thoughts, as he always did. Henry, he wrote, "was our leader in the measures of the Revolution. . . . More is due to him than to any other person. . . . He left us far behind."[16]

But Jefferson was not far behind firebrand Patrick Henry in the call to arms. Richard Henry Lee rose to second Henry's resolution and then Jefferson, despite his dislike for public oratory, stood and in the words of Edmund Randolph, argued "closely, profoundly and warmly" in support of Henry, in favor of armed preparedness. Thomas Nelson, Jr., of coastal York County, abandoned the conservative majority and emotionally called on God as his witness that if British troops landed in his county, he as county lieutenant would call up the militia and defy any orders forbidding him to "repel the invaders at the water edge."[17] At this point, a rumor circulated that, if Patrick Henry and the radicals could not have their resolutions for defense, Henry would move that the Convention seize the entire apparatus of provincial government, appoint magistrates, collect taxes, and arm themselves. The conservatives gave ground slowly, allowing the radicals' resolution to slip past. The secret vote was extremely close, Henry's motion being carried by the western counties and new, younger delegates such as Jefferson. The motion passed by a dangerously small majority, 65–60. Henry was placed in charge of defense; Jefferson was appointed to the committee of twelve to plan Virginia's defense, and its report, submitted to the Convention on March 25, was in his handwriting. Mostly the plan was simple enough: to expand the rapidly forming independent militia companies and bring them under county militia discipline. Virginia's uniform was to be the fringed hunting shirt; its skilled marksmen were to carry rifles and tomahawks. All too soon, Jefferson would join their ranks.

Jefferson himself may have drafted the rules for "each troop of light horse"[18] of thirty men and their officers. Every horseman was to be provided with a good horse, bridle, saddle with pistols, a carbine or other firelock, a bucket, a tomahawk, at least one pound of gunpowder and four pounds of ball. Each trooper was to use the utmost diligence in training and accustoming his horse to stand the discharge of firearms, and in making himself acquainted with the military exercise of cavalry. For the moment, Jefferson was more concerned that all the sister colonies close ranks against the more immediate common threat. The New York House of Representatives had broken with the Association boycott recommended by the Continental Congress. Jefferson wrote a resolution denouncing New York's desertion from "the Union with the other American colonies" formed in Congress for "the preservation of their just rights." It was new language, this "Union."[19]

Amid the excitement over arming the colony, Jefferson helped to steer the Convention to debate a threat of another kind. On May 21, as the protesters met in Richmond, in Williamsburg the royal governor had issued a proclamation ordering that all vacant lands in the colony and the mineral rights to them or any lands with imperfect patents should be auctioned immediately for cash to the highest bidders and that their buyers pay annual quitrents to the Crown. If Lord Dunmore had succeeded in carrying out this particularly odious British mandate, Jefferson, Washington, and many of the men crowded into the church in Richmond would have suffered sharp curtailment of their future fortunes. After Patrick Henry attacked the proclamation and Richard Bland upheld him, Jefferson, the most highly regarded land-tenure lawyer in the colony, rose to challenge Lord Dunmore on March 27. He denounced the Crown order as a dangerous "innovation on the established usage of granting lands"[20] and recommended that a boycott be declared on purchasing the lands and accepting any further land grants.

On the last day of the Convention, the five former delegates to Philadelphia were reelected to a second Continental Congress. The perennial Peyton Randolph was chosen again, but Jefferson was chosen as his deputy in case the ailing Randolph could not attend. At age thirty-two, Jefferson, undoubtedly because of his beautifully written rejection of royal, ministerial, and parliamentary pretensions in his *Summary View*, had risen to number six on Virginia's list of respected leaders. He went home to Monticello, to await the British response.

While the Convention was still debating in Richmond, radical committees all over Virginia were suppressing dissent against the continental protest movement. Orange County committeemen had learned that an Anglican priest, the Rev. John Wingate, had secreted in his parsonage a supply of pamphlets printed in New York on the royalist press of James Rivington. Demanding that Reverend Wingate surrender the Loyalist tracts, the committee menaced the clergyman until he turned them over. Four of the five pamphlets they found had been written anonymously by A. W. Farmer, pseudonym for a Connecticut Loyalist, the Rev. Samuel Seabury, who was later to become the first Episcopalian bishop in America. The fifth was by a Westchester County, New York, Loyalist who had been terrorized by radicals from Connecticut who later packed Rivington's printing press in New York City. All of the pamphlets vehemently attacked the First Continental Congress. One of Wingate's parishioners, young James Madison, wrote that Parson Wingate was known for his "insolence" and "had like to have met with sore treatment, but finding his protection to be not so much in the law as in the favor of the people, he is grown very supple and obsequious." What may have pacified the parson was a second visit on May 27, by the Orange County Committee, which, milling around his house, loudly denounced the Loyalist writings as "a collection of the most audacious insults on that august body [the Grand Continental Congress]" which had been printed "to excite divisions among the friends of America" and "deserved to be publicly burnt." The sentence "was speedily executed in the presence of the Independent Company"[21]—a spontaneously formed unofficial militia—"and other respectable inhabitants," as the *Virginia Gazette* duly reported on April 15.

For many of the delegates, it was probably as important a matter as the boycott on British trade. George Washington had already turned his mind to making the Potomac River navigable to Fort Cumberland in western Maryland to open a "channel of commerce between Great Britain and that immense territory tract of country which is unfolding"[22] to the west. He had recently purchased fifteen thousand acres of land in the Ohio country and had studied lands in Florida and actively pursued a land bounty from Lord Dunmore. An early exponent of larger-scale scientific farming, he had already switched from tobacco to wheat farming, which required vast acreage but did not deplete it so badly. Only the death of his stepdaughter a year before had prevented him from making a land-hunting trip into the Ohio Valley in July 1773. As late as the autumn of 1774, as the Boston Port Act filled newspapers with vituperation, Washington was busy planning to survey lands in the Scioto River country over the mountains in Ohio.

The Convention having approved his resolution and appointed him chairman of a committee to study whether the king had the right to grant or sell further Virginia lands, Jefferson began assembling the documents at once. His committee never reported, but when the Convention reassembled later that year, on his recommendation, the delegates resolved that, until the crisis permitted Jefferson to meet with his colleagues, the ban on buying or accepting lands would be extended. All surveyors were ordered to make no further surveys and to ignore the royal governor's proclamation. Writing the resolution must have given Jefferson a moment's amusement. The first draft was in the handwriting of Robert Carter Nicholas, heavily edited in Jefferson's handwriting. On the reverse side, Jefferson wrote his own much clearer resolution. Six years earlier, when Jefferson had attempted his first public paper as a fledgling burgess, it had been Nicholas who had cut up and completely rewritten his answer to the governor. Forty-six years later, as an elder statesman, Jefferson still remembered Nicholas rewriting his work, but he made no comment on rewriting Nicholas.

In the eye of the gathering colonial storm, Jefferson paused to plant his favorite vegetable, peas, and to reap the harvest of another passion of his, horsebreeding, now made an irresistible temptation by his greatly increased wealth. In Virginia, where horsemanship was a mania, young Jefferson was distinguishing himself as a "bold and fearless rider" who "rode and drove magnificent horses,"[23] a man who rarely drew rein for the roughest terrain and preferred to dash through the Rivanna even when it was dangerously rain-swollen instead of crossing at a safer ford. Finicky about the treatment of his horses, he took out a white linen handkerchief when a stableboy led out a mount. If there was a spot on the horse that did not shine, he rubbed it and, if it came off soiled, reprimanded the groom.

Jefferson's stables at Monticello were to play an important part in the improvement of Virginia thoroughbred horses from this time. Along Mulberry Row, on the south side of the first roundabout down from the crest of his mountain, Jefferson had built a large stable where he bred some of the finest horses in the South.

Until shortly before Jefferson's birth, the general stock of horses in Virginia had consisted of northern European stock from England, France, and Holland, of mixed bloods and Chickasaw stock that had come to the Indians from the Spanish conquistadors. Planters used these quarter horses for farmwork and hunting, but they were suited only for racing short distances. Course racing had been introduced, in 1737, requiring horses of much greater endurance. Breeders began to import fine horses from England and Spain. Peter Jefferson had been bitten by the bug just before his death. Thomas Jefferson's first mare, Allycrocker, born within months of the death of the adolescent boy's father, was a tribute to Peter Jefferson's acumen. The mare's sire was Silver Eye, who was highly esteemed by Virginia breeders and her dam was Patty Banister, foal of a noted stud, Spanker, imported from Spanish Andalusia by William Nelson, president of the Governor's Council, and a fast-running mare named

Jenny Morris, bred by Secretary of the Colony John Carter. The Jeffersons' interest in horsebreeding helped to establish them among Virginia's elite.

Jefferson used his wife's newly inherited money to indulge his passion for fine horses. He began the practice of breeding and buying horses retired after distinguished racing careers. The best of the former English turf champions was Fearnought, who was put to Calista, a Virginia-born mare, to produce Young Fearnought, sire of several of Thomas Jefferson's best horses. Late in 1774, Young Fearnought was put to Jefferson's first mare, Allycrocker, now sixteen years old, just as Jefferson began to keep careful records of his stables. On May 7, 1775, Jefferson noted in his *Account Book,* "Allycrocker's colt by Young Fearnought was foaled." The gentle winter had turned into a cold spring and a late frost had killed off Jefferson's year-old hillside orchards. He rejoiced at the arrival of this thoroughbred colt in his new stable.

It must have been later that same day when a courier swung down from the saddle at Monticello to bring him the worst of all possible news. The British had marched out of Boston and killed American militiamen at Lexington and Concord in a running battle that only ended after heavy casualties. America was at war with its mother country. While Jefferson refused ever to discuss or write about his descent from Welsh kings, he had steeped himself in the laws and customs of his Celtic ancestors and had read in Latin Tacitus's history of the conquest of the British Isles by Roman legions. Before the day was over, Jefferson named the fine young colt teetering on spindly legs in his stable Caractacus—the name of the Celtic king of the Silures who had fiercely resisted the first-century invasion of Britain by imperial Rome.

That same day, Jefferson turned his agitated mind again to England, to a time closer to 1775 and happier years when he had been a student of the effervescent Dr. Small at William and Mary. Civil war with England would cut off correspondence with his old mentor indefinitely. Jefferson had recently heard from a British factor in Bristol that Small was living in Birmingham. When would Jefferson see or hear from him again? With war breaking out, Jefferson wrote to Dr. Small one of his most honest and painful letters, ostensibly to accompany a long-intended gift of three dozen bottles of Madeira port wine, "being the half of a present I had laid by for you." The rest would come later; the captain of the *True Patriot* "was afraid to take more on board lest it should draw upon him the officers of the customs." Jefferson hoped "you will find it fine, as it came to me genuine from the island and has been kept in my own cellar eight years." Sending off a reminder of cultivated evenings of dinner table conversation at the Governor's Palace, Jefferson poured out his anxiety to his distant father-figure. It was a careful letter, composed in draft and then rewritten, and Jefferson artfully worked in his latest political thinking with apparent casualness.

"Within this week we have received the unhappy news of an action of considerable magnitude between the King's troops and our brethren of Boston." Details were sketchy, Jefferson reported, but five hundred Redcoats and the young Earl of Percy, heir to the Duke of Northumberland, had been killed, plunging America into a state of open rebellion against England. This "accident," as Jefferson characterized one of the armed clashes outside Boston, has cut off our last hopes of reconciliation, and a frenzy of revenge seems to have seized all ranks of people."

Jefferson abhorred violence. To Dr. Small, who had taught him to reason, and to dispute only with words and ideas, he wrote on the eve of war, "It is a lamentable circumstance that the only mediatory power acknowledged by both parties"—and here he injected his theory that the only link between England and America was the king, not Parliament—

> instead of leading to a reconciliation of his divided people, he should pursue the incendiary purpose of still blowing up the flames, as we find him constantly doing in every speech and public declaration.

Jefferson's blunt tone as he excoriated the king as if he were an intemperate and dangerous firebrand was calculated to shock. Jefferson went on to question the motives and the good judgment of the young king and his advisers in their heavy-handed treatment of American colonists:

> This may perhaps be intended to intimidate [us] into acquiescence, but the effect has been most unfortunately otherwise. A little knowledge of human nature and attention to its ordinary workings might have foreseen that the spirits of the people here were in a state in which they were more likely to be provoked than frightened by haughty deportment.

Jefferson's angry criticism of the king and his ministers was risky. With open hostilities in America, the king's ships would no doubt begin to blockade all communication between Americans and England. Already it was well known that English censors were reading mail from prominent protesters of British policies. A conspicuous Virginia radical, Jefferson was once again boldly exposing himself to the charge of treason by criticizing the royal person. Jefferson deliberately defied the new British coercive tactic of ordering the arrest of protesters and transporting them to England for trial and almost certain execution. "To fill up the measure of [our] irritation," the seizure and "proscription" of rebellious Americans now was to take the place of a just trial by a jury of American peers without all the unnecessary expense and hardship of deportation to England and what Jefferson was certain would be an unjust trial. "Can it be believed," he asked Small, "that a grateful people" would allow their spokesmen "to be consigned to execution" when their "sole crime has been developing and asserting their rights?" If Parliament had "possessed the liberty of reflection, they would have avoided a measure as impotent as it was inflammatory."

For a moment Jefferson had considered softening his words to Small, writing in the draft letter,

> But for god's sake, where am I got to? Forever absorbed in the distresses of my country, I cannot for three sentences keep clear of its political struggles.

But then he decided to keep the gloves off. He scratched out these more informal lines and let his caustic attack on Parliament stand.

Years before, William Pitt had provided England with an enlightened vision of a British Empire with America as a jewel in its crown and at the same time had championed American rights. From his seat in the House of Lords, he had unleashed his powerful eloquence against Lord North's anti-American policies for the past five years. After a series of meetings with Benjamin Franklin, Chatham had introduced a conciliatory bill in the Lords on February 1, 1775. He called for an imperial union similar to that proposed by Benjamin Franklin at Albany in 1754 and, in 1774, by Franklin's son, the royal governor of New Jersey, and by Joseph Galloway in the First Continental Congress. Chatham's plan proposed making the Continental Congress official and permanent and empowering it to make voluntary grants for imperial expenses once Parliament suspended its primitive acts against New England. But the man who had galvanized the British for their greatest victories in 1760 fifteen years earlier had no following in 1775. After a bitter debate in the Lords, his plan for reconciliation was noisily voted down by an overwhelming and bellicose ministerial majority. The king's chief minister was Lord North, a sure politician who risked holding an election because he was certain he could strengthen the anti-American majority in Parliament supporting his hard-line policies. North shepherded each anti-American measure through the House of

Commons and the Lords while the king himself bitterly complained of "violences of a very criminal nature" in Boston, assuring Parliament, in his opening speech from the throne, of its continued "supreme authority." Jefferson was still stinging from the rejection of British and American attempts at last-minute peacemaking:

> When I saw Lord Chatham's bill, I entertained high hope that a reconciliation could have been brought about. The difference between his terms and those offered by our Congress might have been accommodated, if entered by both parties with a disposition to accommodate. But the dignity of Parliament, it seems, can brook no opposition to its power. Strange that a set of men who have made sale of their virtue to the Minister, should yet talk of retaining dignity!

Pretending this had been an accidental outburst instead of a carefully phrased message intended to be passed around if not published by a man extremely well connected in England, Jefferson now shifted to a friendly tone to his old teacher:

> But I am getting into politics, though I sat down only to ask your acceptance of the wine and express my constant wishes for your happiness. This, however, seems secured by your philosophy and peaceful vocation. I shall still hope that, amidst public dissension, private friendship may be preserved inviolate. . . .

There was much more to Jefferson's draft than he finally sent off, and what he cut out gives an uncommon portrait of his thoughts amid the turmoil produced by the news of actual fighting in Massachusetts. Revolutionary committees raided the homes and warehouses of suspected Loyalists. Illegal militias formed spontaneously all over Virginia. Jefferson had to march briefly as a common soldier with a Charlottesville independent company until he could persuade his neighbors to go home and leave this latest crisis to their leaders. Amid these alarms, Jefferson had started to write to Small more of his theory, contained in his *Summary View* and shared by other American leaders, that the king had been ill-advised by an administration that kept the monarch under "constant delusion." He had decided it was wrong to absolve the king for bloodshed sanctioned by his ministers and their army. Royal governors, too, were complicit in a British conspiracy against America for laboring

> to make the Ministry believe that the whole [American] ferment has been raised and constantly kept up by a few hot headed demagogues —[he crossed out the words] principal men in every colony, and that it might be expected to subside in a short time either of itself or by the assistance of a coercive power. The reverse of this is most assuredly the truth. The utmost efforts of the more intelligent people [have] been requisite, and exerted, to moderate the almost ungovernable fury of the people.[24]

The portion of the Small letter Jefferson excised provides a rare indication of Virginia gentry keeping the lid on what promised to become a far more dangerous revolutionary threat. Jefferson was the colonel and therefore commander of the official Albemarle County militia, even if Dunmore's dissolution of the Assembly had stripped the militia of the entire colony of legal sanction. By joining the extralegal independent militia of Albemarle County, he had coopted it and helped to make it respond to traditional county authority. Like other key county gentry, he had thus helped to prevent his countymen from swelling the ranks of Patrick Henry's militia, which had gathered, six hundred strong, at Fredericksburg and attracted thousands more as they marched within sixteen miles of the colonial capital at Williamsburg. Had they been able to reach Lord Dunmore, Jefferson was sure His Lordship would have perished. Only more moderate Virginians such as Jefferson had saved Dunmore. The anger all over America could have resulted in a bloodbath if so-called hotheads like Jefferson had not intervened:

> To these men, those very governors who have so much traduced them, are indebted that there is this day one man of them left in existence. Within this week past, there have been at least 10,000 men in arms in this colony.

Governor Dunmore's life had been "in the last danger" when Captain Henry's militia "had got within sixteen miles of the capitol before the intercessions of the principal people could prevail on them to return to their habitations." At length, Henry's impromptu army had broken up "and, at present, we appear to possess internal quiet."[25]

But Jefferson apparently thought he was provided evidence of an armed rebellion. He rewrote the letter, toughening it in places, and hurried it to the ship captain. He never heard back from Dr. Small. Just who drank all that Madeira is not known. William Small, Thomas Jefferson's beloved teacher, had died three months earlier.

✧ ✧ ✧

After more than a year of prorogations, Lord Dunmore chose this especially inauspicious moment to summon the Assembly into session on June 1, 1775. After the Second Virginia Convention in March, Dunmore had forbidden Virginians to elect delegates to the Second Continental Congress, but the Convention ignored him. Since then, as the colonists had just learned, on March 30 King George had given his assent to parliamentary bills to restrict the trade of the New England colonies and, on April 13, had assented to a companion bill restricting the trade of New Jersey, Pennsylvania, Maryland, South Carolina, and Virginia. Dunmore's removal of gunpowder from the capital had touched off an angry confrontation with Williamsburg's city government. On April 30, news of Lexington reached Williamsburg. Capt. Patrick Henry's Hanover County armed volunteers had called off their march on the capital only after the provincial receiver general paid Henry the £330 value of the purloined gunpowder. The crisis in the Virginia capital had flared anew when Capt. George Montagu of HMS *Fowey* threatened to bombard Yorktown if anyone tried to hinder the movement of his sailors and marines into Williamsburg; shielded by royal troops, Lord Dunmore denounced Henry's "outrageous and rebellious practices."[26] On May 8, 1775, as the Second Continental Congress in Philadelphia elected Peyton Randolph president and Benedict Arnold and Ethan Allen marched on Fort Ticonderoga on Lake Champlain in New York, Lord Dunmore drafted his proclamation calling for a new General Assembly to consider Lord North's resolution of reconciliation. As the governor published the proclamation on May 12, radicals issued a call for speedy election of delegates to a third Virginia Convention.

Before Thomas Jefferson and his fellow burgesses answered Dunmore's summons, fast-moving events further complicated their task. On May 15, the Continental Congress, following Virginia's and New England's lead, unanimously resolved that the united colonies be put in a state of defense. George Washington had scarcely been appointed to a committee intended to decide which New York posts "ought to be guarded" when Benedict Arnold went on the offensive and sailed up Lake Champlain, seizing the British fort at St. John's in Quebec Province and seizing all British shipping in the region. Writing home from Congress in Philadelphia on May 21, Richard Henry Lee told his brother, Francis, "There never was a more total revolution at any place than at New York," where "the Tories have been obliged to fly" and "the Governor dares not call his prostituted Assembly to receive Lord North's foolish plan."[27]

One week before the Virginia Assembly answered Lord Dunmore's royal summons, Peyton Randolph retired as president of the Continental

Congress, turned the gavel over to John Hancock, and hurried to Williamsburg to preside as Speaker of the House of Burgesses. As Randolph's carriage rattled south, and Jefferson's east toward the Virginia capital, Williamsburg's independent company of volunteers pledged its services anywhere on the continent in defense of American liberty. They turned out again and Jefferson was in the large crowd to greet Randolph with huzzahs and volleys of muskets on May 31, the day before Virginia's last provincial Assembly convened in the turreted red-brick Capitol. A cavalry detachment had met Randolph's carriage thirty miles out of town, possibly to shield him from royal arrest. Two miles out, an infantry company joined the column. At sunset the cavalcade passed the Governor's Palace and, amid tolling churchbells and cheers, marched up to the Speaker's house. Randolph had made the usually ten-day drive in only six. That night candlelight illuminated every patriot's window and the volunteers, "with many other respectable gentlemen, assembled at the Raleigh, spent an hour or two in harmony and cheerfulness, and drank several patriotic toasts."[28]

If Lord North, by feigning reconciliation, had intended to divide the American opposition and unite the British people against them by demonstrating that no terms were acceptable to the majority, he succeeded brilliantly. The king viewed any attempt at reconciliation a sign of weakness, and North's resolution ran so contrary to his usual policies that it confused both his own followers and the loyal opposition. After a week of feverish political jockeying and bitter debate, Lord North was able to ram his plan through Commons by a 261–85 majority.

The arrival in Williamsburg of North's olive branch—"so fair an opening for reconciliation," Lord Dunmore called it—was ominous. Between 2 and 4 P.M. on May 22, 1775, three storms lashed the Tidewater with hailstones "as big as pigeon's eggs, some much larger."[29] Some four hundred panes of glass in the windows of the Governor's Palace were shattered. Within three days of publication in the newspapers, the North plan was branded by the Richmond County Committee as "no kind of redress"; the Fredericksburg Committee recommended to its burgesses that they "oppose such proposal to the utmost of their abilities."[30]

When the General Assembly convened June 1, Lord Dunmore summoned the burgesses to the Council Chamber. He must have been shocked to find that many of the usually elegant lawmakers came armed with swords and were, he reported to London, wearing shirts of "coarse linen or canvas over their clothes and a tomahawk by their sides."[31] Pretending not to notice, Governor Dunmore, in his tactless opening speech, insisted on calling Parliament "supreme legislature of the Empire," urged "reverence"[32] for it. He called on the House to pay the veterans of his war on the Indians and to reopen the courts to civil cases. The burgesses returned to their chamber. Speaker Randolph appointed Thomas Jefferson to draft a reply.

On June 5, Jefferson read aloud to the House that reply. He blamed "the alarming situation of our country and the evils brought upon us" entirely on the unconstitutional policies of the king's ministers. He accused Lord Dunmore of "strangely" misrepresenting the colony to London. He accused the royal governor and ministry of impeding justice and hampering trade. And he bluntly pointed out that the cause of some of Virginia's economic difficulties was not shrouded in mystery. Debts owed by Virginians to British factors were piling up and the British were not able to collect so long as the courts stayed shut. Jefferson spoke for his fellow planters when he said that debtors should not be sired by British creditors, since the halt of all commerce by the British made it impossible for them to sell their tobacco. "Money, my Lord, is not a plant of the native growth of this country."[33] When Jefferson's caustic answer to the governor finally came to a vote, the House approved it unanimously.

But before the vote, a bitter exchange occurred between burgesses and governor. On Saturday night, June 3, a number of youths breaking into the powder magazine to steal guns tripped a string that fired a gun loaded "eight fingers deep with swan shot." Three were hurt, one terribly. On Monday, Speaker Randolph appointed Jefferson to a committee "to inspect the public magazine," that is, to investigate the incident. A curt exchange of accusations flew between the governor and the Speaker of the House. That same day, rumors also flew that a British warship was sailing upriver "with a number of boats in which there was said to be a hundred armed men at least."[34] Agitated townsmen marched to the brick powder magazine and passed out muskets hand to hand.

On the evening of June 7, Jefferson's cousin, Attorney General John Randolph, who had boycotted many of the House sessions, opened his front door to find the short, swarthy Scottish royal governor standing there alone. It had taken considerable courage for him to walk there, through the armed citizenry, despite the fact that John Randolph had recently assured His Lordship that "his person was in no danger."[35] The two royal officials talked late into the night before the governor walked back to his boarded-up palace. At 1 P.M. the next day, the House, waiting for an answer to several questions sent to the governor, was stunned to learn that, in the night, Lord Dunmore and his family had gone aboard HMS *Fowey*, moored twelve miles down the York from the capital. His

Lordship sent word through his Council that he was "fully persuaded" that his family was in constant danger of "falling sacrifices to the blind and immeasurable fury" of "great numbers of the people." When the House sent a committee to ask the governor to return, Dunmore sent off an ultimatum: restore the guns to the magazine, adopt Lord North's plan, reopen the courts entirely, and end the "ferment."[36]

For a week, Speaker Peyton Randolph had been chafing for an opportunity to have someone draw up an address to the governor and the people that "harmonized" with the sentiments and wishes of the Continental Congress he had just left. There would be no more mealymouthed conservative hemming and hawing. He "pressed" Jefferson to "undertake the answer."[37] That same day, Jefferson startled the House by presenting his nine-paragraph answer. Despite a "dash of cold water" from conservatives such as John Mercer and Robert Nicholas which "here and there enfeebled it" with "long and doubtful scruples," Jefferson's address sailed quickly to passage; its language clearly and bluntly declared that Virginia's burgesses had overwhelmingly come around to Jefferson's point of view. There could no longer be any distinction between internal and external regulation. As Britons, the colonists were equally entitled to a "free trade with all the world." Parliament had no constitutional authority "to intermeddle" with any facet of colonial life.

> For us, not for them, has government been instituted here. . . . We conceive that we alone are the judges of the condition, circumstances and situation of our people, as Parliament are of theirs."[38]

Jefferson, thumpingly supported by Virginia, rejected Lord North's olive branch and would wait to learn the wishes of the representatives of all the other American colonies. "Bound in honor as well as interest to share one general fate with our Sister Colonies," Virginians "should hold ourselves base deserters of that union to which we have acceded" to agree to any measure "distinct and apart from them."[39]

Jefferson was not present when the final vote was taken. Peyton Randolph pressed him again, this time to go to Philadelphia to take his place at Virginia's table in the Second Continental Congress.

Thomas Jefferson began his first public mission outside his native Virginia in June 1775 almost hesitantly, taking nearly two weeks to travel from Charlottesville to Philadelphia, slow even for those days. Stopping along the way to buy a retired six-year-old champion racehorse—aptly named the General—he made his entrance into continental politics in his own coach and four. Even as he went off to represent Virginia at the Second Continental Congress because of, he later wrote, "emergencies which threatened our country with slavery but ended by establishing it free,"[40] he traveled with two liveried slaves: Jesse, who rode postilion and whipped along the horses, and Richard, Jefferson's body servant of the moment. Covering the four hundred miles from Monticello to Philadelphia in twelve days, he hired a guide in Wilmington, Delaware, to lead his equipage the last thirty miles on June 21, 1775: he had forgotten the route since his last visit nearly ten years earlier. When he reached Philadelphia, he found a scene succinctly described by Benjamin Franklin: "All ranks of people in arms, disciplining them morning and evening."[41] Jefferson's arrival in Congress caused a stir. "Yesterday the famous Mr. Jefferson arrived," wrote Governor Samuel Ward to his brother in Rhode Island. "I have not been in company with him yet. He looks like a very sensible spirited fine fellow and by the pamphlet which he wrote last summer, he certainly is one."[42] John Adams commented on Jefferson's "reputation for literature. Writings of his were handed about, remarkable for their felicity of expression, science and a happy talent for composition."[43] James Duane of New York was impressed by Jefferson's penchant for researching the law, calling him "the greatest rubber off of dust"[44] he had met.

Boarding with Benjamin Randolph, a cabinetmaker—Jefferson spent only a little more for his lodgings than to put up his horses—he joined the other Virginians at Smith's elegant City Tavern, which had probably the best table fare in Philadelphia, took the place of Peyton Randolph at table as well as in Congress, with George Washington, Richard Henry Lee, Benjamin Harrison, John Alsop of New York, Samuel Chase of Maryland, George Read and Caesar Rodney of Delaware, and other delegates. Congress was midway through its three-month session, its sixty members meeting roughly from nine to five each day, with committee work in the evenings or early in the morning. But Jefferson found time to shop in the tempting stores that abounded in the crowded two square miles of the century-old Quaker city; on the backs of pages of his pocket-size *Virginia Almanac* he recorded his purchases—Jefferson loved to shop—including books and music, noting that he paid a barber by the week to shave him daily at his lodgings.

There was to be little time to relax, however. A few days before Jefferson arrived, Washington had been elected commander-in-chief of all continental forces: since then, he had been exchanging his cumbersome carriage for a small, fast phaeton and carefully selecting his aides-de-camp. In the heavy, humid half-light of a Philadelphia summer dawn, Washington led an impromptu parade of mounted gentlemen and carriages up the Old

York Road, the socialites of the shakoed Philadelphia Troop of Light Horse prancing ahead of his carriage. John Adams reported to Abigail that the generals were all mounted and all the delegates

> with their servants and carriages attended. . . . delegates from the Congress, a large troop of light horse in their uniforms, many officers of militia, besides, in theirs. Music playing, etc., etc. Such is the pride and pomp of war.[45]

At eleven o'clock the next night, an express rider from Connecticut raced down Second Street and turned up Arch to the boardinghouse of Miss Jane Port. He swung down off his horse and asked to see John Adams. He had an urgent message for the Massachusetts delegates to Congress. In a pitched European-style battle outside Boston, hundreds of British and Americans had died in a clash that transformed the men with legalistic grievances in this cautious debating society into an incredibly tense and frequently frightened aggregation of American revolutionaries. Giving few details, the messenger told the Massachusetts men that, on the night of June 15, the Provincial Congress had learned from spies inside British-occupied Boston that the English intended to break the rebel siege a few days later. The Massachusetts Committee of Safety decided to occupy the heights north of Boston, to build a fort the next night, and to man it on June 17, 1775 with twelve hundred New England troops led by veteran officers from the French and Indian Wars. The next morning, the British commander, Gen. Thomas Gage, ordered an immediate attack on the main American position, an earth redoubt only shoulder high atop Breed's Hill. As cannon aboard Royal Navy ships bombarded the American positions, twenty-eight landing barges of heavily uniformed Redcoats with backpacks weighing 100 to 125 pounds crossed the Mystic River and thumped ashore at the beach at Moulton's Point in searing summer sunlight.

In the first attack the British elite light infantry charged with bayonets fixed toward New Hampshire militia behind a rail fence and a quickly set stone wall. Wave after wave of England's best troops, trying to get at the farmers as they reloaded, were mowed down at close range by the incessant fire from John Stark's crack shots, who had orders to aim low at the grenadiers and aim high at the gorgets that marked the throats of British officers. By the time the men of the British Fifty-second Regiment refused to attack further, ninety-six Redcoats lay dead on the beach, one in five who had landed.

In a simultaneous attack against the low redoubt atop seventy-five-foot-high Breed's Hill, Americans held their fire until the British were fifty yards away, causing confusion among the attackers as one wave got mixed up with the last and violated their orders for a bayonet attack by pausing to return the deadly American fire. Fifteen minutes after the survivors of this first frontal assault retreated, the British attacked up the hill again. Murderous fire that seemed to come from all directions left scores more British thrashing and screaming in the high grass.

But the Americans were running out of ammunition and most reinforcements were frightened off by continuous naval gunfire. The third and final British assault had succeeded. The Redcoats were finally allowed to drop their heavy packs. Field artillery, this time with the right ammunition, raked and cleared the forward American breastwork. The British infantry, learning to withstand the devastating musket fire, swarmed into the redoubt from two sides. Thirty trapped Americans, throwing rocks or swinging their muskets as clubs, were bayoneted inside the small fort as they tried to escape; no exit had been provided. American casualties were bad, twenty percent: of 2,000 Americans who had fought, 140 were killed and 301 wounded. British casualties were terrible, double the Americans', forty percent. More than 100 officers and 1,000 men had been killed or wounded in one of the bloodiest hours in the history of the British Empire.[46]

One week later, as word had spread through Philadelphia that a courier had brought grave news of a battle at Boston, "an hundred gentlemen flocked to our lodgings to hear the news," John Adams wrote to James Warren, president of the Massachusetts Provincial Congress. At one in the morning, John and Samuel Adams and John Hancock roused Philadelphia radical leaders "with great politeness." Before morning, ninety small casks of gunpowder were in wagons and rolling north to overtake Washington's column. After three hours' sleep, the Adamses and Jefferson and the other delegates crowded into the State House to vote more aid for beleaguered Massachusetts. John Adams wrote Warren that ten companies of riflemen, including five from Virginia, were on their way north.

> These, if the gentlemen of the Southern colonies are not very partial and much mistaken, are very fine fellows. They are the most accurate marksmen in the world; they kill with great exactness at 200 yards. . . . They have sworn certain death to the ministerial officers. May they perform their oath.

Adams spoke for other congressmen whose "hopes and fears are alternately very strong."[47]

It was three more days before Jefferson had time to write home to Virginia his first letter since his arrival in Philadelphia. Patty Jefferson had taken Patsy and baby Jane to The Forest, her family home, to stay with her half sister, Elizabeth, and Francis Eppes, her brother-in-law. To Eppes, Jefferson wrote, "You will before this have heard that the war is now heartily entered into." The report from Breed's Hill banished Jefferson's pacifism for the moment. There was no "prospect of accommodation" except by the "interposition of arms." While he still did not know precisely how heavy the casualties had been (he underestimated by half), he was sure the battle "was considerably in our favor." But Jefferson was worried by accounts that American positions were dangerously close to the British lines, and he filled his letter with armchair generalship. He was worried whether Washington would arrive safely, concerned about the threat of an invasion from Canada. It was evidently up to Eppes to decide how much detail to pass on to Patty Jefferson. It was another week before Jefferson learned more. His heady letter to Eppes of July 4, 1775, was filled with more news of the battle at Boston. The British officer who had led the march on Lexington had died on Breed's Hill, "at which everybody rejoices." New Englanders were fitting out "light vessels of war" to pounce on British ships "to clear the seas and bays here of everything below the size of a ship of war." Especially after the lukewarm patriotism of so many of his Virginia friends, Jefferson admired "the adventurous genius and intrepidity" of the "amazing" Yankees. Part of what amazed him was the avowed determination of delegates from Massachusetts to burn Boston "as a hive which gives cover to regulars . . . none are more bent on it than the very people who come out of it."[48] Euphorically patriotic, Jefferson had already cast aside any thought that the war could be settled quickly by negotiation. Gunpowder "seems now to be our only difficulty," he wrote Eppes, hoping there would be plenty for the next year's fighting.

Only a few days after Jefferson had taken his seat at Virginia's green-baize-covered table in the State House, he had been appointed vice chairman of a committee writing a Declaration on the Necessity of Taking Up Arms to be published by Washington as soon as he arrived at camp outside Boston. The document was much more important now after the lethal affair at Boston. On the original committee were fiery John Rutledge, a South Carolina planter, whose first draft had been rejected by the full Congress; William Livingston, a Yale-educated New York radical lawyer representing

New Jersey; lawyer John Jay of New York (the only delegate younger than Jefferson); Benjamin Franklin, returned only six weeks earlier after a decade in England, and Thomas Johnson of Maryland. To prepare a new and stronger draft, Jefferson, who had the draft of his Virginia answer to Lord North in his pocket, and John Dickinson of Philadelphia, one of the early spokesmen of American liberties, were now added.

Of all the delegates in the crowded, hot, fly-infested chamber, the man with whom Thomas Jefferson was most likely to clash was the cool, conservative Philadelphia lawyer Dickinson, an opulent Quaker in a gray satin suit. When Jefferson was asked to draft a new version, Dickinson objected to his wording as too harsh. It was the first clash between the reserved middle-of-the-road Middle Colonies lawyers and the passionate southerners that would so divide and protract the Congress over the next year. William Livingston summed up the problem on July 4, 1775:

> We are now working on a manifesto on arming. The first was not liked by the Congress and recommitted. The second was not liked by the committee. Both had the faults common to our Southern gentlemen. Much fault-finding and declamation, with little sense or dignity. They seem to think a reiteration of [the words] tyranny, despotism, bloody, etc., all that is needed to unite us . . ."[49]

The objections of Dickinson and the conservatives must have puzzled and irritated Jefferson. Writing two careful drafts, he had toned down his searing directness in the *Summary View.* He had been chosen of all the Virginia delegates because of his skill and polish as a writer, as he was well aware. There was nothing new here; everyone had read his views before. But, whatever Jefferson's choice of words, Dickinson would have disagreed with Jefferson's views on the relationship of England and America. Finally, Dickinson was asked to write a final draft, which was accepted by Congress. He deleted Jefferson's appeal to "supreme reason" instead of God or king, deleted any hint of the Lockean doctrine of social contract or link between the king and the parts of his empire. Weakening Jefferson's prose, he nonetheless incorporated many of Jefferson's resounding thoughts and words:

> Our forefathers, inhabitants of the island of Great Britain, left their native land to seek on these shores a residence for civil and religious freedom. . . . Our attachment to no nation upon earth should supplant our attachment to liberty.[50]

When Dickinson went on to write another humble petition to the king, the lines were drawn between the conciliatory conservative lawyers and the radical southerners. When Dickinson rose to endorse his petition, he said, "There is but one word in the paper, Mr. President, of which I disapprove, and that is the word 'Congress.' " Rotund Benjamin Harrison rose and responded: "There is but one word in the paper, Mr. President, of which I approve, and that is the word "Congress."[51]

When it came time for the Continental Congress to respond to Lord North's conciliatory message, Jefferson fared better. This committee was more to his liking: Franklin, John Adams, Richard Henry Lee, and Jefferson, the obvious choice as author because of his earlier reply to North on behalf of Virginia. Except for one paragraph by Franklin, the document was Jefferson's Virginia paper made even more powerful. North's plan, so long as it was "accompanied with large fleets and armies seems addressed to our fears rather than to our freedom." Calling the Ministry's proposal "altogether unsatisfactory," labeling Parliament's recent measures "indiscriminate legislation" and a "high breach" of the privileges of Americans, he declared that "nothing but our own exertions may defeat the ministerial sentence of death—or abject submission."[52] Americans, he agreed with Patrick Henry, must fight. Jefferson's ringing rhetoric at first irked understated Puritan lawyer John Adams, but he came to realize their importance: "These things were necessary to give popularity to our cause, both at home and abroad."[53] Working closely with Jefferson for the first time in what was to become a historic friendship, Adams preached patience with the slow pace of the radical agenda in the face of entrenched conservatism in Congress. "Its progress must be slow," he explained to his wife. "It is like a large fleet sailing under convoy. The fleetest sailors must wait for the dullest and the slowest." Or, using an image more suited to Jefferson's tastes, "Like a coach and six, the swiftest horses must be slackened, and the slowest quickened, that all may keep an even pace."[54]

Adams's verdict on his new colleague from Virginia was unequivocal. Like Washington, Jefferson avoided disputatious debates before the full Congress. "Though a silent member of Congress," Jefferson, according to Adams, was "prompt, frank, explicit and decisive upon committees and in conversation." Jefferson, concluded Adams, had "soon seized upon my heart."[55]

After six weeks in hot, muggy Philadelphia made worse by the need to close the State House windows to keep deliberations secret in the heavily Loyalist town, Jefferson drove south on August 1, 1775. But he still could not go home to Monticello: his personal wishes had to be set aside again in favor of his new duties. There was still another Virginia Convention at Richmond. There, delegates took the reins of provincial government into their own hands, setting taxes and troop quotas, electing regimental officers (Patrick Henry was chosen commander-in-chief). The delegates at Richmond voted for the delegates to the next Continental Congress. Jefferson received the third-highest tally, eighty-five votes, behind only Peyton Randolph, eighty-nine, now too ill to attend, and Richard Henry Lee, eighty-eight votes.

Jefferson spent only a week in Richmond. On his way home he performed a painful personal errand. He stopped off to visit Carter Braxton, who was helping to dispose of the property of another Virginia Loyalist going into exile. This time, it was his cousin, Attorney General John Randolph. So many times Jefferson had spent evenings of music and literature with his talented and literate, if stiff-necked, older cousin, who remained very close to Jefferson's mother. Jefferson had no doubt learned of Randolph's plan when his son Edmund, who was in effect Jefferson's junior law partner, stopped off in Philadelphia in July on his way north to join Washington's army at Boston. Invoking his agreement of four years earlier, Jefferson paid Braxton thirteen pounds for Randolph's fine violin and all his music. John Randolph had broken with his brother, Peyton, and his son Edmund, becoming an outspoken critic of American resistance to British policies. In May, just after Henry's march on the capital and before the semiannual meeting of Virginia merchants in Williamsburg, Randolph had allegedly said, in a private conversation in his home, "Do you think the merchants will be such fools to come to Williamsburg with money in their pockets when Patrick Henry or any other set of men might come and take their money from them?" Randolph supposedly also had accused Henry of "extortion" of money from the royal government to replace the gunpowder seized by Lord Dunmore. Week after week, Randolph's denials and the affidavits of witnesses were carried in the newspapers, collectively making Randolph one of the more conspicuous and detested symbols of royal viewpoint. On July 27, the *Virginia Gazette* carried a letter from "a Surry volunteer" to "J-n R-ph, *esquire,* advising him that,

> If your principles are incorrigible, if you are rooted in wrong, pray abscond yourself, push for some remote corner of the globe, where the imprecations of your countrymen and the invectives of a much injured people, cannot assay your adamantine ears.

The week Jefferson had left for Congress, Randolph had announced in the *Gazette* that he intended to "leave the colony for a few months" and was authorizing the sale of his "estate, both real and personal." After Governor Dunmore had fled to a royal warship, Randolph, the last man in Virginia to give up hope of reconciliation, had shuttled back and forth between Dunmore and the House of Burgesses for weeks, trying to resolve their differences.

Shortly after Jefferson reached home at Monticello, Jefferson wrote John Randolph another of his carefully concealed political letters. Ostensibly, he wrote on a trivial subject, the purchase of his cousin's violin. But, as in his recent letter to Dr. Small, Jefferson was trying to open the eyes of any Englishman who saw the letter to the true state of affairs in America. The etiquette of letters in the eighteenth century was different from today's. It was assumed that a letter, especially one crossing an ocean, would be read aloud or handed around, as it was such a precious instrument. But there was a code to be observed with such a private letter between relatives, especially across enemy lines. Jefferson gave his cousin permission to circulate his letter; he hoped he would. He believed it might help ease the crisis:

> I am sorry the situation of our country should render it not eligible to you to remain longer in it. I hope the returning wisdom of Great Britain will e'er long put an end to this unnatural contest. There may be people to whose tempers and dispositions contention may be pleasing and who may wish a continuance of confusion, but to me it is, of all states, but one, the most horrid.

Jefferson told Randolph he had two wishes, "a restoration of our just rights" and that he might withdraw "totally from the public stage," banishing "every desire" of even "hearing what passes in the world." Already, Jefferson yearned to stay at Monticello in "domestic ease and tranquility" without any pretext to try to persuade his cousin to do his utmost to influence the British Ministry. Again, Jefferson blamed the royal governor for deceiving the Ministry into believing that American opposition was "a small faction in which the body of the people took little part." Jefferson reminded Randolph that he knew this "of your own knowledge to be untrue."

> They have taken it into their heads, too, that we are cowards and shall surrender at discretion to an armed force. The past and future operations of the war must undeceive them . . . I wish they were thoroughly and minutely acquainted with every circumstance relative to America.

Jefferson then confided to Randolph that, since blood had already been spilled, Congress was considering accepting foreign aid (which could only mean from England's hated rival, the French). Moreover, he wrote in a passage he later cut out of the finished copy, that if England succeeded in subduing New England, the southern colonies were prepared to secede from the British Empire and to establish themselves independently, annexing the lands west of the mountains. Before Jefferson would submit to Great Britain, he told his cousin he would "lend my hand to sink the whole island [of Great Britain] in the ocean." Whether Britain remained "the head of the greatest empire on earth" or returned to her smaller former importance, depended on Parliament's decisions that winter. Affecting a lighter personal tone in an otherwise austere lecture, Jefferson proposed to buy Randolph's extensive library. But it was too late for personal dealings: Jefferson did not offer to meet with Randolph. He would deal with any agent Randolph appointed to handle the sale. Randolph indeed showed the letter to Lord Dunmore, whom Jefferson had so soundly denounced for misinforming London, before he sailed for England.

Before John Randolph left America, perhaps forever, he wrote a sad, kind letter to his cousin on August 31, 1775, apologizing that there was no case for the violin. Their differences were only political, he reminded Jefferson. Privately their esteem for each other should remain intact. "Should any coolness happen between us, I'll take care not to be the first mover of it. . . . We, both of us, seem to be steering opposite courses. The success of either lies in the womb of time."[56] Jefferson had lost another father figure.

On September 9, 1775, the *Virginia Gazette* tersely reported that John Randolph's son Edmund had been appointed one of Washington's aides-de-camp in Massachusetts, removing all question of where his personal loyalty lay. On the same page, there was another announcement: "Yesterday morning John Randolph, Esq., his Master's Attorney General for this colony, with his lady and daughters, set out from this city for Norfolk to embark for Great Britain."

Jefferson's return to Monticello was marred by yet another sad loss; he arrived just in time to bury his second daughter, seventeen-month-old Jane. He did not record the event in his record books. The exact date is unknown. He did not mention the child in any surviving correspondence. In his *Garden Book,* however, he recorded the untimely arrival of winter. For all of September, he made only one entry, on the 21st, "This morning the northern part of the Blue Ridge (to wit from opposite to Monticello northwardly as far as we can see) is white with snow."[57]

He wrote little of planting that first year of the Revolutionary War, despite the fact that there were ten family members, twenty-four free men and women, and eighty-three slaves to feed from crops that certainly were sown and reaped. Yet, in his *Account Book,* he now recorded a conversation with his old neighbor.

> Dr. Walker says he remembers that the years 1724 and 1741 were great locust years. We all remember that 1758 was, and now they are come again. . . . They come out of the ground from a prodigious depth. . . . The females make a noise well known. The males are silent.[58]

Did Jefferson connect these pestilential locust years with times of death? All had occurred in wartime. The year 1758 he must have remembered: Dr. Walker's frequent visits to Shadwell, and his father's death. Silently, without other comment, Jefferson buried his child, the first of his children to die, near his sister Jane and his friend Dabney Carr.

Before he left Monticello again that autumn to attend the adjourned session of the Second Continental Congress, the Virginia Committee of Safety commissioned Jefferson County lieutenant and commander of Albemarle militia, but Colonel Jefferson had no time for military service. He remained within sight of the mountains as long as he could, riding north by way of Orange Courthouse and Culpeper, taking along only one slave, a boy, Bob Hemings. He made faster time, returning to his old digs at carpenter Benjamin Randolph's house. Jefferson forwent the services of his slave to have him inoculated against smallpox and helped keep a common kitchen with several fellow delegates, including Peyton Randolph and his wife, and his fat, lively friend, Thomas Nelson, Jr., of Norfolk. Jefferson was with Speaker Randolph and his wife at the country estate of Henry Hill on an October evening when the Speaker died of a stroke at dinner. Jefferson helped to carry him back to their boardinghouse, sent off a messenger to notify his cousin Edmund, and helped to prepare for an elaborate funeral service at Christ Church attended by all of Congress. Wearing a black crepe band around his arm, he paid his respects to the kind old cousin who had brought him along.

Most of the news in Congress in the autumn of 1775 seemed auspicious for the American cause. A New England–New York joint expedition under Philip Schuyler invaded Canada. While Schuyler's forces besieged St. Jean-sur-Richelieu in southern Quebec, Benedict Arnold led a secret expedition up the Kennebec River through Maine toward Quebec City. By late November, Montreal had fallen. When word arrived from London that the king had declared the colonies in a state of rebellion and threatened dire treatment to the traitorous rebel leaders, Jefferson took the opportunity to write John Randolph again, this time in England, under the pretext of telling him what he doubtless already knew, that his revolutionary brother, Peyton, had died six weeks before. Once more, Jefferson lambasted the king:

> . . . We are told, and everything proves it true, that he is the bitterest enemy we have. . . . To undo his empire, he has but one truth to learn; that, after the colonies have drawn the sword, there is but one step more they can take. That step is now pressed upon us. . . . Believe me, dear sir, there is not in the British empire a man who more cordially loves a union with Great Britain than I do. But by the God who made me, I will cease to exist before I yield to a connection on such terms as the British Parliament propose; and, in this, I think I speak the sentiments of America.[59]

At this early date, Jefferson hinted to his Loyalist cousin in a letter intended for British government eyes, that there was now a growing movement in Congress to separate America from the British Empire. In his *Notes on Virginia,* Jefferson wrote expressly in 1784: "It is well known that, in July, 1775, a separation from Great Britain and establishment of republican government, had never yet entered into anyone's mind." America lacked

> neither inducement nor power to declare and assert a separation. It is will alone that is wanting, and that is growing apace under the fostering hand of our king. One bloody campaign will probably decide, everlastingly, our future course, and I am sorry to find a bloody campaign is decided on.[60]

There was little left for Congress to do now but wait for battles and to prepare for them. Jefferson had become the writer of Congress. When word arrived that Ethan Allen had been captured in a premature attack on Montreal and was being taken in irons to England, Jefferson was given the task of drafting a stiff protest. He demanded that England respect the "right of nations" that insisted on humane treatment—but he threatened retaliation "blood for blood" against British prisoners in American hands. Jefferson was immersing himself in the details of planning and policing the Canadian invasion: he was convinced that Arnold's surprise attack on Quebec had "determined the fate of Canada." The protest, never sent, shows that Jefferson had thoroughly come around to the necessity of "success in the fields of war." To punish a few hapless prisoners for "pretended treasons" which were "created by one of those laws whose obligations we deny and mean to contest by the sword" would be "to revive ancient barbarism."[61] Jefferson had resorted to war when peaceful attempts to reform the law had failed. But he expected the British to follow the rules.

In late December 1775, Jefferson took a leave from the Continental Congress. The other delegates could cast Virginia's single vote. Jefferson was becoming increasingly anxious about the health and safety of his wife, who had gone to stay at The Forest with her sister while Jefferson was in Philadelphia. For months, despite the fact that he had taken a day each week to write and post letters south, he had heard nothing from her.

As soon as Jefferson had arrived in Philadelphia for the resumed Continental Congress, he had learned alarming news from England and, on October 10, he had transmitted it in a long, detailed distillation of British strategic plans to his brother-in-law Francis Eppes, at The Forest. The immediate British response to the Battle of Bunker Hill was to send off ninety brass cannon from the Tower of London and two thousand Redcoat reinforcements, which were now en route to New York City while ten thousand more Redcoats were rotated to America from Gibraltar and Ireland and replaced on garrison duty by hired Hessian mercenaries. Worse still for Virginia,

> Commodore Shuldam was to sail about the same time with a great number of frigates and small vessels of war to be distributed among the middle colonies. He comes at the express and earnest intercessions of Lord Dunmore. . . . The plan is to lay waste all the plantations on our river sides. . . . If any defense could be provided on the rivers by fortifications or small vessels it might be done immediately.[62]

By October 31, a full two months after he had left Monticello, when letters to his wife and brother-in-law drew no reply, a worried Jefferson complained at the end of a military-news-laden letter to John Page,

> I have set apart nearly one day in every week, since I came here, to write letters. Notwithstanding this, I have never received the scrip of a pen from any mortal breathing.[63]

By November 21, Jefferson wrote to Eppes that he had "written to Patty a proposition to keep yourselves at a distance from the alarms of Lord Dunmore." To his brother-in-law, Eppes, whom Jefferson assumed was also at The Forest, Jefferson wrote again on November 7:

> I have never received a scrip of a pen from any mortal in Virginia since I left it, nor been able by any inquiries I could make to hear of my family. . . . The suspense under which I am is too terrible to be endured. If anything has happened, for God's sake, let me know.[64]

All the news Jefferson had received from Virginia came through Congress and it was disturbing. In early November, Lord Dunmore, on board HMS *Fowey* off Norfolk, had declared martial law in Virginia. On November 11, Page wrote back a long letter detailing a British attack on Hampton and Jamestown by a flotilla of small boats with some runaway slaves in the redcoated ranks. In Norfolk harbor, which was the colony's principal seaport and the most strategic location for a naval base between New York and Charleston, South Carolina, Lord Dunmore had gathered a small navy of some two hundred Loyalist merchant ships and three British sloops of war. Hundreds, perhaps thousands, of Loyalists from coastal Virginia and the Eastern Shore of Maryland along with Scottish merchants, their employees, and their families had gathered for the king's protection. Up to a thousand Scots had operated general stores in the region and had boycotted patriot attempts to enforce the trade embargoes of the past ten years. For two years, patriot committees had raided Scottish stores and auctioned off their goods as part of the Association's boycott on imports. Along with his proclamation of martial law, Dunmore had issued a proclamation emancipating all slaves and indentured servants "appertaining to rebels" who fled their rebellious Virginia masters and rallied to the king's colors.

On Friday, December 1, 1775, the fourth Virginia Convention ended its rump session in Richmond and its delegates rode over rutted mud roads to the old capital at Williamsburg. The next day, the Second Regiment of Virginia's new militia, reinforced by North Carolina troops, marched to Great Bridge, which crossed the southern branch of the Elizabeth River twelve miles southeast of Norfolk. There, the patriots, who were mostly overmountain frontiersmen called "shirtmen" by the English, found that Lord Dunmore had built a wooden fort flanked by earthworks and protected by small cannon. There, nine hundred buckskin-clad riflemen from western Virginia and North Carolina dug in to await orders from the

Virginia Convention, throwing up a 150-foot-long breastwork seven feet high.

Jefferson's close friend John Page, now vice president of the Virginia Committee of Safety and second-in-command to Patrick Henry, lived in Gloucester County, within reach of British sailing parties at Great Bridge. He was alarmed by the confrontation and wrote to Richard Henry Lee in Congress at Philadelphia that the Virginia troops had been blocked "by a body of Negroes headed by Scotchmen and a few regulars." Page's letter, which hinted at the slaveowner's fear of a general slave insurrection, certainly was shown to Jefferson and may have prompted his anguished letter to Francis Eppes. But Eppes was not at The Forest to receive it; he was now a major leading three companies of the First Virginia Regiment toward the coast to reinforce the troops gathering at Great Bridge. The British were also strengthening the fort there. The night of December 8, Lord Dunmore dispatched two companies of the British Fourteenth Regiment, including 153 grenadiers, light infantry, gunners from British ships in Norfolk harbor, and sixty Loyalist volunteers. In all, there were 672 in the British force inside the fort by the morning of December 9, when the first land battle in the south of the American Revolution broke out.

Lord Dunmore had taught his troops, according to prisoners, to believe that the western riflemen wouldn't stand up to attack and would undoubtedly be frightened away by the first British charge, but also that any man falling into their hands would be scalped. At dawn on December 9, 150 British grenadiers, tall men who stood seven feet in their pointed hats, attacked the American breastwork with bayonets fixed. They were followed by more than three hundred Loyalists and blacks with two field artillery pieces dragged along to rake the American breastworks.

Confined by a narrow causeway, the British regulars, resplendent in red uniforms with pale buff facings, marched six abreast in perfect parade formation to the beat of two drums toward the silent patriots.

The regular vanguard alternated volleys of musketfire by platoons, and paused the fifteen seconds it took these skilled professionals to reload. The patriot troops waited behind the sagging-**M**-shaped breastwork, which blocked the Norfolk road. They had been ordered to hold their fire until the British came within fifty yards. When they did, "bullets whistled on every side."[65] The two British officers in the lead went down. Capt. Charles Fordice, commander of the grenadiers, wounded in the knee, rose, brushed his leg lightly, swept off his tricorn, and, waving it, shouted, "The day is our own."[66] As American riflemen reinforced a trench fifty yards off to the west, their commander, Adj. Gen. Thomas Bullitt, told them to "keep an eye"[67] on the British captain. Fourteen men did. When

Fordice and his grenadiers came within fifteen yards of the American breastwork, Fordice went down again, this time with fourteen bullets in him. Only two or three British regulars reached the breastwork before they fell. As the Americans poured out of their trenches to counterattack, the British, their officers all dead or wounded, retreated pell-mell, dragging their wounded, some of them yelling, "For God's sake, do not murder us."[68] One hundred Culpeper Virginia Minutemen dashed to an opposite line of trenches and began pouring a deadly fire into the British retreating over the bridge. Ordering all his men and guns back inside the fort, the ranking British survivor, Capt. Samuel Leslie, cradled his dying nephew in his arms. A single Virginian had been wounded in a finger in the battle; 102 British lay dead or dying, an even worse casualty rate than at Bunker Hill. The battle had lasted less than thirty minutes. An American officer, Capt. Richard Kidder Meade, reported a "vast effusion of blood. . . . Too much . . . I then saw the horrors of war. . . . Worse than can be imagined. . . . 10 and 12 bullets through many, limbs broke in two and three places, brains turning out . . . Good God, what a sight."[69]

The American victory threw Virginia's Loyalists into a panic. Hundreds more Scottish merchants and their families crowded onto two hundred vessels, large and small, in the harbor as the British commander ordered his troops aboard ships and said he would sacrifice no more men to Dunmore's "whims." On December 12, Norfolk's magistrates, including Dr. Archibald Campbell, Jefferson's friend and client from the smallpox riots case, met with Virginia officers to plead that the patriots do nothing to provoke Dunmore to carry out his latest threat, which was to bombard the town. But that night, one thousand patriot militia marched to Norfolk, taking up positions out of sight of the Royal Navy gunners.

The arrival of the news of the victory at Great Bridge prompted the first open debate on a declaration of independence from England, held on December 13. Once so conservative, old and stooped Robert Carter Nicholas introduced a resolution that denounced Lord Dunmore as a champion of "tyranny," a monster "inimical and cruel" for pronouncing martial law and assuming "powers which the King himself" could not exercise. Dunmore had broken "the bonds of society."[70] Nicholas's declaration was adopted unanimously.

The Convention instructed Nicholas's committee to draw up a second document. The next day, Woodford reported an attack by white troops on a stockade manned by black troops under British sergeants. One black Loyalist had died when the whites set fire to the outpost; three others were captured and twenty-six blacks and nine whites had escaped. Woodford had to restrain his officers from killing the black prisoners "to make an immediate example of them."[71] On December 4, Colonel Woodford had informed the Convention that large numbers of slaves led by regulars and Scotch Loyalists had joined the British, including the entire garrison of 250 inside the fort at Great Bridge. The news that black slaves had not only run away and joined the king's standard but were willing to take up arms to fight and kill their former masters sent a shudder down the backs of the slaveowning Convention. Colonel Woodford had also sent to the Convention a specimen of a scored lead bullet that would later be called a dumdum bullet. He said the bullet had been taken from the cartridge of a black troop he had just captured. Lines had been cut into it to assure that it would split into two or more pieces that would cause a ghastly wound when it struck flesh. All the black prisoners were carrying such bullets, he said, and had been taught how to make them by their Scottish sergeants.

Nicholas was instructed to write a document "offering pardon to such slaves" as should voluntarily return "to their duty"—to their Virginia masters. The very next day, Nicholas's resolution was adopted by the full Convention, which "earnestly" recommended "to all humane and benevolent persons in this colony to explain and make known this our offer to mercy to those unfortunate people" who had been "deluded" by Lord Dunmore's "base and insidious acts."[72] The resolution also asked the Committee of Safety to instruct Colonel Woodford to exchange his prisoners with the British, but if that were not possible he was to send all captured slaves to Williamsburg to be dealt with. By this time, fourteen runaway slaves had been captured and were in irons.

Two days later, on December 15, 1775, Captain Squire of the British sloop *Otter*, commander of the British squadron off Norfolk, complained about "several musket balls" fired on his ships even as he demanded the patriots supply the British ships with food and water. Further, if the patriots failed to return a brig loaded with precious salt that they had captured, Squire said he would "most assuredly fire the town."[73] For their part, the patriots would also have welcomed an excuse to burn Norfolk, which these western troops considered a Tory haven filled with "a contemptible set of wretches"[74] to whom many Virginians owed considerable money. John Page was among many patriots who were damning the townspeople of Norfolk. To Jefferson, he wrote that they "deserved" to have "their town burnt," deserved "to be ruined and hanged. . . . Rather than the town should be garrisoned by our enemies and a trade opened for all the scoundrels in the country, we must be prepared to destroy it."[75]

On December 21, in a snowstorm, cheers of joy came from the shivering, hungry Loyalists and Scots crowded into the "near 200 sail, large and small,"[76] as the twenty-eight-gun British frigate *Liverpool,* carrying 250 men and accompanied by a supply ship, came up the bay. Its skipper, Capt. Henry Bellew, considered Norfolk the key to navigation in two colonies. He quickly decided to clear and retake or destroy the town to flush out its rebel defenders. He deployed a line of five British warships interspersed with armed merchantmen, all riding broadside and ready for action and again demanded fresh provisions from the patriot commander ashore. They hastily sent a courier to the Convention at Williamsburg for instructions. Before any answer could come from Williamsburg, newly arrived patriots, apparently without orders, fired into a British guardboat. Gunners on the *Liverpool* answered with cannonfire. On December 30, *Liverpool's* captain sent a message that the "honour" of his commission made him demand that the rebels "leave the town."[77]

On New Year's Day, 1776, as if to goad the British, the patriots paraded in the streets down to the wharves, waving their hats on their bayonets. They were clearly visible to British spyglasses and they gave the British every possible "mark of insult." At 3 P.M., *Otter* sent three cannonballs whizzing across Norfolk's wharves and into the rebel guardhouse; soon every armed vessel in the harbor opened fire. Lord Dunmore "sent some boats on shore to burn some detached warehouses" and hailed Captain Bellew on *Liverpool* to destroy a rebel-captured salt ship. Bellew complied. Every ship in the harbor sent its crews ashore to burn the wharves and nearby houses that could shelter snipers. As the winter wind picked up, "the fire soon became general." Ships sporadically fired their cannons for eight hours. By the time the gunfire ceased shortly before midnight, "the clouds," wrote one witness on an approaching ship, "appeared as red and bright as they do in an evening at sun setting."[78]

To the dismay of property-owning Loyalists aboard the fleet, the town burned for five days. A Virginia patrol venturing into Norfolk the morning of January 2 found "the lower streets and wharves in a great measure extinguished." But Sgt. Henry Henly of the First Virginia, a Norfolk resident, was startled to find Virginia troopers "with firebrands in their hands, destroying the houses of the inhabitants." When Sarah Smith approached Col. Robert Howe of the North Carolina militia and asked him, in the presence of the Virginia commander, Colonel Woodford, whether they intended to burn her house, Howe paused, then answered yes, adding that they would "burn up the two counties"—Norfolk and Princess Anne—suspected Loyalist strongholds. For hours patriot troops, after looting a rum distillery, roared through Norfolk's streets, burning, hooting, and shouting, "Keep up the Jigg!" A barber protested when he saw the outhouse of a prominent merchant being torched: this seemed beyond the call of duty. He saw Virginia troops "roll a pipe of wine out of the house of Capt. Cornelius Calvert and, after beating in the head, they drank of the wine and filled their canteens and bottles with it." By January 3, when Colonel Woodford forbade further sacking and burning, he judged that "nine-tenths"[79] of the town was destroyed. A Virginia General Assembly investigating commission found in 1777 that, of the 1,331 Norfolk structures destroyed, 51 had been burned under British orders and 863 by patriot militias.

Amid rumors that an army of five thousand Scottish Highlanders was about to attack from the east while Cherokees struck from the west and south, Jefferson gathered up his family at The Forest and took them to Monticello, as far as he could from the coastal fighting. He gave no hint in his surviving writings why he had heard nothing from his wife in so many anxious months, and he destroyed all correspondence between them at her death. He celebrated their safe arrival home by opening a pipe of 1770 Madeira and then he helped to foal the sorrel mare Ethelinda, another proud descendant of Young Fearnought. For the next four months, Jefferson tried to live the life of a country squire, among other things concerning himself with stocking his deer park. He was still worried about his wife and tried to put off the problem of what to do when he returned to Congress. Thomas Nelson, writing about congressional business on February 4, urged Jefferson to bring Patty to Philadelphia, where he had brought his own wife. "Mrs. Nelson shall nurse her," Nelson promised, "and take all possible care of her."[80] Was Patty once again pregnant? There is no way of knowing. Jefferson's most personal letters were later weeded out and destroyed by the family.

Then, on March 31, 1776, Jefferson's mother, who had been living at Tuckahoe, died unexpectedly of a stroke. Jefferson could barely have reached her side before she died. She was fifty-seven. Even though Jefferson had never been close to her and had, as an adolescent, strongly resented being cast back under her control at his father's death, her death added to all his other anxieties over the war and his wife's health and apparently triggered a five-week bout of migraine headaches that left him bedridden and unable to write about her death or anything else. He had suffered these headaches at least once before in a period of great stress when he had come back to Shadwell from college for the first time. Every six or eight years, at periods of heightened tension, he was felled by severe headaches for about a month. Writing to a friend about one bout he experienced while president, he described "a severe attack of periodical headache which came on every day at sunrise and never left me till sunset." Whatever Jefferson thought about during one of these daily attacks he did not attempt to write down until evening. "What had been ruminated in the day under a paroxysm of the most excruciating pain was committed to paper by candlelight."[81] The bout of headaches Jefferson suffered in April 1776, was among his worst.

In all that long winter and spring at home at Monticello, Jefferson apparently wrote almost nothing. Sometime after January 19, 1776, he researched and wrote a long historical refutation of the British Ministry's assertion that the American colonies had been established at the expense of the British nation, a favorite Loyalist argument. Concentrating on the voyages of Sir Humphrey Gilbert and Sir Walter Raleigh, and drawing his "narration of facts" principally from the writings of Hakluyt, Jefferson ended with another broadside at George III, again warning the king, because he heeded "wicked favorites," that he would once again become "a petty king":

> . . . Vice is a foul blemish, not pardonable in any character. A king who can adapt falsehood and solemnize it from the throne, justifies the revolution of fortune which reduces him to a private station.

Was Jefferson prophesying the overthrow of the king? If so, he was certainly holding both the king *and* his ministers accountable for the "weakness" and "vices" of his character, and he wished upon them "perpetual execrations."[82] This tirade, entered in a notebook but never published, rivaled in antimonarchical invective the writings of Tom Paine, whose *Common Sense* was published about the same time. Jefferson's angry blast, apparently dictated to a secretary, did not seem to help vent his anxiety or ease his headaches.

It was three months after Jefferson buried his mother before he wrote a word about her death. He waited until he returned to Congress in May before he sent word to his mother's brother, William Randolph, for many years a merchant in Bristol, England, that Jane Randolph was dead. In his only surviving scrap of writing about her loss, he wrote, totally unemotionally,

> The death of my mother you have probably not heard of. This happened on the last day of March after an illness of not more than an hour. We suppose it to have been apoplectic.[83]

Jefferson's coldness toward his mother may have been exacerbated by her alleged sympathies to the Loyalist viewpoint of cousin John Randolph. Whatever the causes of his hostility toward her from childhood on, he owed her much beyond their strong physical resemblance. He had learned his reticence and refinement from her and his taste for reading, music, and dancing. His granddaughter wrote that it was from Jane Randolph that Jefferson "inherited his cheerful and hopeful temper and disposition." According to all other family sources (save Jefferson himself), his mother was a woman of "clear and strong understanding,"[84] but Thomas Jefferson seems never to have forgiven his mother for living on after his father died. Perhaps they were too much alike.

While Jefferson was convalescing at Monticello, his fellow congressman Thomas Nelson sent him from Philadelphia a copy of an explosive new pamphlet—"a present of two shillings worth of *Common Sense.*" Written by Tom Paine, the hired pen of Benjamin Franklin and editor of the *Pennsylvania Magazine,* its angry antimonarchical rhetoric was clear and blunt, not classical or refined. It came on the heels of the news of the king's denunciation from the throne of Americans as "traitors" who were in "open and avowed rebellion,"[85] a declaration that helped to convince many Americans, including Jefferson, that the time for talking was over. Like most Virginia leaders, Jefferson had been unwilling to face the idea of a total break with England when he wrote his *Summary View* in July 1774. The king had still linked England and the American colonies. But by the time Jefferson had written to John Randolph in late November 1775, he had shifted his ground and was blaming the king as well as Parliament for the out-of-control pace of attacks and reprisals. He had inched closer to talking about independence when he warned Randolph that America lacked "neither inducement nor power to declare and assert a separation."[86] The shocking news that Norfolk had been shelled and burned made Jefferson pay close heed to Tom Paine's insistence that the period of debate was over, that Americans must

> . . . Stand forth! Every spot of the world is overrun with oppression. Freedom hath been hunted round the globe. Asia and Africa have long expelled her. Europe regards her like a stranger . . . and England hath given her warning to depart. O! receive the fugitive and prepare in time an asylum for mankind.

Jefferson wrote nothing about *Common Sense* at the time, but when most people had forgotten Paine, Jefferson praised his contribution to the

American cause in the early months of 1776. A commoner newly arrived in America, Paine was a professional revolutionary who saw the absurdity of her imperial connection to England, and foresaw the import of the American cause for everyone and not just for Americans:

> The sun never shone on a cause of greater worth. 'Tis not the affair of a city, a county, a province or a kingdom, but of a continent —of at least one eighth part of the habitable globe. 'Tis not the concern of a day, a year, or an age. Posterity are virtually involved in the contest, and will be more or less affected even to the end of time by the proceedings now.[87]

Paine's new, naked, powerful prose appealed to the "simple voice of nature and reason," and Jefferson agreed with most of it. As Paine's propaganda swept the colonies closer to independence, Jefferson polled his Virginia neighbors and constituents. "When at home," Jefferson wrote to Thomas Nelson on May 16, "I took great pains to enquire into the sentiments of the people on that head. In the upper counties, I think I may safely say nine out of ten are for it."[88] But the ambivalence of many other Virginians was still hindering efforts even to defend the American colonies, let alone appeal for foreign aid to fight the British. By May 1776, Jefferson left his trusted Jupiter to help his wife and took fourteen-year-old Bob Hemings north with him. Jefferson's old friend John Page spoke for many Americans when he wrote, "For God's sake declare the colonies independent and save us from ruin!"[89] Page lived in Gloucester County, Virginia, not far from the British ships. Page's letter awaited Jefferson at Benjamin Randolph's boardinghouse in Philadelphia when he arrived to resume his seat with the Virginia delegation. Another letter from a William and Mary schoolmate, Dr. James McClurg, declared: "The notion of independence seems to spread fast in this colony, and will be adopted, I dare say, by the next convention."[90]

On May 6, 1776, the day Jefferson left Monticello to ride north, forty-five elegantly dressed Virginia gentlemen marched to the seventeenth-century red-brick Capitol at the east end of Duke of Gloucester Street in Williamsburg. Filing into the long dark rows of facing benches in the ornate first-floor chamber of the 157-year-old House of Burgesses, they gazed for the last time at the gilded emblems of royal authority decorating the walls. The burgesses had assembled according to an order of adjournment voted on March 7 for a meeting of the royal Assembly of Virginia. But assemble was all they did. The Speaker, Peyton Randolph, was dead.

House Clerk George Wythe was in Philadelphia, where he was now a member of Congress. No members of the colony's Council of State had assembled in the upstairs chamber. The governor was aboard a warship in the Elizabeth River and could not summon them upstairs to greet them. Nothing could happen. The oldest royal government in America was dead. As Edmund Pendleton later wrote, the "several members met, but did neither proceed to business, nor adjourn, as a House of Burgesses." The burgesses simply "let that body die."[91]

Inspiriting these rebellious former burgesses was the knowledge that they had the backing of the revolutionary delegates to the Continental Congress. On December 4, after a messenger had brought them Lord Dunmore's proclamation, Jefferson and his Virginia colleagues had urged the Congress to send troop reinforcements. Many of Virginia's best riflemen had been killed or captured in the invasion of Canada; most of the others were opposing Dunmore's force on the coast. Three companies of western Pennsylvania riflemen had been sent south into Northampton County. In the same resolution, Congress had urged Virginia "to resist to the utmost" Dunmore's "arbitrary" government. Congress declared that to impose martial law was to tear up the "foundations"[92] of Virginia's civil government. Congress then had taken the historic step of encouraging Virginia to establish a new government and call free elections. Jefferson had at once turned his mind to drawing up a new constitution for Virginia.

The wealthy representatives of Virginia's established counties who had witnessed the death of the House of Burgesses later joined a second group outside the Capitol. Together they reentered the downstairs chamber. This time they took their seats as delegates to the fifth (and last) revolutionary Virginia Convention: Many carried instructions from their county freeholders to declare Virginia independent of Great Britain. Of 108 members who presented credentials, one-third were novices who had never before served in a province-wide assembly. They were, wrote one, "not quite so well dressed, nor so politely educated, nor so highly born" but were "full as honest, less intriguing, more sincere."[93] The Commonwealth of Virginia had been born. After two years of suspended government, the Fifth Convention now moved quickly to take up an agenda crowded with issues unresolved by the first four—reinvigorating the court system, drafting a new form of government, collecting taxes, punishing "the enemies of America," producing salt, and protecting the frontiers.

But nothing else could be resolved before the question of independence. Day after day, the debate had to be put off as the house divided over the evacuation of Norfolk and Princess Anne counties, where Lord Dunmore's foraging parties carried out daily raids for provisions for his troops and Loyalist refugees. Finally the debate over independence came on. Meriwether "Fiddlehead" Smith of Essex County proposed "that the government of this colony as hitherto exercised under the Crown of Great Britain be dissolved and that a committee be appointed to prepare a Declaration of Rights."[94] He also called for a "plan of government" to "maintain peace and order" and "secure substantial and equal liberty to the people." Bartholomew Dandridge then introduced a resolution that went a giant step further:

> That the Union that has hitherto subsisted between Great Britain and the American colonies is hereby totally dissolved, and that the inhabitants of America are discharged from any allegiance to the Crown of Great Britain.[95]

A resolution written by Patrick Henry recited the colonies' grievances against the king and "absolved" Virginians of allegiance "to the crown of G.B."[96] Only a month earlier, in a rare fit of caution, Henry had worried that to declare independence before receiving foreign aid would guarantee disaster. But he now argued that no foreign power would offer aid unless the colonies declared themselves independent states. Preparing a composite resolution from all the proposed resolutions, a committee appointed by the Convention's new Speaker, Edmund Pendleton, resolved that Virginia's delegate to the Continental Congress be instructed to propose that Congress "declare the United Colonies free and independent states absolved from all allegiance to or dependence upon the Crown or Parliament of Great Britain." On May 15, 1776, all 112 delegates at Williamsburg shouted aye.[97]

On May 14, on the eve of Virginia's daring resolution, Jefferson had arrived in Philadelphia. It was twelve more days before the Virginia resolution instructing her delegates to ask Congress to "declare the United Colonies free and independent states" and to give Virginia's assent "to whatever measures may be thought proper and necessary by the Congress for forming foreign alliances and a confederation of the colonies"[98] arrived. While North Carolina had already instructed its delegates to vote for independence if the majority favored it, Virginia boldly became the first to instruct its delegates to lead the way and propose a break with England. On May 27, 1776, Jefferson and his Virginia colleagues presented their new instructions for independence to President John Hancock. As they did, ardent Virginia patriots in Williamsburg, considering Virginia's independence a fait accompli, struck the Union Jack over the Capitol and hoisted up the new flag of the Continental Union.

For nearly two weeks, the Virginia resolution sat on the pile in front of Secretary Thomson, all the delegates assuming that Virginia had declared her independence from England. In fact, it was not for another month, until June 28, that Virginia would adopt a new constitution, the vital step in convincing all Virginians that their patriot leaders had severed all imperial ties and embarked on independence. When it was finally adopted, Virginia's new constitution contained a long preamble that reflected Jefferson's feverish efforts in the weeks after he returned to Philadelphia and heard of the plan to draw up a new frame of government for Virginia. To allow more quiet and privacy, Jefferson had left his digs with cabinetmaker Randolph on May 23 and moved to a handsome new three-story house on the edge of town at the southwest corner of Seventh and Market streets, only a short walk from the State House. On May 16, he wrote that, "as the excessive heats of the city are coming on fast," he was moving to lodgings "on the skirts of the town where I may have the benefit of freely circulating air."[99] The house belonged to a newlywed German bricklayer, Jacob Graff, who rented Jefferson two spacious rooms on the second floor, a bedroom, and a parlor flanking the stairwell. He spent long hours in the bright, airy parlor, using a folding writing box made according to Jefferson's plans by his former landlord. On this "plain, neat, convenient" portable desk, which Jefferson used until his death fifty years later, he wrote almost all his most famous state papers and thousands of letters. "It claims no merit of particular beauty," he then wrote. "Taking no more room on the writing table than a moderate quarto volume, it yet displays itself sufficiently for any writing."[100] Here, in late May and early June, he finished writing a series of bleak reports and recommendations that resulted from the failed American invasion of Canada, which had ended only weeks before in defeat and retreat of a decimated, smallpox-riddled army to Fort Ticonderoga, New York.

His migraine finally banished, Jefferson brooded about being in Philadelphia, which to him seemed far from the scene of actions to the north, where a British invasion force of ten thousand Redcoats and Hessians now threatened to descend on New York, and on his beloved Virginia, where his friends were drawing up the documents of a new and independent state. To Thomas Nelson, now in Williamsburg, he wrote,

> Should our Convention propose to establish a new form of government, perhaps it might be agreeable to recall for a short time

> their delegates. It is a work of the most interesting nature and such as every individual would wish to have a voice in. In truth, it is the whole object of the present controversy, for should a bad government be instituted for us, in future it had been as well to have accepted at first the bad one offered us from beyond the water without the risk and expense of contest.

Once again Jefferson was lonely and worried, "in the same uneasy anxious state in which I was last fall without Mrs. Jefferson, who could not come with me."[101] Even as war menaced all around him, Jefferson, in a letter to John Page full of a naval battle at the mouth of the Delaware River and "disagreeable news of a second defeat" in Canada, daydreamed of buying more books. The Anglican rector of Bruton's Parish in Williamsburg had fled, leaving his books "with Molly Digges for sale." Jefferson recalled two books the parson had, Pelloutier's two-volume *Histoire des Celtes* and Payne's eight-volume *Observations on Gardening.* Would Page purchase them and send him a catalogue of the rest? And he reminded Page he could use his Committee of Safety position to ship via "military or commissary's wagon a box of books you have of mine" to Monticello.[102]

By the time Jefferson had returned to Congress, events were pushing the politicians in Philadelphia ever closer to a showdown over independence. Congress had authorized privateering and thrown open American ports to all the nations of the world except Great Britain. Secret negotiations were taking place with French agents in hopes of procuring foreign aid. French arms were already being smuggled into American coves and ports from the Caribbean. While Arnold's and Montgomery's invasion into Canada had failed, the guns of Ticonderoga, dragged to Dorchester Heights overlooking Boston, had forced the last British army to evacuate the thirteen rebellious colonies. But the three Middle Colonies—New York, New Jersey, and Pennsylvania—still were controlled by pro-British governors, and the majority of their citizens still opposed independence outright or hesitated to risk declaring it. "The novelty of the thing deters some," Franklin wrote from Congress in April, "the doubts of success, others; the vain hope of reconciliation, many."[103] As Jefferson had returned to Congress on May 15, the day that Virginia passed its resolution calling for a congressional vote on independence, Congress was just winding up a debate on John Adams's motion calling on all colonies to establish their own governments. Adams was exultant: "Every post and every day rolls in upon us, Independence like a torrent."[104]

11

"An Expression of the American Mind"

The sentiments of men are known, not only by what they receive, but what they reject also.

—THOMAS JEFFERSON, 1776

ON FRIDAY, JUNE 7, 1776, Thomas Jefferson was in his usual Windsor armchair near the oak-paneled fireplace in the State House when Richard Henry Lee, tall, patrician, eyes piercing, stood to deliver Virginia's challenge, proposing a resolution that "These united colonies are, and of right ought to be, free and independent states."[1] At this historic moment, Jefferson made the most accurate and complete surviving record of the proceedings. The colonies, Jefferson noted, were to be

> absolved from all allegiance to the British crown, and that all political connection between them and the state of Great Britain is and ought to be totally dissolved; that measures should be immediately taken for procuring the assistance of foreign powers, and a Confederation be formed to bind the colonies more closely together.[2]

The next morning the Committee of the Whole took up the debate on independence. Instantly the conservative opposition—Jefferson listed James Wilson and John Dickinson of Pennsylvania, Robert R. Livingston of New York, and Edward Rutledge of South Carolina—attacked the timing of the resolution. Congress should wait. While New England heartily supported Virginia's initiative, the Middle Colonies (Delaware, Maryland, New Jersey, New York, Pennsylvania) as well as South Carolina were unready for independence, the conservatives argued. Jefferson recorded "that they were friends to the measures themselves" and realized it was impossible "that we should ever again be united with Great Britain," but they were against declaring independence "till the voice of the people drove us into it." The people of the individual colonies were "our power." The Middle Colonies "were not ripe for bidding adieu to their Great Britain" but were "fast ripening." A congressional resolution passed on May 15 suppressing all royal governments in the individual colonies had "thrown the middle colonies" into "ferment." Several colonies, such as New York and New Jersey, had expressly forbidden their delegates to consent to such a declaration. The Pennsylvania Assembly was "now sitting above stairs; in the next four days, conventions would gather in three other colonies to "take up the question of Independence." If the Continental Congress voted before hearing the "voice" of each state, these colonies "might secede from the Union," Jefferson wrote, a weakness more than offsetting "any foreign alliance."[3]

But the proponents of independence were just as determined. "The people wait for us to lead the way. . . . The declaration of independence alone could satisfy European delicacy before European powers would negotiate or even receive an ambassador. . . . That they would not receive American vessels into their ports. . . . That no time should be lost in opening trade with foreign countries. . . . People will want clothes and money, too, to pay taxes." At the close of the second day, Congress voted to postpone the vote for independence until July 1 to allow delegates to send home for new instructions. Meanwhile, "a committee was appointed to prepare a declaration of independence. The committee [members] were J. Adams, Dr. Franklin, Roger Sherman, Robert Livingston and myself."[4]

The independence committee appointed on June 11 in turn asked Adams and Jefferson to be a subcommittee to draw up the declaration. When the two men got together two days later, "Jefferson proposed to me to make the draught," Adams wrote years later.

I said, "I will not. You shall do it."
"Oh no!"
"Why will you not?"
"You ought to do it."
"I will not."

"Why?"

"Reasons enough."

"What can be your reasons?"

"Reason first: you are a Virginian and a Virginian ought to appear at the head of this business. Reason second: I am obnoxious, suspected and unpopular. You are very much otherwise. Reason third: You can write ten times better than I can."

"Well," said Jefferson, "if you are decided I will do as well as I can."

"Very well, when you have drawn it up we will have a meeting."[5]

According to Jefferson, writing to Madison forty-seven years later, the exchange never took place, as charming a recollection as it was. Citing his notes, Jefferson told Madison,

> The Committee of 5 met, no such thing as a subcommittee was proposed, but they unanimously pressed on myself alone to undertake the draught. I consented; I drew it.[6]

As soon as Jefferson learned that the Virginia Convention had declared independence from Great Britain and was creating a commonwealth, he set about writing the first of three drafts of a new constitution for his home country. Brooding that he could not be there at such a crucial moment, he labored over it before Congress convened in the morning and late at night. Even before he had left Virginia, he knew that the task of setting up a new government there was imminent and, as soon as he had reached Philadelphia, he wrote back to Thomas Nelson, who had gone to Williamsburg to report on recent developments in Congress, as gingerly as possible, hinting that the Virginia Convention should bring him home and use his writing skills. Congress could function without him. "In the other colonies who have instituted government, they recalled their delegates, leaving only one or two to give information to Congress," he wrote Nelson. Repeating the point as if to underscore its importance, he urged Nelson to have him recalled if even "for a short time" should the Convention establish a new form of government. He had little doubt that Congress would vote independence. The outcome in his native Virginia was far more important to him:

> It is a work of the most interesting nature and such as every individual would wish to have his voice in. . . . In truth, it is the whole object of the present controversy, for should a bad government be

> instituted for us in future, it had been as well to have accepted at first the bad one offered to us from beyond the water without the risk and expense of contest.[7]

Jefferson had an additional reason to wish he were in Williamsburg, not Philadelphia. He questioned whether the Virginia Convention had the right to create a permanent constitution until deputies were elected expressly for that purpose. A government, he had argued in his *Summary View,* had to be based on the will of the people. He sent his practical as well as theoretical misgivings south with his young cousin, Edmund Randolph. But the Virginia Convention had already forged ahead, drafting a declaration of rights. By May 27, George Mason had finished a draft constitution declaring that "all men are born equally free and independent and have certain inherent natural rights . . . among which are the enjoyment of life and liberty, with the means of acquiring and possessing property and pursuing and obtaining happiness and safety."[8] In uninspired prose, Mason's Declaration of Rights called for free and regular elections, a free press, and trial by jury, and denounced taxation without representation, general warrants, excessive bail, cruel and unusual punishment, and hereditary office, while preserving these largely traditional English liberties. This Virginia document was deeply conservative, keeping power in the hands of the planter oligarchy that had dominated Virginia for a century and a half and upholding the status quo. Mason's declaration retained the property-owning qualification for voting, keeping power in the hands of fewer than one-tenth of one percent of the population.

Unable to present his objections and amendments in person, Jefferson, deeply worried that the old gentry-dominated system would only be perpetuated, took advantage of an uncharacteristic cool spell in Philadelphia to write draft after draft of his own version of the constitution for Virginia. Jefferson's differed principally in granting a far wider voting franchise, preserving the property qualification by requiring a fifty-acre freehold of land but then granting enough free acreage from Virginia's vast public landholdings to any white male who lacked it to make up the fifty acres. In this revolutionary scheme, on a given day in the near future, every white male Virginian would own at least fifty acres of land and have the right to vote, extending the franchise to tens of thousands, making Virginia far more democratic. Jefferson was right to be alarmed by the Mason constitution. Jefferson's version had been aimed, one participant wrote years later, at preventing "the undue and overwhelming influence of great landholders in elections"[9] by disenfranchising landless tenant farmers and retainers who depended "on the breath and will" of great men by ensuring that only men who owned their own land could vote, but it actually would shrink the vote: in other colonies, property counted for voting purposes included horses, tools, furniture, leases. Mason's favored continuing dominance of a small, wealthy landowning elite in power, a fact that alarmed Jefferson that summer of 1776. He continued to rail against the American fascination for "founding great families" all his life. In his autobiography, he wrote nearly fifty years later that the "transmission" of property "from generation to generation in the same name" had, at the moment of revolution, only ratified the long-term tendency in Virginia to "raise up a distinct set of families" who remained "privileged by law in the perpetuation of their wealth" and constituted "a patrician order distinguished by the splendor and luxury of their establishments."[10]

In three distinctly different drafts, Jefferson laid down many of the philosophical principles that he was to articulate in the Declaration of Independence and in the reforms he later championed in Virginia. Those three lonely, frustrated weeks in a rented room in Philadelphia formed the cornerstone of his political career. His draft constitution propounded a coherent system of individual rights. The people were the true source of authority. "Public liberty" and individual rights were to be protected against authoritarian control. The right to vote was to be broadened and the distribution of representation in the legislature equalized, granting a greater voice to the western frontier counties. Jefferson proposed using unappropriated lands to establish a society of independent farmers who would hold their land "in full and absolute dominion of no superior whatever." He provided for just and equitable treatment of Indians. In a document envisioning wise use of Virginia's vast western lands, Jefferson's revolutionary Declaration of Rights for Virginia foresaw the use of these untapped territories as a means of removing friction with neighboring states with conflicting claims and of providing vast wealth and security for the new nation. He also encouraged immigration and the lowering of barriers to naturalization, the elevation of civil over military authority. He endorsed the abolition of privilege and prerogative and the regulation of descent of property to eliminate primogeniture. He discouraged the importing of any more slaves. The document also provided for eliminating capital punishment and thoroughly revising Virginia's penal code as well as proposing land tenure reform. It was a revolutionary document. Jefferson's plan most resembled Mason's in that it split the government into three branches.[11]

On June 13, Jefferson sent his third draft south with George Wythe. As he had drawn a blueprint for a new nation—Virginia—he had confided to Wythe one of his greatest desires, one based on his years of practicing law, that an independent and highly educated judiciary replace the amateur and capricious gentlemen-judges he had so come to loathe. Whether he also wrote Wythe a letter outlining his anxiety over the need for judicial reforms remains a cause of controversy among Jefferson scholars. Appearing in a Charleston, South Carolina, newspaper in 1803, during a violent attack on Jefferson's administration of the presidency, it was never refuted by Wythe, who was still alive and the preeminent law scholar in America:

> The dignity and stability of government in all its branches, the morals of the people and every blessing of society depend so much upon an upright and skillful administration of justice. . . . The judicial power ought to be distinct from both the legislative and executive, and *independent* upon both, that so it may be a *check* upon both, as both should be checks upon that. The judges, therefore, should always be men of learning and experience in the laws, of exemplary morals, great patience, calmness, coolness and attention. . . . Their minds should not be *distracted with jarring interests;* they should not be dependent on any man, or body of men. To these ends, they should hold *estates for life* in their offices. . . . Their commissions should be *during good behavior* and their salaries . . . established by law. . . . For *misbehavior,* the grand inquest of the colony, the house of representatives, should impeach them. . . . If convicted, [they] should be *removed* from office . . .[12]

The extract from the Charleston *Courier* was printed in the latest edition of Jefferson's papers with the caveat that the original was missing and that the provisions for judges contained in it "do not wholly conform to any known plan" proposed by Jefferson. But the editors nonetheless published the "highly suspect" letter. If genuine, the letter reflects Jefferson's attitude toward the law as the revolution gathered momentum. The arrival in Williamsburg of yet another version of Virginia's proposed constitution came too late. If anything, Jefferson's proposal caused consternation among the hot, tired delegates to the fifth convention in two years. The convention already was in turmoil. As Edmund Randolph reported, a "very large committee" had been nominated to prepare "the proper instruments" and "many projects" had surfaced that, collectively, had "discovered the ardor for political notice rather than a ripeness in political wisdom." Among the pile of proposals, George Mason's Declaration of Rights "swallowed up all the rest by fixing the grounds and plan"[13] which, after great discussion and correction, were finally ratified.

By the time Jefferson's proposed constitution arrived, the Mason committee's plan was already being revised and prepared for a final vote. The Mason plan contained elements suggested by John Adams, who had made written suggestions to Wythe and Richard Henry Lee in January, which were then published as a widely read pamphlet, *Thoughts on Government.* Lee had sent copies to Patrick Henry, Charles Lee, and others. In effect, Mason's plan was an adaptation of Adams's except for its narrow electoral base. Without consulting Jefferson, Wythe decided not to press the Convention even to consider Jefferson's plan; instead, he lined up support to cut it up as a series of amendments and to add Jefferson's denunciation of the king as a preamble justifying Virginia's need for a new constitution. Jefferson's preamble was added to the Mason document almost verbatim. To Jefferson, his friends in Williamsburg wrote, almost apologetically, anticipating his bitter disappointment. "Some of your friends have no doubt given you a history" wrote Will Fleming, of the "spirit (evil spirit I had almost said) and general proceedings of our convention."[14] "For God's sake, be with us quickly," wrote Edmund Randolph. "Our counsels want everything to stamp value on them." Young cousin Randolph related there had been "vast affection" for the honor of "being the manufacturers of the new government."[15] That John Adams had made himself "obnoxious," as he said to Jefferson, does not seem an exaggeration, at least from the point of view of the scant minority of conservatives who were so outmaneuvered by the energetic Boston lawyer in the first months of 1776. On May 15, just as Jefferson had returned to Philadelphia and the Virginia Convention had decided to push for independence, Adams had capped a year-long campaign in Congress to drag less bellicose colonies into the fray that had originated, in the minds of many, in troublesome Massachusetts. Lee's May 15 congressional resolution called for an end to all royal government in America and virtually made independence inevitable. Moderates such as New York lawyer James Duane, who did not share the extremism of Adams and Jefferson, angrily told Adams that he considered the May 15 resolution "a machine for the fabrication of independence."[16] The move that Adams had "pursued for a whole year" and "contended for" through "a scene of anxiety, labor, study, argument and obloquy,"[17] led to each colony's revolutionary government assuming all the powers of government. By May 17, Adams could write to his wife, "Great Britain has at last driven America to the last step, a complete separation from her, a total absolute independence not only of her Parliament but of her crown." Adams predicted the speedy formation of "a whole government of our own choice, managed by persons whom we love, revere and can confide in," a government "for which men will fight."[18]

Even Adams was astonished by the sweeping and radical revolution that was "taking place in the minds and hearts of the people." He found the sudden onset of antimonarchical tendencies "astonishing" and wrote that "idolatry to monarchs and servility to aristocratical pride was never so totally eradicated from so many minds in so short a time."[19]

On May 27, as Congress planned with General Washington the summer season of war, the arrival of Virginia's instructions sped along events, which were already moving so swiftly that, as Adams wrote on June 3, "I fear we cannot proceed systematically." Out of order "we shall be obliged to declare ourselves independent states before we confederate and, indeed, before all the colonies have established their governments."[20] Bad news came to Philadelphia with almost every dispatch rider. A British fleet had sailed up the St. Lawrence. Ten thousand Redcoats had retaken Quebec and were driving the Americans out of Canada. By the time Adams and Jefferson decided just who should draft a declaration of independence, there was an urgent drumbeat to the political rhetoric that charged the air around the State House. It had been unseasonably cool, "as cold as New England and very wet,"[21] wrote an ailing Massachusetts delegate, but now the usual Philadelphia summer heatwave began. As British ships bore down on the American coast and many members of Congress rode home to harvest the authority to vote for independence, Thomas Jefferson closed himself in the second-floor parlor—and began to write the Declaration of Independence.

It was years before most people knew Thomas Jefferson wrote the Declaration of Independence, so carefully was it locked up in the secret anonymity of a revolutionary committee. It was years later before he became really proud of it. It was just a document among many Jefferson wrote, early and late, in the tumultuous summer of 1776 around the edges of days of committee meetings and worried debates about the latest developments of a spreading civil war. For two weeks, between June 13 and 28, on cool mornings, on sultry evenings, Jefferson sought the relative peace and quiet of his airy parlor and, spreading his papers about him, worked away on his new portable desk. Four leaders and an age of revolutions later, Jefferson, at age seventy-seven, recalled a few details erroneously, insisting that he had "turned to neither book nor pamphlet" while writing "the declaration." But while Jefferson did not need to consult any book, he did have at least one document, the final draft of his precious constitution for Virginia: from it he would copy the long list of grievances against the king. Jefferson was not expected to create something original: quite the opposite. Yet he was free to glean his words from a hundred writers from the time of ancient Greece to the day-before-yesterday's charged rhetoric of

Tom Paine. He was free to select and shape and rewrite, but what he had in mind, often tortured by modern scholarly interpretations, he expressed frankly enough himself in a letter shortly before he died. His task, he wrote, was

> not to find out new principles, or new arguments, never before thought of, not merely to say things which had never been said before, but to place before mankind the common sense of the subject in terms so plain and firm as to command their assent, and to justify ourselves in the independent stand we are compelled to take. Neither aiming at originality of principle or sentiment, nor yet copied from any particular and previous writing, it was intended to be an expression of the American mind, and to give to that expression the proper tone and spirit called for by the occasion.[22]

The genius of Jefferson's writing lay in his ability to take deep and complicated concepts of history, law, and philosophy and clothe them in easy, graceful, direct, almost simple language—what Adams admired as a "peculiar felicity of expression"[23]—that was concise, clear, crafted perfectly to his purpose. His writing style gives us the sharpest self-portrait of the man: soft in tone and gracious in sentiment when he spoke of human rights, angry and self-righteous when he catalogued the crimes of a tyrannical king, magnificently Roman when, forging in a single paragraph the creed of a new nation, he pledged "our lives, our fortunes and our sacred honor"[24] to the glorious cause.

But Thomas Jefferson was also an Enlightenment man of science. He saw as the essence of his task the choice of the right device to express "the American mind."[25] He had the mind of an inventor who knew what innovations had come before him, which to adopt, which to adapt, which to throw away. He realized that all the "authority" his declaration could carry rested on "the harmonizing sentiments of the day, whether expressed in conversation, in letters, printed essays or the elementary books of public right, [such] as Aristotle, Cicero, Locke, Sidney, etc."[26] Jefferson was acquainted with virtually all of the ideas of fundamental law from Plato to Blackstone. It was not his task to educate, but to use the power of reasoning to argue, to persuade, and to justify a revolution, a tearing away from the womb of an empire that allowed the birth and survival of a new and sharply different type of nation.

For his complicated task, Jefferson dipped back beyond all the years of reading and rewriting dusty legalisms to the clarity of thinking and prose of his college years. As the format for his declaration, he decided to use the powerful, succinct logic his Scottish schoolmaster, Dr. Small, had brought to Virginia from Aberdeenshire. Small was the student of William Duncan of Aberdeen, whose two-volume *The Elements of Logic* had been published in London with an introduction by Samuel Johnson. The book was the best-known work of logic in English at the time of the education of members of the Second Continental Congress. Jefferson owned a copy of Duncan's work, which appeared in the second volume of Thomas Dodsley's *The Preceptor* in 1748. Professor of moral, natural, and experimental philosophy at Marischal College, Aberdeen, from 1752 to 1760, Duncan was honored by his graduates in 1755, who chipped in to buy him experimental equipment for his physics classes. William Small had been one of these students. Later, instructing Jefferson in rhetoric and logic at William and Mary, Small very likely taught from Duncan's book of logic, "the proof of a proposition," by "a series of syllogisms, collecting that proposition from known and evident truths," as well as "self-evident propositions."[27] Duncan's *Logic* was widely used when arguments were supposed to achieve the authority of science. Jefferson's structure for the Declaration of Independence follows Duncan's dictum, "If therefore evident and allowed truths are disposed in a syllogistic order, so as to offer a regular conclusion, that conclusion is necessarily true and valid."[28]

Knowing his audience in Congress and beyond the walls of the Pennsylvania State House, he chose the form and rhetoric of science, of syllogistic logic, self-evident premises, and deductively established conclusions to achieve persuasion. Jefferson's Declaration of Independence was one of the early attempts at political science. He managed masterfully to construct an elegant political argument according to the rules of science. Brilliantly he settled on a short set of powerful logical arguments, as if summing up a defense of revolution to justify a new nation or trial before a world dominated by a jury of kings and princes. The form was as important as the content: an argument using syllogisms that every educated reader would easily recognize.

To begin, to clear his throat and rivet attention, Jefferson wrote a preamble that laid out his major premise: that the step America was about to take was necessary, coming reluctantly after many attempts at reconciliation had failed. It was "in the course of human events," not a dangerous invention. His meaning would be clear from events; his case so clear and irrefutable that he did not have to demean the jury by telling them what he meant by "the laws of nature and of nature's god." Still, this document was necessary out of "a decent respect to the opinions of mankind." The United States must declare itself equal among the nations of the earth, and here it would "declare the causes" that impelled them to separate from England.

In his major premise, an appeal to reason and all reasonable people, based on moral truths he did not feel compelled to prove, Jefferson spelled out "self-evident" truths. "All men are created equal . . . they are endowed by their creator with certain inalienable rights." What all of them were, he did not detail, only those he held most important . . . "life, liberty, and the pursuit of happiness." His choice of words "pursuit of happiness" over John Locke's "property" marked a sharp break with the Whig doctrine of English middle-class property rights. It was a felicitous, memorable turn of phrase, the most succinct expression ever of American political philosophy. Beyond any man's ability to prove with fact or legal precedent, Jefferson transcended events and precipitated them at the same time, depending only on the self-evidence of his prose to prove his major premise:

> . . . that to secure these rights, governments are instituted among men, deriving their just powers from the consent of the governed; that whenever any form of government becomes destructive of these ends, it is the right of the people to alter or abolish it and to institute new government.

In marvelously few words, Jefferson asserted the rule of right reason that philosophers since Thomas Aquinas had taken volumes to argue, then plunged on to posit the doctrine, at the heart of the English revolutions of the seventeenth century, that, as John Knox had put it, "resistance to tyrants is obedience to God."

As his minor premise, Jefferson flatly asserted that the king and his ministers had carried out a plot, "a design to reduce" the American colonies "under absolute despotism." Thus, it was, their right . . . their duty "to throw off such government." And here he launched into an angry, burning assault on "the present king of Great Britain"—not his Parliament. Here he was shifting the argument boldly to new ground. Steering clear of any overt antagonism to the idea of kingship in general before this tribunal of the world, Jefferson held that Americans had the right to revolt against a bad king, a tyrant. And then, in eighteen withering charges intended to prove George III's guilt of tyranny over America, he drew up a bill of indictment against the king. If he had proved his case, there was no question that the Americans had to absolve themselves of their allegiance to a wicked Crown and that, therefore, "these United Colonies are and of right ought to be free and independent states."[29]

According to notes he cited in a letter to Madison in 1823, when Jefferson finished his rough draft, "before I reported it to the committee, I communicated it separately to Dr. Franklin and Mr. Adams, requesting their corrections." Franklin and Adams made minor changes of a word here, a word there—"merely verbal,"[30] wrote Jefferson—interlining the draft in their own hands. The committee, making no substantive changes, presented a clean copy to Congress on June 28. Four days later, on July 2, Congress, after voting in favor of the resolution on independence, sat as a committee of the whole, and began two and a half days of debate on every line and provision of the Declaration of Independence. No one took notes except for a few details Jefferson himself wrote down.

The debate was one of the more painful ordeals of Jefferson's long political career. He sat there, beside Franklin, silent in his humiliation at the number, extent, and importance of the changes. He mostly maintained this silence for years, but what little he wrote indicates his mounting disgust at the timidity of the conservatives in Congress, their slashing deletions of at least two major clauses in Jefferson's draft declaration. In his notes, he derided the conservatives:

> The pusillanimous idea that we had friends in England worth keeping terms with, still haunted the minds of many. For this reason, those passages which conveyed censures on the people of England were struck out, lest they should give them offense.[31]

Jefferson wrote to Robert Walsh in 1818, expanding on this theme: "The words 'Scotch and other foreign auxiliaries' excited the ire of a gentleman or two of that country."[32] A shipload of Highlanders bound for Boston to reinforce the British had blown offshore and fallen into American hands. Scots under the Scottish Lord Dunmore had burned Norfolk. All mention of Scottish mercenaries was nevertheless suppressed.

But of far greater consequence was Congress's "immediate" agreement to cut out Jefferson's latest attempt to condemn the slave trade. In his draft Declaration, approved by the Committee of Five, Jefferson charged that the king personally was to blame for the slave trade:

> . . . He had waged cruel war against human nature itself, violating its most sacred rights of life and liberty in the persons of a distant people who never offended him, captivating and carrying them into slavery in another hemisphere, or to incur miserable death in their

> transportation thither. This piratical warfare, the opprobrium of *infidel* powers, is the warfare of the *Christian* king of Great Britain. Determined to keep open a market where *MEN* should be bought and sold, he has prostituted his negative for suppressing every legislative attempt to prohibit or to restrain this execrable commerce, determining to keep open a market where *MEN* should be bought and sold: and that this assemblage of horrors might want no fact of distinguished die, he is now exciting those very people to rise in arms among us, and to purchase that liberty of which *he* has deprived them, by murdering people upon whom *he* also obtruded them: thus paying off former crimes committed against the *liberties* of one people with crimes which he urges them to commit against the *lives* of another.[33]

In all of Jefferson's hammer-blow indictments of the king of England written in language usually reserved for a common criminal, this was his strongest attack. It echoed seven years of his determined attempts to curtail the slave trade in Virginia and the spread of this murderous institution. In his first bill before the Virginia House of Burgesses in 1769, in *Howell v. Netherland* before the General Court, in his "instructions" to the Virginia Convention of 1774, rejected so quickly out of hand, Jefferson had used his every forum to attack slavery. Now he used his first national office to renew the attack—and he was rebuffed. In his *Notes on the State of Virginia*, Jefferson recorded that the clause he had written into the Declaration

> reprobating the enslaving the inhabitants of Africa was struck out in complaisance to South Carolina and Georgia, who had never attempted to restrain the importation of slaves and who, on the contrary, still wished to continue it. Our northern brethren also, I believe, felt a little tender under those censures, for, though their people have very few slaves themselves, yet they had been pretty considerable carriers of them to others.[34]

Writing four decades later, Jefferson added that

> severe strictures on the British king in negativing our repeated repeals of the law which permitted the importation of slaves, were disapproved by some southern gentlemen whose reflections were not yet matured to the full abhorrence of that traffic.

Yet Jefferson was still smarting that his antislavery "expressions were immediately yielded" by north *and* south. He noted that the same "gentlemen," especially Rutledge of South Carolina, were encouraged by this easy victory to continue "their depredations on other parts of the instrument."[35]

Before the brutal editing process ended, Jefferson's theory of empire had also been cut out and numerous charges against the British had been watered down, including his version of a clause about the closing of courts in Virginia and Massachusetts: "He had suffered the administration of justice totally to cease in some of these states" became a tamer "he has obstructed the administration of justice." Lawyer Jefferson wrote that the king was guilty of "depriving us of the benefits of trial by jury." Congress qualified this, adding, "in many cases." All hint of Lord Dunmore's offer to give land he seized from rebels to any Loyalists who supported him disappeared from the Declaration. Congress deleted the clause, "He has incited treasonable insurrections of our fellow citizens, with the allurements of forfeiture and confiscation of our property."[36]

Through three days and late into the evening of July 4, Jefferson later recalled, he was "sitting by Dr. Franklin, who perceived that I was not insensible to these mutilations." In his longest comment on the brutal editing process, he said that Franklin tried to reassure him by whispering a parable to him:

> "I have made it a rule," said Franklin, "whenever in my power, to avoid becoming the draftsman of papers to be reviewed by a public body. I took my lesson from an incident which I will relate to you. When I was a journeyman printer, one of my companions, an apprentice hatter, having served out his time, was about to open a shop for himself. His first concern was to have a handsome signboard with a proper inscription. He composed it in these words: 'John Thompson, hatter, makes and sells hats for ready money,' with a figure of a hat subjoined. But he thought he would submit to his friends for their amendments. The first he showed it to thought the word 'hatter' tautologous, because followed by the words, 'makes hats' which show he was a hatter. It was struck out. The next observed that the word 'makes' might as well be omitted, because the customers would not care who made the hats. If good and to their mind, they would buy, by whomever made. He struck it out. A third said he thought the words 'for ready money' were useless, as it was not the custom of the place to sell on credit. Everyone who purchased expected to pay. They were parted with, and the inscription now stood: 'John Thomp-

> son sells hats.' 'Sells hats?' says his next friend. 'Why, no one will expect you to give them away. What, then, is the use of that word?' It was stricken out; and 'hats' followed it, the rather as there was one painted on the board. So, his inscription was reduced ultimately to 'John Thompson' with the figure of a hat subjoined."[37]

That same day, Congress adopted the Declaration of Independence. Four days later, on July 8, the sheriff of Philadelphia County, Col. John Nixon, read it to a cheering crowd in the courtyard behind the State House, later renamed Independence Hall. Jefferson took no prominent part in the ceremony if he attended at all, and no one except the members of Congress even knew he had written the Declaration. It would be many years before his authorship became well known. By the time he died, he was proud of even the eviscerated version, listing it first among his achievements when he wrote his epitaph as he lay dying half a century later.

At the time he was humiliated and disappointed by Congress's cuts although, at first, he characteristically committed little about his pain to paper. He was too private a man for that. He did write to Lee in Virginia, sending him the approved version and a copy "as originally framed"—by Jefferson. "You will judge whether it is the better or worse for the Critics."[38] His strongest reaction was to let it be known that he did not want a second term in Congress. When he did write home to his good friend, Page, he at first never reported the struggle over the Declaration. "There is nothing new here,"[39] he reported. And when he wrote his scientifically objective, deliberately unemotional autobiography at age seventy-seven, he gave the episode short shrift. "The sentiments of men are known not only by what they receive but what they reject also," he wrote—and then published in full "the form of the Declaration as originally reported" with "the parts struck out by Congress"[40] underlined in black.

Disappointed that he had played no important role in writing the new Virginia constitution, smarting from the hack-and-slash editing of his Declaration of Independence, Jefferson was determined to quit Congress and return to Virginia politics. He could not see into the future. To him, the Continental Congress was but a temporary meeting in convention of delegates from the new states, of little importance in the long run. Even if there were a permanent confederacy of states, it would be of far less importance than the reshaping of the weak old English colonies into strong independent countries. Jefferson wanted to join his friends in Virginia in an effort that mattered far more to him than writing of declarations and resolutions in Philadelphia. He wanted to carry out a revolution that would last, a revolution in the laws of his native country, Virginia, a sweeping legislative reform movement that would transform the old semifeudal Tidewater aristocracy into a democratic republic. Whatever he accomplished in Virginia would be seen and studied and could be emulated in other states, but someone must take the lead. Jefferson's frustrations as a delegate to the Second Continental Congress were compounded by his misgivings about the new Virginia constitution. George Wythe had written from Williamsburg that he had arrived too late to submit Jefferson's draft constitution in time to be considered seriously by the Virginia Convention. He warned that the new constitution, adopted in Jefferson's absence, "required reformation." Writing Jefferson on July 27, Wythe told his protégé, "I hope you will affect it."[41]

His one-year term in Congress to expire in August, Jefferson wrote to Edmund Pendleton, now the president of the Virginia Convention, insisting that he did not wish another term in Congress. Mounting a virtual campaign against his reelection, Jefferson also wrote to his friend and physician, George Gilmer, who was temporarily filling Jefferson's Albemarle seat, and then he wrote to ask Edmund Randolph to make a speech on the floor of the Convention to appeal that he be relieved. Despite his pleas, he was reelected. He polled fourth out of the five delegates elected to the new contingent. He would have ranked higher, wrote a friend, had he not opposed his own reelection. Others had voted for him anyway because they thought he was jesting about leaving Congress.

Even before the mauling of his draft Declaration, however, Jefferson had been worried about rumors of his disloyalty to the revolution in Virginia. In his *Summary View,* he had not gone so far as to recommend independence; nothing he had written since then for public consumption had gone any further. In Congress, he had coauthored key documents with the conservative John Dickinson. When instructions to vote for independence had arrived in Philadelphia, Jefferson had been among those opposed to introducing them for a vote until there was stronger support from the more timid states. Were there whisperings that he, in fact, had Loyalists in his family? Had his daring correspondence with John Randolph become known? Randolph certainly had made no secret of it among the king's party. To Will Fleming, he wrote,

> It is a painful situation to be 300 miles from one's country, and thereby open to secret assassination without a possibility of self-defense. . . . It cannot be easy. . . . If any doubt has arisen as to me, my country will have my political creed in the form of a "Declaration, etc." which I was lately directed to draw. This will give decisive

> proof that my own sentiment concurred with the [May 15] vote [for independence] they [the Convention] instructed us to give.[42]

Jefferson's friend Fleming explained to him that he had polled fewer votes for reelection to Congress than he might have had because of his request to be replaced. Delegates had swung their votes to more willing candidates. But Jefferson still worried that he was suspected of being less than staunch in the cause of independence, that his desire to resign had marked him as a Loyalist. Will Fleming hastened to reassure Jefferson that he should make himself "perfectly easy, for you are as high in the estimation of your countrymen as ever."[43] Edmund Pendleton praised Jefferson for his ringing indictment of George III in the Declaration and asked him to consider "one of the most important posts" in Virginia's new judiciary, "where I most fear our deficiency."[44]

Even if reelection to Congress must have assuaged some of Jefferson's anxiety about the imagined whisperings against him, he remained less than thrilled that he had to stay on as the lone Virginia delegate in Philadelphia for several more months. Thanking the Virginia Convention for its "continued confidence," Jefferson tried again to be relieved of his post immediately, citing his wife's frail health.

> I am sorry the situation of my domestic affairs renders it indispensably necessary that I should solicit the substitution of some other person here in my room. . . . The delicacy of the House will not require me to enter minutely into the private causes which render this necessary.[45]

But Virginia did not free Jefferson from his congressional seat until September. Meanwhile he was to help settle an old boundary dispute with Pennsylvania, quietly and diplomatically. His days were full with committee work. A large British expeditionary force landed at New York City only days after the Declaration of Independence was signed. The remnants of the smallpox-infested Canadian army had retreated to Fort Ticonderoga. Jefferson felt the American withdrawal to the dilapidated fortress was "against everything which in my eye wears the shape of reason."[46] Jefferson had spent long weeks exploring the "ill successes in Canada"[47] and had written new rules for the Continental Army at Washington's request. He worried about a British armada off the Carolinas, hoped that Virginia "is perfectly safe now"[48] and personally wrote to Virginia to summon two regiments of Virginians to reinforce Washington and to turn around wagonloads of lead shot headed south that were now needed to repulse what was obviously the main British onslaught at New York City. His personal knowledge of the lead mines in western Virginia made him confident that Virginia would comply. And every time a week went by without a letter from Patty, he worried more about her.

By July 29, his pleas to go home were shrill. To Richard Henry Lee, he wrote,

> For god's sake, for your country's sake and for my sake, come. I receive by every post such accounts of the state of Mrs. Jefferson's health that it will be impossible for me to disappoint her expectation of seeing me at the time I have promised. . . . I pray you to come. I am under a sacred obligation to go home.[49]

What ailed Patty Jefferson remains unknown and maddeningly elusive. Was she possessive to the point of deep depression whenever he went away? One can only speculate. On September 3, Lee finally arrived in Philadelphia and Jefferson was free to go home. Happily he made a last-minute shopping trip: clothes for Patty, a doll for Patsy. After what must have been a record drive south—Jefferson took a new route, the Philadelphia-to-Lancaster Turnpike, America's first toll road, newly opened—six days later he was in his wife's arms. A week short of nine months later, their first son was born. There could be no greater pleasure for Jefferson than domestic bliss at Monticello, but once again his time at home was short-lived. From the heat haze atop Monticello he would have to descend all too soon. Edmund Pendleton, Speaker of the newly created House of Delegates, had written to Jefferson before he left Philadelphia to urge him to come to Williamsburg to help rewrite Virginia's sanguinary penal code. Jefferson's legal talents were needed "much in the revision of our laws and forming a new body, a necessary work for which few of us have adequate abilities and attention."[50] It was an undertaking Jefferson relished.

12

"With a Single Eye to Reason"

I am certainly not an advocate for frequent and untried changes in laws and constitutions, but I know also that laws and institutions must go hand in hand with the progress of the human mind.

—THOMAS JEFFERSON TO SAMUEL KERCHEVAL,
JULY 12, 1816

THREE WEEKS AFTER arriving home at Monticello, Jefferson, his wife Patty, their three-year-old daughter Patsy, and their servants rolled down the mountain in a little cavalcade of carriage, wagons, and mounted slaves for the one-hundred-fifty-mile journey to the capital, where Jefferson was to take his seat in the new Assembly of 1776. He had decided not to leave his family behind while he took up his duties in the House of Delegates as it began inventing a new state. In Williamsburg, the Jeffersons crowded temporarily into two rented rooms. Jefferson did his best to furnish them, borrowing a wagonload of furniture from their old friend Mrs. Drummond, who was Patty Jefferson's constant companion during Jefferson's long days at the Capitol. The rooms proved unsatisfactory; within a week Jefferson was dashing off an urgent request to George Wythe at the Continental Congress in Philadelphia. Could they borrow his house on Palace Green until the Wythes returned to Williamsburg? Wythe wrote back promptly: "Make use of the house and furniture. I shall be happy if anything of mine can contribute to make you and Mrs. Jefferson's residence in Williamsburg comfortable."[1]

On October 8, the day after the Jeffersons reached the capital, a dispatch rider from Philadelphia rushed up to Jefferson with an urgent message from the Continental Congress. Jefferson tore open the familiar seal of John Hancock:

> The Congress [has] appointed you to fill a most important and honorable department. . . . Doctor Franklin, Mr. [Silas] Deane and yourself are appointed commissioners at the Court of France.

For months, Jefferson and Franklin had worked closely together. With Adams, Jefferson had made common cause in convincing the Congress of work to be done making a complete break with England and splicing new trade and military ties in Europe. Indeed, Jefferson had asked to go to France with Franklin, and Franklin had prevailed on Congress for his appointment. Now, Hancock was not making it easy for Jefferson to refuse. A ship was waiting. Appealing for him to leave at once, Hancock made it clear that Congress did not expect Jefferson to decline. "To promote the interest and happiness of your country" was well known to be Jefferson's wish. Congress "expects from you" the "great services" which were "fully in your power."[2]

For three days, Jefferson kept the messenger waiting while he agonized. Did he try to win Patty over? Congress was offering to send him *and* his family to Paris. He must have held out some hope that he could persuade her. But he could not bring himself to leave her alone again so soon after such a long absence. If he went without his wife, he could be away from her for months, even years, trying to win aid from France and her allies and establishing diplomatic relations for the new American states. He would also be away from Virginia at the crucial time of reforming the laws of his native country. Finally, however, on October 11, he wrote his demurrer to Congress:

> . . . No cares for my own person nor yet my private affairs would have induced one moment's hesitation to accept the charge. But circumstances very peculiar in the situation of my family such as neither permit me to leave nor to carry it [to Europe with him], compel me to ask leave to decline . . .[3]

There were others in Congress who could replace him at Franklin's side, argued Jefferson. For most of a year, he had been away. He could not tell Congress, but he knew that, for much of that time, Patty had been depressed, staying in her room atop Monticello, not even writing to him. His place now was closer to his family. On October 11, the day he sent back the messenger with his refusal, Jefferson turned his full attention to reforming Virginia into a republican commonwealth.

In the Virginia House of Delegates, meeting in the former House of Burgesses chamber on the ground floor of the Capitol, Jefferson established himself as the principal and most prolific writer of a radical legislative revolution, one that was as sweeping as the English Revolution of the 1640s, which in many ways Jefferson used as his model. In two and a half years, he almost single-handedly invented a country, the Commonwealth of Virginia, the largest of the new American states and a closely watched example for the others. Setting out to dismantle the old English system and replace it with a solidly democratic state, he drafted and submitted for legislative approval some 126 laws in three years, in what the distinguished legal scholar Julian Boyd has called "one of the most far-reaching legislative reforms ever undertaken by a single person."[4] Many of the legal principles he forged out of his years of research into ancient and early English and American law have become the guiding principles of the United States.

Jefferson built his personal crusade to republicanize Virginia through its laws on four cornerstones: repeal of the laws of entail and primogeniture that had built up a Tidewater aristocracy; establishing religious freedom; providing for "the more general diffusion of knowledge"[5]—for the rest of his life he proselytized free public education—and streamlining Virginia's judiciary system while liberalizing her brutal penal code. He was often at odds with the majority of his fellow delegates. His proposals were sometimes set aside and not adopted for years: only his proposal for primary schools of all his educational proposals was adopted, and not for twenty years. His call for a state library was among many others never enacted. But no one drafted more bills that were introduced and adopted in any state during the American Revolution. So many were passed that he managed to succeed in overhauling Virginia's laws root and branch. No one was more influential or played a more critical role in the legislation that followed Virginia's declaration of independence from England in 1776 and in the creation of a commonwealth that was the forerunner of the federal government of the United States.

Thoroughly shaken by the conservative tone and intent of George Mason's Declaration of Rights of May 1776, Jefferson had been working secretly in Philadelphia before his return to Virginia, meeting with his radical congressional colleagues and preparing the first round of a legal counterattack that would completely overthrow the English legal system that had chained Virginia for 170 years. Jefferson was quickly coming to distrust the ebbs and flows of political fervor: no orator, he believed in the law, and he believed he must move quickly before the military stage of the Revolution ended (like many radicals, he believed it would be over in a matter of months). He must act before the pressure for change dissipated. He did not want his ideas to vanish with the morning mists of revolutionary enthusiasm. As soon as he took his seat in the new House of Delegates, he sounded the alarm. As Jefferson explained forty years later in his *Autobiography*:

> The first question was whether we should propose to abolish the whole existing system of laws and prepare a new and complete institute, or preserve the general system and only modify it to the present state of things.

No work could be more necessary or important. "When I left Congress in '76," he recalled in his autobiography, "it was in the persuasion that our whole code must be reviewed, adapted to our republican form of government and, now that we had no negatives of councils, governors, and kings to restrain us from doing right, that it should be corrected in all its parts with a single eye to reason and the good of those for whose government it was framed."

To Jefferson's surprise, Edmund Pendleton, Speaker of the House, "contrary to his usual disposition in favor of ancient things," favored a completely new code of laws. Jefferson, George Wythe, and George Mason took the opposite view: better to create new laws where absolutely needed, but to leave intact the vast bulk of laws based on English Common Law: "to abrogate our whole system would be a bold measure and probably far beyond the views of our legislature,"[6] said Jefferson. Jefferson hatched out an array of ideas that could easily qualify as bold in the midst of this already bold revolution, but he was ever careful, as Adams had advised, to keep the flock together and not step out too far in front of it. This was not a conservative revolution, but in the Virginia Assembly it was a revolution of conservatives. On October 14, at the opening of the second week of Virginia's most revolutionary session, Jefferson won the House of Delegates' assent to bring in an omnibus bill that would provide for general revision of Virginia's laws. The House then passed an act that gave a committee, to be headed by Jefferson, full authority to revise, amend, or repeal any or all of the laws of Virginia or to draft new ones subject to the approval of the General Assembly. This brilliant blue-ribbon committee also included Pendleton, Wythe, George Mason, and Thomas Ludwell Lee. Jefferson was by far the youngest. His appointment at age thirty-three to the head of this distinguished group of revisors was a tribute not only to his political stature but to his ability to bring understanding and order to the chaos of centuries of often-contradictory laws. He set out to create a whole new universe of legal reason, in the end doing most of the work himself. Mason, no lawyer, was too ill and soon resigned. Lee, also no lawyer, died a year later without taking up his share of the work. Jefferson, Pendleton, and Wythe divided the immense undertaking at first, in the end agreeing on a plan to revise and modify essential laws but to leave intact most of the great body of English Common Law.

Jefferson was not willing to wait until they completely revised the laws. Month after month he bombarded the legislature with bills he considered the most urgent reforms. His first assault was on primogeniture and entail, the companion laws that had perpetuated inequalities at all levels of society. Jefferson's bill to abolish entail, submitted on the very day the revolutionary House of Delegates met for the first time, was his opening shot in a war of agrarian reform that would give broader popular access to the land and sharply cut back the political, social, and economical influence of landed aristocrats. In his recently ignored draft Virginia constitution, he had proposed that white males not owning fifty acres of land be granted enough land to raise their holdings to this minimum amount required to qualify to vote. Now, the first bill he introduced in the House of Delegates abolished entail. Until then, in Virginia as in England, recipients of large land grants had been required by law to perpetuate their vast family estates by conveying intact their lands to their descendants. This limited the inheritance of property to an unalterable succession of specified male heirs. Only an act of the legislature could change the succession.

To break the hereditary power of this landed aristocracy was at the top of the agenda of Jefferson's bloodless revolution, as he related in his autobiography:

> In the earlier times of the colony, when lands were to be obtained for little or nothing, some provident individuals procured large grants and, desirous of founding great families for themselves, settled them on their descendants. . . . The transmission of this property from generation to generation in the same name raised up a distinct set of families who, being privileged by law in the perpetuation of their wealth, were thus formed into a patrician order, distinguished by the splendor and luxury of their establishments.

The king had drawn his council from this elite corps, the hope of royal preferment bending its members to "the interests and will of the crown."

Jefferson wanted to "annul this privilege" and the "harm and danger" of this "aristocracy of wealth."[7]

The huge hereditary estates of eastern Virginia blocked the access of newer arrivals to the rich alluvial soils of the Tidewater and its vital transportation network of rivers. New migrants had to trek into the hardscrabble backcountry where, as land-tenure lawyer Jefferson well knew, they were also blocked from owning their own land by feudal quitrent laws and by greedy speculators who had engrossed much of the west country without paying for it.

Primogeniture and entail had concentrated wealth and power in approximately 140 families in Tidewater Virginia, whose property by law descended to the eldest son unless a will provided for other sons, daughters, or heirs. Most of the gentry wrote wills but many were already sympathetic to the problem of entails. Jefferson had, like many others, gone to court to break his wife's father's entail. John Wayles had left no sons and Jefferson himself had as yet no male heir. His own progressive father had refused to entail his lands and had provided for all his children by will: the girls had received property and cash dowries and Thomas Jefferson had divided his father's lands with his younger brother. In the Assembly of 1776, Jefferson wished the law to "authorize the present holder to divide the property among his children equally, as his affections were divided." Jefferson considered the abolition of primogeniture and entail "essential to a well-ordered republic," opening the way for

> the aristocracy of virtue and talent which nature has wisely provided for the direction of the interests of society and scattered with equal hand through all its conditions.

Jefferson believed that merit would surface unless it was stifled by a system of class and privilege upheld by law. A class system, he argued, was against the laws of nature. In its place, he intended to erect "a system by which every fibre would be eradicated of ancient or future aristocracy and a foundation laid for a government truly republican." Submitting a bill to abolish entail only three days into the Assembly of 1776, as he later told Adams, he "laid the axe to the root of pseudo-aristocracy."[8] Even if there had been no hint of such a bold attack on privilege in Virginia's earlier revolutionary assemblages, Jefferson sensed the revolutionary fervor of his compatriots in the Assembly of 1776 and he decided to seize the moment, to take the first swing of the axe himself. On October 14, he introduced his bill to end entail, which promptly passed the House of Delegates and was accepted by the Senate on November 1, the same day Jefferson was appointed to head the committee of revisors. Following Jefferson, Virginia led the way in abolishing hereditary privilege in America.

Within a week of joining the revolutionary Assembly and leading the attack on the laws of inheritance and land descent, Jefferson clashed again with the powerful landed oligarchy, this time over western land speculation. During the previous two years, his statements in the *Summary View,* in correspondence with Edmund Pendleton, and in his drafts of a constitution had revealed his liberal attitude to the potential of the vast and fertile overmountain region. He envisioned a land of independent farmers who divided up all unappropriated lands, with each settler given up to fifty acres. He would have barred any further purchase from the Indians without the approval of the legislature. To Jefferson, the new lands in the west belonged to all the people; to many of the most powerful leaders in Virginia and other states, the west offered limitless opportunities for private profit. Jefferson considered it absolutely essential to confront their schemes of land-capitalism before undertaking the creation of new laws and legal institutions. He carried off an important success in behalf of western settlers, helping them to triumph over profit-seeking land proprietors in the October 1776 partition of Fincastle County into three distinct counties.

The Transylvania Company claimed lands between the Cumberland and Kentucky rivers, purchased from the Cherokees in March 1775. Judge Richard Henderson and his self-styled "true and absolute proprietors" tried to have their princely colony, mostly within Virginia, admitted to the Union as a separate and fourteenth state. Jefferson and Wythe had conferred with their emissary in Philadelphia at about the time Henderson, in Williamsburg, was blandly assuring the Virginia Convention he didn't want a separate government. The Convention, drawing on Jefferson's draft constitution in late June, barred private purchases from the Indians without legislative approval. The would-be proprietors had run into stiff anti-Transylvania resistance by overmountain settlers led by John Gabriel Jones, nephew of Jefferson's first law associate in Augusta County. They organized at Harrodsburg and petitioned Virginia to set off West Fincastle as a separate county and chose as their delegates to go to Williamsburg John Jones and George Rogers Clark, whom Jefferson had known since boyhood. After the two groups' representatives applied for seats in the Assembly on October 11, the House barred seats to Jones and Clark. At the same time, the House Committee on the State of the Country reported that, because the settlers wanted a separate county, a committee should be formed to consider drafting a bill establishing it. Jefferson was appointed to the committee to draw up the bill along with two powerful spokesmen of the Tidewater aristocracy, Robert Carter Nicholas and Carter Braxton. When the bill was brought to the floor, the House deadlocked and recommitted it to a new committee. This time Jefferson won the chairmanship, at the head of representatives from the western counties, and the conservatives were dropped. Two days later Jefferson's amended bill was argued to a standstill by Tidewater opponents and, ten days later, was defeated. Jefferson's sense of triumph at mobilizing western opinion and taking the bill away from the landed interests had been premature. When he took a two-day leave from the House over the weekend of October 18–19, Henderson and his friends were busy lobbying; the following Tuesday, Braxton, the Tidewater planters, and the land speculators took away the bill to the much more Tidewater-dominated Committee on Propositions and Grievances, where there were few western sympathizers. Returning, an aroused Jefferson created so much resistance that, two days later, the House took the bill away from the Grievance Committee. On October 26, Jefferson, not Braxton, reported out a new bill to divide Fincastle. Now there were two nearly identical bills, only differing in their backers. Four days later, Jefferson shrewdly backed Braxton's bill on the House floor—then thoroughly amended it, striking out everything after "Whereas" in Braxton's bill and substituting a completely new bill! Jefferson's "amendments," dividing Fincastle into three instead of two counties, including Kentucky and Washington counties, passed the House and the Senate in two days. The conservatives had stirred Jefferson into uniting the west behind him and winning six new western seats in the House.

In the opening days of that extraordinary revolutionary assemblage, Jefferson was also appointed to a nineteen-member committee on religion. Article 16, the last article of the Virginia Declaration of Rights, had affirmed the principle of religious freedom. Cautious George Mason had done nothing more than update the English Act of Toleration, and he had kept silent on the vital question of disestablishing the Church of England. In his notes for a speech in 1776, Jefferson estimated there were fifty-five thousand dissenters, or one in seven adult white males. In his 1784 *Notes on Virginia,* Jefferson stated that two-thirds of Virginians had become dissenters since 1776. In the debate over Article 16, Patrick Henry was asked point-blank if a more liberal wording by young James Madison precluding "peculiar emoluments" was designed as a prelude to an attack on the established church. Henry answered with an emphatic no. The constitutional convention had merely guaranteed that evangelists would no longer be arrested for preaching without a license, but this had not gone nearly far enough for the majority of Virginians, who were by now religious dissenters. They included Presbyterians in the western counties, many of whom had emigrated from Northern Ireland to escape Anglican bishops and exactions, as well as evangelical Methodists and the Baptists, in the southwest and central Piedmont, the severest critics of paying taxes to support the Anglican establishment.

When the House of Delegates convened in October 1776, it was greeted by a mound of protests on the subject, including a "ten-thousand name" petition, circulated by the Baptists, declaring the signers' "hopes have been raised and confirmed by the declarations" of "equal liberty." After long having "groaned under the burden of ecclesiastical establishment, they pray that this as well as every other yoke may be broken and that the oppressed may go free."[9] The dissenters demanded a law granting full equality in carrying out religious beliefs and actual disestablishment of the Church of England. Nearly half a century later, Jefferson attested that the dissenters "brought on the severest contest in which I have ever been engaged." In 1776, Jefferson identified himself as the foremost proponent of separation of church and state, declaring freedom of religion "one of the natural rights of mankind."[10] He launched a ten-year legislative battle with a new lieutenant at his side, twenty-five-year-old freshman legislator James Madison of Orange County, who had pressed the May Convention for a more liberal statement on religious freedom.

To Jefferson, religion was a private matter, like marriage, and in 1776 he said little about his private views on the subject. He did not attend church frequently, eschewed religious dogma, and believed in a supreme being who had set the world on its foundation and stepped aside. But he respected all honest men and their moral beliefs and believed firmly in tolerating all religions, not only Anglicanism or Christianity. He had studied Hebrew and the Koran. "It does me no injury for my neighbor to say there are twenty gods, or no god. It neither picks my pocket nor breaks my leg,"[11] he wrote in his *Notes on Virginia* in 1784. Neither Jefferson nor Madison was prepared to compromise on less than complete disestablishment and religious freedom. Jefferson held that no one had the right to dictate the faith of another: there *must* be freedom of religion to be free at all. Jefferson drew up six resolutions to carry out disestablishment and religious freedom late in 1776, but the reform movement could get no further than the repeal of all acts that oppressed dissenters plus a bill that exempted dissenters from having to pay taxes that contributed to the support of Church of England clergy. The preamble stating that "reason and justice" opposed such enforced support was left standing. The controversial act postponed a decision on the question of a general assessment to support the clergy of all religious denominations: "Although the majority of our citizens were dissenters," Jefferson wrote later, "a majority of the legislature were Churchmen."[12] Jefferson opposed even this as a deprivation of individual liberty. The state should neither support nor oppose any religion. Jefferson made this an American article of faith.

That the Anglicans should lose their favored status, according to Jefferson, was at least in part because they deserved to—the people had lost respect for their clergy. To John Adams, he wrote, "Our clergy, before the Revolution, having been secured against rivalship by fixed salaries, did not give themselves the trouble of acquiring influence over the people."[13] Jefferson, who believed in a free society, objected to an established clergy on the rational grounds that their special status and privileges had made them think of themselves as an artificial aristocracy that had usurped authority. The dissenters shared Jefferson's view that the respect demanded by the Anglicans was no longer warranted. The legislature voted to freeze temporarily the salaries of Anglican clergy, but a general clergy support bill for all denominations was never accepted. Jefferson led the legislative attack on a number of "tyrannical spiritual laws"[14] still stipulating, for example, capital punishment for heresy against the doctrines of the Church of England and imprisonment for denial of the Trinity or the divine authority of the Scriptures.

Despite Jefferson and Madison's efforts in 1776, backed by such political forces as Pendleton and the Lees, the establishmentarian majority of Anglican landed gentry still dominated the Assembly, which insisted on maintaining the principle of establishment by asserting that "proper provision should be made for continuing the succession of the clergy and superintending their conduct" in the face of the wartime cutoff from the bishops of the church in England and the Declaration of Independence. A set of six resolutions passed by the House in November also reaffirmed the determination of slaveowning gentry to maintain their legal right to control, by force if necessary, popular meetings within their counties. Unspoken was the deep-seated fear of fire-and-brimstone sermons by the evangelical Baptists and Methodists stirring up a slave revolt. The House declared the principle that "public assemblies of societies for divine worship ought to be regulated."[15] In his autobiography Jefferson wrote of the vitriolic struggle, of "desperate contests" in his Religion Committee "almost daily."[16]

The House staved off Jefferson's religious reforms for nearly three years before he was allowed to bring forward for consideration his "Bill for Establishing Religious Freedom" on June 4, 1779. He knew by now not to rely on fickle flickerings of revolutionary fervor. He believed in law. He wrote a statute that would blot out the authority of the established church and leave absolute freedom of choice in religious beliefs. Here is its core passage:

> We the General Assembly of Virginia do enact that no man shall be compelled to frequent or support any religious worship, place or ministry whatsoever, nor shall be enforced, restrained, molested, or burthened in his body or goods, or shall suffer otherwise on account of his religious opinions or belief; but that all men shall be free to profess, and by argument to maintain, their opinions in matters of religion, and that the same shall in no wise diminish, enlarge or affect their civil capacities.[17]

Still, the House refused to vote on the bill, allowing its authority to lapse. Jefferson and his associates anonymously had it printed for general circulation, setting off another round of controversy in the press and in pulpits. When the House reconvened in the fall of 1779, Jefferson found that the effort had backfired. Petitions in favor of his bill for disestablishment were far outnumbered by petitions calling for a bill providing tax support for the clergy of all denominations. One petition, from Lunenburg County, pointed out that, in the past three revolutionary years, voluntary support of religion had been weak and would predictably remain "very inadequate," discouraging "men of genius and learning" from "engaging in the ministerial office," depriving the state of "one of the best means of promoting its virtue, peace and prosperity."[18] This backlash discouraged Jefferson and his faction from bringing his bill to a vote. Indeed, a bill providing a general tax levy for all organized churches was introduced on October 25, 1779. While Jefferson's bill mentioned no particular religions and extended freedom to all, this bill declared that "the Christian Religion shall in all times coming be deemed the established religion of this commonwealth." Clergy must adhere to four articles: the existence of God, "a future state of rewards and punishment,"[19] the truth of the Scriptures, and the social duties of Christians. Ministers must be formally educated, a pointed attack on homegrown Baptist and Methodist preachers. The reactionary antidisestablishment bill created such an uproar among Jefferson's followers that it, too, was shelved after a second reading on November 15, 1779.

Forced to take some action, the House voted a bill that confirmed legal title to ecclesiastical properties and endowments, abolished the tax-subsidized salaries of parish rectors and curates and all other tax levies to support "the former established Church."[20] This was an oblique but important disavowal of the established state religion, but the Church of England in Virginia was far from being dismantled, its vestries continuing to act as the public bodies entrusted with the care of the poor, its membership still confined to baptized and communion-taking churchmen, its clergymen, those who had not fled into Loyalist exile, still enjoying special privileges. Jefferson's defense of the Bible-beating hillfolk against the rational Anglicans horrified some of his friends like John Page, who said the "bigoted and illiberal" Baptists were stealing the flocks of the "rational" sect. "Nothing but a general assessment can prevent the State from being divided between immorality and enthusiastic bigotry,"[21] Page wrote Jefferson. But Jefferson would not heed Page's warning. "Reason and free inquiry are the only effectual agents against errors," he countered.[22]

It was five more years before a general concern for lapses in morality brought the clergy-support battle out into the open again. This time, in 1784, Patrick Henry was leading a drive for a "general assessment" to support all clergy. Urged on by Presbyterians decrying the decline of "public worship of the deity," the House voted 47–32 a progressive tax "for the support of the Christian religion." A companion bill to pay for the training of clergymen only passed 44–42, but with so little support that the House referred it to public discussion in the counties before final passage. The public outcry from tax resisters joining forces with Enlightenment reformers, such as Madison, who wrote his famous "Memorial and Remonstrance," one of eighty papers printed on the question, buried the Christian Teachers Bill and resurrected Jefferson's Act for Establishing Freedom of Religion, which now sailed to final passage with overwhelming support, eleven thousand signatures to one thousand on the petitions.[23] Madison informed Jefferson, who was in Paris at the time, that its enactment had "in this country extinguished forever the ambitious hope of making laws for the human mind."[24]

Jefferson celebrated this victory after a ten-year struggle as a triumph of the Enlightenment spirit in the new republic of Virginia. He promptly translated the act's final text into French and Italian and paid to have it printed and distributed all over Europe. To Madison, he exulted at the "infinite approbation" of Europeans for their labors. "I do not mean by the governments, but by the individuals which compose them." He had it inserted in the new edition of Diderot's *Encyclopédie* and sent it to "most of the courts of Europe." The Virginia statute, he reported, was considered "the best evidence of the falsehood of those reports which stated us to be in anarchy." Jefferson found it "comfortable" to see

> the standard of reason at length erected, after so many ages during which the human mind has been held in vassalage by kings, priests and nobles. . . . It is honorable for us to have produced the first legislature who had the courage to declare that the reason of man may be trusted with the formation of his own opinions.[25]

Forty years later, when he wrote his own epitaph, he included the Virginia Statute on Religious Freedom among the three great accomplishments of his long life, second only to his writing the Declaration of Independence.

As if to prove his conviction that each person should voluntarily support his own local church and clergyman, in 1777 Jefferson subscribed six pounds—more than double any other parishioner's contribution—to the annual support of the Rev. Charles Clay of St. Anne's Anglican Parish, Charlottesville, and wrote the subscription petition *gratis.*

The struggle over freedom of religion had, like the revolution, come on suddenly in 1776, before Jefferson was able to organize the systematic revision of Virginia's laws that he had left Congress to carry out. Before the opening session of the Assembly of 1776 was over, however, Jefferson was able to settle down to a thorough study of those laws. On November 1, the same day the Senate ratified his bill to eliminate entail, the Assembly passed his Bill for the Revision of the Laws. Edmund Pendleton appointed a five-man House committee, with Jefferson its chairman, to revise the commonwealth's entire law code. All statutes were to be revised, digested, and altered wherever necessary. It was to be a mammoth undertaking that required nearly three years. In the house of George Wythe and with the books he had once studied as a law clerk, Jefferson began a two-month preliminary study to recommend to his committee how to proceed even as he began to write and introduce emergency law-reform bills.

By November, he was hard at work remodeling Virginia's judiciary instead of the old General Court made up of the Governor's Council. In his draft constitution, he had recommended an independent judiciary. Between November 25 and December 4, 1776, building on his frustrated constitutional drafts, Jefferson drafted five separate bills to establish a court of appeals, a high court of chancery, and a general court, and to regulate the proceedings in the county courts. His goal was to bring the judicial system into harmony with republican principles, but Jefferson must have known he faced strenuous opposition from many Virginia planters who feared ruin if the courts were reopened and British creditors were able to sue for payment of long-overdue debts. Overriding the opposition of the conservative Braxton on his committee, Jefferson introduced the first three judicial reform bills on November 25. He immediately ran into stiff resistance. While his admiralty court bill sailed smoothly to passage because Virginia was a maritime state with one eye turned to prize vessels captured from the British, all four of the other bills were stalled. Jefferson tried to push county-court reforms by reassuring debtors in an amendment that suspended all executions for debts, but this attempt failed. County courts had a ten-pound debt limit and most debts to the British were many times that. His bill creating a court of appeals was blocked for more than two years. His bills establishing chancery and general courts were delayed until January 1778. And his bill to regulate county-court proceedings never was enacted. Jefferson wanted to elevate the character of the lawyers and the judicial process and to require the courts to remain open at all times to accommodate the people, instead of only during infrequent, short sessions with years of delays. Although those reforms were far less radical than some of the changes he brought about in other spheres, he had taken on the entrenched county social, economic, and political system wrapped in its sacrosanct robes of custom and tradition, and he was as doomed as any criminal dragged before it. It would not be until after the Civil War that these local bulwarks of the landed gentry's power could be reformed.

With his earlier attempts at religious and judicial reforms qualified successes, Jefferson retrenched. At the end of the legislative session, he took his family home to Monticello, where his prodigious labors were eased by the scope of his extensive law library. On January 13, 1777, he met at Fredericksburg with the other revisors—Wythe (recalled from Congress), Pendleton, Mason, and Thomas Ludwell Lee—for a preliminary conference to divide the vast labor of overhauling the law.

"At the first and only meeting of the whole committee," wrote Jefferson, "the question was discussed whether we would attempt to reduce the whole body of the law into a code, the text of which should become the law of the land." Pendleton was backed by Lee and the others, except Jefferson, in favoring a new and complete institute of the laws, but chairman Jefferson opposed them emphatically:

> To compose a new Institute like those of Justinian . . . or that of Blackstone, which was the model proposed by Mr. Pendleton, would be an arduous undertaking.

Jefferson was still recovering from his last solitary writing project, the Declaration of Independence and its heavy-handed editing. He feared that every word of such a new code of laws would be subjected to test lawsuits and "render property uncertain until, like the statutes of old, every word had been tried and settled by numerous decisions and by new volumes of reports and commentaries."[26] Years later, Jefferson confided to John Tyler that he had feared that, had Pendleton prevailed in a complete rewrite, the overwhelming task would have fallen to Jefferson himself. Worse, "we should have retained the same chaos of law-lore from which we wished to be emancipated, added to the evils of the uncertainty which a new text and new phrases would have generated."[27] Jefferson hated the "vain declaration of the lawyer" and he despised legalistic verbiage, but he feared that, if every word of every law were recast, if the Common Law were

> reduced to a text, every word of that text, from the imperfection of human language and its incompetence to express distinctly every shade of idea, would become a subject of question and chicanery.

For the same reasons, Jefferson yielded also in his desire to rid the law of the clumsy language of ancient law courts:

> I thought it material not to vary the diction of the ancient statutes by modernizing it . . . the text of these statutes had been so fully explained and defined.

But he steadfastly refused to allow the dust and cobwebs to accompany any new laws into Virginia statute books. Not only were all statutes to be revised, digested, and altered where necessary, but their diction was to be modernized. Nevertheless, Jefferson wanted to make as few changes as necessary. In all new laws, he vowed to rid the lawbooks of "their verbosity, their endless tautologies, their involutions of case within case and parenthesis within parenthesis . . . their multiplied efforts at certainty by *saids* and *aforesaids*, by *ors* and by *ands*—to make them more plain."[28]

When the revisors divided their labors, Jefferson agreed to study crime and punishment, the Common Law, the statutes of England to the settlement of Virginia, and the laws of the English Commonwealth of 1641–60, including the laws of descent and religion. Wythe took on the remainder of the statutes. Pendleton undertook Virginia's laws. Lee was to study the laws of property and slavery. Mason was assigned the laws of the other colonies: "any good ones were to be adopted."

The committee next disagreed on the first major body of laws it considered. Wrote Jefferson in his autobiography,

> Mr. Pendleton wished to preserve the right of primogeniture, but seeing at once that that could not prevail, he proposed we should adopt the Hebrew principle and give a double portion to the elder son. I observed that, if the eldest son could eat twice as much, or do double work, it might be a natural evidence of his right to a double portion. But, being on a par in his powers and wants, with his brothers and sisters, he should be on a par also in the partition of the patrimony.[29]

Jefferson's argument prevailed. The committee recommended a bill to the Assembly abolishing primogeniture. In fact, Jefferson's attempts to liberalize Virginia's penal code proved one of his greatest frustrations. Seeking a rational proportion between crime and punishment, Jefferson's bid for penal reform very nearly wrecked his entire reform package. Between 1776 and 1779, Jefferson gave more time to researching the criminal laws than to any other segment of revisions. He systematically studied Anglo-Saxon laws, medieval authorities like Bracton, and the chief foreign writers, including Beccaria. The finely crafted bill he submitted in advance to George Wythe was a model of elegant, plain writing. He wrote footnotes in Anglo-Saxon characters, in Latin, and in old French and English.

In clear, beautiful language, Jefferson relaxed the severity of the law, making punishments not only more rational but also more humane. He took his well-known Enlightenment views further than many of his colleagues were willing to go. In 1776, Pendleton had warned him,

> I don't know how far you may extend your reformation as to our criminal system of laws. That it has hitherto been too sanguinary, punishing too many crimes with death, I confess, and could wish to see that changed for some other mode of punishment in most cases. But if you mean to relax all punishments and rely on virtue and the public good as sufficient to prompt obedience to laws, you must find a new race of men to be the subjects of it . . .[30]

In Section I of his bill, Jefferson borrowed from Beccaria to argue that harsh punishments were neither right nor effective. He argued that a citizen "committing an inferior injury does not wholly forfeit the protection of his fellow citizens, but after suffering a punishment in proportion to his offense, is entitled to their protection from all greater pain." He described capital punishment as "the last melancholy resource against those whose existence is become inconsistent with the safety of their fellow citizens."[31] He urged reforming criminal offenders instead of executing them; a thoroughly modern theoretical criminologist, he believed that the punishment should fit the crime, and he set up a table of punishments. In essence he followed ancient traditions, the Roman *lex talionis,* the law of the claw, and the Mosaic law, an "eye for an eye and a tooth for a tooth," as he put it.[32]

Some of his suggestions, when set in cold print, made it easy for his critics to attack the reform bill:

> Section XIV. Whosoever shall be guilty of rape, polygamy or sodomy with man or woman, shall be punished, if a man, by castration, if a woman, by boring through the cartilage of her nose a hole of one half inch in diameter at the least.[33]

His penal reform bill was thoroughly repudiated in a stampede of public opinion arising from a wave of horse stealing.

Later, after Jefferson was severely criticized by Europe's intelligentsia for his principle of retaliation and his rape law was derided in Paris as indecent and unjust, an older and wiser Jefferson suggested altering the rape section of the bill to take into account "the temptation women would be under to make it the instrument of vengeance against an inconstant lord and of disappointment to a rival."[34]

He ignored more unnatural sexual acts. He divided "buggery" into sodomy—which he punished severely—and bestiality, until then a capital crime, which he now chose to ignore:

> Bestiality can never make any progress; it cannot therefore be injurious to society in any great degree, which is the true measure of criminality . . . and will ever be properly and severely punished by universal derision. It may, therefore, be omitted.[35]

Jefferson aimed at enacting criminal laws that were clear enough for laymen to understand without the help of lawyers. To Wythe, as he sent him his draft code of laws, he confided,

> In its style I have aimed at accuracy, brevity and simplicity, preserving however the very words of the established law, wherever their meaning has been sanctioned by judicial decisions or rendered technical by usage. Indeed, I wished to exhibit a sample of reformation in the barbarous style into which modern statutes have degenerated from their ancient simplicity. . . . I have thought it better to drop

> in silence the laws we mean to discontinue, and let them be swept away by the general negative words of this than to detail them in clauses of express repeal.[36]

All the revisors agreed on the necessity for one reform: the death penalty should be abolished except for treason and murder. Under the English-style criminal code followed in Virginia, more than one hundred offenses, many of them minor, brought the death sentence. "For other felonies," wrote Jefferson, "hard labor in the public works"[37] should be prescribed. He had in mind putting criminals to work digging canals, building roads until they repented, but in an attempt to be more humane Thomas Jefferson had fostered the chain gang. By 1796, when his criminal code was finally enacted after twenty years of controversy, he favored the construction of penitentiaries to administer justice, and he never was very proud of his penal code.

On his mountaintop of Monticello, as slave laborers proceeded with the construction of his rural retreat, Jefferson the lawgiver turned his research and writing skills to the problem of slavery. In 1781, much of his legislative reform package still pending before the House, Jefferson wrote his best statement of his views on slavery in answer to queries from a French diplomat, and it deserves to be quoted here at length:

> . . . There must doubtless be an unhappy influence on the manners of our people produced by the existence of slavery among us. The whole commerce between master and slave is a perpetual exercise of the most boisterous passions, the most unremitting despotism on the one part, and degrading submissions on the other. Our children see this, and learn to imitate it; for man is an imitative animal. . . . If a parent could find no motive either in his philanthropy or his self-love, for restraining the intemperance of passion towards his slave, it should always be a sufficient one that his child is present. But generally it is not sufficient. The parent storms, the child looks on, catches the lineaments of wrath, puts on the same airs in the circle of smaller slaves, gives a loose to the worst of his passions, and thus nursed, educated, and daily exercised in tyranny, cannot but be stamped by it with odious peculiarities. The man must be a prodigy who can retain his manners and morals undepraved by such circumstances. And with what execrations should the statesman be loaded, who permitting one half the citizens thus to trample on the rights of the other, transforms those into despots, and these into enemies, destroys the

> morals of the one part, and the amor patriae of the other. . . . With the morals of the people, their industry also is destroyed. For in a warm climate, no man will labour for himself who can make another labour for him. This is so true, that of the proprietors of slaves a very small proportion indeed are ever seen to labour. And can the liberties of a nation be thought secure when we have removed their only firm basis, a conviction in the minds of the people that these liberties are of the gift of God? That they are not to be violated but with his wrath? Indeed I tremble for my country when I reflect that God is just: that his justice cannot sleep forever; that considering numbers, nature, and natural means only, a revolution of the wheel of fortune, an exchange of situation, is among possible events: that it may become probable by supernatural interference! The Almighty has no attribute which can take side with us in such a contest.[38]

Jefferson's opposition to slavery was well known when he was appointed chairman of the revisors. He had not only denounced the African slave trade in the Declaration of Independence, but in his draft state constitution he had urged the prohibition of holding in slavery any person who came into Virginia. He did not need to write a separate law banning slavery. By the time his complete set of revisions was delivered to the legislature, the revolutionary Assembly had already taken a first step and banned any further importation of slaves in 1778. Jefferson claimed authorship of the act, but it was a heavily edited version of his original bill. It encouraged the private manumission of slaves, which Virginia law still made virtually impossible. When Thomas Ludwell Lee died in 1778, the task of studying and revising the slave code devolved on Jefferson. His first task was to codify the colonial slave laws, and he wrote a digest of them.

Jefferson favored gradual emancipation of slaves and he presented no sweeping antislavery law as part of the revisors' reforms. He proposed not only banning foreign importation of slaves but limiting slavery henceforth to the descendants of female slaves currently in Virginia.

According to Jefferson, he offered the revisors a plan under which all of the slaves born in Virginia after the passage of the act would be freed, should remain with their parents and be trained at public expense in a useful trade and, at the age of adulthood (twenty-one for men, eighteen for women) be colonized outside Virginia "as a free and independent people." While the revisors remained silent on the subject, Jefferson was apparently overruled by the rest of his committee. He later wrote that "it was thought better that this should be kept back and attempted only by way of amendment, whenever the bill should be brought up."[39] Jefferson later wrote that he was prepared to argue why freed slaves had to leave Virginia: "Deep-rooted prejudices entertained by the whites, ten thousand recollections, by the blacks, of the injuries they have made; and many other circumstances will divide us into parties and produce convulsions which will probably never end but in the extermination of the one or the other race."[40] Jefferson was abroad in France by the time the bill was brought before the House in 1786, and by then revolutionary fever had cooled to the point that no prominent Virginia politician would risk his friends, his office, or his influence to speak up for the slaves. As Jefferson put it, "the public mind would not yet bear the proposition."[41]

Jefferson's own mind on the subject can only be judged fairly in the context of his times. Most of his neighbors, it can be fairly said, considered both Afro-Americans and Indians inferior to whites. Jefferson attempted to study the question scientifically and unemotionally:

> . . . I advance it, therefore, as a suspicion only, that the blacks, whether originally a distinct race, or made distinct by time and circumstances, are inferior to the whites in the endowments both of mind and body. It is not against experience to suppose that different species of the same genus, or varieties of the same species, may possess different qualifications. Will not a lover of nature history then, one who views the gradations in all the races of animals with the eye of philosophy, excuse an effort to keep those in the department of man as distinct as nature has formed them? This unfortunate difference of colour, and perhaps of faculty, is a powerful obstacle to the emancipation of these people. Many of their advocates, while they wish to vindicate the liberty of human nature, are anxious also to preserve its dignity and beauty. Some of these, embarrassed by the question, "What further is to be done with them?" join themselves in opposition with those who are actuated by sordid avarice only. Among the Romans emancipation required but one effort. The slave, when made free, might mix with, without staining the blood of his master. But with us a second is necessary, unknown to history. When freed, he is to be removed beyond the reach of mixture.[42]

But Jefferson was writing this only as a philosopher for the eyes only of his close friends, the French in the circle of his friend, Condorcet. Jefferson never intended it to be published. Yet more than most southerners of his time, he attempted to keep an open mind. In 1791, shortly after Jefferson returned from France, he learned of a "very respectable" Afro-American mathematician, Benjamin Banneker. After perusing some of Banneker's work, Jefferson wrote him,

> Nobody wishes more than I do to see such proofs as you exhibit that nature has given to our black brethren talents equal to those of the other colors of men, and that the appearance of a want of them is owing merely to the degraded condition of their existence, both in Africa and America. I can add with truth that nobody wishes more ardently to see a good system commenced for raising the condition both of their body and mind to what it ought to be, as fast as the imbecility of their present existence . . . . will admit.[43]

When Jefferson was nearly eighty, he wrote what may have been his final judgment on the subject of slavery. "Nothing is more certainly written in the book of fate than that these people are to be free. Nor is it less certain that the two races, equally free, cannot live in the same government."[44]

When Jefferson rewrote the laws of citizenship in Virginia as part of his revisal, he included all whites who migrated into the state, asserting the "natural right which all men have of relinquishing the country in which birth or other accident may have thrown them, and seeking subsistence and happiness wheresoever they may be able, or may hope to find them."[45] Jefferson was governor by the time the bill came before the House, and he was always proud of its passage, but it did not include slaves.

To bring about the humanitarian changes in society that his new laws required, Jefferson put his faith in the education of posterity and planned for it. Writing a bill "for the more general diffusion of knowledge," Jefferson told George Wythe he considered it the most important piece of legislation he proposed in the omnibus "Report of the Committee of Revisors."[46] One biographer, Dumas Malone, has called him "the chief prophet of public education in the first half-century of the Union."[47] Education became his most enduring crusade, and he went about it by trying to inspire others to action. To Wythe, he wrote from Paris as he waited anxiously for the passage of his education bill, "Preach, my dear sir, a crusade against ignorance, establish and improve the law for educating the common people."[48]

In the Virginia of the eighteenth century, Jefferson offered a sweepingly revolutionary proposal: dividing each county into "hundreds" and building a school in each hundred located so that all free boys and girls could attend daily. All children would receive free schooling for three years, being taught reading, writing, and simple arithmetic and being exposed to Greek, Roman, English, and American history through the books they read. Any child could go on to higher grades at private expense. In addition, from each group of ten elementary schools, one boy, "of the best and most promising genius and disposition" whose parents couldn't afford to pay for his schooling, was to be chosen each year to go to one of the state-run secondary grammar schools serving several counties, his board and tuition paid by the state. Qualified students whose parents could pay for their education would be admitted to these schools to study Latin, Greek, English grammar, geography, and advanced arithmetic. The least promising third of the state scholarship students would be cut after one year, and after two years, only one—"the best in genius and disposition"—from each grammar school would be allowed to go on at taxpayers' expense for four more years. Proposing twenty grammar schools, Jefferson wanted "twenty of the best geniuses raked from the rubbish annually." Each grammar school in alternate years would send its most promising student to the College of William and Mary for three years, to be educated, boarded, and clothed at public expense. Thus, ten meritorious "public foundationers"[49] would annually reach the college level.

To reform the pinnacle of this system, William and Mary, he attacked his alma mater, asserting that it had not measured up to public "expectations" as a superior educational institution. Originally the college had come under Pendleton's charge as a revisor, but Jefferson explained in his autobiography that the revisors decided his "systematical plan of general education" should include the college. Jefferson wanted the college to provide "an ultimate grade for teaching the sciences generally and in their highest degree." In his bill "for amending the constitution of the college" and providing public revenues for it, Jefferson asserted that "the experience of near an hundred years," although amply endowed "by the people," had not answered their expectations." Jefferson found, however, "reason to hope that it would become more useful." To improve the college he depended upon to train "the future guardians of the rights and liberties of their country," he urged appointment of a five-member board of visitors and a three-man board of chancellors, voted by the Assembly each year, who would choose the rectors of the college for one year only, assuring accountability and tight public control. The "ultimate result" of his whole "scheme of education would be

> teaching all children of the state reading, writing and common arithmetic; turning out ten annually of superior genius . . . turning out

> ten others annually of still superior parts who . . . shall have added such of the sciences as their genius shall have led them to."[50]

At a time when there was no public education at all in the south, Jefferson's proposal to recruit an "aristocracy of virtue and talent"[51] from all classes was truly revolutionary. Jefferson believed that, to preserve republican government and prevent tyranny, it was necessary "to illuminate, as far as practicable, the minds of the people at large."[52]

Jefferson the lawgiver came down from Monticello twice a year between October 1776 and February 1778, to attend Assembly sessions, but for the most part, as the Revolution raged to the north and south, there was virtual calm in Virginia, and he was free to carry on his legislative revolution with the aid of his formidable library at Monticello. The committee of revisors came down to two active members, Jefferson and Wythe, and they exchanged drafts of laws and reform bills frequently.

For all but two weeks in the first half of 1777, he worked at home. Patty was pregnant again. He journeyed to Williamsburg without her in May 1777, but requested a leave of absence after only sixteen days to return to Monticello in time for the birth of their first and only son on May 28. But the infant did not live long enough to be christened, appearing only as "our son" in Jefferson's records. One evening in his third week, the baby died. For a year longer, Patsy remained an only child. Then, on August 1, 1778, Jefferson made another precise entry in his *Account Book,* "our third daughter born." They named her Mary. She became known as Maria to some, Polly to others. She lived into her twenties, and the servants said she was as pretty as her mother. The year of her birth was a busy one atop Monticello. Some ninety thousand good bricks were laid and three stone columns set in front of the main house, making its façade similar to the main entrance to the Wren Building at William and Mary. Jefferson worked grafting the laws of America and England in his study and grafting scores of trees—cherries, apples, pears, quinces, plums, apricots, almonds—in his orchard on the south side of the mountain. He planted an olive tree given him by Philip Mazzei that fall, noting it would take ten years to bear fruit. While fellow Virginian George Washington held the Continental Army together at Valley Forge, Jefferson raced to finish his law revisions.

By the time he finished his vast undertaking in February 1779, a British army had surrendered at Saratoga and the French had joined in an alliance with the United States against Great Britain. The war had turned.

Jefferson and Wythe met in Williamsburg, combined their drafts, and revised their work, reading every bill line by line to each other and suggesting final revisions. Returning to their homes to supervise the copying out of final drafts, they each carried a large bundle of bills into the capital on June 18, 1779. Filling ninety folio pages when printed, the combined total of 126 bills included several already adopted, the completely revised criminal code, and a modernized, digested version of British statutes and Virginia laws that thoroughly republicanized the state's legal system. But they were not to be adopted en masse; indeed, each was to be brought forward, debated, and voted in a grindingly slow process that dragged on until 1786. The "Report of the Committee of Revisors" was never brought to a vote; its provisions were enacted or defeated piecemeal. By 1785, only half the proposed laws had been passed and the revolutionary reform movement seemed spent when Madison, who had all along supported Jefferson's reform efforts in the legislature, led a final drive that carried to victory many of Jefferson's most important programs, including freedom of religion, by the end of 1786. By that time, of Jefferson's fondest reformations all but the regulation of the county-court system, the abolition of slavery, and his plan for "the more general diffusion of knowledge" were adopted into law by the Virginia revolutionaries. And, in the end, after twenty years of legislative struggles, the law he considered the most important to the success of all the others, his law to establish a democratic system of education, was finally adopted in 1796. In three years of backbreaking studies that, according to Madison, "exacted perhaps the most severe of [Jefferson's] public labors," Jefferson had almost single-handedly provided "a mine of legislative wealth"[53] that provided Virginians with a modern republic built on the foundations of Greece and Rome. It became a model for other states and the pattern after which the federal republic of the United States was modeled. Jefferson, in short, in his legal laboratory atop Monticello, invented the United States of America.

For nearly three years, while Jefferson liberalized the laws of Virginia, war left him in peace and he was able to maintain a detached, fantastic notion of it. He spent long weeks in the legislature at Williamsburg. In one spell, in the fall of 1777, he attended seventy-seven days in an unbroken row, voting supplies and requisitions for troops and seizing Loyalist property. But he had not yet seen anyone killed. For the most part he could pursue not only his research and writing but his loves of music and natural philosophy. While hundreds of Virginians with Washington at Valley Forge were shivering and starving, Jefferson calculated the angles of the Blue Ridge mountains to the horizon and took daily measurements of the climate and temperature at Monticello. And he corresponded with amateur scientist David Rittenhouse of Pennsylvania about making him an accurate clock, chiding him about wasting his time as a revolutionary: "Nobody can conceive that nature ever intended to throw away a Newton upon the occupations of a crown."[54]

Jefferson abhorred war and was tired of politics by the spring of 1779. He aspired to nothing more than the life of a philosopher tending his books and his orchards atop Monticello, where he had a gardener, a weaver, and a stonecutter and dreamed of having his own orchestra. He actually wrote to a friend of Mazzei's in Paris to see if he could find him skilled artisans and a vigneron who could also play the clarinet, the oboe, the French horn, and the bassoon. He talked to his friend Edmund Pendleton about retiring from the House of Delegates. Pendleton would not hear of it: "You are too young to ask that happy quietus from the public life."[55]

Pendleton's mild rebuke came shortly after war intruded for the first time into Albemarle County in January 1779. Some five thousand British and Hessian troops who had surrendered after Saratoga marched nearly seven hundred miles from Boston to Charlottesville. The Convention Army was sent to Virginia when Congress refused to honor the terms of surrender. Suddenly the county lieutenant of Albemarle was busy building barracks in winter and pleading with Patrick Henry, who was serving out his third term as governor, for supplies and humane treatment for his prisoners:

> I would not endeavor to show that their lives are valuable to us, because it would suppose a possibility that humanity was kicked out of doors in America and interest only attended to. . . . But is an enemy so execrable that, though in captivity, his wishes and comforts are to be disregarded and even crossed? I think not. It is for the benefit of mankind to mitigate the horrors of war as much as possible. The practice, therefore, of modern nations of treating captive enemies with politeness and generosity is not only delightful in contemplation but really interesting to all the world, friends, foes and neutrals.[56]

Busily mitigating the horrors of unsuccessful war for British and Hessian officers in the spring of 1779, Jefferson arranged for the ranking Hessian, Maj. Gen. Baron de Riedesel, and his wife and daughters to rent Colle from Philip Mazzei, who was away in Europe on a diplomatic errand for

Virginia. Inside a week, the Baron's grazing horses brought an end to Mazzei's five-year experiment in wine culture.

Welcoming cultivated company, Jefferson and his family accepted invitations to his captives' rented houses:

> Major General Phillips sends his compliments to Mr. and Mrs. Jefferson, requests the favor of their company at dinner on Thursday next at two o'clock to meet General and Madame de Riedesel. Major General Phillips hopes Miss Jefferson will be permitted to be of the party to meet the young ladies from Colle [the Baron's daughters].[57]

The Riedesel girls became Patsy Jefferson's playmates; the vivacious baroness her close friend. Among the Hessian prisoners there were good musicians. Young Baron de Geismar played his violin with Jefferson. When de Geismar was paroled, he left Jefferson all the music he had brought with him when he came to invade America.

Opening his home and his library to his fallen enemies, Jefferson exhibited hospitality that prompted this warm tribute from a German officer in a Hamburg newspaper, which also provides a rare glimpse inside Monticello at this early period:

> . . . My only occupation at present is to learn the English language. It is easier for me, as I have free access to a copious and well-chosen library of Colonel Jefferson's [who] possesses a noble spirit of building. He is now finishing an elegant building, projected according to his own fancy. In his parlor he is creating on the ceiling a compass of his own invention by which he can know the strength as well as direction of the winds. I have promised to paint the compass for it. He was much pleased with a fancy painting of mine. . . . As all Virginians are fond of music, he is particularly so. You will find in his house an elegant harpsichord, pianoforte and some violins. The latter he performs well upon himself, the former his lady touches very skillfully and who is in all respects a very agreeable, sensible and accomplished lady.[58]

By joint ballot of the Virginia House of Delegates and the Senate on June 1, 1779, Thomas Jefferson, at age thirty-six, was elected governor. Patrick Henry had served the maximum number of terms, three in a row, and whoever succeeded him, given the short one-year terms, faced the same prospect. Pulled away from the "private retirement to which I am drawn by nature with a propensity almost irresistible,"[59] Jefferson was elected in a closely contested election over two close friends, John Page and Thomas Nelson, Jr. Put up by his backers almost against his will on the first ballot, Jefferson had a plurality but not a majority, with Page, president of the Assembly's Council of State, and Nelson, commander of the state militia, running closely behind him. On the second ballot, Nelson's backers swung most of their ballots to Page, but Jefferson picked up enough votes to edge out his old college roommate, 67–61. The outcome, Jefferson hastened to write his close friend, gave him "much pain that the zeal of our respective friends should ever have placed you and me in the situation of competitors."[60] For his part, Page wrote that the contest had left him no "low dirty feelings."[61]

13

"It Is Not in My Power to Do Anything"

I have no intimate acquaintance with Mr. Jefferson, but from the knowledge I have of him, he is in my opinion as proper a man as can be put into office, having the requisites of ability, firmness and diligence.

—John Jones to James Monroe, March 1780

The streets of Williamsburg, once so full of fashionably dressed Tidewater gentlemen when the royalist Assembly convened, now were packed with men in fringed linen shirts carrying flintlocks and rifles. There was a ragged, martial air about the place as Governor Jefferson, driving his Monticello-made phaeton, led his family, in a coach and four with outriders, around the curving drive and up to the Governor's Palace, where he had first come as a college student to play his fiddle nearly twenty years earlier. Patrick Henry had already left for Leatherwood, his home in the western forests. There was no transitional briefing, but Jefferson already knew the state of things. Most of all, he knew he had no real executive power.

Jefferson's monumental legal labors, his devotion to the revolutionary cause, his reputation for being disinterested and widely trusted had made him the chief executive of the largest independent nation in the confederation of United States, which all seemed very imposing except for the fact that he had helped to give Virginia exactly what it wanted, an Assembly with all power and a governor who was only a figurehead. He might in the past have thrived for the maximum three terms, meeting every day with his friends on the Assembly's all-powerful Council of State—John Page, James Madison, John Walker, Jacquelin Ambler (who had won Rebecca Barwell from him so long ago), but, more and more, the war intruded. A pacifist could only survive in a war so long as the fighting remained far removed. Before his second term was over, Jefferson's pacifism was in tatters. He was to go back to Monticello, frustrated, disappointed, facing charges of cowardice under fire. His two terms as governor of Virginia ended in failure only months before the final Franco-American victory of the American Revolution, an event that took place on Virginia soil and could have solved Jefferson's problems.

As governor of the largest of the new United States from 1779 to 1781, Thomas Jefferson presided over a vast territory. Virginians claimed rights to the west all the way to the Mississippi. New counties had been settled not only west of the Blue Ridge but in Kentucky and along the Ohio River. Jefferson had secretly backed his former Albemarle neighbor, George Rogers Clark, in planning the conquest of the Illinois country from the British. Clark's expedition was under way when Jefferson took office. Virginia, even without its western claims, comprised, east of the Blue Ridge, a settled land area and a population of five hundred thousand whites and nearly as many black slaves, greater than that of all of New England, but nearly half of its population was enslaved.

Except for Lord Dunmore's brief depredations early in 1776, Virginia had been largely spared from war. This was fortunate, because the state was basically indefensible. With no factories capable of mass-producing heavy weapons and a neglected state navy of only three undermanned ships and four armed row galleys, its vital tobacco exports were effectively bottled up on Chesapeake Bay by British warships offshore. While Jefferson believed "that our only practicable defense was naval," he concluded within his first six weeks in office that "we should be gainers were we to burn our whole navy."[1] As British armies marched and countermarched around New England and the Middle States and attacked Georgia and South Carolina, Virginia's eleven battalions of willing warriors had been siphoned off by the Continental Army, taking with them most of Virginia's usable firearms. Resupply from imperial French, Spanish, and Dutch trading partners on Caribbean islands had been all but stifled by an ever-tightening British navy blockade.

A month before Jefferson's election, as if to demonstrate their absolute control of the seas and the weakness and vulnerability of vast Virginia, a British fleet had sailed into Hampton Roads, landed eighteen hundred

Redcoats, seized Portsmouth, raided the Eastern Shore, put several towns to the torch, captured 130 vessels, destroyed three thousand barrels of tobacco and large quantities of military and naval stores, in all wreaking £2 million pounds' worth of damage. After sixteen days of leisurely plunder and rape, the British had returned to their ships and, prizes in tow, had sailed back to New York without the loss of a single man. Patrick Henry's governorship had begun with bravado in the heady revolutionary days of 1776, but Henry had failed disastrously to defend Virginia's shores against repeated British attacks ever since Lord Dunmore's first depredations at Norfolk. Now, his three years as governor ended as they had begun, in military disgrace.

Despite its hypothetical fifty-thousand-man all-white militia, many of the best-armed Virginians were far from the coast in the overmountain region. The militia in coastal counties had proven useless against well-armed and well-disciplined British Regulars. Even if all of Virginia's able-bodied white men between ages sixteen and fifty had turned out armed, trained, and led well, they would have amounted to only one soldier per square mile. But, except for the mountain men, they were not armed well.

The 1779 invasion of Portsmouth had demonstrated all of the weaknesses of the state's defenses only two weeks before Jefferson moved into the Governor's Palace. Without a powerful and ruinously expensive navy, the Chesapeake and the rivers running into it, navigable far into the interior, lay open to British raids. The enemy blockade was rapidly ruining Virginia's economy, the richest in America at the outbreak of war, by closing off the export of its principal cash crop, tobacco. In an attempt to finance a cash-and-carry war and avoid piling up debts, Virginia had imitated the Continental Congress in adopting its own paper money. In 1777, Jefferson had confidently endorsed the plan: "I think nothing can bring the security of our continent and its cause into danger if we can support the credit of our paper."[2] But by the time he was sworn in in mid-1779, Virginia's paper money had depreciated dangerously. As government printing presses created more and more money, its purchasing power shrank. In May 1779, constitutionally prohibited from propping up the paper as legal tender, the Assembly had lifted the ban on discounting its paper money. As civilian and soldier suffered the ravages of currency depreciation, they stoutly refused to pay taxes to finance a war that had started with protests against taxation. Half the counties reneged on their tax quotas in 1779. Widespread tax resistance necessitated even higher tax assessments, which led to even fewer tax payments, which in turn were made with steadily depreciating currency. Even when the Assembly reluctantly resorted to limited taxation on enumerated imported goods in May 1779, the new Governor Jefferson had to admit by June that "taxation has become of no account."[3]

"In a virtuous government, and more especially in times like these," Jefferson wrote to well-wishing Richard Henry Lee shortly after his election, "public offices are what they should be—burdens to those appointed to them which it would be wrong to decline, though foreseen to bring them intense labor and great private loss."[4] Jefferson made few more accurate predictions. His new job brought him no glory other than to be called "His Excellency" and much trouble and expense after long days of enormous labor. With no real authority and few resources to draw on, he depended on his powers of personal ingenuity and the respect accorded the representative of scarcely half the people, and those the less wealthy and powerful upland voters. Under the constitution of 1776, the governor depended totally on the legislature, which was no longer made up of the illustrious revolutionaries of 1776. Jefferson had none of the power of a royal governor. He could neither veto a law he did not like nor dissolve the legislature that passed it. He was enjoined only to "exercise the executive powers of government according to the laws of the Commonwealth and . . . not, under any pretense, exercise any power or prerogative by virtue of any law, statute or custom of England."[5] He could not make appointments, call out the militia, or grant pardons without the approval of a majority of the eight-man Council of State, elected by the Assembly. While he commanded the militia, he had to rely on its county officers for its actual operations. What little real authority Jefferson held, he had to exercise with the advice and consent of the Council of State, which often lacked a quorum. Yet Jefferson was responsible for carrying out all the laws of Virginia. To do so, he developed a close working relationship with each Council member. Within six months, he had learned to employ this so skillfully that the members yielded to him the power to act when the Council was not sitting and when its approval of a measure could have been expected. He was able to accomplish this largely because of his increasing reliance on Council member James Madison, whose influence on fellow Assemblymen was growing. Jefferson's reforms of Virginia law had extended to some innovations in the Constitution of 1776. In May 1779, the month before his election as governor, he had won Assembly approval for two bills, one establishing a Board of War and the other a Board of Trade. While extending the governor's power, these boards were to prove too cumbersome. Jefferson kept up his attempts to enhance the efficiency of the office of governor; the following year he replaced the boards with a commercial agent, a commissioner of the navy, and a commissioner for the war office. All three were appointed by the governor with the advice and consent of the Council and were placed under the control of both. But for all intents and purposes, the Council of State, a revolutionary committee, ran Virginia throughout the Revolutionary War. While Jefferson had a staff of clerks, he had to involve himself in myriad petty details in matters large and small, and write many of the letters himself. From Philadelphia, Congress pressed him for money and troops for the Continental Line. He corresponded regularly with leaders in Congress and with Washington. As the fighting shifted south, he received desperate calls for arms, supplies, clothing. Governor Jefferson was expected to know the proper width of cloth and to estimate how many uniforms it would make. When he provided 1,495 yards to make uniforms for four hundred recruits, the Virginia militia quartermaster, Colonel William Davies, wrote back to correct him. This was only enough to make 370: "I should have imagined that the width as well as the length of the cloth would have been reported to your Excellency."[6] He seemed calm as he went about his new duties, unperturbed that most other Virginians did not seem to be working harder to prepare for the possibility that the war would ultimately come to Virginia. "Mild laws, a people not used to war and prompt obedience, a want of the provisions of war and means of procuring them render our orders often ineffectual, oblige us to temporize and, when we cannot accomplish an object in one way, to attempt it in another," Jefferson explained to Lafayette.[7]

From June to August, 1779, Jefferson met daily with the Virginia Council of State at the Governor's Palace. By 10 A.M., when the meetings began, he had already drawn up the day's detailed proposals for Council debate and decision. He had once again packed off Patty and the girls to The Forest in nearby Charles City for a long visit with her sister, leaving his days and nights clear for the kind of concentration he needed to tackle the state's problems. Jefferson's capacity for long days at his desk served him well: he had to deal with countless detailed transactions. He promulgated and executed new laws, maintained relations with Congress and the Continental Army, handled Indian affairs and prisoners-of-war, was in charge of public works, including lead mines and a gun factory, was responsible for trade, taxes, buying supplies, disbursing government funds, raising and writing orders for troops, and commanding the militia, the state navy, and fortifications. While all decisions were made "in Council," he set the agenda and carried out the Council's decisions, doing much of the paperwork himself.

He brought two major plans to the Council almost at once, drawing on proposals he had made the month before in the legislature as an assemblyman. To "increase the annual revenue" of the state and to "create a fund for discharging the public debt,"[8] Jefferson had proposed the Land Office Act, which allowed the sale of Virginia's vast unsettled western lands to back the state's war efforts. So successful was this device that thousands of landless pioneers began streaming into western Virginia, especially along the Ohio River into Kentucky, buying for small change homestead lands far from the coastal depredations of the British. Jefferson's bold plan worked so well that it set off a howl of protests from landless states. By the end of 1779, Congress strongly protested Jefferson's Land Office, charging that the prospect of "savage freedom" on the Ohio had "afforded both an asylum and a temptation for desertion"[9] from the war. As some states worried about the increased costs of frontier defense caused by Jefferson's land policy, pressure against it also mounted at home, as land sale revenues proved inadequate to offset currency depreciation.

The governor's second executive attempt to raise money proved more popular but even less successful. He had long led a legislative movement to seize the estates of Loyalists, most of whom had left long ago with Lord Dunmore, and in June, he ordered the confiscation of all the property of British subjects in Virginia, presumably including his own cousin's. The estates of British subjects involved not only valuable lands and buildings but an estimated £3 million in planter debts. Ironically, Jefferson's championing of personal liberty and the rights of citizens to their property did not include the Loyalists, to whom he was quite ruthless. It seems he regarded them only as legal abstractions: under the law, they were enemy aliens. Their estates represented money that the state badly needed.

In January 1778, when he had first introduced his Land Office bill, he had successfully pushed through the Assembly his Bill for Sequestering British Property. The act had temporarily left intact titles to land held by Loyalist exiles, but it provided for patriot administrators to pay the income of Loyalist property into the state treasury. His bill also provided that planters' debts owed to the British could be discharged by paying the money in Virginia currency into Virginia's coffers, a confiscatory measure at best. Jefferson himself paid off his father-in-law's debts to British merchants in this manner. But few Virginians rose to this tempting bait and sequestration yielded little revenue.

Jefferson began his next anti-Loyalist law, an act confiscating all British property, with an apologetic preamble stating that he knew it was "a departure from that generosity which so honorably distinguishes the civilized nations of the present age," but this did not stop the new wartime governor from invoking the feudal Norman law of escheat and forfeiture so often applied by English monarchs against rebels. Bending English law to the needs of the new commonwealth, Jefferson simplified it. Only an inquest was now needed to establish whether a property owner was a British subject or a Virginia citizen in good standing. A ruling of "British" meant that the property was automatically forfeited to Virginia by escheat. In the words of Jefferson scholar Merrill Peterson, Jefferson's anti-Loyalist law represented

> an astounding revival of feudal practice. . . . All the states confiscated British property and some of them were far less generous than Virginia, but there alone, ironically by the most enlightened statesman of his time, was confiscation carried out in the shadow of feudalism.[10]

It also failed. If Jefferson conceived of Loyalist confiscations as a means of refilling the revolutionary coffers by redistributing wealth among Virginia's landless pioneers, the money netted from confiscations was being paid in depreciated dollars, and the effort was a fiasco. While Jefferson's confiscation policy remained on the lawbooks until 1784, by September 1779, he had already given up on it and was looking elsewhere for revenues. He admitted he had been beaten in the law courts, where legal pleadings on behalf of British subjects were "throwing the subject into a course of legal contestation which . . . may not be terminated in the present age."[11]

Governor Jefferson's desperate fiscal measures belied his personal philosophy, that government solvency could only be established by winning aid from some foreign power or by a complete embargo of Virginia's foreign trade, which would force the state to become completely self-reliant. In theory, Virginia tobacco should bring hard money to pay taxes to finance the war effort, and the alliance with France was to bring French naval protection of Virginia trade from the British navy. But by the time Jefferson took office, more than a year after the French alliance had been signed, the British attack on Portsmouth demonstrated how easily the tobacco trade could be blockaded in the Chesapeake. Jefferson could see no other means to avert financial disaster but to send an emissary to Europe to seek "a plentiful loan of hard money."[12] Jefferson sent several agents abroad, including his friend, Philip Mazzei, who was to seek aid from Italian princes.

When this scheme also failed, Jefferson concluded that the only alternative would be a formal embargo to force the development of domestic manufacturing. Indeed, several key manufacturers—cloth weaving, leather tanning, iron forging, and gunpowder making—were beginning to make rapid strides in wartime Virginia, and Jefferson did all he could to further them, increasingly making Virginia the arsenal of the revolutionary cause. His most ambitious project as governor was to establish a large arms factory at Westham, far from British coastal raiders, seven miles above the Falls of Richmond on the James River. Jefferson wanted to end American reliance on arms imports from beyond the Atlantic. To depend on "the transportation of arms across an element on which our enemies have reigned for our defense," he explained to Benjamin Harrison, the new Speaker of the House of Delegates,

> has already been found insecure and distressing. The endeavors of five years aided with some internal manufactures have not yet procured a tolerable supply of arms. To make them within ourselves, then, as well as the other implements of war, is as necessary as to make our bread within ourselves.[13]

The Assembly authorized Jefferson to enter into a contract with the French munitions firm of Penet, Windel and Company to build and operate a cannon factory at Westham. It also authorized Jefferson's plan to dig a canal around the falls at Richmond to enable transportation between Westham and the lower James. Jefferson's grand project was doomed, however, by the French government's refusal to allow munitions workers to emigrate. Jefferson had to content himself with using the public foundry already operating at Westham to turn out a few cannons, small arms, and ammunition.

One of the more controversial episodes of Jefferson's long political career took place during his early days as governor. One evening early in June 1779, an armed troop of dragoons rode up the governor's driveway leading a bedraggled Sir Henry Hamilton, the British lieutenant governor of Canada, and two other captives in handcuffs on horseback to Jefferson's doorstep after a harrowing three-month, twelve-hundred-mile wilderness trek from Vincennes, Indiana, where they had been captured by a Virginia army led by George Rogers Clark. Governor Jefferson did not come out to welcome the captured royal governor as a high-ranking British dignitary deserved, nor did he invite him inside out of the rain. The arrival of Hamilton "the Hair Buyer" was Jefferson's first shock over the horrors of war. Usually an objective lawyer who demanded evidence, he accepted frontier rumors of Indian atrocities as proof of Hamilton's guilt. In peacetime, he had honored the noble savage, the Native American; now he decried Hamilton for inciting the Indians and accepted the stories of their barbarity as commonplace. With both Hamilton and the Indians, he exhibited a strong belief in the principle of retaliation.

According to Hamilton's later testimony, Jefferson treated him insultingly when they arrived. Hamilton and his aide, Capt. Guillaume La Mothe, "were conducted to the Palace where we remained about half an hour in the street at the governor's door, in wet clothes, weary, hungry and thirsty, but had not even a cup of water offered to us." When Jefferson sent out orders that the governor and his men were to be confined in the Williamsburg jail, they were followed through the streets by a jeering, menacing crowd. At the jail, Hamilton and La Mothe were put in a crowded cell with five common criminals and another of Hamilton's aides. "The next day," wrote Hamilton, "we three were taken out about 11 o'clock and, before a number of people, our handcuffs taken off and fetters put on in exchange." These were heavy chains. "I was honored with the largest, which weighed eighteen pounds, eight ounces." In irons, Sir Henry and his aides were paraded back to jail under orders from Jefferson and his Council that they be "confined in the dungeon of the public jail, debarred the use of pen, ink and paper and excluded all converse except with their keeper."[14]

Ordinarily, any officer and gentleman, and especially one of such high rank as Hamilton, could expect to be paroled to a private house and be able to walk, ride around, and talk to whom he pleased, all on his honor as a gentleman. But Jefferson clearly felt Hamilton was beyond the bounds of the laws of nations after a military career that Jefferson had secretly helped to bring to an inglorious end. Jefferson, like so many other people in Virginia, knew and hated Governor Hamilton as the "Hair Buyer," who had allegedly paid Indian warriors for the scalps of Virginia frontier settlers. His cruel treatment of Hamilton, so sharply in contrast with his humane concern for the prisoners of Burgoyne's Convention Army, can be explained in part at least by his deep anxiety over British power in the west.

When Hamilton was forced to stand in humiliation before Jefferson and his Council, the only witness against him was John Dodge, a Connecticut trader to Detroit's Indians before the war who had been imprisoned by Hamilton for disloyalty to the Crown. Escaping after two years, Dodge had published a sensational account of his treatment by Hamilton and his lieutenants, vowing to hang them "without redemption."[15] Jefferson and his Council accepted Dodge's testimony, relying on frontier rumor, a pile of Hamilton's printed proclamations and propaganda, and a highly biased, uncorroborated witness. It was undoubtedly Jefferson who uttered the Council's verdict that condemned Hamilton to unusually harsh treatment. The Council members found

> that Governor Hamilton has executed the task of inciting the Indians to perpetuate their accustomed cruelties on the citizens of these states without distinction of age, sex or condition, with an eagerness and activity which evince that the general nature of his charge harmonized with his particular disposition.[16]

Jefferson would go on for years trying to justify the Council's judgment on Hamilton on "the general principle of national retaliation."[17] But there was something more to his acquiescence in the Council's judgment than he let reach the public eye in June 1779. For two years, he had been deeply involved in the expedition that had ended with Hamilton's ignominious appearance on Jefferson's doorstep.

So active in building up the western counties of Virginia over the past dozen years, Jefferson had drawn up the bill and led the fight to create three new Kentucky counties on the Virginia frontier in 1776. And so it had been natural for Jefferson's old friend, George Rogers Clark, to seek him out again when the British in 1777 began to send Indian scalping parties against Virginia's frontier settlements.

In June 1777, the British Ministry had sent orders through Canada to Hamilton, its commandant at Detroit, to exploit Indian hostility to the encroaching Virginia settlers. Detroit was to serve as the base for attacks on frontier settlements. Raids into Kentucky had already begun, with settlers there retreating into two besieged towns. Daniel Boone had been captured. After an attack on Wheeling, in present-day West Virginia, on September 1, 1777, twenty-four-year-old George Rogers Clark traveled to Williamsburg to seek arms and orders for an expedition to wipe out the British forts in the northwest. He met with then-Governor Henry, who favored the expedition but knew that Virginia could not officially sponsor such a march, which came under the jurisdiction of the Continental Congress. Henry secretly appointed three "select men,"[18] Jefferson, George Mason, and Richard Henry Lee, to see that Clark and his men secured not only Kentucky but Virginia's disputed claims to the entire northwest country, opening the Ohio River to permit trade with the Spanish at New Orleans and renewed settlement beyond the Appalachians. Jefferson had thrown himself behind the illegal expedition, in January 1778 writing Clark's secret instructions. Clark was allowed to draw money and supplies and raise seven companies of volunteers in Virginia, ostensibly for the defense of the Kentucky counties. Jefferson and his committee also promised that, if Clark and his men succeeded, they would receive generous grants of western land.

Clark's volunteer army of two hundred marched over one thousand miles and seized the old French forts in southern Illinois and Indiana. Since the French defeat in 1763, the *habitants* had accommodated themselves to British rule, but they had little affection for Governor Hamilton and his Eighth Regiment of Regulars, who alienated them by failing to conceal their contempt for the conquered French. Clark and his officers knew the French well and had been careful to treat them respectfully, a stroke of diplomatic brilliance that was to prove decisive. Clark's makeshift army shot the rapids of the Ohio River at Louisville in flatboats on June 26, 1778, during a total eclipse, built a fort and then rowed day and night to Fort Massac, marching 120 miles overland in four days, two without eating. On July 4, 1778, they marched up to Fort Gage at Kaskaskia. Its French militia commander surrendered without firing a shot. Cahokia and other towns in a two-hundred mile swath quickly surrendered.

News of the Franco-American alliance of February 1778 had just arrived on the Mississippi. Clark skillfully played this card. Frenchmen, including priests, accompanied Clark's troops as they rode on. By July 10, Frenchmen brought back word that Vincennes had shifted its allegiance to Virginia. Clark's men promptly occupied the place. Flattering, cajoling, and threatening neighboring Indian tribes, Clark won them over in time to frustrate a British counteroffensive from Detroit. Wading through icy armpit-deep waters in February 1779, he besieged Governor Hamilton at Fort Sackville, tomahawking five captured Indians in full view of the garrison to show what would happen unless Hamilton surrendered unconditionally. The French *habitants* inside the fort refused to fight on. Hamilton surrendered with seventy-nine Redcoats and began the long forced march to Williamsburg, where news of the victorious campaign reached Jefferson shortly after he was elected.

Virginia's annexation of the vast Illinois country had far-reaching consequences. First, it nearly wrecked the ratification of the Articles of Confederation, the first United States constitution, delaying the process for nearly four years after land-poor Maryland objected to the huge unauthorized acquisition by her already dominant neighbor, Virginia. Jefferson finally broke the logjam by ceding Virginia's claims to the United States government in 1780.

Jefferson's clandestine role in Clark's western expedition, his harsh treatment of Loyalists and of Governor Hamilton proved he could act illegally and quite ruthlessly if he perceived a serious threat to one of the causes he considered sacred. As governor, he was beginning to display a growing and lifelong hatred for the British and for anyone who sympathized with them in the bitter civil war that was beginning to engulf the revolution. The Hamilton affair was one of the few episodes of Jefferson's forty-year public life in which he let his personal emotions win out, and his conduct is hard to fathom. Had his long-ago fears of accusations of disloyalty for his continued correspondence with the Loyalist former attorney general, his cousin, John Randolph, made him overcompensate in meting out punishment to Hamilton? Or, as he found himself all but powerless to help Virginia's deteriorating defenses, did he vent his frustrations by dealing harshly with Virginia's enemies? Was the public perception of firmness necessary to offset rumors of his hobnobbing with the Convention prisoners before he came to Williamsburg? Or was the rational ruler simply punishing a corrupt official who had put himself beyond the bounds by breaking the rules of civilized warfare? In the western campaign, Jefferson had both the greatest achievement of his governorship, extending Virginia authority as far as the Illinois country, and the worst notoriety.

Three weeks after the drumhead proceedings at Williamsburg, Gen. William Phillips, the commander of British prisoners in Virginia, interceded from his house arrest in Charlottesville on behalf of Hamilton. He told his former dinner guest he found it hard to believe Jefferson's about-face in dealing with a captured officer: "It must have been very dissonant to the feelings of your mind to have inflicted so severe a weight of misery and stigma of dishonor upon the unfortunate gentleman in question." Phillips threatened Jefferson that, unless Hamilton were given more lenient treatment, the British would retaliate on Virginia prisoners. If Hamilton had been the beast Jefferson believed he was, Clark should have put him "to the sword." But Clark had accepted Hamilton's capitulation. "It matters not how barbarous the disposition of the lieutenant governor might have been prior to the surrender, the capitulation was assuredly sacred, and should remain so."[19] Jefferson responded that the capitulation had not included any mention of the terms of confinement (he enclosed a copy for Phillips's perusal). Under the law of nations, Hamilton's treatment was at Jefferson's discretion, "gentle and humane, unless a contrary conduct in an enemy, or individual, renders a strict treatment necessary."[20]

Jefferson had carefully researched international laws binding treatment of prisoners of war in the summer of 1776 when he had drawn up the congressional protests over the treatment of Ethan Allen, captured at Montreal, and prisoners taken at the Cedars during the American retreat from Canada. Forwarding Phillips's letter to General Washington and asking his decision in the case, Jefferson said he had at first supported the Council's decision in the case but, regardless of the terms of capitulation, now felt he should have put Hamilton "upon a different footing from a mere prisoner."[21] Washington urged Jefferson to keep Hamilton under arrest but to publish a full report of Hamilton's cruelties. In Jefferson's absence, the Council had taken off Hamilton's leg irons, but Jefferson never revised his opinion of Hamilton, nor did he publish the report.

Hamilton continued to make trouble for Jefferson until October 1780, refusing the terms of his parole until Jefferson finally had him shipped to British headquarters at New York to await his exchange for another high-ranking prisoner. The whole business badly embarrassed Washington and Congress in their delicate prisoner negotiations with the British and revealed an intransigent side to Jefferson's complex personality. He refused to accept captive American officers offered in exchange for Hamilton, and, only two weeks before Hamilton's departure, told Washington that he supported the Council's determination to hold Hamilton, warning him in a letter that helped to explain his own conduct during the fifteen-month-long controversy:

> You are not unapprised of the influence of this officer with the Indians, his activity and embittered zeal against us. . . . You also perhaps know how precarious is our tenure of the Illinois country, and how critical is the situation of the new counties on the Ohio.[22]

The west continued to mesmerize Jefferson, and worry about Virginia's immeasurable frontier was a constant drain on his time and the state's meager resources. One thousand scarce Virginia troops were tied down under Clark. Frontier traders refused Virginia's increasingly worthless currency. Congress kept up its pressure. Unless Virginia yielded its claims, Maryland would not sign the Articles of Confederation, which had been stalled in Congress since 1776. Finally, in January 1780, Jefferson yielded. Virginia's western lands would become part of the United States. He ordered Clark "to withdraw to the [southern] side of the Ohio" all but vital defense forces and to build a fort there to secure the Mississippi trade. Then he won Assembly approval for a force of Virginia regulars—a first—to attack and subdue the Shawnees: "The same world will scarcely do for them and for us,"[23] he wrote. Clark and his Virginians defeated the

Shawnees that summer and built Fort Jefferson just below the mouth of the Ohio, honoring Jefferson's perseverance in his first western policy, the crowning achievement of his troubled governorship. The fort, undersupplied and repeatedly attacked, held out for another year, then had to be abandoned. Jefferson's efforts to supply Clark with precious lead, gunpowder, men and supplies created friction with Friedrich Wilhelm, Baron von Steuben and the southern Continental Army commander, Nathanael Greene, who considered their mission paramount. In the end, however, Jefferson's bold support of Clark and the continued presence of a skeleton force of Virginians strengthened the hand of American negotiators at the Paris peace talks. And it helped to bring the vast midwest into the Union when the treaty defined the new western boundary of the United States as the Mississippi River, doubling American territory.

Expanding Virginia into territories that it could not defend or hold characterized Jefferson's grand efforts that did not at the time seem to work. He did bring to fruition his longtime ambition to create a new capital. He had proposed Richmond, which was closer to the geographical and population center of the state. Because this hill town on the James was accessible to western farmers, it would have the effect of making Virginia more democratic by weakening the power base of the Tidewater aristocracy. As early as 1776, Jefferson had drawn up a bill in the Assembly to move the capital away from the old royalist town of Williamsburg to Richmond, chiefly on the argument that it would be safely out of reach of British warships. The devastating raid on Portsmouth in May 1779 finally convinced the Assembly it was time to move. Jefferson set the date for April 1780. Taking his family home to Monticello during the dog days of August 1779, Jefferson brought them back to Williamsburg for the fall session of the Assembly. After 160 years, it was the last assembly of Virginia's elective officials in the town Jefferson had so disliked as a schoolboy and had come to despise as a symbol of entrenched English-style aristocracy.

His grand design for the new capital at Richmond would hold Jefferson's interest for a decade. He envisioned six great public squares, each accommodating a handsome porticoed brick building. But on a cold spring day in 1780, Richmond was a picturesque village of eighteen hundred people, and nothing had been done to make it resemble a capital when the Jeffersons arrived with a train of carriages and rented wagons carrying forty-nine crates of books, furniture, wine, china, chandeliers, framed portraits—all the accoutrements Jefferson believed necessary to uphold the elegant style of a governor. He rented a brick house with a garden atop Shokoe Hill, with a splendid long view of the Great Falls of the James, and he directed the slaves as they carefully unpacked his books, his best bottles of thirteen-year-old Madeira, and the records of Virginia's young government.

By the time Jefferson moved Virginia's capital to Richmond, the American Revolution was in jeopardy. The Continental war effort was foundering and the vaunted French Alliance had produced little more than false hopes and inflation. In March, Madison wrote Jefferson from Congress in Philadelphia, aptly summing up the dismal state of the Continental union:

> Our army threatened with an immediate alternative of disbanding or living on free quarter [by confiscation and foraging]; the public treasury empty; public credit exhausted. . . . Congress complaining of the extortion of the people, the people of the improvidence of Congress, and the army of both; our affairs requiring the most mature and systematic measures, and the urgency of occasions admitting only of temporizing expedients, and those expedients generating new difficulties.[24]

Omitted from Madison's depressing list was a shift in British policy that had led to increasingly frequent attacks in the south. The entry of France into the war had made the struggle a worldwide contest between European superpowers, with the mainland of North America but one theater. The British army had been ordered on to the defensive in the north and into support of England's more potent weapon, the Royal Navy. In 1778 shifting army operations to New York City, the new British commander-in-chief, Sir Henry Clinton, began to spin off military operations in the south. The British Ministry had come to believe that the south could be split off from the north, where the war appeared deadlocked, by supporting the supposed majority of Loyalists. The devastating raid on Portsmouth in May 1779 signaled the launching of this new British policy.

Ever since the British had failed to capture Charleston, South Carolina, in 1776, they had waited for the right moment to return. The failure of a Franco-American joint attack to dislodge Georgia Loyalists from Savannah in October 1779 encouraged Clinton to commence full-scale operations in the south, where patriot morale and troop strength were at an ebb. Jefferson's comment on Virginia's pitiable defenses summed up the south in general. The British were strong, there was "not a regular within our state nor arms to put into the hands of the militia."[25] Loyalist activity and armed resistance had picked up correspondingly in the Carolinas and Georgia. The French had removed their navy from American waters and, overall, seemed more concerned with winning back territories in the Caribbean and Canada lost in the Seven Years' War than in actively engaging in the American Revolution.

Sailing from New York City the day after Christmas, 1779, with one hundred ships and 13,700 regular soldiers, sailors, and marines, Clinton besieged a ragtag force one-third his number at Charleston, but it took the British until May 12 to surround the city and reduce its starving populace to surrender. So slow were communications that Jefferson did not know of the final attack on Charleston until one day after the city fell, making reinforcement impossible. He only learned of the fall of Charleston three days after he began his second term as governor. Nearly the entire Virginia Line of Continentals was killed or captured. Two weeks later, the last American resistance in the deep south ended with a victory by Banestre Tarleton's Loyalist Dragoons at the Waxhaws. A Continental army of reinforcements rushed south by Washington under the command of Horatio Gates was routed at Camden, South Carolina, on August 16, 1780, its commander retreating eighty miles in one day onto Virginia soil and pleading to Jefferson for reinforcements. But all of Virginia's three thousand Continentals had already been shipped out. Jefferson had only the infirm under arms. He could not call up the militia himself and he could not even ask the Council of State to mobilize the militia unless Virginia was attacked directly. He told Gates he would send more Continentals "as soon as they come out of the hospital."[26]

Now Jefferson learned of plans for an impending British invasion of Virginia: a British army under Earl Cornwallis, marching from the south, was to link up with a fleet under Clinton, arriving from New York. Jefferson, while he was no soldier, grasped the strategic threat. Once again, he skirted the impotent Continental Congress and appealed for aid directly to Chevalier de la Luzerne, French commander in America, in a cleverly diplomatic letter:

> The interest of this state is intimately blended so perfectly the same with that of the others of the confederacy that the most effectual aid it can at any time receive is where the general cause most needs it. Of this you yourself, Congress, and General Washington are so perfect judges that it is not for me to point out. . . . If their action in the north will have more powerful influence toward independence, they ought not to be wished for in the south, be the temporary misfortunes there what they will.[27]

But Jefferson resented Virginia's naked exposure to British forces and he feared imminent invasion unless help came soon from the north.

He did not have to wait long for his fears to be realized. On October 20, 1780, a British fleet of sixty ships brought five thousand troops into the Chesapeake and landed them at Newport News, where they seized Hampton and, on the opposite shore, Portsmouth, without any militia opposition. Only Cornwallis's summons for reinforcements to the south in mid-November ended this latest incursion before the British had time to attack the capital at Richmond. Desperate, Jefferson put aside his loathing of ancient British institutions and appealed to the Council of State to create a standing Virginia army. The Council refused him. Disheartened, Jefferson, who had just begun serving a second term as governor, threatened to resign, but his friend Page urged him not to desert his post in such a crisis. Jefferson stayed on, but he warned that they must find a replacement, that he would not accept a third term.

Jefferson was rebuffed again, in December 1780, when he appealed to the Assembly to follow a recommendation from the Virginia Board of War and appoint a general officer to take charge of the militia. He had to content himself with improving the transmission of intelligence from the south that would give the militia earlier warnings of British attacks and allow more time for them to mobilize. He secretly set up a relay of horsemen every forty miles between the Carolina coast and Richmond, calculating that a warning could travel 120 miles every twenty-four hours. At the sensitive southern terminus of the line, he appointed a twenty-two-year-old Virginian, a Continental Army veteran recommended by Washington himself, Col. James Monroe, whom Jefferson had recently taken on as his personal law student.

Jefferson got no respite from a year of dismal news about men, money, and supplies. On November 30, 1780, Patty gave birth again, this time to another girl, her fourth, christened Lucy Elizabeth. The child thrived, but frail Patty Jefferson never fully recovered from her fifth difficult birth in seven years. Jefferson stayed close to her at Richmond, occupying himself in the temporary winter lull from fighting by beginning the first scientific project he had permitted himself since the war began. He had purchased Diderot's *Encyclopédie* for the new state library; he had recently been elected a councillor of the American Philosophical Society. But he had kept strictly to his gubernatorial duties. Then, in December, one of Virginia's delegates to Congress brought him an extensive set of questions on the state of Virginia drawn up by François Barbé-Marbois, secretary to the French Legation in Philadelphia. From bales of notes and manuscripts he had gathered over the years, Jefferson diverted himself for hours. "I am at present busily employed for Monsieur Marbois without his knowing it," he wrote a friend, "and have to acknowledge to him the mysterious obligation for making me much better acquainted with my own country than I ever was before."[28] To friends, he confided once again his urgent desire to resign, something he had done once before when he was deeply worried about Patty's health. He pleaded with John Page to succeed him. Page humored his old friend, suggesting he stay on for eighteen months more until his third term would run out:

> I know your love of study and retirement must strongly solicit you to leave the hurry, bustle and nonsense your station daily exposes you to . . . 18 months will soon pass away. . . . Deny yourself your darling pleasures for that space of time . . .[29]

On October 7, 1780, nine hundred Loyalists detached from the main British army under Cornwallis were surrounded on a rocky, treeless ledge called King's Mountain, North Carolina, by an equal number of hard-riding, Kentucky-rifle-bearing overmountain men from southwestern Virginia and the Carolinas. The only combatant who was not American was Maj. Patrick Ferguson, a brave and most innovative British officer, who made but one mistake: attempting to defend his position by bayonet charges signaled with his silver whistle. Firing from behind the border of evergreens and giving out inhuman yells, the patriot marksmen decimated the Loyalists. One mountain man named Robert Young, firing a deadly long rifle he called "Sweet Lips," later claimed he killed the British commander, but there were actually eight slugs in Ferguson when he fell. At least a hundred died before white flags went up; still the firing continued. After one Loyalist, returning from a foraging patrol, fired at a patriot officer, the mountain men continued to fire into the clustered prisoners. Soon another hundred Loyalists lay dead. The next day, thirty Loyalists were tried for war crimes and eighteen were lynched on the battlefield.

News of the bloody American victory reached New York City shortly after a new British brigadier general was commissioned to lead a Loyalist corps. He was Benedict Arnold, who had recently crossed over to the British from American lines after a brilliant career as a field commander at Ticonderoga, Quebec, and Saratoga, a less successful term as the politically controversial military governor of Philadelphia, and a failed plot to capture Washington with his entire staff while surrendering the key stronghold at West Point.

Arnold was itching for an opportunity to prove himself as a key player among the British generals. The Loyalist defeat at King's Mountain gave him an opportunity for a quick, bold stroke. He had just asked Clinton for permission to attack Philadelphia to round up the Continental Congress and burn the American capital. Refused, he now successfully pleaded with Clinton the necessity of reinforcing the southern army before the momentum of victory shifted to the Americans. Clinton authorized him to sail to Portsmouth, Virginia, to join forces with Maj. Gen. Alexander Leslie and the Loyalists he had gathered in the Chesapeake region to compel the Americans to withdraw from the Carolinas and defend Virginia, thus easing pressure on the weakened army under Cornwallis. Arnold took along his newly minted Loyalist regiment, the American Legion, made up entirely of foreign soldiers-of-fortune and deserters from the Continental Army, and an elite regiment of dragoons, the Queen's Rangers, made up entirely of Virginia-born Loyalists who had already killed or captured twice their number. A British officer was also to lead Scottish and Hessian companies. On December 11, 1780, Arnold ordered his sixteen hundred troops aboard ships. Despite the bitter-cold weather at sea, morale was high: the men were looking forward to avenging the Loyalists killed at King's Mountain. On the first day of 1781, after surviving a wild snowstorm at sea, Arnold's Loyalist legion arrived off the Chesapeake capes. Their arrival was a complete surprise to Jefferson, despite the fact that, as long ago as September, Gates had forwarded intelligence from British deserters that the British planned a concerted invasion at Portsmouth involving Clinton and Cornwallis. And Washington had warned him, at the end of a long letter of December 9, "I am at this moment informed from New York, another embarkation is taking place . . . destined southward."[30] But Jefferson had become preoccupied with helping the new southern commander of the Continental Line in the Carolinas, Gen. Nathanael Greene, who had replaced the disgraced Gates, and he had stripped Virginia's coastal defenses bare. There had been so many false alarms in recent weeks that, if he had mobilized the militia at every warning, the militia would have been in the field and eating constantly at Virginia's expense.

On Sunday morning, December 31, 1780, an aide to Gen. Thomas Nelson dismounted at the Governor's House in Richmond as the chief executive emerged for a morning stroll. He could give Jefferson few details, but two days earlier twenty-seven ships had entered the Chesapeake. Whether they were British or the long-awaited French was unknown. Jefferson immediately ordered General Nelson, head of the state militia, to the scene to do whatever he deemed necessary. He also sent word to Baron von Steuben, Continental commander for Virginia, at nearby Chesterfield Courthouse, and House Speaker Benjamin Harrison, and he awaited further developments. William Tatham, the aide to General Nelson who carried Jefferson's orders, had happened to be in Richmond that Sunday and had been riding to the Governor's House when a dispatch rider arrived from the coast. Did His Excellency intend to call out the militia? No: the ships' appearance probably portended nothing more than a foraging expedition ashore before they sailed on. Jefferson also decided not to call a special meeting of the Council. They were to meet the next day anyway. He did write to militia officers in coastal counties, warning them of a possible emergency and asking for the latest intelligence. The next morning, when the Council met, it endorsed Jefferson's conservative measures and adjourned.

On January 2, Jefferson received another message from the coast that a British fleet had entered the James River and had reached Jamestown a full two days earlier. In considerable consternation, Jefferson tried to call an emergency Council meeting, but he could not find a quorum of four. Years later, according to his autobiography, Jefferson was still stinging at the communications breakdown. "From a fatal inattention to the giving us due notice of the arrival of a hostile force, two days were completely lost in calling out the militia."[31] After refusing to believe the worst for two days, Jefferson now called up militia from six counties. He also ordered all official records and military supplies carted to Westham, seven miles above the falls, although he believed Williamsburg was the British objective. He also ordered the Convention Army of British prisoners at Charlottesville marched as quickly as possible to Frederick, Maryland, in case ships had been sent to rescue them. Still Jefferson, the perpetual optimist, could not fathom the worst—that the British target was Jefferson himself, his government, and the new capital at Richmond. Yet fortuitously he sent his wife and children to stay with the Randolphs at Tuckahoe.

At 5 A.M. on January 4, a servant answered a banging at the front door, lit candles, and awoke Jefferson. The ships were still on the move. They had landed at Westover, the Byrd estate, which was only twenty-five miles from Richmond. Jefferson ordered militia from surrounding counties assembled at Westham and further ordered that the Convention prisoners be hurried off north without their baggage with two ships, the sloop-of-war *Swift* and the troop transport *Hope*. Now Jefferson learned the worst. Helped by following winds and the tide, the arch-traitor Benedict Arnold, a skilled mariner, had sailed seventy miles upriver, shelled and destroyed Hood's Fort, built to guard the approach to the upper James. Arnold had landed a force of fifteen hundred at Westover, on the north side of the river at its bend, and was marching with nearly one thousand men toward

Richmond. Jefferson could not know that Arnold had decided completely on his own initiative to teach both the Americans and the British a lesson in his style of scorched-earth warfare, ignoring strict orders to take no unnecessary risks and speedily reinforce Portsmouth. He intended to destroy the cannon foundry at Richmond, the small-arms depot at Westham, the warehouses crammed with vital supplies in both places, in the process bagging the author of the Declaration of Independence. Disgusted, Jefferson later wrote, "Winds favoring them in a remarkable degree, they almost brought the first news themselves of their movements."[32] But, for the moment, he sent off an urgent message to Steuben to hurry and take charge of the militiamen. He assumed they would come pouring in to defend Richmond. But they did not. Only two hundred gathered to protect the governor, council, and government offices by 11 A.M. on January 5.

Putting his least able men ashore at Westover, Arnold had marched his best nine hundred men, including mounted light infantry and field artillery, through a driving rain thirty-three miles overnight. Jefferson, too, was in the saddle, riding about Richmond, directing the evacuation of military stores by Conestoga wagon to Westham. At one o'clock that morning, he had ridden to Tuckahoe to spend the rest of the night with his family, trying to calm his anxious, ailing wife. Early the morning of the fifth, he had taken his family even farther from the capital to refuge at Fine Creek, his father's first farm, which Jefferson had sold to "Tuckahoe Tom" Randolph a dozen years before. Then he galloped the twenty miles to Westham to urge on the evacuation of vital supplies. His horse finally stumbled and fell under him. Jefferson yanked off the saddle, commandeered another at a farmhouse, and rode on. All that day he supervised the rescue effort, inspiriting workers with his tireless energy to save fifteen tons of vital gunpowder and supplies. Only six tons fell into enemy hands.

Benedict Arnold was disappointed to find that Jefferson and his council had fled. At first the townspeople of Richmond did not recognize the troops marching up from the river. Dressed in green, not red, they appeared to be Americans. Arnold's field artillery was able to get within point-blank range of Jefferson's house before opening fire with a three-pounder. The militia on Shokoe Hill scattered at the first deadly volley of grapeshot. Arnold, establishing his headquarters at City Tavern on Main Street, offered to spare the capital if Jefferson consented, in writing, to allow the British ships to remove all stored goods—especially the valuable tobacco—that belonged to rebels or their government. Loyalist goods, ships, and real estate were to be spared. Arnold later reported to British headquarters that he had offered to pay half price for the "thirty to forty ships full of tobacco, West Indies goods, wines, sailcloth," he found at the crowded wharves and in the warehouses along the James, but he insisted that local merchants find Jefferson and procure "the approbation of the nominal governor." Arnold gave the merchants "until the next morning to obtain an answer from Mr. Jefferson, who was in the neighborhood."[33] But neither the merchants of Richmond nor Arnold's legionnaires could find Jefferson.

They did find his house. A troop of Arnold's American Legionnaires aimed straight for it, rounding up the servants and demanding to know where Jefferson was. When they could not find him, they ransacked the house. In the basement wine cellar, they slashed open his best bottles of Madeira and other wines with their cutlasses, drank all they wanted, destroyed the rest. Upstairs they looted his library, carrying off or burning three-quarters of his state and personal papers for the entire war period along with the irreplaceable records of the Council of State and the War Office, making a mess of state government and the intricate business of supplying revolutionary armies. They also marched off ten of Jefferson's slaves as prizes of war to be sold for cash.

When they could not find Jefferson, Arnold lost his small supply of patience. "As Mr. Jefferson was so inattentive to the preservation of private property," he wrote General Clinton, "I found myself under the disagreeable necessity of ordering a large quantity of rum to be [staved in], several warehouses of salt to be destroyed, several public warehouses and [blacksmith and gunsmith] shops with their contents consumed by the flames." The flames quickly spread in the rising wind to a sailcloth factory and to the powder magazines. Arnold reported that "a printing press and types were purified by the flames." Meanwhile, Arnold's American Legionnaires were busily stripping the tobacco warehouses, rolling half-ton hogsheads down to their flotilla of ten captured ships and thirty-four open boats. Then Arnold ordered the warehouses burned, sparing only those belonging to Loyalists to whom he had already issued safe-conduct passes to escape downriver with their shiploads of wine.

He had detached the Queen's Rangers to march up the James to Westham and destroy Jefferson's fledgling munitions plant and ammunition warehouses. "They burnt and destroyed one of the finest foundries for cannon in America,"[34] reported Arnold proudly to Clinton. He enumerated their destruction: 26 cannons, 310 barrels of gunpowder (ordered it rolled off cliffs into the river), several warehouses of oats. That John Simcoe's Virginia-born Rangers also burned the town of Chesterfield with its mills, clothing depot, and warehouses. Richmond and the surrounding towns were steaming in the rain as Arnold's commandeered ships slid downriver and his tireless plunderers trudged back to Westover. The next day, he occupied Portsmouth on Albemarle Sound and dug in.

It is unclear where Jefferson spent the night of January 5, 1781: his political enemies said he spent it cowering in a barn. Others say he slept in a house along the evacuation route to the west. Jefferson was forever silent on the point. All day on the fifth, he had been searching futilely for Steuben, he later said, riding up and down the river road, trying to find the Continental inspector general to organize resistance. He took time out to record his movements on scraps of paper. Stuffing them back in his pockets, he rode all day and half the night. He visited Westham to see if the records had arrived safely. He left just before the Queen's Rangers arrived and destroyed the foundry. In eighteen hours on horseback, Jefferson was never far ahead of the enemy. He crossed the river to Manchester; there he had a clear view through his telescope of the enemy on the heights of Richmond. He rode out again to meet with Steuben, now attempting to stiffen the militia on the south side of the James. Steuben did not know Nelson was attempting the same thing on the north shore. When a deputation from Richmond finally reached Jefferson at Manchester and told him that Arnold had proposed the ransom of the city, Jefferson rejected the plan. He knew Arnold had no intention of holding Richmond, even if he could. When he was finally able to reenter Richmond on the sixth, Jefferson surveyed the ruins and printed a full report in the *Virginia Gazette:* three hundred precious muskets, several artillery pieces, (he understated the number), several tons of gunpowder, large quantities of tools, clothing, stores, one hundred twenty sides of leather. The boring mill and the roof of the cannon foundry had burned, but the furnace and chimneys were intact. The Council of State papers were lost.

Worst of all, Arnold had demonstrated how ill-prepared Virginia was, even after five years of declared war. Still, he correctly argued that Arnold was only creating a diversion in favor of Cornwallis. The Continental Army must not rush to respond. The worst damage from the raid was psychological. "When they came, we were getting in a fair way of providing both subsistence and men," Jefferson wrote. In less than forty-eight hours, Arnold had "amazingly interrupted both operations."[35] At a time when a year of drought had ruined the wheat crop, the militia had devoured precious provisions gathered for Greene's army in the Carolinas, had tied up the quartermaster wagons, had stopped all recruiting for the Continental Army, worsening Jefferson's relations with the imperious Continental Army commander, Steuben. Worse, Arnold's raid helped Jefferson's critics in their attempt to discredit the Virginia governor. Charges began to fly that Jefferson had been incompetent in not calling up the militia in time, negligent for not providing in his term in office for the state's defense, cowardly for running away from the enemy and not personally leading the militia in active resistance. Urged on by Patrick Henry's lieutenants, proud Virginia, supine and violated, pointed the finger of blame at Governor Jefferson. Attempting to reconvene the Council in ruined Richmond, he called a meeting every morning from the sixth to the nineteenth, when there were finally four members on hand to constitute a quorum.

Thomas Jefferson believed so deeply in the rule of law, the use of reason, and the gentleman's code of honor that he could not tolerate anyone whose behavior went beyond the bounds he set for himself and others. Now, he took out all his frustrations in office on Benedict Arnold. To Gen. J.P.G. Muhlenberg, the frontier parson who had shed his clerical robes to join the Continental Army, Jefferson wrote an angry letter, asking his help in capturing or killing Arnold:

> You will readily suppose that it is above all things desirable to drag Arnold from those under whose wing he is now sheltered. . . . Having peculiar confidence in the men from the Western side of the mountains, I . . . ask you to pick from among them proper characters, in such number as you think best, to reveal to them our desire and engage them to seize and bring off this greatest of all traitors. . . . I will undertake, if they are successful in bringing him off alive, that they shall receive five thousand guineas among them and . . . their names will be recorded with glory in history.

From his rough draft, Governor Jefferson, the guardian of Virginia's laws, crossed out the possibility that "circumstances" might oblige them to put him to death:

> If this happened and America was deprived of the pleasure of seeing him exhibited as a public spectacle of infamy and of vengeance, I must give my approbation to their putting him to death. . . . In the event of his death, however, I must reduce the reward to 2,000 guineas, in proportion as our satisfaction would be reduced.[36]

No amount of dreaming of revenge would wash away what Steuben called, in a letter to Jefferson in February, "the disgrace" at Richmond. While the Council at first endorsed Jefferson's conduct in the affair, more and more Jefferson's friends seemed willing to let him take full responsibility.

For three months, until Arnold was replaced and sailed back to New York, Jefferson was consumed by demands for food, clothing, arms that did not exist, giving orders that were rarely obeyed. From April 26 to May 10, his daily calls for Council never once produced a quorum. On April 15, when his five-month daughter, Lucy Elizabeth, died, he had no time to spare to mourn her.

That bleak morning, Jefferson could not stand the thought of leaving his stricken wife for yet another futile meeting with the Council, which probably would fall short of a quorum anyway. Outside it was unseasonably cold, pouring rain. He sent off a servant with a note to David Jameson, a member of the Council. "The day is so very bad that I hardly expect a council, and there being nothing that I know of pressing, and Mrs. Jefferson in a situation in which I would not wish to leave her, I shall not attend today."[37] Then he went off with his family to Monticello to bury his third child.

It seems to be exactly at this time that Jefferson decided not to seek a third term as governor. His wife had been ill for all nine months of her pregnancy in 1779. Whenever she suffered from long bouts of depression, she stopped serving as mistress of the household. Patsy never served as an official first lady, despite the fact that other wives of revolutionary leaders, including Martha Washington and Abigail Adams, had served. (When Martha Washington once asked her to spearhead a ladies' drive to raise money for uniforms, she declined on the grounds that "I cannot do more for promotion."[38]) One indicator of her periods of depression was when she stopped spending money for the myriad household accounts. Jefferson's account books show these times: there are no transfers of funds to her, but there are notations for medical bills. For most of 1779, and from April 15, when Lucy Elizabeth died, Jefferson's account books are silent on her purchases. There were frequent calls for Dr. Gilmer to come up from Charlottesville. On April 17, as soon as the Jeffersons arrived home from Tuckahoe, Dr. Gilmer summoned "for medicine—£108," a sizable expenditure. Nearly four weeks later, only two weeks before Jefferson resigned, he paid Dr. Gilmer nine pounds after another visit. Later, Jefferson would couple his public and private sufferings in a letter to his protégé, James Madison: "I think public service and private misery inseparably linked together."[39]

But Jefferson had become exhausted and embittered by this time. He left Richmond that day, joined his family at Tuckahoe, and took them home to Monticello, moving what was left of his government to Charlottesville. From May 24 to May 30, he never could find more than two councilors out of the eight on the Council. By May 30, only Jefferson's old friends

John Page and Will Fleming regularly attended the meetings; the others had fanned out across Virginia to their plantations. On June 1, at the end of his second term, Jefferson officially resigned as governor.

Lord Cornwallis, the British commander in the south, did not know that Governor Jefferson was considering resigning. With some seventy-two hundred regulars, Cornwallis invaded Virginia in late May 1781. Under the threat of war, the Virginia General Assembly could not raise a quorum of members for its scheduled meeting at Richmond, adjourning to meet in Charlottesville. Most of its members were also unaware that Jefferson had quit.

Jefferson had already arrived at Monticello by the time that Cornwallis, obviously bent on capturing Jefferson and the Assembly, had attacked the capital at Richmond the second time. As he had prepared to wind up his governorship, Jefferson had made one last futile attempt to bolster the defense of Virginia. Maj. Gen. Marquis de Lafayette, who had been sent belatedly by Washington with a feeble nine-hundred-man force of Continentals, pleaded with Jefferson to call up the entire state militia. Abjectly, Jefferson admitted the impotence of his situation: "It is not in my power to do anything." Unless the Assembly provided "more effectually" for the governor's powers to carry out the state laws, "it will be [in] vain to call on militia."[40] As Cornwallis crossed the James River to attack Richmond, Lafayette retreated northward.

Humiliated, Jefferson wrote to the Assembly that he thought it would be better for Virginia if Gen. Thomas Nelson, commander of the state militia, should become governor. "The union of the civil and military power in the same hands at this time would greatly facilitate military measures," Jefferson wrote, adding that he considered himself "unprepared by his line of life and education for the command of armies."[41] He wrote one more despairing letter, this time to Washington, imploring his "personal aid."[42] If only Washington would come home to Virginia and take command, Virginians would rise up, drive out the British. Washington replied, too late to help Jefferson, that he must remain in New York, but that he hoped that his attacks on the British in the north would require the British to pull their army out of Virginia to defend New York City. Washington's last paragraph must have consoled the frustrated, deeply disappointed Jefferson in the months ahead:

> Allow me to express the obligations I am under for the readiness and zeal with which you have always forwarded and supported every measure which I have had occasion to recommend.[43]

14

"I Tremble for My Country"

If pride of character be of worth at any time, it is when it disarms the efforts of malice.

—Thomas Jefferson, 1781

On the last day of May 1781, shortly after the Jeffersons returned to Monticello, Lord Cornwallis gave up his pursuit of Marquis de Lafayette and the retreating Continentals and unleashed his most daring leader of dragoons, Col. Banastre Tarleton, to make a raid on Charlottesville with 250 of his British Legion of cavalry, to seize Jefferson, the Virginia Council of State, and the General Assembly. Cornwallis's plan might have succeeded if it hadn't been for Capt. Jack Jouett. A Virginia militia officer, Jouett had been sipping a drink in a roadside tavern at eleven o'clock at night when Tarleton's horsemen rode up. Jouett knew that Jefferson had moved the capital to Charlottesville, supposedly out of Cornwallis's reach, and he also knew where the road outside the tavern door led. While Tarleton gave men and mounts a three-hour rest, Jouett slipped out the back door and lashed his horse on a forty-mile dash over a shortcut to warn Jefferson. He barely managed to reach Charlottesville in time. As Tarleton's legionnaires rode through the night, Captain Jouett had just enough time to flog his exhausted horse up the steep, coiling driveway to Monticello by dawn. A red-coated giant, Captain Jouett scooped off his plumed hat, bowed deeply before His Excellency, and blurted out his warning.

Unfazed, Jefferson woke his wife and their houseguests, including several assemblymen. They ate breakfast and the legislators rode down to Charlottesville, as if they had all morning to wake their colleagues. Meanwhile, Jefferson went around the house indicating to the servants what silverware and other valuables to hide. A few minutes later, a neighbor named Hudson breathlessly warned Jefferson that the British had almost reached the foot of the mountain. In a wild scramble, Jefferson crowded his wife, their daughters, eight-year-old Patsy and two-year-old Polly, and two maids into his fastest phaeton and sent them off to nearby Blenheim on Carter's Mountain to await him. It was the third nerveracking flight from the enemy for Patty in as many months. As the phaeton disappeared down the mountain, Jefferson ordered his fastest horse, Caractacus, brought from the blacksmith's shop to a road that led from Monticello to nearby Carter's Mountain. He calmly walked down the drive, then led his horse into the woods over a path that came out at Carter's, five miles away, pausing occasionally to peer through his telescope toward Charlottesville. At first he saw nothing unusual. When he didn't see any cavalry, he decided to go back to the house to gather up some of his papers, fearing the British would burn Monticello. Kneeling to focus his telescope, he unknowingly dropped his short sword. He rode on but once he realized the sword was missing he went back to look for it. He took another squint through the telescope and suddenly saw that the streets of Charlottesville were filled with men in the green-and-white uniforms of the Loyalist light infantry. An instant later, much closer, a troop of horsemen loomed into focus, filling up the round lens as they charged up Monticello after him. Jefferson sprang on Caractacus and galloped into the dense woods toward Carter's Mountain. Going back to look for his dress sword may have saved Jefferson's life.

Five minutes after Jefferson had left the house, British Capt. Kenneth McLeod and his dragoons reached the top of Monticello as two slaves, Martin Hemings, the butler, and Caesar, hid sacks of silver under the planks of the front porch floor. As the dragoons appeared, Hemings pushed back the planks, trapping Caesar under them in the dark, where he remained for eighteen hours without food or water. A dragoon shoved a pistol against Hemings's chest, cocked it, demanded to know where Jefferson was. If he refused, he would be shot. "Fire away, then,"[1] answered Martin Hemings. The dragoons went inside. For eighteen hours, they occupied Monticello, drinking only a few bottles of wine, disturbing nothing else. Tarleton had given strict orders against looting.

Jefferson arrived on Carter's Mountain in time for the noonday meal with his hosts and his family before the Jeffersons continued their ninety-mile flight southwest to Poplar Forest in Bedford County, a summer retreat Patty had inherited from her father. But the Jeffersons did not completely escape British depredations. As Lord Cornwallis advanced up the James, he occupied Elk Hill, Jefferson's best income-producing plantation at the Point of Fork. Jefferson later related "that spirit of total extermination with which he [Cornwallis] seemed to rage over my possessions":

> He destroyed all my growing crops of corn and tobacco. He burned all my barns containing the same articles from the last year, having first taken what corn he wanted. He used, as was to be expected, all my stock of cattle, sheep and hogs for the sustenance of his army, and carried off all the horses capable of service. Of those [horses] too young for service, he cut the throat; and he burned all the fences on the plantation, so as to leave it an absolute waste.[2]

The cash value of Jefferson's loss amounted to £3,700, more than $350,000 in 1993. Tarleton captured seven assemblymen that day and captured Jefferson's neighbor, Thomas Walker, home from the Continental Congress.

For eight days during the British invasion, Virginia was without a governor until the Assembly, reconvening at Staunton, could elect one. Jefferson's flight on horseback still haunted his political career three and a half decades later when the ex-president penned his autobiography:

> Would it be believed, were it not known, that this flight from a troop of horse whose whole legion, too, was within supporting distance, has been the subject with party writers of volumes of reproach on me, serious or sarcastic? That it has been sung in verse, and said in humble prose that, forgetting the noble example of the hero of La Mancha, and his windmills, I declined a combat singly against a troop in which victory would have been so glorious? Forgetting, themselves, that I was not provided with the enchanted arms of the knight nor even with his helmet. . . . These closet heroes forsooth would have disdained the shelter of a wood, even singly and unarmed, against a legion of armed enemies.[3]

The pillorying of Thomas Jefferson over his flight from the British began little over a week after he fled. On June 12, 1781, George Nicholas, son of Robert Carter Nicholas and Jefferson's old nemesis at the bar of the General Court, moved, in the House of Delegates, that a formal inquiry be made at the next session of the Assembly into the conduct of Governor Jefferson in his last year in office. Shortly after Jefferson arrived at Poplar Forest, he experienced another embarrassing fall. He was out for what must have been one of his breakneck morning rides on his high-spirited Caractacus when the horse suddenly reared. Jefferson fell hard. Trying to break his fall, he shattered his left wrist. Six weeks passed before he could ride again. His injury came at a time when his political rivals were willing to believe that he had fallen while running away from the enemy as fast as he could.

Housebound at Poplar Forest for six weeks, Jefferson was able to keep abreast of politics. The news from the east continued to be bad for weeks. The Assembly at Staunton still lacked a quorum. The British had built up their base at Portsmouth and were attracting significant numbers of Loyalists and runaway slaves: twenty-two of Jefferson's slaves had run away to join the British. Jefferson tried to refrain from politics while he convalesced. In four months, he wrote only seven letters. One of them was to decline another tantalizing nomination to go to Europe with John Adams and John Jay to join Benjamin Franklin as ministers plenipotentiary to negotiate peace. Obviously his friends in Congress did not regard his resignation or his narrow escape in the same harsh light as did some Virginians. Racked again by indecision, this time he wavered for four weeks, but he could not bring himself to leave his ailing wife, and she apparently was in no condition to go with him. A voyage across an Atlantic infested with British warships and Loyalist privateers would, no doubt, be too much for her nerves. To be sure, John Laurens, who went instead of Jefferson, *was* captured and sent to the Tower of London, a fate that might have befallen Jefferson. Jefferson also wanted to clear his name in Virginia. He had written Nicholas for a detailed list of the specifications of misconduct against him and he may have already made up his mind to run for his old seat in the House of Delegates so that he could defend himself on the floor of the Assembly. But he yearned to go to Europe, where he might better use his talents in bringing this awful war to an end. To Lafayette, who was pressing Congress's offer on him and dangling all sorts of alluring prospects for introductions to *philosophes,* and visits to the men of science in different countries, he finally wrote that he was painfully aware that he was missing the opportunity he had wanted for twenty years "of combining public service with private gratification, of seeing countries whose improvement, in arts and in civilization it has been my fortune to admire at a distance but never see." To refuse Lafayette, whose warm friendship was heartening to Jefferson, now caused him "more mortification than almost any occurrence in my life."[4] Lafayette, who carried Jefferson's letter of refusal to Philadelphia, could not completely understand Jefferson's malaise. Nonetheless, the young marquis wrote to Washington, "Jefferson has been too severely charged."[5] In Paris, when Benjamin Franklin learned that Jefferson had been named, but would not come, he wrote Jefferson, "I was in great hopes." He had looked forward, five years after they had served so closely in Congress, to "the happiness of seeing you here and of enjoying again, in this world, your pleasing society and conversation. But I begin now to fear that I shall be disappointed."[6] For forty years, Franklin had been sure he was dying when it was all his friends who were.

To his young cousin Edmund Randolph, Jefferson tried on September 16, long after he had returned to Monticello, to explain his latest rejection of a diplomatic appointment that would have honored him:

> I have taken my final leave of everything of that nature, have retired to my farm, my family and books, from which I think nothing will ever more separate me.[7]

He had made up his mind to attend one more meeting of the Assembly, but only long enough to vindicate his official behavior and then resign.

To distract himself from his misery, Jefferson began, at Poplar Forest while his left arm was still in a sling, to write a book with his good hand. He had sent for Barbé-Marbois's set of twenty-three questions about the state of his state—something that the French diplomat had sent to every governor and which no one else took very seriously—and before the year was out, Jefferson had compiled the first two-hundred-page draft of his first book, *Notes on the State of Virginia.* For years he had taken careful notes on his law cases, farming operations, travels, astronomy, the weather, political debates. He now had stacks of "these memoranda bundled up without order,"[8] and he decided that Marbois's questionnaire was the perfect excuse to organize them into some useful form. What he wrote during his self-imposed exile is probably the best portrait compiled until then of America's flora, fauna, geology, and natural history, an Enlightenment study that used his powers of reasoning to portray beautifully and exactly the history of Virginia, its Indians and the story of its Afro-Americans and their relations with whites, its customs, its farming, manufacturing, and government. A piece of prose that was both romantic and patriotic, the manuscript, not intended for publication, was eventually printed in Europe. It was written at a time when he felt the need to prove his knowledge and his love of America. He crammed the manuscript with useful facts—

on the flooding of the Missouri River, on how long it took a boat to pass from the mouth of the Ohio to the mouth of the Mississippi in the summer and get back up again: three weeks down, three months back—and from the mouth of the Ohio to Santa Fe: forty days.

But his most brilliant display of learning was about nature and natural resources. He was taking this opportunity to respond to the assertion by the French naturalist Comte Georges de Buffon that there was a "degenerative" process in the species of North America that made men and animals smaller in size and less vital than in Europe. The arrogance of the French king's botanist set off Jefferson on a years-long campaign to make Buffon eat a large American crow. He totaled up the weights of American versus European animals: an American elk weighed three hundred pounds more than the puny European. An American cow could be almost a ton heavier. Even the American otter was precisely 2.3 pounds heavier on average than its European cousin. Buffon had never been to America and had not solicited Jefferson's opinion, but Jefferson was stinging after reading Buffon's comments in his multivolume natural history. Buffon had suggested the natural inferiority of blacks and American Indians. Rising to the defense of the American Indian, Jefferson was almost lyrical about his attributes: "He will defend himself against a host of enemies, always choosing to be killed rather than surrender." Did the words trouble Jefferson as he wrote them? The Indian was "affectionate to his children" and loyal to his friends. He quoted the noble speech of Logan to make his point. Jefferson attacked the institution of slavery at length, showing its depravity in great detail. Slavery had produced an indolence among whites that was tempting to consider in the light of so many Virginians' failure even to defend themselves: "With the morals of the people, their industry is also destroyed. For in a warm climate, no man will labor for himself who can make another labor for him." Jefferson's condemnation of slavery, printed and circulated in Europe against his wishes and then distributed in America, incensed many of his slave-holding neighbors:

> Can the liberties of a nation be thought secure when we have removed their only firm basis, a conviction in the minds of the people that these liberties are the gift of God? That they are not to be violated but by his wrath? Indeed, I tremble for my country when I reflect that God is just, that his justice cannot sleep forever.[9]

Jefferson resigned five months before the war in Virginia ended with Cornwallis outnumbered four-to-one and surrendering at Yorktown. Quitting office so close to the end of the Revolution only heightened Jefferson's humiliation. He remained at Monticello as a twenty-nine-vessel French fleet under de Grasse sailed at last into the Chesapeake capes with three thousand French troops. Washington quick-marched sixty-five hundred Continentals and Frenchmen south to bottle up Cornwallis. By September 28, as Virginia militia rallied behind the new governor, Gen. Thomas Nelson, Jr., some twenty-eight thousand Franco-American troops were bombarding Cornwallis while the French fleet beat off a belated British rescue attempt from the sea. Jefferson only learned fragments of news, mostly from Monroe, who wrote to criticize Washington's decision to keep the militia on duty around Yorktown along with the regulars. Monroe thought it was costly and unnecessary: the Continentals had the situation well in hand. "I think with you," Jefferson wrote back to Monroe,

> that the present force of regulars before York might admit of the discharge of the militia with safety. Yet, did it depend on me, perhaps I might not discharge them. As an American, as a Virginian, I should covet as large a share of the honor in accomplishing so great an event as a superior proportion of numbers could give.[10]

Jefferson seemed acutely aware that scarcely one-fourth of the allied force at Yorktown was American, that it was really a French victory, certainly not a victory for Virginia, facts that he would long remember.

When Jefferson learned of Washington's greatest victory, he did not go in person to congratulate him. He would have, "notwithstanding the decrepitude to which I am unfortunately reduced," but he said he thought Washington had better things to do than take time for "a private individual."[11] In fact, his wrist had healed months ago, but he obviously was so distressed at the time that the draft of his letter to Washington was a mass of scribbled changes. When the Assembly convened in December 1781, the members had so forgotten the panic of the year before that they overwhelmingly elected Jefferson a delegate to the Continental Congress.

Jefferson's supporters wanted the House to rescind the resolution calling for an inquiry into Jefferson's second term but, behind the scenes, they insisted that the inquiry be conducted and that public hearings be held. "The inquiry was a shock on which I had not calculated," Jefferson wrote in his memoirs. Jefferson felt "suspected and suspended in the eyes of the world without the least hint then or afterwards made public which might restrain them from supposing I stood arraigned for treasons of the heart and not mere weaknesses of the head." The inquiry, he said, "inflicted a wound on my spirit which will only be cured by the all-healing grave."[12] To George Nicholas, he had complained of his evident desire "to stab a reputation by a general suggestion under a bare expectation that facts might be afterward hunted up to bolster it."[13]

There was an added reason for Jefferson's anxiety in late 1781. Patty was pregnant again. Each pregnancy had threatened her frail health, but this time, in her sixth pregnancy in eight years, something was terribly wrong. She began to gain a great deal of weight. She was unable to endure sitting politely when visitors came. She turned all the household affairs over to the slaves again. Certainly her husband could not think of leaving her to go see Washington or rejoin the Assembly.

But Jefferson's repeated refusal to participate in the burdens of government was only bringing more censure on him. When Edmund Randolph learned of Jefferson's resolve to reject a seat in Congress and abandon politics, he fired back, "If you can justify this resolution to yourself, I am confident that you cannot to the world."[14] Jefferson temporarily stemmed the criticism when he went to Richmond in December. Sending off his draft *Notes on Virginia* to Marbois in Philadelphia, he took his seat in the House of Delegates. On December 19, 1781, he rose and declared that he was ready to answer any and all charges. His accuser, George Nicholas, discreetly left the chamber, but Patrick Henry, who Jefferson was sure was the real instigator of the inquiry, remained. Nicholas had sent Jefferson a list of five charges, all regarding Arnold's invasion. Why had the governor not acted on the receipt of Washington's notice that British troops were embarking for the south? Why hadn't he stationed lookouts at the capes? Why were there no carriages prepared for heavy artillery? Why had the forts at York and Portsmouth been abandoned? All autumn, Jefferson had sought affidavits, and he also had written his own answers.

By the time Jefferson stood up to address the House, his comments were out of proportion and anticlimactic. House Speaker John Tyler had appointed a committee to conduct an inquiry but neither Nicholas nor anyone else came forward to offer information, so there was no inquiry. The committee reported to the full House that the charges had stemmed from "rumors which were groundless."[15] But Jefferson insisted anyway on rising and reading the paper he had prepared. Later he would argue the necessity of the spectacle: "If pride of character be of worth at any time, it is when it disarms the efforts of malice."[16] He believed he owed it to his friends as well as to himself and his family to restore his honor. He could not stand the whisperings. He never could. Typical was a comment in a letter by Betsy Ambler, daughter of Rebecca Burwell, once his "fair Belinda." She saw nothing wrong in the fact that her father, a member of the Council of State, had run from the siege. "The public office which he holds makes it absolutely necessary for him to run no risk of falling into the hands of the enemy." But she found nothing "more laughable than the accounts we have of our illustrious governor," she had written, "who, they say, took neither rest nor food for man or horse till he reached Carter's Mountain."[17]

By December 1781, only Thomas Jefferson really cared any longer about his vindication. The war was over. Virginia was safe. But the House listened and, when Jefferson sat down, unanimously passed an already drafted resolution that thanked him "for his impartial, upright, and attentive" use of executive power as governor and noted

> Popular rumors, gaining some degree of credence by more pointed accusations, rendered it necessary to make an inquiry into his conduct and delayed that retribution of public gratitude so eminently merited.

Wiping from the record the "unmerited censure" by Jefferson's enemies, the Assembly praised "Mr. Jefferson's ability, rectitude and integrity."[18]

His honor placated, Jefferson still refused to serve another minute in the House of Delegates. He went home to Monticello. When he was reelected, he refused to take his seat. That Christmas Eve, in a private letter, he vented his contempt for George Nicholas and for Patrick Henry, who had insisted that he be brought up on charges. Calling Nicholas a "trifling body" who was "below contempt," an "object of pity," whose "natural ill-temper was the tool worked with by another hand"—Patrick Henry's. Nicholas was "like the minners that go in and out of the fundament of the whale. . . . But the whale himself was discoverable enough by the turbulence of the water under which he moved."[19]

Jefferson's rage at what he perceived as betrayal by Patrick Henry, to whom he had been sincerely affectionate long ago, only grew with time. He made Henry a traitor, a Henry Hamilton or a Benedict Arnold worthy of his implacable hatred. Fully three years later, he told Madison, "While Mr. Henry lives, another bad constitution would be formed and saddled forever upon us. . . . What we have to do, I think, is devoutly to pray for his death."[20]

In the spring of 1782, one of the French friends of the Marquis de Lafayette stopped over to meet Thomas Jefferson at Monticello. The Marquis de Chastellux, a general in the French army based in Rhode Island, was a cultivated nobleman and a member of the French Academy who frequented some of the most distinguished literary and scientific salons in Paris. Before going home to France, he was traveling widely and recording his observations on America and Americans. Chastellux was not impressed with Monticello in its first incarnation: the house was "rather elegant and in the Italian taste, though not without fault." He declared that "Mr. Jefferson is the first American who has consulted the fine arts to know how he should shelter himself from the weather." He admired some details of the as-yet-incomplete mansion, especially its "very large lofty saloon which is to be decorated entirely in the antique style."

He found Jefferson at "first appearance serious, nay even cold," but "before I had been two hours with him, we were as intimate as if we had passed our whole lives together." The four days Chastellux visited left him deeply impressed with the master of Monticello:

> Let me describe to you a man not yet forty, tall, and with a mild and pleasing countenance, but whose mind and understanding are ample substitutes for every exterior grace . . . an American who, without ever having quitted his own country, is at once a musician, skilled in drawing, a geometrician, an astronomer, a natural philosopher, legislator and statesman.

Chastellux detected something far above the ordinary in Jefferson's mind. "It seemed as if from his youth he had placed his mind, as he had done his house, on an elevated situation, from which he might contemplate the universe." He envied Jefferson's splendid country prospects: it was spring in full flower, and Monticello was one great blossoming garden with white-tailed deer running right up to flick Indian corn from the palm of his hand. He found Patty Jefferson, nine months pregnant, "mild and amiable;" nine-year-old Patsy and three-year-old Polly "charming," swirling around him with the six children of Jefferson's sister Martha and the late Dabney Carr.

One evening, Patty excused herself and the children and left Jefferson to entertain the marquis. They quickly discovered their mutual love for the poems of Ossian. A bowl of punch between them, as Chastellux described the scene in his travel book,

> . . . A spark of electricity . . . passed between us rapidly from one to the other. . . . We recollected the passages in those sublime poems which particularly struck us. . . . In our enthusiasm, the book was sent for, and placed near the bowl where, by their mutual aid, the night far advanced imperceptibly upon us.

The next morning, Chastellux's last at Monticello, Jefferson was ready to escort him partway on a ride to Staunton and the Natural Bridge, which Jefferson insisted he visit. He could only ride to the Mechum River with him. Chastellux recorded that "his wife [was] expected every moment to lie in."[21]

At one o'clock on the morning of May 8, 1782, only a few days after Chastellux left, Patty Jefferson gave birth to another girl. Family lore holds that the second Lucy Elizabeth, named to replace the first, weighed more than sixteen pounds. Two days before her birth, Jefferson, terribly anxious, refused to take his seat in the House of Delegates, even though Speaker John Tyler said the House "may insist upon you to give attendance" and warned him that he could have the "censure of being seized"[22] and forced to come to Richmond by the sergeant-at-arms. When even his young friend Monroe wrote to berate him that he should "not decline the service of your country,"[23] Jefferson, nearly unhinged by worry as his wife's health sank by the hour, unleashed a salvo at Monroe that was obviously intended for his old friends in the House:

> If we are made in some degree for others, yet in a greater, we are made for ourselves. . . . Nothing could so completely divest us of that liberty as the establishment of the opinion that the state has a *perpetual* right to the services of all its members.

He had, he reminded the House, "dedicated to them the whole of the active and useful part of my life." Could he not now, as his wife lay dangerously ill, be "permitted to pass the rest in mental quiet."[24]

His letter to Monroe was his last to anyone for seventy-nine days; all that summer, he made only a single entry in the *Garden Book*. From the beginning of May, Jefferson nursed his wife, watching helplessly as she wasted away. His daughter, Patsy, three weeks short of her tenth birthday, became "his constant companion." She was to write years later about those days with her father and dying mother:

> He nursed my poor mother in turn with Aunt Carr and her own sister—sitting up with her and administering her medicines and drink to the last. For four months that she lingered, he was never out of calling. When not at her bedside, he was writing in a small room which opened immediately at the head of her bed.[25]

Only ten years before, Jefferson and his fragile Patty had read their favorite novel, *Tristram Shandy*, aloud to each other. Laurence Sterne had written the book for someone he loved as he lay dying. One day in early September 1782, ten years after Patsy had been born to them, Patty grew tired of the stifling polite refusal of everyone around her to confront aloud what was so obviously happening. She took a pen from her bedside table and began to write on a piece of paper from memory,

Time wastes too fast: every letter
I trace tells me with what rapidity
Life follows my pen. The days and hours
Of it are flying over our heads
Like clouds of windy day, never to return—
More everything presses on—

Her strength was gone. Her handwriting stopped. But Jefferson knew the words as well as she did. In his strong, bold hand, he wrote on

—and every
Time I kiss thy hand to bid adieu,
Every absence which follows it, are preludes to
that eternal separation
Which we are shortly to make![26]

In her last hours, Patty Jefferson asked her husband never to marry again. She did not want her little girls raised by a stepmother, as she had been. Jefferson agreed.

But he could not stand to face the truth. As she sank into a coma and began to breathe in gasps, he blacked out and had to be carried, unconscious, from the room by his thoroughly alarmed sister and sister-in-law, who thought he, too, was dying. He was unconscious for an hour. On September 6, 1782, Patty Jefferson died. She was thirty-three. It was three weeks before Jefferson came out of the library, and then all he could do was ride hour after hour, day after day, five and six miles at a time over back roads and through deserted woods, with ten-year-old Patsy at his side. Fifty years later, she wrote, "The violence of his emotion to this day I dare not describe to myself."[27]

Five days after Patty's death, Jefferson made the first entry in nearly three months in the *Garden Book*. He described, in grisly detail, a "method of preserving birds."[28] It was a month before he could write a letter, and then it was to Patty's sister, Elizabeth, at The Forest. "This miserable kind of existence is really too burdensome to be borne. . . . Were it not for the infidelity of deserting the sacred charge left me, I could not wish its continuance a moment. For what could it be wished?"[29]

Ten weeks after her death, he described to Chastellux his "stupor of mind"[30] but, after that, he only mentioned Patty three or four times in his letters to others, and he destroyed all the letters that had passed between them.

Near the bodies of her children, near Dabney Carr and Jefferson's sister Jane, he now placed a plain white marble slab. Beneath the inscription that told how she had been "torn from him by death," Jefferson had two lines from the *Iliad* carved in Greek:

> *If in the house of Hades men forget their dead,*
> *Yet will I ever there remember you, dear companion.*[31]

After the summer of "dreadful suspense" and "the catastrophe which closed it,"[32] Patty's death changed everything for Jefferson. Devoted to his wife and family and their home at Monticello, he was cut off, at her death, from his fondest pleasures. The music, the joy went out of his life. He was left, not quite forty, with three children and six nieces and nephews to raise, but he would never remarry, and it would be a dozen years before he could stand their home, Monticello, again. In a self-described stupor of mind, he lost all interest in *their* interests. In the entire year of her death, he made only one notation in the *Garden Book.* On September 2 and 3, a few days before she died, he noted the unseasonable "white frosts which killed vines in this neighborhood" and "killed tobacco in the North Garden."[33]

To Chastellux, his new friend, he wrote, "Before that event, my scheme of life had been determined. I had folded myself in the arms of retirement, and rested all prospect of future happiness on domestic and literary objects." It was months before he knew what to do with himself. "A single event wiped away all my plans and left me a blank which I had not the spirits to fill up."[34]

But needs of his family first and then his country swept him back into the main currents of life. Smallpox was once again raging in Virginia. Jefferson's friend Archibald Cary had arranged private inoculations at Ampthill, near Richmond, and invited him to bring his children. As the mansion became a hospital, Jefferson became its chief nurse. He was there on November 25 when he received a letter from Robert R. Livingston, the president of the Continental Congress. His young friend Madison, worried about him, had arranged to have Jefferson once again appointed minister plenipotentiary to go to France to negotiate peace with Great Britain: the vote had been unanimous. Madison had suggested to his fellow delegates in Congress that "the death of Mrs. Jefferson had probably changed the sentiments of Mr. Jefferson with regard to public life."[35] This time, Jefferson did not hesitate. The very next day, he wrote Congress,

> I will employ in this arduous charge, with diligence and integrity, the best of my poor talents, which I am conscious are far short of what it requires.[36]

Jefferson wrote the same day to Chastellux, who was soon to sail home to France; Jefferson hoped to reach Philadelphia in time to make the voyage with him. "Your letter recalled to my memory that there were persons still living of much value to me." He hurried to assure his new friend,

> If you should have thought me remiss in not testifying to you sooner how deeply I had been impressed with your worth in the little time I had the happiness of being with you, you will, I am sure, ascribe it to its true cause. . . . In this state of mind an appointment from Congress found me, requiring me to cross the Atlantic and, that temptation might be added to duty, I was informed at the same time from his Excellency the Chevalier de la Luzerne that a vessel of force would be sailing about the middle of December in which you would be passing to France. I accepted the appointment and my only object now is so to hasten over those obstacles which would retard my departure . . . to join you in your voyage, fondly measuring your affections by my own and presuming your consent.[37]

Something of his old exuberance restored, Jefferson, as soon as it was safe for the children to travel after the inoculations, arranged to have the baby and Maria taken to his sister-in-law's at Eppington, sending his sister Martha and the Carr children back to Monticello. He turned his business affairs over to two trusted friends, his brother-in-law, Francis Eppes, and his neighbor Nicholas Eppes. Ten-year-old Patsy was to come with him. By December 19, he set out from Monticello for Philadelphia and France. A week later, Jefferson was poring over the papers of the congressional Secretary of Foreign Affairs. He took the time to renew his old friendship with Madison, attend a meeting of the American Philosophical Society, of which he was a newly elected councilor. Then he dashed south to Baltimore to overtake the French frigate *Romulus*. But it was locked in thick harbor ice.

More than winter barred his departure. British warships still blockaded the mouth of the Chesapeake. Jefferson waited and fumed about his lodgings, "the most execrable situation in point of accommodation and society which can be conceived."[38] When the *Romulus* managed to get free and drop down the bay, Jefferson pursued it in a small rented vessel until the ice-laden tide reversed and "closed on us on every side and became impenetrable to our little vessel, so that we could get neither backwards nor forwards."[39] This seemed to describe his predicament. Locked in by ice and twenty-five British warships which "have made our little fleet the sole object," Jefferson requested further instructions from Congress and worried that the war would end before he could leave to negotiate peace. He could travel, winter notwithstanding, to Boston and leave from there: "Would to God I had done this first, I might now have been halfway across the ocean." He could "stay here with patience till our enemies think proper to clear our coast," but that "may not be till the end of the war." Or he could request a flag of truce from the British allowing safe conduct to France, which seemed to him the best course. "I shall acquiesce in anything."[40]

In his enforced idleness, Jefferson studied the conditions in the United States as they prepared for peace. In a long letter to Edmund Randolph, Jefferson confided a deep fear that, unless the confederation of states was strengthened, the individual states would turn against each other and resort to civil war. In his recent visits to Congress in Philadelphia and to Maryland, he had found

> the pride of independence taking deep and dangerous hold on the hearts of individual states. I know no danger so dreadful and so probable as that of internal contests. And I know no remedy so likely to prevent it as strengthening the band which connects us.
>
> We have substituted a Congress of deputies from every state to perform this task, but we have done nothing which would enable them to enforce their decisions. . . . They will not be enforced. The states will go to war with each other in defiance of Congress. One will call in France to her assistance, another Great Britain, and so we shall have all the wars of Europe brought to our doors. Can any man be so puffed up with his little portion of sovereignty as to prefer this calamitous accompaniment to parting with a little of his sovereign right and placing it in a council of states.

Jefferson, who still considered himself in retirement from the legislature, wished that Randolph would join Madison and Monroe to "bring into fashion principles to the form of government we have adopted." He urged Randolph to "lay your shoulders to strengthening the band of confederacy and averting those cruel evils to which its present weakness will expose us."[41]

For the third time, Jefferson was disappointed in his hope to go to Europe. While Congress reconsidered Jefferson's instruction, Madison wrote back that John Adams was to join him and Dr. Franklin in the peace negotiations. In a coded message, Madison confided to his mentor his misgivings about Adams's appointment: "Congress yesterday received from Mr. Adams several letters . . . not remarkable for anything unless it be a display of his vanity, his prejudice against the French Court and his venom against Dr. Franklin."[42] Madison shared Jefferson's view that Virginia politics was deteriorating as the revolution wore on into its ninth year.

From Baltimore, where he was by February 15 "feeling the effects of my idleness here," Jefferson wrote to his young cousin Edmund Randolph, who was considering a run for the Virginia General Assembly:

> Indeed, I hear it with as much pleasure as I have seen with depression of spirit the very low state to which that body has been reduced. I am satisfied there is in it much good intention but little knowledge of the science to which they are called. I only fear you will find the unremitting drudgery, to which any one man must be exposed who undertakes to stem the torrent, will be too much for any degree of perseverance.[43]

Jefferson hoped that Madison would soon return from Congress to join Randolph in the Assembly. By March 1783, Jefferson had heard the rumors that a peace treaty had been signed in Paris, but it was not until April 4 that Congress finally decided that Jefferson was no longer needed in Paris. Disappointed, he bundled Patsy into the phaeton and whipped the matched grays off toward Monticello.

They stopped off for two weeks in Richmond where Jefferson, still a delegate from Albemarle County, renewed his old acquaintances in the Assembly. It had been nearly two years since he had retreated from the revolutionary capital with the British in pursuit, setting off rumors that had led to his humiliation. Now, a pleasant surprise awaited him. The College of William and Mary, where his old mentor George Wythe had been appointed the first law professor in America, awarded him the honorary degree of doctor of civil laws, honoring him as

> most skilled both in private and public law, if exceptional love for his country, illustrious . . . in championing American liberties, and so imbued with letters . . . that all the fine arts seem to foregather in one man. . . . These arts are adorned by the greatness of his mind, which proposes nothing with regard to ostentation, everything with regard to conscience, and for a deed well done, he seeks his reward not from popular acclaim but from the deed itself.[44]

George Wythe's name, and obviously his paeans of praise, appeared on the handsome document that quite publicly honored Jefferson at this pivotal moment. Wythe and Madison had helped to pull Jefferson out of his grieving isolation. Surprised at his own popularity, he found he could return to his much-needed work of defending, revising, and strengthening Virginia's laws.

Since 1776, when his draft constitution had arrived too late to alter substantially Mason's Bill of Rights, Jefferson had been worried about the weakness of the state constitution. In the 1782 session of the Assembly, radical young lawmakers had been pushing to scrap many of Jefferson's law revisions and renounce all laws that had existed before the revolution, leaving Virginia literally in a "state of nature." This radical interpretation already was the law of Vermont and would have made possible the passage of laws excluding the return of the Loyalists to the state, their claiming of their property and the debts owed them. Writing to Edmund Randolph from Baltimore in February, Jefferson had cited these young radicals' insistence that the change of government in 1776 had dissolved the social contract. Jefferson considered this "a doctrine of the most mischievous tendency. For my part, if the term *social contract* is to be forced from theoretical into practical use, I shall apply it to all the laws."[45] In other words, he would make them sorry they had brought up the subject. Jefferson had recently wrestled with the question of what to do about returning Loyalists and their confiscated lands. While he was in Baltimore, he had been visited by Gov. Abner Nash of North Carolina, who had offered to make Jefferson a partner in a land scheme. The governor and his friends were buying escheated Loyalist lands at rock-bottom prices for resale at high prices once the war ended. On March 11, Jefferson rejected the offer as "one of those fair opportunities of bettering my situation which in private prudence I ought to adopt." If he had been "a private man" he would have accepted the offer "without condition or hesitation." But he was well aware that the Loyalists could be a topic on the table for negotiations with the British. He would do nothing that could "lay my judgment under bias."[46]

Returning in mid-April 1784 to Monticello to await further developments with Congress, Jefferson, obviously reinvigorated by his warm reception in Richmond, threw himself with his old fervor into the task of writing a new constitution—even if no one had asked him—to turn over to a growing number of zealous young followers he had found in the Assembly. In the six years since his first model constitution had arrived too late to influence the commonwealth's Bill of Rights, he had thought long and hard, especially in his impotent years as governor, of how better to govern a state. In the summer of 1783, he proposed a constitution that prefigured the founding documents of republics in Europe, Asia, Africa, and South America as well as the Confederate States of America and the 1787 revision of the Constitution of the United States. At its heart was the doctrine of the balance of powers:

> The powers of government shall be divided into three distinct departments, each of them to be confided to a separate body . . . those which are legislative to one, those which are judiciary to another, and those which are executive to another.

As he revised his *Notes on the State of Virginia* into his first book, Jefferson wrote his model constitution, laying out his guiding principles of political philosophy. His legislature would have two houses, their members chosen by the vote of all free male citizens. The governor would be chosen by both houses of the Assembly: he would serve a five-year term and be ineligible for reelection. The governor would have a power of veto that he would share with a Council of Revision that included two members of the Council of State and a judge from each of three superior courts. He also provided for freedom of religion and freedom of the press: "Printing presses shall be subject to no other restraint than liableness to legal prosecution for false facts printed and published." The death penalty would be banned except for "treason or murder or military offenses." The "military shall be subordinate to the civil power." The legislature would be forbidden to introduce "any more slaves to reside in this state" or continue slavery "beyond the generation which shall be living on the thirty-first day of December, 1800: all persons born after that day being hereby declared free."[47]

Many of the reforms Jefferson proposed would be pushed through the

Assembly by Madison and his faction. Many others would never be enacted. But, when Jefferson's *Notes* were published in 1784, they were widely circulated in America at a time when the Founding Fathers, too, were becoming worried about the problems Jefferson had observed in his two years on the political sidelines between 1782 and 1784.

On June 6, 1783, Madison led a successful drive to have Jefferson elected to his old seat in the Virginia delegation to Congress. He was to reach Philadelphia by the time the new Congress met on November 1. In his six-month stay at Monticello in 1783, Jefferson filled his days with revising his constitution and his first book, immersing himself in work to blot out his loneliness. He catalogued his rapidly growing library of 2,640 books, one of the largest in America, according to a system used by Francis Bacon, following his intellectual interest: I, Memory (history and biography); II, Reason (philosophy, both "moral and mathematical"); III, Imagination (the fine arts). He also took a careful inventory of his other possessions, including 204 slaves.

On October 15, as the Blue Ridge reached the peak of its rust-and-gold autumn colors, Jefferson and Patsy, now a tall, solemn eleven-year-old, set out for Philadelphia. There, they found that Congress, threatened by mutinous soldiers demanding back pay, had fled to Princeton, New Jersey, where they were being protected by Washington's army, encamped nearby. Jefferson, not worried about her safety, left his daughter temporarily with Mrs. Trist, whose boardinghouse he had frequented, while he arranged for her to live and study with Mrs. Thomas Hopkinson and her son. Mrs. Hopkinson was a close friend of Franklin's, and her husband had been the first president of the American Philosophical Society. Their son, Francis, a signer of Jefferson's Declaration of Independence, was Philadelphia's leading literary figure and composer. He left Patsy with a rigorous regimen of study to follow:

> From 8 to 10, practice music. From 10 to 1, dance one day and draw another. From 1 to 2, draw on the day you dance, and write a letter next day. From 3 to 4, read French. From 4 to 5, exercise yourself in music. From 5 till bedtime, read English, write, etc.[48]

If the program was rigorous for a girl of Patsy's age, it would not have been for a boy at a good boarding school of the period. He prescribed a similar burden for his nephew, Peter Carr, now studying at Walker Maury's school. His advice to the teenage boy was as self-revealing as his instructions to his young daughter. To Peter, he wrote,

> It is for you now to begin to learn those attentions and that complaisance which the world requires should be shown to every lady. The earlier you begin the practice of this, the sooner it will become habitual, and with the more ease to yourself will you enter on the public stage of life and conduct yourself with ease through it. You will find that, on rendering yourself agreeable to that sex will depend a great part of the happiness of your life. . . . The way to do it is to practice to every one all those civilities which a favorite one might require.[49]

In a letter to Barbé-Marbois at the French Legation, Jefferson justified his instructions to his daughter by reasoning that he was preparing her to be "at the head of a little family of her own." He told Barbé-Marbois he thought her education needed to be "considerably different from what I think would be most proper for her sex in any other country than America." He did not want her to have to depend on a husband to solve all her problems and educate her own children: "The chance that in marriage she will draw a blockhead I calculate at about fourteen to one, and of course [conclude] that the education of her family will probably rest on her own ideas and direction without assistance. With the best poets and prose writers, I shall therefore combine a certain extent of reading in the graver sciences."[50] She was, at eleven, to read one of his favorites, *Don Quixote.*

Since he also had to take on giving some of the guidance of a mother to an adolescent girl, Jefferson tried to do it scientifically, consigned to a letter that she could save and reread over and over so that it would impart the kind of strict guidance he thought Patsy needed to become a lady. The result seems fussy, heavy-handed and at times, dictatorial by modern standards,

> Never do or say a bad thing. . . . Our maker has given us all [a] faithful internal Monitor, and if you will always obey it, you will always be prepared for the end of the world, or for a much more certain event, which is death. This must happen to all: it puts an end to the world as to us, and the way to be ready for it is never to do a wrong act. . . . Above all things and at all times let your clothes be clean, whole and properly put on. Do not fancy you must wear them till the dirt is visible to the eye. You will be the last who will be sensible of this. Some ladies think they may under the privileges of the *déshabillé* be loose and negligent of their dress in the morning. But be you from the moment you rise till you go to bed as cleanly and properly dressed as at the hours of dinner or tea. A lady who has

> been seen as a sloven or slut in the morning will never efface the impression she made with all the dress and pageantry she can afterwards involve herself in. Nothing is so disgusting in our sex as a want of cleanliness and delicacy in yours. I hope therefore the moment you rise from bed, your first work will be to dress yourself in such a style as that you may be seen by any gentlemen without his being able to discover a pin amiss.

However successfully Patsy was able to satisfy her demanding father, there was one admonition she dutifully followed: "Keep my letters and read them at times, that you may always have present in your mind those things which will endear you to me."[51]

No sooner did Jefferson overtake Congress in the crowded, battle-scarred town of Princeton than Congress, responding to intense political pressure to hold its sessions at a more central location, adjourned for three weeks before reassembling at Annapolis, where it met in 1783 and 1784. Jefferson headed south, sad that Madison, who was becoming his best friend, had gone home to Virginia. He tied a message to Madison to take back with him: he was to find out whether George Mason would help him bring pressure for a stronger federal constitution. "Is he determined to sleep on, or will he rouse and be active?"[52]

It was at this time, as he concluded revising his *Notes on Virginia*, that Jefferson warned, "It can never be too often repeated that the time for fixing every essential right on a legal basis is while our rulers are honest and ourselves united. From the conclusion of this war, we shall be going downhill."[53]

When Congress reconvened at Annapolis, only six of the seven states' representatives needed for a quorum had arrived. Internal bickering and indifference had already jeopardized the peace treaty, which carried a clause requiring its ratification by March 3. Nine states out of thirteen must ratify, but by December 11, when there were still not seven delegations in Annapolis, Jefferson worried that the "other states" arrival was "as unsusceptible of calculation as when the next earthquake will happen."[54] Jefferson, who had been appointed chairman of the committee on ratification, had cause for alarm. The British peace negotiators had given the Americans extremely favorable terms, so much so that the government that had caved in to Franklin's demands had since fallen. The British had not only granted American independence but agreed to greatly expand American borders, including the vast Northwest Territories so central to Jefferson's philosophy of cheap and abundant western lands as the cornerstone of a philosophy of democracy and prosperity for all Americans. In exchange, the Americans had only promised that all debts owed to British creditors would be paid and that the states would restore the property and rights of Loyalist Americans. Jefferson well knew there was growing opposition to both these pledges in the state legislatures. The British still had not abandoned their forts on the northwest frontier, and the new British Ministry might not again grant such lopsided terms.

In his first month at Annapolis, Jefferson was also preoccupied with arranging public ceremonies to honor George Washington, who was resigning his commission as Continental Army commander-in-chief on December 23, 1783, after eight and one-half years. At a ball given by the American Assembly, Washington drank thirteen toasts proposed by the states and was still standing to return one of his own. Then he danced every dance of the ball, as James Tilton wrote to Gunning Bedford, so that "all the ladies might get a touch of him."[55] Jefferson did not attend. He pleaded illness; in fact, he was suffering again from headaches. At noon the next day, Washington came to the Maryland State House to resign formally. The twenty congressmen present, including Jefferson, sat with their tricornered hats on as Washington, escorted by Congressional Secretary Charles Thomson, in his blue-and-buff uniform, marched down the center aisle to the front of the Assembly Chamber. When the doors opened, gentlemen flooded the chamber below as ladies crowded the gallery above. The president of Congress, Thomas Mifflin, who had ridden north to war at Washington's side in 1775 as his aide-de-camp, presented Washington, who bowed. The congressmen, following Jefferson's lead, doffed their hats to salute him. Putting on the wire-rimmed spectacles he now needed to read, Washington haltingly, almost inaudibly, read his final address as commander-in-chief. The United States had become "a respectable nation," he said; he "resign[ed] with satisfaction the appointment I accepted with diffidence." Toward the end of the short speech, Washington's voice broke, but he regained his composure. "Having now finished the work assigned me, I retire from the great theater of action."[56] After the public ceremony, Washington met privately with the delegates to shake hands with them individually. There were few dry eyes in the room: Jefferson described his own farewell to Washington as "affecting."[57]

As Washington rode off to keep his promise to Martha to be home at Mount Vernon on Christmas Day, Jefferson was alone, reporting to Governor Benjamin Harrison his "extreme anxiety at our present critical situation."[58] He had skipped the farewell ball for Washington and, by New Year's, wrote Madison, "I have had very ill health since I have been here and am getting rather lower than otherwise."[59] He worried because there were still not enough delegates for ratification; the day-to-day waiting for a quorum must have brought back haunting memories of his wartime governorship. When one member went home, all business stopped. When would others arrive? "Chance may bring them in, and chance may keep them back,"[60] he wrote Harrison. Now, as 1784 began, only two months remained for ratification.

In his autobiography, Jefferson admitted he was finding it more and more difficult to abide the new breed of congressmen, virtually all of them lawyers, which "talks much and does nothing:"

> Our body was little numerous but very contentious. Day after day was wasted on the most unimportant questions. A member, one of those afflicted with the morbid rage of debate, of an ardent mind, prompt imagination and copious flow of words, who heard with impatience any logic which was not his own, sitting near me on some occasion of a trifling but wordy debate, asked me how I could sit in silence, hearing so much false reasoning, which a word should refute? I observed to him, that to refute indeed was easy, but to silence impossible: that in measures brought forward by myself, I took the laboring oar . . . but that in general, I was willing to listen . . .

But Jefferson had come to think that congressional time-wasting was inevitable so long as the people sent lawyers, "whose trade it is to question everything, yield nothing and talk by the hour."[61]

Jefferson did not sit in silence when some delegates began to argue that the peace treaty could be ratified by the seven states represented and the British never informed of the shortfall. Jefferson opposed the proposal as "risky and dishonorable."[62] Jefferson counterproposed that, if one more delegate arrived, Congress should move to Philadelphia to the bedside of a delegate too ill to travel to Annapolis, but before the motion could come to a vote, on January 14, delegates from two more states arrived and the peace treaty was promptly ratified.

Jefferson's two years in Congress at Annapolis in 1783 and 1784 marked a fruitful climax to a fifteen-year period as a legislator. He became the dominant figure in Congress in those years as well as its driving force. As head of many key committees, he wrote some thirty-one major reports and documents that helped to strengthen and stabilize the American Confederacy, and he laid the foundation for the new nation's gradual and orderly expansion to the west. In his two final years as a lawmaker, Jefferson helped to play out many of the last acts of an era that had begun in his first Congress in 1775. He personally rounded up the votes to ratify the Treaty of Paris, bringing the United States independence and peace. He handled the arrangements for Washington's peaceful departure from military power and wrote Congress's response to Washington's farewell address.

In his last months in the Confederation Congress, he shunned Annapolis society and he drove himself hard. A young Dutchman touring America, G.K. van Hogendorp, managed to crack briefly Jefferson's "cool and reserved behavior" and to observe Jefferson at work:

> Retired from fashionable society, he concerned himself only with affairs of public interest, his sole diversion being that offered by *belles lettres*. The poor state of his health, he told me occasionally, was the cause of this retirement, but it seemed rather that his mind, accustomed to the unalloyed pleasure of the society of a lovable wife, was impervious since her loss to the feeble attractions of common society, and that his soul, fed on noble thoughts, was revolted by idle chatter.[63]

The illness Jefferson complained about was not contrived to put off congressman's wives or tourists, but marked a persistent return of his migraine headaches. Privately he complained of them to Patsy and he wrote of "an attack of my periodical headache" to William Short, one of his young protégés in Virginia politics, lamenting that he was "obliged to avoid reading, writing and, almost, thinking."[64] Almost.

But Jefferson was still suffering from grief, as an exchange with the nervy young Dutchman revealed. "I pitied your situation," wrote van Hogendorp,

> for I thought you unhappy . . . Why, I did not know, and though you appeared insensible to social enjoyments, yet I was perfectly convinced you could not have been ever so. One evening, I talked of love, and then I perceived that you still could feel, and express your feelings.[65]

Jefferson was developing a way of opening himself up and revealing himself to people he had no reason to dislike or fear. To van Hogendorp, he replied,

> Your observation on the situation of my mind is not without foundation, yet I had hoped it was unperceived, as the agreeable conversations into which you led me often induced a temporary inattention

> to those events which have produced the gloom you remarked. I have been happy and cheerful. I have had many causes of gratitude to heaven, but I have also experienced its rigors. I have known what it is to lose every species of connection which is dear to the human heart: friends, brethren, parents, children—The sun of life with me [has] already passed the meridian.[66]

Jefferson still could not bring himself to add Patty to the list of his lost.

Despite or because of his deep and enduring sadness and his terrible headaches, Jefferson drove himself restlessly. Jefferson wrote Governor Harrison that he had a five-point agenda. He gave priority to "authorizing our foreign ministers to enter into treaties of alliance and commerce": America needed new trading partners to shake its economic dependence on England. He wanted to arrange "the domestic administration": for the new nation to be able to conduct government business only when a quorum of delegates allowed Congress to function seemed ludicrous to this former state executive. He set out to establish "arsenals within the states and posts on our frontiers," the only objective in which he failed. His other goals were "disposing of the Western territory," arranging "treaties of peace and purchase with the Indians," and "Money."[67]

On this last note, he was most successful. Financier Robert Morris of Philadelphia had submitted a totally unworkable plan for units of currency. A loaf of bread would cost 1/20 of a dollar, or 72 units on a scaled 1,440; a pound of butter, 20 cents or 288 units; a horse, 80 dollars or 115,200 units. Jefferson wrote that this may have been fine for a merchant but not for a farmer. Jefferson reacted by proposing his own system, one he had been thinking about off and on since 1776. In *Notes on Coinage,* he proposed a decimal system, the world's first, based on the dollar. "Everyone knows the facility of decimal arithmetic," he wrote. "The bulk of mankind are schoolboys through life. . . . In all cases where we are free to choose between easy and difficult modes of operation, it is most rational to choose the easy."[68] His simple system was a currency based on tens: one-hundredth of a dollar; the dime; the dollar; ten dollars; etc. Congress agreed with his clear mathematician's logic and adopted his system in 1785. Jefferson's brilliantly uncomplicated system of coinage was one of his great legislative achievements.

On March 1, 1784, he presented a committee report on a plan to establish the government of western territories ceded by the original thirteen states to the United States. Jefferson's Land Ordinance of 1784, although superseded by the Northwest Ordinance of 1784, laid the groundwork for that more famous act, establishing the fundamental principles of American territorial policy. Jefferson had helped to organize the George Rogers Clark expedition to the Illinois country, which had vastly extended Virginia's western lands by conquest. Then, to unite and strengthen the new nation, as governor he had ceded an unprecedented amount to the Confederation's western reserves. But he was worried about British troops in Canada and in the northwestern forts at Detroit and Michilimackinac, Spanish troops along the Mississippi River.

Now, Jefferson drew on his own plan for revising the Virginia state constitution to draft a model federal ordinance for the Northwest Territories. He saw them as a valuable buffer zone against the colonies of the European powers in North America. To induce settlement by hardy pioneers, many of the Revolutionary War veterans buying the land with their veterans' land bounties, he proposed that the old northwest be divided into distinct republican states that were to be admitted into the United States on an equal footing with the original thirteen states, a clear result of his studies of the inequality between England and her American colonies that had helped to precipitate the Revolution. Once before, he had made the suggestion in his draft 1776 Virginia constitution. The persevering Jefferson proposed dividing the northwest into fourteen new states. Had his plan succeeded, the power in Congress would have shifted rapidly to the west. He suggested names for ten new states, odd names, somewhat Indian-sounding names with classical-sounding endings: Metropotamia, Assenisipia, Cherronesus, Pelisipia. Fortunately Congress ignored these particular suggestions. There were to be two tiers of new states, between the Atlantic Ocean and the Mississippi River; they were to be created north and south of the Ohio River, each state comprised of two degrees of latitude.

Jefferson's plan of western government, adopted by Congress in 1784, provided for stages of orderly territorial development. Settlers had the right to form a temporary territorial government by adopting the constitution and laws of any of the original states. When population reached twenty thousand, Congress was to authorize the calling of a constitutional convention to establish a permanent government. Once population equaled the least populous of the original states, the new state was to be admitted to the Union on an equal footing with existing states. There was to be universal male suffrage and local republican rule at every step, allowing great freedom to the settlers in establishing their own self-government. They were to elect their own lawmakers, make their own laws, lay out their own townships and counties. They, in return, were to pay their share of the federal debt and to participate in the common defense.

It was the unprecedented final article of the Ordinance of 1784 that set off fiery debate in Congress: "That after the year 1800 of the Christian era, there shall be neither slavery nor involuntary servitude."[69] The abolition of slavery had become a matter of overriding national urgency to Jefferson. It must not be left to settlers hungry for land and wealth; it must be decided by Congress in advance of settlement. The northern states were all with him, but when this article came to a vote on April 19, 1784, Jefferson's southern friends, including his fellow Virginians, abandoned him. Only one other southerner, Hugh Williamson of North Carolina, voted to abolish slavery in the new territories. When the vote came, the delegates were deadlocked. Jefferson needed only one more vote. He needed seven states to carry the vote; he got only six. John Beatty of New Jersey had a cold. He stayed in his lodgings. Jefferson's defeat by a single vote sickened him: "South Carolina, Maryland, and !Virginia! voted against it,"[70] he wrote Madison. Years later, Jefferson was still disgusted by his southern neighbors' refusal to extend their freedoms across the new land. To the French historian Démeunier, he wrote sadly,

> The voice of a single individual would have prevented this abominable crime from spreading itself over the new country. Thus we see the fate of millions unborn hanging on the tongue of one man, and Heaven was silent in that awful moment![71]

Channeling his disgust with Congress into his *Notes on Virginia,* Jefferson grieved for his fellow Americans:

> What a stupendous, what an incomprehensible machine is man! Who can endure toil, famine, stripes, imprisonment or death itself in vindication of his own liberty, and the next moment be deaf to all those motives whose power supported him through his trial, and inflict on his fellow men a bondage, one hour of which is fraught with more misery than ages of that which he rose in rebellion to oppose. . . . We must await with patience the workings of an overruling providence.[72]

He waited for "a god of justice" to free his own slaves, but he could no longer stand to serve in Congress. Two weeks later, when word arrived that John Jay was resigning as one of the American ministers in France, Jefferson claimed the post. On May 7, 1784, Congress appointed him to join Franklin and Adams to negotiate treaties of amity and commerce with sixteen European states as well as the Barbary powers to carry out a policy formulated by Jefferson himself. Four days later, with no regrets, he left

Annapolis and Congress behind and hurried to Monticello to pack his and Patsy's trunks and recruit his favorite servant, James Hemings. Then he headed for Philadelphia—and France.

Jefferson's mission as minister plenipotentiary to France was to represent the interests of the thirteen states, yet he had never seen two-thirds of them. He decided to depart for France from Boston: "I mean to go through the Eastern states in hopes of deriving some knowledge of them from actual inspection and inquiry,"[73] he told Edmund Pendleton. He drew up a questionnaire covering every imaginable aspect of trade for merchants and local officials to fill out on his route. At Philadelphia, he met with the French Legation, watched three balloon ascensions (the latest rage from Paris), and measured a four-year-old bull as part of his evidence against Buffon's claims of diminutive American fauna. At New York City, he lodged on Maiden Lane next door to Hector St. John de Crèvecoeur, the French consul-general famous for his *Letters of an American Farmer.* The two became fast friends, and Crèvecoeur wished Jefferson "your health and plenty of friends. . . . I hope you'll be pleased with our social scene, which is the shining side of our nation."[74] In New Haven, Jefferson toured the Yale College campus with President Ezra Stiles, who proclaimed him "a most ingenious naturalist and philosopher, a truly scientific and learned man, and every way excellent."[75]

A year later, borrowing from Montesquieu's theory of climate and personality, Jefferson drew on his notes of the journey and wrote Chastellux that he had worked out a typology of northerners versus southerners:

In the North they are	In the South they are
cool	fiery
sober	voluptuary
laborious	indolent
persevering	unsteady
independent	independent
jealous of their own liberties, and just to those of others	zealous for their own liberties, but trample others'
interested	generous
chicaning	candid
superstitious and hypocritical in their religion	without attachment or pretentions to any religion but that of the heart

He told Chastellux that Pennsylvania, because of its temperate climate, had formed "a people free from the extremes both of vice and virtue."[76]

Journeying to Portsmouth, New Hampshire, before going to Boston, he missed his chance to sail with Abigail Adams and her daughter. In Boston he took the time to confer with the Massachusetts General Court on trade problems. "No small part of my time," he wrote to his congressional ally, Elbridge Gerry of Massachusetts, "has been occupied by the hospitality and civilities of this place, which I have experienced in the highest degree."[77] Boston newspapers hailed Jefferson. One praised "Governor Jefferson, who has so eminently distinguished himself in the late glorious revolution," as a "mathematician and philosophic . . . the memorable declaration of American independence is said to have been penned by him."[78]

At last he booked passage on a new merchant ship, the *Ceres,* owned by Nathaniel Tracy of Newburyport, himself at the helm for the maiden transatlantic voyage to France. Could Jefferson have known that it was Tracy who had rounded up a flotilla of ships for Benedict Arnold's expedition to Maine at the beginning of the Revolution? The ocean crossing, after three becalmed days on the Grand Banks off Newfoundland, took only a remarkable sixteen days. Much of the time Jefferson pored over *Don Quixote* with the aid of a new dictionary he had purchased in New York City, teaching himself Spanish. "We had a lovely passage in a beautiful new ship," recorded Patsy in the journal she was to keep faithfully, writing down things her father would never enter in his meticulous account books. "There were six passengers, all of whom Papa knew, and a fine sunshine all the way, the sea as calm as a river." Captain Tracy obviously was following the Gulf Stream; every day at noon, Jefferson made notations of the latitude and longitude, the winds, the distance covered, describing whales and sharks as well.

By the time the *Ceres* approached Cowes on England's western tip, Patsy was running a high fever and Jefferson had to be rowed ashore to fetch a doctor. By June 30, she was well enough for them to face a rainy, turbulent Channel crossing in a tiny, windowless cabin to Le Havre. Thomas Jefferson had never been farther from his Appalachian home than Philadelphia before this voyage. Now, he went ashore in France, in horrible Scottish-accented schoolboy French attempting to superintend the unloading of their trunks by opportunistic porters. Jefferson's account books show that the porters reduced this English-looking gentleman's fortune by about as much as the transatlantic voyage had cost. Patsy reported indignantly, "It is amazing to see how they cheat the strangers." But, at last, they were on French soil.

Jefferson entrusted James Hemings with the baggage and seventy-two pounds to hire a wagon to get their baggage safely to Rouen, their first stopover. Hemings, who spoke no French, went ahead of them and managed to return half the money. Did he know, did Jefferson tell him, that, as soon as he landed on French soil, he was a free man?

After three days of rest and an early morning coffee at Le Havre, father and daughter went rattling off in Jefferson's best Monticello-made phaeton. "The singularity of our carriage," Patsy recorded, "attracted the attention of all we met." But it was not just the carriage. To the French, the tall, freckled, redheaded American looked exotic. To Patsy's dismay, the shining black carriage, surrounded front, back, and on the sides by isinglass, also attracted hungry beggars at every stop: Patsy counted nine at one time as they changed the rented horses. They followed the meandering Seine River toward Paris, through, as Patsy described it, "the most beautiful country I ever saw in my life, a perfect garden." To Monroe, Jefferson wrote that it was the beginning of the harvest season as they went "through a country than which nothing can be more fertile, better cultivated and more elegantly improved." Passing chalk hills, they entered Rouen, the capital of Normandy, where Joan of Arc had been tried and burned at the stake. They remained for two days of rest and sightseeing. One ancient cathedral, Patsy noted, had been built by William the Conqueror. Patsy had never seen stained glass: "All the windows are dyed glass of the most beautiful colors that form all kinds of figures." Another church had to be approached "by as many steps as there are days of the year."[79] On August 6, 1784, they stopped one last time on the outskirts of the capital to change horses. No one but a determined Jefferson himself would have cracked the whip over them for an incredible seventy miles in one day. Jefferson and Patsy stretched their legs and looked out from the terrace of the royal chateau of Saint-Germain-en-Laye, birthplace of Louis XIV, home in exile of Stuart kings of England and now the haunt of English tourists. From the high terraces of Saint-Germain, Jefferson could see Paris through his telescope, its palaces, its new walls, and the pall of smoke above a million chimneys. Jefferson was far more interested in the next stop, the hundred-year-old hydraulic machine of Marly, great dragonflylike blades and screws that pulled up water from the Seine to create cascades adjoining Versailles, where Jefferson was to succeed or fail, in the next five years, to secure the survival of his weak, new nation at the pleasure, at the feet of the king of France and his ministers.

15

"I Do Love This People"

Life is of no value but as it brings us gratifications. Among the most valuable of these is rational society. It informs the mind, sweetens the temper, cheers our spirits and promotes health.

—Jefferson to Madison, February 20, 1784

When Thomas Jefferson crossed the Seine on the gracefully arching Pont de Neuilly—he proclaimed it the most beautiful bridge in the world—and rolled down the Champs-Elysées on August 6, 1784, it was every bit his Rubicon. For the next five years, he was leaving behind parochial American society and politics and a provincial view of innate American superiority to eventually become America's best-traveled, most cosmopolitan president until the late twentieth century. In his years as an American diplomat in Europe, he metamorphosed into a respected statesman who brought back to the infant United States a sophisticatedly radical world view that was to lead him to found the Democratic Party, win two terms as president, and, in one of the great diplomatic coups of all time, bloodlessly double the territory of the United States. In his years in Europe, Jefferson became an intimate of kings and revolutionaries, Enlightenment *philosophes,* artists, writers. He met with Louis XVI and Frederick the Great, who admired him; George III, who snubbed him; the princes of Italy; and the envoys of the Barbary states. With his close friend Lafayette, he took an active and quite illegal part in the early events of the French Revolution. His years as a diplomat in Paris were the watershed of his mature political life, recharging him and revitalizing him until he was ready to accept appointment as the first American Secretary of State. But first he had to endure a period of exile, what he described, in another context, as "seasoning,"[1] a time of alternating extremes of exhaustion and exhilaration at his spreading fame. The most obscure ambassador to Versailles when he first arrived, he became one of the most sought-after. "Although Mr. Jefferson was the plainest man in the room and the most destitute of ribbons, crosses, and other insignia of rank," wrote one American visitor toward the end of Jefferson's tenure, "he was the most courted and most attended to, even by the courtiers themselves, of the whole diplomatic corps."[2] Surviving passionate love and severe personal losses, injury and lingering illness, he emerged from his European years a supple and successful diplomat whose creativity and resourcefulness helped to open new trade routes and bring new riches to his young country. He became more patient and self-possessed. He liked his new ability to move from private aloofness to public celebrity, preferring diplomacy to the buffetings of elective office. What he liked best about his new office as minister plenipotentiary to France was "that I can do its duties unseen by those for whom they are done."[3]

Like so many Americans in Paris since those days, Jefferson lived a lonely existence on the fringes of French society for several months, clinging to the friendship of whatever compatriots he could find and compensating for his isolation by a frenzied round of shopping. With his daughter, he took up lodgings briefly at the Hôtel d'Orléans on rue de Richelieu near the King's Library, a few yards from the birthplace of Voltaire. He paid six days' rent in advance for rooms in this noisy carriage hotel, but found the place intolerable.

Jefferson dashed off word of his arrival to Franklin and Chastellux. Four days later, he moved to another Hôtel d'Orléans on present-day rue Bonaparte, thus becoming the first famous American writer to live on the Left Bank. Here, he was a short walk away from the quayside chateau of Chastellux, who lived at rue de Bac and the river. Jefferson's hotel advertised itself as a haven for foreign gentlemen, and here Jefferson installed himself. He hired a major domo named Gaspard and a *valet de chambre,* Marc, and he began to line up all the little affairs that came with his first days in office. Soon there were urgent visits by the tailor Dubuquay, by a staymaker, and by a milliner. Two years earlier, Adams had written Jefferson what to do when he finally reached Paris: "The first thing to be done in Paris is always to send for a tailor, a [wig] and shoemaker, for this nation has established such a dominion over fashion that neither clothes, wigs nor shoes made in any other place will do in Paris."[4] To which Abigail had added, "to be out of fashion is more criminal than to be seen in a state of nature, to which the Parisians are not averse."[5] Abigail's advice proved to be crucial. When Jefferson arrived, she wrote to him, "There is now court mourning and every foreign minister with his family must go into mourning for a prince of eight years old, whose father is an ally of the King of France." The period of mourning was for eleven days; Jefferson had only two days to have a "whole new black suit"[6] made up. Patsy loved every moment of it. "We were obliged to send immediately for the staymaker, the mantua maker, the milliner and even a shoemaker before I could go out." She drew the line at following the hairdresser's advice: she turned down her own hair "in spite of all they could say."[7] He sallied forth to the chic boutiques in the porticoed Palais Royal and the main shopping street of the Right Bank, the rue St. Honoré. Jefferson taught Patsy to follow the traveler's guidebook advice to *Garde! Garde les voitures!* and avoid being spattered with execrable filth by the hurtling carriages and speeding one-horse cabriolets jockeying for position in the crowded, narrow streets.

Jefferson bought himself a new sword and belt (something he would have to wear more often), shaving brushes and razor, silver buckles, and lace ruffles for his sleeves. He began to haunt the bookstalls around rue St. Jacques, in his first visit spending the equivalent of five days' food and lodging on leather-bound volumes and a Paris street map before he extricated himself from the shop of the happy Monsieur Molini. "While residing in Paris," he wrote Madison, "I devoted every afternoon I was disengaged, for a summer or two, examining all the principal bookstores, turning over every book with my own hands and putting by everything related to America, and indeed whatever was rare and valuable to every science,"[8] a word Jefferson freely defined to include all useful or natural sciences. But if he seemed omnivorous, Jefferson's taste was quite particular, according to instructions he sent a London dealer. "When I name a particular edition of a book, send me that edition and no other. . . . I disclaim all pompous editions and typographical luxury, but I like a fine white paper, neat type and neat binding, gilt and lettered in the modern style."[9] To a friend in Virginia, he confessed his "malady of bibliomania," which he treated by submitting "to the rule of buying only at reasonable prices, as a regimen necessary to that disease."[10] By the time he returned to America, he had acquired enough books to fill two hundred fifty running feet of shelves.

He began to take noontime rides to the Bois de Boulogne, afternoon strolls in the Jardin de Luxembourg and the gardens of the Tuileries

Palace, where he mixed with elegantly dressed nobles and their ladies, listened to concerts and could observe the impromptu trysts of young *gentilhommes* with prostitutes in the thorny marginal shrubbery.

He sought out everything new. In the Palais Royal, he found the Café Mécanique, where dumbwaiters connected with the kitchens below and were concealed in the large, columnlike legs of the tables. No waiters needed be present. Jefferson would design his own and build it into Monticello. He took a dimmer view of the craze for the "animal magnetism" of Mesmerism that had enchanted Paris society, including Queen Marie Antionette and Jefferson's friend Lafayette, who had actually lectured on the subject before the American Philosophical Society in Philadelphia the year before. Like Franklin, who was at the time a member of a royal commission investigating these early psychological experiments, Jefferson considered Friedrich Anton Mesmer a quack, but he could hardly avoid the excitement over the séance in the salons around Place Vendôme. Rages or discoveries, he kept his American correspondents up to date on them. Phosphorous matches were "a beautiful discovery" which allowed "the convenience of lighting a candle without getting out of bed, of sealing letters without calling a servant, of kindling a fire without flint, steel, punk, etc." He touted the new Argand cylinder lamp which gave "a light equal as is thought to six or eight candles." He was fascinated by the screw propeller "which takes hold of the air and draws itself along by it."[11] Jefferson proposed its use for moving vessels through water.

By the end of August, before his official business began, Jefferson had enrolled Patsy in an aristocratic convent school, the Abbaye Royale de Panthemont at 106 rue Grenelle, with the aid of a letter from his friend, the Marquis de Chastellux. Chastellux apparently had assured Jefferson that, while it was a Catholic school, religion was not pressed on Protestants. Among sixty pensioners and students in the handsome domed abbey near the Seine, Patsy found three princesses wearing distinctive blue ribbons over one shoulder. Patsy was outfitted with a crimson uniform and rapidly learned her new routines, turning her already proficient written French into happy chatter with new friends who dubbed her "Jeffy." At first Jefferson visited her almost daily. After three years of observation, he termed the Panthemont the best school in France. But poor motherless Patsy, yanked from a boardinghouse in Philadelphia after two years, was still recovering from yet another jolt a year later when she wrote an American friend,

> I was placed in a convent at my arrival and I leave you to judge of my situation. I did not speak a word of French. . . . Speaking as much

> as I could with [the pensioners], I learnt the language very soon. At present, I am charmed with my situation.[12]

Jefferson's official duties could not begin until John Adams, his fellow minister, arrived from London, where he had gone to meet his family, just arrived from Boston. On August 30, the American commissioners were to confer with Franklin, who was in charge of the delegation in France. The American envoys gathered over a three-week period. Jefferson's closest collaborator in Europe was to be John Adams, who arrived with his family on August 17. The two had last worked together in the Second Continental Congress in 1776, most notably on the committees that investigated the Canadian debacle, and produced the Declaration of Independence. They had corresponded infrequently during the war. Jefferson's appointment to Paris pleased the crusty Adams. "He is an old friend," he told James Warren of Massachusetts, "with whom I have often had occasion to labor at many a knotty problem, in whose abilities and steadiness I have always found great cause to confide."[13] When the envious Arthur Lee wrote of Jefferson, "His genius is mediocre, his application great, his affection greater, his vanity greatest of all,"[14] Adams defended him as "an old friend and coadjutor whose character I studied nine or ten years ago and which I do not perceive to be altered. . . . The same industry, integrity and talents remain without diminution."[15] Yet Jefferson was more circumspect about Adams. A year earlier, he had written Madison, "His dislike of all parties and all men, by balancing his prejudices, may give the same fair play to his reason as would a general benevolence of temper."[16]

Their personal styles were polar opposites. Jefferson was cool, reserved but amiable, a born diplomat able to avoid confrontation without ceding the main point. Adams, also a brilliant lawyer, was irascible, aggressive, blunt, and uncompromising, the Boston bull terrier to Jefferson's sleek Andalusian riding horse. Adams fitted almost exactly Jefferson's typology of a northerner in his letter to Chastellux; he was sober, laborious, persevering, independent, jealous, ungenerous, and chicaning. He was also a pessimist by nature who had a dark view of men and nations, sure that both were actuated by greed and ambition and never to be expected to act from generosity or benevolence. Jefferson deeply believed in

> but one code of morality for man, whether acting singly or collectively. . . . The best interest of nations, like men, was to follow the dictates of the conscience. I think, with others, that nations are to be governed according to their own interest, but I am convinced that it is their interest, in the long run, to be grateful, faithful to their

> engagements, even in the worst of circumstances, and honorable and generous always.[17]

Jefferson believed the United States owed France an undeniably great debt of gratitude for its national existence. When he had first learned of the Franco-American alliance in 1778, he had written to an Italian correspondent, "If there could ever have been a doubt before as to the event of the war, it is now totally removed by the interposition of France and the generous alliance she had entered into with us."[18] Not only did the United States owe France a great debt for its existence as a nation, but still the United States depended totally on its military and trade alliance with the most powerful nation in Continental Europe. These, for Jefferson, were sufficient reasons to support wholeheartedly the continued Franco-American alliance. But Adams, who had been brought up in anti-French, anti-Catholic Boston, resented any notion of moral obligation to the French and considered the alliance only a matter of temporary expediency. Both men shared, nonetheless, what historian Merrill Peterson has called a "puritanical indignation toward European vice and luxury,"[19] but Jefferson did not share Adams's personal austerity.

On August 10, 1784, only four days after he arrived in Paris, Jefferson rode out to Passy to present himself to his old friend Benjamin Franklin, the senior American diplomat in Europe. According to Jefferson's *Account Book,* he rode along the high Left Bank and paid to be ferried across to the foot of the even-higher bluff at Passy, then rode up the steep incline to the handsome, aristocratic village that looked down over Paris, some three miles away. It was easy to get directions to the Hôtel Valentinois, Franklin's headquarters in Passy, across the Seine from Paris on a high, wind-swept bluff midway between Paris and the court of Louis XVI at Versailles. The retired Philadelphia printer and former lobbyist for American colonies in London had created a sensation when he had landed in France nearly eight years earlier. He had arrived in Paris on Christmas Day 1776, the very day Washington had routed the Hessians at Trenton. He not only personified the American Revolution to Frenchmen who detested the English, but he embodied the Enlightenment as if he had been *le Quaker* himself. He was self-aware of his roughcut American *philosophe* image. "Figure to yourself," he wrote to a woman friend in London, "an old man with grey hair appearing under a martin fur cap among the powdered heads of Paris."[20]

At Passy, Franklin had taken up residence in a wing of the vast, baroque Hôtel de Valentinois recently purchased by a wealthy businessman, Jacques Leray de Chaumont, who had visions of cornering the American trade in exchange for free rent and meals for Franklin and his entourage. Eventually, Franklin insisted on paying rent and, also eventually, Chaumont bellied-up a bankrupt, but their mutual exploitation had its charming elements. Chaumont's son introduced Franklin's grandson to French society, invited him to country-house weekends, to hunts and balls at the family's estate in the Loire. Franklin himself had never lived in such high style as at Passy, surrounded by handsome villas, terraces, gardens, and stairs that cascaded down to the Seine, where he liked to bathe in a riverside spa. Much of Franklin's success in obtaining treaties and crucial loans in Europe flowed from his enormous popularity among people at all levels. He instantly received the respects of "the first people,"[21] he reported in 1776 to his sister. In a letter to his daughter, Sarah, in Philadelphia, he described in 1778 the rage for Franklin souvenirs in France, "some to be set in the lids of snuff-boxes and some so small as to be worn in rings. . . . These, with the pictures, busts and prints (of which copies upon copies are spread everywhere), have made your father's face as well known as that of the moon."[22]

To French government leaders such as Turgot, the king's finance minister, he was an Enlightenment hero who had "snatched the lightning from the heavens, the scepter from tyrants."[23] To John Adams, sent to Paris to join the American mission in 1780 to help negotiate peace, it was evident, even if Adams begrudged the fact, that Franklin's personal diplomacy had played a tremendous role in winning and holding French support:

> His name was familiar to government and people, to kings, courtiers, nobility, clergy and philosophers, as well as plebeians to such a degree that there was scarcely a peasant or a citizen, a *valet de chambre,* coachman or footman, a lady's chambermaid or a scullion in the kitchen who was not familiar with it and who did not consider him a friend to human kind.[24]

Yet it was Franklin's cult of personality that aroused the ire of Adams and a passing cast of fellow commissioners who stood in his shadow and carped about his personal habits, his amoral social circles, his ungrammatical French. Adams made common cause with the Lees of Virginia in an unsuccessful effort in Congress to have Franklin recalled. "On Dr. Franklin the eyes of all Europe are fixed," wrote Adams. "Neither [Arthur] Lee nor myself are looked upon of much consequence."[25] When Franklin printed his collection of bagatelles partly in French, partly in English, on his own press at Passy, Adams complained that the *bon mots* were not really that good and were ungrammatical anyway. The petty backstabbing within the

American mission had led to Franklin's resignation when it turned on the minor post he had given his grandson, Temple, who made longhand copies of the Franco-American treaty between his sexual escapades. Franklin doted on his handsome, playful grandson, whom he had wrested from his Loyalist father in a tug-of-war. When submitting his resignation, Franklin succeeded in dissolving any more effective opposition to him with a heart-wrenching letter to Congress:

> I know these gentlemen have plenty of ill-will to me, though I have never done to either of them the smallest injury. . . . It is enough that I have lost my *son,* would they add my *grandson?* An old man of seventy, I undertook a winter voyage at the command of Congress . . . with no other attendant to take care of me. I am continued here in a foreign country where, if I am sick, his filial attention comforts me and, if I die, I have a child to close my eyes and take care of my remains. . . . I am confident the Congress will never think of separating us.[26]

Congress had rejected his resignation. But now, Franklin was seventy-nine, increasingly infirm, with severe gout and a large kidney stone, in those days inoperable. He had once gone every Tuesday to the levees at the Foreign Ministry at Versailles, and he had chatted with the king, an amateur locksmith, about Franklin's inventions and he had been permitted to watch Marie Antoinette at her gaming table. Now, it was painful for him to leave Passy. The American delegation and most French friends had to come to him, and he had given out word that he would go home soon.

It did not seem to bother Franklin at all that, as soon as Jefferson arrived, the rumor went around that he had been sent to relieve him, although it was not technically true. Jefferson was only one of three ministers plenipotentiary assigned to Europe, and the others, Franklin and Adams, were his seniors. Many in Congress considered Adams the logical successor. Franklin had once vowed to remain in France "among a people that love me and whom I love,"[27] until peace with England was restored. Now that Congress had sent out new commissioners in an attempt to proliferate trade treaties, he seemed tired at the prospect of even more work that he had no reason to believe would succeed. He had already negotiated a commercial treaty with Sweden, signed in 1783, and talks were under way in Denmark, Portugal, and Austria. His chores, as Jefferson would learn soon enough, were already daunting. He referred to himself as a "perfect drudge."[28] He had acted for years as a kind of American secretary of the treasury in Europe. Since this treasury was invariably in arrears, he was perpetually bombarded by merchants or bankers with bills of credit drawn on Congress before he had managed to borrow the money to cover them for Congress. He also served as the mediator between French and American merchants. As head of American naval operations in Europe, he had spent countless hours arranging for the exchange of men captured at sea. The British were no longer taking Americans prisoner, but the Barbary states were. And most time-consuming of all were the soldiers-of-fortune, travelers, merchants, adventurers, scientists, artists, and well-connected sycophants who expected a letter of introduction and a free meal.

Franklin's appearance must have surprised Jefferson. He had the same ample forehead and big frame, of course, and the old glimmering smile, but he no longer felt compelled to wear the pretense of a wig and had only a few thin wisps of hair. His face was drawn and tired and he wore his homemade bifocals low on his nose. A stout cane braced him, but his weight certainly had sunk, adding to the painful pressure on his gout-stricken foot. Franklin, for all the stories of his sexual exploits in the salons of Paris, was an old man who could barely move. In his recent *Dialogue between Franklin and the Gout* which he had printed for his worshipping neighbor, Madame Brillon, he had taunted himself for his decrepitude: "You have praised the fine view and looked at the beauties of the garden below, but you have never stirred a step to descend and walk about in them."[29]

Franklin was happy to see Jefferson again. Franklin had thrice requested that Jefferson come to France with him, and now he had finally come. Together, the bent old man and his tall, erect protégé stared across the Seine toward the massive Royal Military Academy with its pristine parade grounds and artillery parks where so many of the officers who had helped to win America's independence had been trained. They talked softly about how much the United States still depended on France for its military and economic independence. Then they went inside for more introductions and a French lunch. Living *en famille* with Franklin were his courtier-grandson, Temple, now twenty-two, and the grandson by his daughter, Sarah, the fifteen-year-old Benjamin Franklin Bache, who was studying with François Didot, the eminent Paris printer. Frequently at Franklin's table were two neighbors, the *abbés* Chalut and Arnauld, whom Jefferson knew to be among the Paris literati. The two *abbés* could always be counted on, even on fast days, to insist on a luxurious table. At a later visit, Jefferson took painter John Trumbull along, who made sketches of Passy and used words to describe in his notebook "the luxury of the table in soups, fish and fruits, truly characteristic of the opulent clergy of the time."[30] Franklin's introductions to the *abbés* were among many he accorded Jefferson in the year they worked together. Not since his college days with Governor Fauquier in Williamsburg had Jefferson feasted on such table talk. Then, when the talk of work was done, a servant brought Franklin's chessboard, and the quiet, contented game between master and novice began.

On August 21, a few days after his arrival from London with his family, Adams met Jefferson at a dinner party at the Paris residence of Thomas Barclay, the American merchant who was consul-general to France. John's sparklingly bright wife, Abigail, a year younger than Jefferson, was there with Nabby, their nineteen-year-old daughter, and their son, John Quincy, already learned at seventeen and an admirer of Jefferson.

Abigail had not seen her husband for four years, but she had affection enough left over to take Jefferson and his daughter under her wing at once. In the next two years, the Adamses and the Jeffersons were to be a close-knit family. Patsy took Nabby to her convent when two young women took their veils; they cried together when the women, wearing white wedding gowns, lay prostrate on their faces for half an hour while priests held a black pall with a white cross over them and nuns holding candles chanted prayers.

Soon afterward, when the Adamses moved to a spacious rented villa in suburban Auteuil, Jefferson was their frequent dinner guest. In cramped quarters for a few more months, Jefferson was unable at first to reciprocate, but he took Nabby and John to orchestral concerts at the Tuileries Palace, where he pointed out Prince Henry, the short, ugly brother of the king of Prussia, and, when Jefferson rented a "small hôtel" at 5 cul-de-sac Taitbout (present-day rue de Helder near the Opéra), John Quincy was so often his guest there that, when he became the sixth president of the United States in 1825, the elder Adams wrote to Jefferson about "our John" and those long-ago days in Paris: "I call him our John because, when you were at the cul-de-sac at Paris, he appeared to me to be almost as much your boy as mine."[31]

Many young men were to try to fill the lonely void Jefferson felt for not having a son. John Quincy Adams was only one. Madison, Monroe, also to become early American presidents after serving as protégés of this Founding Father, had their turns close to him. Others tried. Tall, dark, and pompous Col. David Humphreys, thirty-two, had been appointed by Congress as secretary to the American commission to supplant Franklin's grandson, William Temple Franklin. Humphreys had arrived in Paris ten days after Jefferson, who had last seen him standing at Washington's elbow when the commander-in-chief had resigned the preceding December. Humphreys, from Connecticut, was an outsize would-be poet who thought a diplomatic sinecure in Paris would be just the thing to advance his literary career. Typically, he had tried to describe his emotions as he crossed the Atlantic toward the Old World. The result was an epistle in verse that he sent to his young evangelist-friend Timothy Dwight in New Haven. Smoothly detaching himself from Washington and attaching himself to Jefferson, the stiff, solemn former staff officer went everywhere Jefferson went, including that first dinner party at Barclay's with the Adamses. In fact, he invited himself along to the Adamses' several times. They never warmed to him and Nabby could never quite figure him out. Out of his own modesty, Jefferson may not have realized the significance of having such a sycophant, but Humphreys had recognized that Jefferson's star was the highest and fastest-rising now and brought him an autographed copy of *McFingal*, the revolutionary epic by his friend John Trumbull, which had been all the rage—a decade earlier. Jefferson was hospitable to the point of misplaced generosity to younger men. He arranged a hotel for Humphreys in advance and, when he rented the house at cul-de-sac Taitbout in mid-October, invited him to live there rent-free.

A month later Humphreys had a rival, Jefferson's dashing young Albemarle neighbor, William Short, a tall, slender twenty-five-year-old graduate of William and Mary and Jefferson's in-law through the Skipwiths. When Short had completed his law studies, Jefferson and George Wythe, Short's law professor at William and Mary, had served as his bar examiners. Jefferson had invited him to set up his practice in Charlottesville but then urged Short to drop everything and come to Paris as his personal secretary. Eventually Jefferson introduced him into French society, where he cut a figure that came close to embarrassing Jefferson on more than one occasion. Yet Jefferson liked and trusted Short and treated him as his own son. He left him behind in Paris as chargé d'affaires whenever he traveled. Short gradually eased Humphreys out of Jefferson's circle of Americans. Humphreys returned to the United States, much later capturing the essence of his political life in his epitaph:

To sum all titles of respect in one—
There Humphreys rests—belov'd of Washington.[32]

Before his first private and unofficial visit to Franklin had ended, Jefferson had gotten Franklin to agree that Adams should be brought back to Paris after the negotiations he was just concluding with the Netherlands. Together, the three American envoys were to begin preparing for a new round of negotiations at Versailles with the ambassadors of some twenty nations called for by the new instructions which Jefferson had carried from Congress and which he himself had written. On August 30, the entire delegation gathered at Passy out of consideration for Franklin's inability to travel into Paris, where all earlier negotiations had taken place. There, every day for the next month, three giants of the American Revolution met and worked together. The last time they had been together on what John Adams called a "knotty problem," the result was the Declaration of Independence. Now that a war had made American independence possible, it was their task to win trading partners to assure that the United States would survive. After six weeks, Adams could report happily, "We proceed with wonderful harmony, good humor and unanimity."[33] Even Adams and Franklin for once got along cheerfully. Between Jefferson and Franklin, there had never been a question: each openly admired the other. No father and son could be closer than they were in those happy, productive days together.

Jefferson quickly spelled out Congress's instructions, which were his ideas, that he had brought to Europe with him to free the United States from the restrictive mercantile policies of rival European powers. He wanted to push the American Revolution into the arena of free trade.

Espousing a dynamic doctrine of international commercial cooperation, of "perfect liberty" of trade, he argued:

> Instead of embarrassing commerce under piles of regulating laws, duties, and prohibitions, could it be relieved from all its shackles in all parts of the world, could every country be employed in producing that which nature had best fitted it to produce, and each to be free to exchange with others mutual surpluses for mutual wants, the greatest mass possible would then be produced of those things which contribute to human life and human happiness. . . . The numbers of mankind would be increased and their conditions bettered.[34]

As an incentive to the European mercantile nations, he held out the bait of American trade. The United States could exchange its vast surplus of raw materials and food with any nation that would accord it a free and equal footing. Jefferson saw trade as a potent American weapon that could help the new nation enhance its own feeble strength in the scales of the European balance of power. No trading nation could ignore the American challenge to the old equilibrium, which had been based on British mercantilist supremacy.

But his motives were not purely international. Jefferson's primary motive, he said, in forging new international trade alliances was "to take the commerce of the states out of the hands of the states and to place it under the superintendence of Congress." He intended his treaty-making in Europe to coerce the consolidation of the American states, to force the states to make a "new compact [that would] make them more perfect." He was certain that "the moment these treaties are concluded, the jurisdiction of Congress over the commerce of the states springs into existence." After his years as Virginia's shackled governor, and two years as a badly handicapped congressman, watching the deteriorating level of politics in his home state, and surveying the growing chaos in trade in the Middle Atlantic and New England states, Jefferson was determined to adduce every power he could to the Confederacy. In his written instructions to the treaty commission, he used the words "one nation" for the first time. The United States was to be considered, not as a loose confederation of neighboring states with different laws, constitutions, and trade regulations, but "one nation."[35] As he reported to Monroe, he had with difficulty convinced Congress to accept this implicit denial of each state's authority over its own trade, but, he assured the other commissioners at Passy, "The majority, however, is for strengthening the bond of Union. They are the growing party, and if we can do anything to help them, it would be well."[36] Jefferson was thus taking the lead in the drive for a stronger central United States government by asserting one of Congress's few clearcut peacetime powers to take commerce out of the hands of the states for the good of all. As he had written privately to his disciple, Monroe, that summer before sailing to Europe,

> I would then say to every nation on earth, *by treaty*, your people shall trade freely with us, and ours with you, paying no more than the most favored nation, in order to put an end to the right of individual states, acting by fits and starts, to interrupt our commerce or to embroil us with any nation.[37]

Jefferson also saw the trade negotiations as the means to propagate his revolutionary philosophy among nations that wanted trade with revolutionary America. Characteristically he had spent much of his time since his appointment by Congress immersing himself in the existing American treaties with France, the Netherlands, and Sweden. Now, he produced a model treaty to offer to some twenty potential trading partners. He attempted to replace with his simple, clear writing style and an orderly systematic structure the tangle of arcane verbiage and conflicting principles in existing treaties. His new treaty expressed in rational Enlightenment terms an entirely new model of diplomacy. Article XXIII, for example, exempted neutral shipping from capture in wartime and protected civilians of either warring party already residing in the other. Merchants caught in a country at the outbreak of a war would have nine months to collect their debts, settle their affairs, and go home. The model treaty also protected from war "all women and children, scholars of every faculty, cultivators of the earth, artisans, manufacturers and fishermen, unarmed and uninhabited unfortified towns, villages or places whose occupations are for the common subsistence and benefit of mankind."[38] Article IV allowed each signatory the right to carry cargo for the other and buy and sell on terms accorded by that country to its most favored trading partner. One key provision opened ports of a neutral treaty partner in wartime to American ships. Jefferson was seeking a safe haven in Europe for America's primary private business enterprise, shipping, in the event of renewed warfare with England, but he was also seeking a base for American warships to prey on the shipping of a potential enemy and to cut off its supply routes. Another provision abolished the notion of contraband of war. Along with Jefferson's principle that "free ships make free goods," this article was to assure the safety of a neutral America in a war involving her treaty partners in Europe.

When the commissioners concluded their weeks of preparation at Passy, negotiations began with several key nations, including Prussia, Denmark, Portugal, and the Kingdom of Sardinia, at once. Jefferson was surprised how little interest in treaty making with the United States the commissioners found, the exact opposite of what he had expected. At Versailles, there was a collective yawn in the diplomatic corps, echoing the same reaction in the capitals of Europe. Jefferson tried later to explain away this initial indifference:

> They seemed, in fact, to know little about us but as rebels, who had been successful in throwing off the yoke of the mother country. They were ignorant of our commerce, which had always been monopolized by England, and of the exchange of articles it might offer advantageously. They were inclined, therefore, to stand aloof until they could see better what relations might be usefully instituted with us."[39]

Jefferson believed that any good will America had enjoyed at the time of its victory over Britain had been since sabotaged by the British press, which consistently portrayed the United States as riot-torn, depressed, and on the brink of collapse. No nation would sign a treaty with a country about to dissolve. And he never forgot that the British still controlled the United States economically, selling Americans eighty-five percent of all imported necessities and luxuries.

Only Frederick the Great, the king of war-wrecked Prussia, the only Enlightenment ruler in Europe, refused to believe the rumors about Americans and recognized the rational genius of Jefferson. Prussia accepted his model treaty, concluding negotiations in 1785. The aged warlord personally endorsed Jefferson's enlightened diplomacy and asked to meet him. Jefferson was gleeful: Frederick's reaction had mattered to him most of all. "Of all the powers not holding American territory, a connection with him will give us the most credit," he had prophesied at Passy. There were a few modifications, including Prussia's reservation of the right to prohibit trade "when reasons of state require it,"[40] but Jefferson had won vital access to Prussian ports in peace or war, when American privateers could prey on British trade in the Baltic and interdict the flow of raw materials from Russia. While other negotiations lagged, Jefferson was far from discouraged.

The weak American economy began to affect Jefferson's personal finances soon after he arrived in Paris. At the same time Jefferson was appointed, Congress, responding to persistent stories that the American ministers in Europe lived in royal luxury, slashed the envoys' salaries twenty percent and stopped paying for a new diplomat's outfit—the customary allowance for furniture, clothing, carriage, rental deposits, etc. Someone had given Congress the impression that the foreign ministers were "eating the bread of idleness" to which Franklin, his meager salary cut $2,000, replied, "that we might not eat too much, our masters have diminished our allowances."[41] Jefferson had to go $4,000 into debt to live in the modest manner he had followed as a congressman in Annapolis, but he had to uphold American honor, even if he considered himself "the lowest and most obscure of the whole diplomatic tribe."[42] On October 16, 1784, after a nine-week search for a house he could afford, he signed a nine-year lease and began to lay out 6,000 livres a year for a small unfurnished *hôtel* on cul-de-sac Taitbout, a relatively newly built neighborhood of villas on the Right Bank between the towering St. Eustache Church and the present-day Opéra Garnier. The lease described a three-section-deep house with a courtyard and two gardens. A visitor could pass through the façade by an entry to the courtyard. Then, turning left, he could seek admission through a more elaborate, pillared doorway that led into the central part of the house. Inside was a rectangular private garden; out back was a garden plot. It was not long before Jefferson had laid out a full year's salary for "all plain" clothing, furniture, and carriage. He needed everything. Except for the "brooms and other small affairs for the house" which he left for his *maître d'hôtel,* he shopped carefully for everything himself—furniture, carpets, an *armoire* he stocked with linen and blankets, clocks, silver, dishes. Disavowing any "ambition for splendor," he nonetheless selected damask for drapes and upholstery, bought expensive Oriental rugs to protect the parquet floors, hired a carpenter to build bookshelves and modify two rooms—"nothing can be worse done than the house-joinery of Paris." He rented a piano, bought music and music stand; he haunted art auctions and bought engravings and oil paintings to cover his lodging's nakedness, including a "heriodade" with Salome holding the head of John the Baptist, copied from Simon Vouet. It was one of five items he purchased at the great Dupille de Saint-Séverin sale in February 1785. (The painting now hangs at Monticello.)

To manage his new house, Jefferson relied on Marc, his *maître d'hôtel,* to handle the household accounts, which he did badly for two and a half years before Jefferson discovered his filchings and fired him. Marc directed the other servants. There was a *valet de chambre* named Hegrand; Saget, the *frotteur,* whose job it was to dance on brushes that buffed the floors; a coachman named Vendôme to maintain and drive Jefferson's rented equipage. To cook for him and do his gardening, there was his faithful servant James Hemings, who fell ill shortly after they arrived and, after Jefferson hired a doctor and nurse, recovered enough to begin an apprenticeship with a caterer named Combeaux to learn some of the secrets of French cooking as well as the language.[43] In Hemings's absence, Jefferson paid the caterer for his meals, wine, and tea.

By mid-November, Jefferson was joined in his *hôtel* and at table by Humphreys and Short. That rainy, wind-swept winter, Jefferson spent most of his time in the new house. "I have had a very bad winter," he reported to James Monroe, "having been confined the greatest part of it. A seasoning, as they call it, is the lot of most strangers, and none, I believe, have experienced a more severe one than myself."[44]

Abigail Adams wrote on December 8 that Jefferson had been sick for six weeks. When he went out, it was often to the Adamses', and when he had company other than his young aides, it was often John Quincy. Jefferson was formally presented at the Foreign Ministry at Versailles in October. Flanked by the short, stocky Adams and the drooping Franklin, the tall, elegant southerner in his new wig and best tailored suit appeared an exclamation point between a period and a question mark. After that, he shunned the weekly *levées* of the beribboned, bowing-and-scraping diplomatic corps, considering their *habitués* a worthless tribe of time wasters.

Among the Americans, only David Humphreys was impressed by them and frequented them.

There were other welcome distractions that first gloomy winter. It took little French for Jefferson to stroll and shop at the Palais Royal or to enjoy concerts or balloon watching in the Tuileries Gardens. Jefferson came to the Tuileries with the rest of the people of Paris on Sunday, September 19, for the twenty-fourth French balloon ascent since the first *montgolfière* had flown over Paris less than a year earlier. In May 1784, shortly before he had left Philadelphia, Jefferson had first seen an air-filled balloon reach three hundred feet, but he had never seen human passengers borne aloft. Jefferson queued up at one of the three gates into the Tuileries that Sunday morning and handed in the two tickets (one for Patsy?) he had purchased earlier at the office of the balloonists, the Robert brothers. So vast a crowd had been expected that the police had published elaborate traffic rules in the *Journal de Paris.* But there was no jostling, no one hurt, and Americans in the audience were impressed at the orderly crowd. "Nobody was even incommoded," wrote a friend of Dr. Johnson. "Some comforts must then be confessed to result from a despotic government."[45] But the crowding did take its toll. When the colorful silk balloon was pulled along the central *allée* at eleven-thirty that morning to the platform specially built over the garden's central fountain, the crowd pressed in for a closer look and crushed the balloon's rudder. Undaunted, the two Robert brothers and a brother-in-law climbed into the basket and, at noon, ascended amid the crowd's roar. Jefferson could watch it through his telescope for nearly two hours before the balloon disappeared over the horizon. After a six-hour, forty-minute flight, it landed in Artois province, establishing a record for manned flight. Jefferson was fascinated and, throughout his years in Paris, followed the latest developments of ballooning, predicting great things for aviation. That winter, he reported to friends in America the first cross-Channel flight on January 7, 1785, and then, in June, the fatal crash of the balloon of "the two first martyrs to the aeronautical art."[46] But he did not lose interest; in June 1785, he hiked to the Luxembourg Gardens on the Left Bank to watch Monsieur Tetu take off. When there was no such excitement, on days when weather permitted, Jefferson formed the habit of strolling in the Tuileries Gardens and sitting silently on a terrace overlooking the Seine. As he later wrote to the turbulent Adams, "I do not love difficulties. I am fond of quiet."[47]

Jefferson's longed-for peace was shattered in January 1785, when Lafayette brought from America the news that Jefferson's two-year-old daughter Lucy had died. Lafayette brought two letters. One, from his brother-in-law, told him Lucy was ill. His brother-in-law's physician, Dr. James Currie, wrote the other:

> I am sincerely sorry my dear friend now to acquaint you of the demise of poor Miss L. Jefferson, who fell a martyr to the complicated evils of teething, worms and hooping cough which last was carried there by the virus of their friends without their knowing it was in their train. I was called too late to do any thing but procrastinate the settled fate of the poor innocent, from the accounts of the family, a child of the most auspicious hopes and having among other early shining qualities an ear nicely and critically musical. . . . Mr. Eppes lost his own youngest child from the same cause.[48]

Tortured and angry, Jefferson wrote to his wife's kin, demanding more details and begging him to write by the fastest mail, the French packet boat, that would bring him an answer in seven weeks. Not until May 6, 1785, did Jefferson receive the two letters posted by Elizabeth and Francis Eppes at the time little Lucy had died. Elizabeth, whose own daughter, Lucy, had also died at the same time, wrote:

> It is impossible to paint the anguish of my heart. . . . A most unfortunate hooping cough has deprived you, and us of two sweet Lucys, within a week. Ours was the first that fell a sacrifice. She was thrown into violent convulsions lingered out a week and then expired. Your dear angel was confined a week to her bed, her sufferings were great though nothing like a fit. She retained her senses perfectly, called me a few moments before she died, and asked distinctly for water. Dear Polly has had it most violently, though always kept about, and is now quite recovered. . . . Life is scarcely supportable under such severe afflictions.[49]

As soon as he learned of Lucy's death, Jefferson tried to decide whether he should send for six-year-old Polly, but he feared the harshness of the long voyage and he believed he would be returning to Virginia soon, his treaty-making completed. But he was worried that, if she did not rejoin him soon, especially in the face of repeated tragedies, she could never be taken away from her aunt, who had taken her mother's place for more than two years already. He could not stand to lose her, too: "It would be unfortunate through life both to her and us," he wrote Elizabeth Eppes, "were those affections loosened which ought to bind us together."[50] Unable to decide, he waited.

While the Adams and Lafayette households went into mourning to help him through his latest ordeal, Jefferson seems to have poured out his grief only to the distant sister of an old Williamsburg friend. He had tried to visit Betsy Blair Thompson in England while he was fetching a doctor for Patsy, but an insouciant servant had said she was not at home; actually she had been upstairs. When she wrote to him in Paris, he poured out his grief to her:

> My history, since I had the pleasure of seeing you last, would have been as happy of one as I could have asked, could the objects of my affection have been immortal. But all the favors of fortune have been embittered by domestic losses. Of six children, I have lost four, and finally their mother.[51]

In his desperate loneliness and grief, Jefferson began to write negative impressions of France and the French in letters home. To Eliza House Trist, his former landlady in Philadelphia, he wrote of the overwhelming poverty he had observed so far in his travels: "Of twenty millions of people supposed to be in France, I am of opinion there are nineteen millions more wretched, more accursed in every circumstance of human existence, than the most conspicuously wretched individual of the whole United States."[52] Presumably, he included the condition of slaves in America, believing them better treated than the majority of Frenchmen. "The truth of Voltaire's observation offers itself perpetually, that every man here must be either the hammer or the anvil."[53]

Increasingly he held the French monarch at fault. To John Jay he wrote, "The king goes for nothing. He hunts one half the day, is drunk the other, and signs whatever he is bid." Marie Antoinette was "detested and an explosion of some sort is not impossible."[54] To Washington, he wrote, "I was much an enemy of monarchy before I came to Europe. I am ten thousand times more so since I have seen what they are."[55] Writing "Hints to Americans Travelling in Europe," he wrote that the royal courts of London or Versailles should be viewed "as the Tower of London or menagerie of Versailles, with their lions, tigers and hyenas and other beasts of prey."[56] He held diplomats in special contempt. Of Foreign Minister Comte de Vergennes, he wrote,

> His devotion to the principles of pure despotism render him unaffectionate to our governments but his fear of England makes him value us as a make weight. He is cool, reserved in political conversa-

> tion, free and familiar on other subjects. . . . It is impossible to have a clearer, better organized head but age has chilled his heart.[57]

He still did not know enough French people well individually, but he felt competent to generalize. He was shocked, he wrote, by the beautiful prostitutes in the streets, the frank, uninhibited behavior at all levels of society. To Eliza Trist, he observed,

> The domestic bonds here are absolutely done away. And where can their compensation be found? Perhaps they may catch some moments of transport above the level of the ordinary tranquil joy we experience, but they are separated by long intervals during which all the passions are at sea without rudder or compass. . . . Fallacious as these pursuits of happiness are, they seem on the whole to furnish the most effectual abstraction from a contemplation of the hardness of their government."[58]

And to young John Banister, Jr., he warned in October 1785 against a European education because a young man is

> led by the strongest of all the human passions into a spirit for female intrigue destructive of his own and others' happiness, or a passion for whores destructive of his health, and in both cases learns to consider fidelity to the marriage bed as an ungentlemanly practice and inconsistent with happiness: he recollects the voluptuary dress and arts of the European women and pities and despises the chaste affections of those of his own country."[59]

He still considered himself a "savage of the mountains," he wrote in September 1785, a man accustomed to "tranquil permanent felicity." He still could not understand the Frenchman and Frenchwoman's pursuit of happiness:

> Conjugal love having no existence among them, domestic happiness of which that is the basis, is utterly unknown. In lieu of this are substituted pursuits which nourish and invigorate all our bad passions, and which offer only moments of ecstasy amidst days and months of restlessness and torment.[60]

In his loneliness, Jefferson was obviously becoming more and more preoccupied with questions of sexual mores. All his friends and colleagues except the Adamses seemed to have abandoned their inhibitions or wanted to. His young secretary, William Short, was soon engaged in a long affair with the beautiful young wife of the Duc de La Rochefoucauld. Lafayette had an affair with Madame de Hunolstein. Franklin was, at seventy-nine, in love with a widow, Madame Helvétius, and had proposed marriage, complaining to a friend that "it seemed ungrateful in her" that, while he had given her so many years of his days, "she has never given [me] a single one of the nights."[61]

Jefferson's "days and months of restlessness and torment" seemed to come to an end in mid-1785 when he received word that Franklin was finally going home and that he, Jefferson, was to replace him as the sole American minister to France, John Adams being sent to England as first ambassador to the Court of St. James. Jefferson would miss Franklin. Yet Jefferson did find some things more to his liking in Paris. "A man might pass a life without a single rudeness. . . . In the pleasures of the table they are far before us, because with good taste they unite temperance. They do not terminate the most sociable meals by transforming themselves into brutes. I have never yet seen a man drunk in France, even among the lowest of the people." And he had "no words capable of expressing how much I enjoy their architecture, sculpture, painting, music. . . . It is in these arts they shine."[62]

In the near-year they had worked together in France, Jefferson benefited enormously from Franklin's company as well as his generous introductions. Jefferson visited the Chaumont villa at Passy often, where he became a regular at the big dining table with Dr. Franklin and his friends, the *abbés,* and stayed for the Doctor's chess matches, which lasted by candlelight late into the night and sometimes until dawn. A grandson of the Chaumonts recalled that Franklin once refused to open an important dispatch from America until their game was finished. Jefferson related that the Doctor, while playing with the old Duchesse de Bourbon, happened to put her king into checkmate and take it.

"No, we do not take kings so," objected the duchess.

"We do in America,"[63] retorted Franklin.

Franklin gradually introduced Jefferson to the pro-American aristocrats and intellectuals in the salons that had made him their prize. The friends of Dr. Franklin became the early friends of Jefferson in Paris. Among the first he had met had been the *abbés* Arnauld and Chalut, whose names have always been mentioned together. Abigail Adams reported that Arnauld was "about fifty, a fine sprightly man who takes great pleasure in obliging his friends."[64] Chalut was seventy-five, and John Adams thought he was a spy who watched Franklin and the Americans for the French foreign minister. He described the two *abbés* as "learned men" who "came early to visit me."

> They had a house in the city, another in the country at Passy, in our neighborhood, where they resided in summer. Whether they were spies of the court or not, I know not, but I should have no objection to such spies, for they were always my friends, always instructive and agreeable in conversation. They were upon so good terms, however, with the courtiers, that if they had seen anything in my conduct or heard anything in my conversation that was dangerous or very exceptional, I doubt not they would have thought it their duty to give information of it.[65]

If Adams distrusted the two *abbés*, Jefferson, at first at least, did not. As early as July 1785, he was writing that the death of another *abbé* "is likely to detain his friends Arnauld and Chalut in Paris the greatest part of the summer. . . . It is a fortunate circumstance for me, as I have much society with them." It was not the dinners with the *abbés* that attracted Jefferson, he wrote; it was "the crumbs of science on which we are subsisting here." Their table talk revolved around a recently propagated theory; that summer, the late friend of the *abbés* had "shaken, if not destroyed," the theories of Descartes and Newton for explaining the phenomenon of the rainbow. . . . We are wiser than we were by having an error the less in our catalogue, but the blank occasioned by it must remain for some happier hypothesis to fill up."[66]

Jefferson's acquaintance with Abbé Chalut was fairly typical of how he intertwined diplomacy and the pleasures of the salons. Their meeting led to his introduction to Abbé Chalut's brother, Chalut de Verin, one of the farmers-general who controlled the French monopoly over the American tobacco trade, a prime interest to Jefferson, the Virginian who hoped to break the monopoly. De Verin also had one of the great art collections in Paris in his apartment in Place Vendôme. Here Jefferson could chat with connoisseurs, artists, and sculptors such as Jean-Antoine Houdon, whom Jefferson soon commissioned to make a full-length statue of Washington for the new Virginia capital at Richmond. He introduced his young friend, John Trumbull of Connecticut, at De Verin's salon and accompanied him again to dinner at Abbé Chalut's in Passy.

Chalut, in turn, introduced Jefferson into the circle of Madame de Laye Epinaye. With letters of introduction from Abbé Chalut, Jefferson was able, among a half dozen stops at the homes of French leaders, to visit the

Château de Laye Epinaye in Beaujolais, where he was enchanted with the statue of Diana and Endymion, "a very superior morsel of sculpture,"[67] on a long trip he made through the south of France in the spring of 1787. It was undoubtedly at Passy that Jefferson met another distinguished secular cleric-without-a-charge, Abbé Morellet, one of the last of the *philosophes.* Fifty-eight and highly respected when Jefferson met him, he had recently been elected one of the immortals of the Académie Française who was known for his conversation—once described as "malignant without being caustic"—and who considered himself a "slightly less bitter"[68] Jonathan Swift. After studying at the Sorbonne and in Rome, he had been taken up by one of the leading women of France, the *saloniste* Madame Geoffrin, whose Tuesday-night dinner guests had included the *philosophes* Voltaire and Montesquieu, Diderot and d'Holbach, Abbé Reynal and Saint-Lambert. She kept the painters separate from the philosophers. She fed the painters on Mondays—Boucher, La Tour, Van Loo, Chardin, and Greuze all feasted at her *petits soupers.* On Wednesdays only men were allowed, except for her rival *saloniste,* Mlle. de Lespinasse, out of deference to her friend, the mathematician D'Alembert. On Wednesdays, she invited distinguished foreigners: Franklin, David Hume, and Horace Walpole had mingled with the likes of the novelist Marmontel and his uncle, Abbé Morellet, who now latched on to the newest member of the charmed American circle, Jefferson.

Since Franklin's arrival in France, Morellet had interested himself in the rights of man and become an ardent follower of the American cause, attempting a history of the American colonies. His translation of Enlightenment-spirited works that favored such ideas as religious toleration, penal reform, and smallpox inoculation, enjoyed a wide audience. He had also recently written a sly parody of Franklin and his devoted Madame Helvétius, whose salon at Auteuil near the Adams villa was one of the most fashionable in France despite the fact that it crawled with eighteen cats and innumerable dogs that urinated at her feet as she presided over the most notable literary circle of France, reading aloud the letters of absent members or the writings of those present. Jefferson had always found the scene charming, even if he left little record of the weekly meetings. The puritanical Adamses, whom he also visited frequently at Auteuil, detested the salon scene. Abigail wrote a friend a description of Madame Helvétius, dressed in "a chemise made of Tiffany, which she had on over a blue lutestring, and which looked as much upon the decay as her beauty, for she was once a handsome woman; her hair was frizzled; over it she had a small straw hat with a dirty gauze handkerchief around it, and a bit of dirtier gauze than ever my maids wore was bowed on behind." The Adamses had stayed for dinner at Madame Helvétius's salon and Abigail was horrified when Madame Hutchins greeted Franklin with "a double kiss, one upon each cheek, and another upon his forehead." Then, during dinner, Abigail looked on at Madame

> frequently locking her hand into the Doctor's, then throwing her arm carelessly upon the Doctor's neck. After dinner, she threw herself upon a settee, where she showed more than her feet. She had a little lapdog who was, next to the Doctor, her favorite. This she kissed and when he wet the floor, she wiped it up with her chemise.[69]

Jefferson, too, was ill at ease in the salons at first, but for another reason. As he confided to Eliza Trist, he had never seen women who had real political power before. But, in France, from Marie Antoinette to Madame Helvétius, women had become highly influential. He haunted the salons of the venerable Duchesse d'Anville (mother of his friend the Duc de La Rochefoucauld), whom Franklin had told him was "a lady of uncommon intelligence and merit,"[70] of Comtesse d'Houdetot, the Julie of Rousseau's *La Nouvelle Héloïse,* and of Madame Necker and her celebrated daughter, Madame de Staël. He became the close friend of the Comtesse de Tessé, the beautiful young aunt of Lafayette, lady-in-waiting to the queen, and a dedicated botanist; later, he sent her many plants for her gardens at Chaville, on the road to Versailles.

Morellet had met Franklin in France before the Revolution and had introduced him into the salons when he arrived in France in 1776. One of their favorite hostesses was Madame Brillon, with whom they dined twice a week and took tea twice a week. She once described one of these tea parties: "My fat husband will make us laugh, our children will laugh together, our big neighbor [Franklin] will quiz, the Abbés La Roche and Morellet will eat all the butter. . . . Père Pagin will play 'God of Love' on his violin, I the march on the piano, and you 'Petits Oiseaux' on the harmonica."[71] Now, another violinist could be called upon, the tall, graceful, attentive, and thoroughly captivated Jefferson.

To his expanding weekly round, Jefferson now added Monday morning visits to the splendid apartment of Abbé Morellet, who lived with his sister and his niece, who was married to the novelist Marmontel. Their apartment looked down on the Tuileries Gardens; from it, as the noonday sun shone in and the distinguished guests arrived for the *abbé*'s famous *matinées,* Jefferson could look down at "the most celebrated walk in Paris," as Abigail Adams called the Tuileries, with "the six large gates by which you enter." Abigail had obviously visited the *abbé* and studied the view, the royal gardens "adorned with noble rows of [chestnut] trees, straight, large and tall, which form a most beautiful shade."[72] In Morellet's sun-filled library, Jefferson basked among "men of unusual gifts [who] brought the tribute of their talents to enrich the life of this sage."[73]

It was at one of Abbé Morellet's weekly salons that Jefferson and the *abbé* agreed that Morellet should be allowed to translate into French Jefferson's privately printed *Notes on the State of Virginia,* which in 1785 earned Jefferson a place of his own in the best literary salons.

In his grief in the spring of 1785, Jefferson had turned inward to his writing. He decided to put into print anonymously his *Notes on Virginia* to circulate privately among influential Europeans as a remedy for the colossal ignorance and prejudice he had found on the subject of America. Jefferson had intended to have the book printed in Philadelphia but found it would cost too much; he brought it to France where, as he had hoped, the cost of printing was one-fourth that in Philadelphia. He ordered two hundred copies printed. When the book came off the press that spring, he began mailing copies to friends in the United States, with some reservations. "I fear the terms in which I speak of slavery and of our constitution," he wrote Monroe, "may produce an irritation which will revolt the minds of our countrymen against the reformation of these two articles, and thus do more harm than good."[74] When Chastellux asked him for permission to make extracts for the *Journal de Physique,* Jefferson gave permission because they would not include his passages on slavery and the constitution:

> In my own country, these strictures might produce an irritation which would indispose the people towards the two great objects I have in view, that is, the emancipation of their slaves and the settlement of their constitution on a firmer and more permanent basis. If I learn from [Virginia] that they will not produce that effect, I have printed and reserved just copies enough to be able to give one to every young man at the College [of William and Mary]. It is to them I look, to the rising generation, and not to the one now in power, for these great reformations.[75]

When Jefferson asked Madison what he thought about distributing copies to students at William and Mary, Madison in turn asked George Wythe, who thought Jefferson's opinions would ruffle the feathers of many of the students' parents, but "we think both the facts and remarks which you have assembled too valuable not to be made known, at least to those for whom you destine them." Wythe urged Jefferson to place the books in the college library rather than to give them directly to the students to avoid offending "some narrow-minded parents."[76] After reading Jefferson's book, Philadelphian Charles Thomson, Secretary of Congress, wrote of his regret "that there should be such just grounds for your apprehension" in the "southern states by what you have said about slavery. However, I would not have you discouraged. This is a cancer that must be got rid of."[77]

Jefferson had not heard back from friends in America before he had to decide whether to publish the book without restrictions. So many of his Paris acquaintance had asked for copies that his fear that the book would fall into the wrong hands was realized. One recipient died, and *Notes* wound up in the possession of an unscrupulous bookseller who was about to publish "a very abominable translation."[78] Abbé Morellet, to whom Jefferson had presented a copy, had asked for permission to make his own translation. Jefferson acceded and he discussed the work during his weekly visits to Morellet's splendid apartment overlooking the Tuileries.

Jefferson welcomed Morellet's "friendly proposition" because "a translation by so able a hand will lessen the faults of the original instead of their being multiplied by a hireling translator. I shall add to it a map," Jefferson wrote to Charles Dumas, "and such other advantages as may prevent the mortification of my seeing it appear in the injurious form threatened."[79] Morellet was to submit his translation for Jefferson's inspection. Jefferson's identity was to be shielded. Jefferson was to pay for engraving the map; Morellet was to be paid from sales of the book and was to pay the printer. Jefferson was not prepared for the liberties a famous translator could take. Morellet followed guidelines for translation laid down by the *philosophe* D'Alembert in 1759 in his "Observations on the Art of Translating in General"; Morellet was interested in disseminating "liberal ideas to a readership which appreciated classical order, clarity and stylistic elegance," a recent study by classical scholar Dorothy Medlin points out. "Probably because he did not understand the translator's point of view and his purpose. Jefferson was surprised, disappointed and disturbed by Morellet's translation."[80]

He certainly was. Thirty-five years later, still stinging at being caught up in a publishing venture he could not control, Jefferson in his autobiography denounced Morellet's translation. "Introverted, abridged, mutilated and often reversing the sense of the original, I found it a blotch of errors from beginning to end."[81] After Jefferson handed Morellet a seven-page list of seventy errors, a certain coolness developed in their relationship. When Morellet listed years later in his memoirs the famous literary figures he had known, he gave Franklin a paragraph, but he did not even mention Jefferson.

Worried now that a pirated English edition in London based on bad translations would be even worse, Jefferson sent John Stockdale a copy in English with the plate of his own carefully drawn copy of his father's map of Virginia. Stockdale's superior edition, published in 1787, carried Jefferson's name on the title page for the first time. But he still did not consider the work important. When he wrote Thomson about the latest literary events, he said there was nothing new in Paris. Despite the fact that the book was making him a literary figure in Paris, Jefferson wrote Thomson that he did "not consider" he had "added anything to that field."[82] But his friends in America and the *salonistes* of Paris ignored his self-deprecation. John Adams wrote, "It will do its author and his country great honor. The passages about slavery are worth diamonds. They will have more effect than volumes written by mere philosophers."[83] The influential *Mercure de France* reviewed it much more enthusiastically than did the *Monthly Review* of London, which found "much to applaud as well as some things to which we cannot afford a ready assent."[84] Jefferson had not set out to write a book, only to answer a questionnaire, then to seek a better opinion of America from influential Europeans. All accidentally, on both sides of the Atlantic, he emerged a noted literary figure, an early political scientist, and a respected man of science.

16

"A Situation Much More Pleasing"

Fancy to yourself a being who is withdrawn from his connections of blood, of marriage, of friendship. . . . Continue, then, to give me facts, little facts. Your letters will be most precious.

—JEFFERSON TO DR. JAMES CURRIE, SEPTEMBER 27, 1785

ON MAY 2, 1785, Paul R. Randall arrived from America with new instructions for American diplomats in Europe. In March, Congress at last had accepted Franklin's resignation. The treaty-making commission was to be dissolved after its two-year contract expired. Jefferson was to become the sole American minister plenipotentiary to France; Adams was to go to England as the first minister to the Court of St. James. Randall himself was to go with Thomas Barclay as agents to Algiers. Jefferson was ecstatic, but he could not have known how much thanks he owed Adams for the appointment. In February 1783, Adams had written Robert R. Livingston, then-secretary of foreign affairs, what he considered to be the qualifications for an ideal minister to France:

> In the first place, he should have had an education in classical learning, and in the knowledge of general history, ancient and modern, and particularly the history of France, England, Holland and America. He should be well versed in the principles of ethics, of the law of nature and nations, of legislation and government, of the civil Roman law, of the laws of England and the United States, of the

> public law of Europe, and in the letters, memoirs and histories of those great men who have heretofore shone in the diplomatic order and conducted the affairs of nations and the world. He should be of an age to possess a maturity of judgment arising from experience in business. He should be active, attentive and industrious and, above all, he should possess an upright heart and an independent spirit and should be one who decidedly makes the interest of his country, not the policy of any other nation, nor his own private ambitions or interest, or those of his family, friends and connections, the rule of his conduct.[1]

If Adams was not describing himself, he certainly was describing Thomas Jefferson at age forty-two. Congress had also learned from other sources that Jefferson "was peculiarly acceptable to the Court of Versailles."[2] Jefferson wrote back at once to accept: "Fervent zeal is all which I can be sure of carrying into their service. . . . The kind terms in which you are pleased to notify this honor to me require my sincere thanks."[3] On May 14, Jefferson went to Versailles to hand the Comte de Vergennes "my appointment as minister plenipotentiary to this court," at the Foreign Ministry in the tall, vast, gambrel-roofed Hôtel de la Marine et des Affaires Etrangères, in the rue de l'Intendance abutting the chapel of the royal palace. He was invited to return that Tuesday, May 17, to deliver "my letter of credence to the King at a private audience."[4]

It had been only eight months since Jefferson, fresh from America, had first driven up the breathtaking avenue de Paris to Versailles, a dazzling royal city with scores of palaces and the villas of courtiers that surpassed anything in the Western World. He could only have been awed, this farmer from the American wilderness, but, as usual, he did not record his emotions, after that first brief visit, only that he had "paid for chair hire at Versailles, 3 francs."[5] It must have seemed incongruous for this American revolutionary, the son of a frontier settler, to pay to be carried in a sedan chair to meet with powdered dignitaries, but as a diplomat, he knew he must do whatever was *de rigueur* to uphold the dignity of his country. Fortunately there were those present that day who preserved scraps of evidence of the scene. David Humphreys, who attended Jefferson as secretary of legation, wrote as proudly to Washington, his former chief, as if he himself were the new ambassador:

> I have passed through the ceremony of going to Court and being presented to the King and royal family. . . . The King, who is rather fat and of a placid, good-tempered appearance, is thought to possess

an excellent heart and to aspire only to the distinction of being considered as the father of his people.[6]

The courtier assigned that day to introduce Jefferson to the royal family was the Comte de Cheverny, who described the usual ceremony for such occasions, whether the new ambassador was from Turkey, mighty England, or the anemic string of former British colonies called the United States of America.

At ten that Friday morning, the carriages of Jefferson and other diplomats due to be presented that day rattled over the wide, cobbled avenue de Paris through the gold-encrusted grillwork gate crowned with the royal fleur-de-lis. Ordinarily Jefferson would have veered off to the right to tie up his rented coach-and-four at the Foreign Ministry. Today, however, he rolled straight ahead through a second gilded gateway flanked by statuary. The carriage began the prescribed turn around the royal courtyard. Franklin must have briefed Jefferson on the elaborate routine. After he steered his carriage beneath the windows of the king's apartments, Jefferson ordered his driver to halt as two gold-liveried *écuries* came to meet the carriage, one to steady the horses, the other to fold down the carriage steps deftly. As the two Americans alighted, Cheverny, carrying the mace of a *maître d'hôtel*, ushered them through the palace entrance, where a double row of white-uniformed French and Swiss guards formed an inner corridor through the Hall of Mirrors, through the row of golden nymphs and cherubs holding gilded candelabra dripping crystal and showering reflections all along the seventy-five-meter-line of march. "We entered the (lower) hall," Cheverny wrote, where he told Jefferson to wait. "I went to the King's apartments to await the hour of the audience and I also passed through the apartment of the whole royal family. At the hour fixed, I went to get the ambassador." Cheverny rejoined Jefferson, fell in at his left. "My footmen were lined up an equal distance apart from the door to the foot of the great staircase." Jefferson's appointed servants-for-the-day in matching royal liveries, waited for him, turned, and preceded Jefferson, with Cheverny at his left as *introducteur* whispering directions in French. Humphreys, as secretary of legation, walked behind them, down the gold-balustraded, marble-inlaid Stairway of the Queen, past one hundred Swiss Guards in full dress white uniforms. Inside the great Hall of the Ambassadors, the entire diplomatic corps was waiting. The Guards officers gave an order for the sentinels and the king's bodyguard to salute Jefferson.

At the doors to the king's bedchamber, two more Swiss guards stood at their posts, looked down at them from the *oeil-de-boeuf*. The captain of the King's Guards walked at the *introducteur's* left now as the gilded doors swung open admitting them to the king's bedchamber. Jefferson faced Louis XVI. "We found him seated surrounded by his leading officers," wrote Cheverny, "the grand chamberlain, all the dukes and people of title —in short, the courtiers." When Louis XVI saw Jefferson approaching, he removed his hat and then stood up. Jefferson, escorted now by a prince and by Cheverny, took off his hat and advanced toward the king, Humphreys following him. Then Louis sat down again and put back his hat, Jefferson's signal to do likewise. All the princes, dukes, and courtiers put their hats on again as Jefferson began to read a prepared speech. Each time he mentioned the name of the king or the queen, Jefferson doffed his hat. "The King did the same and the courtiers followed faithfully," wrote Cheverny. When Jefferson was finished, the king, eleven years Jefferson's junior, made a few brief formal comments before the Americans "retired as they had come in, again making three deep bows." Then Jefferson and his staff went to the queen's apartment, repeating the ritual, "and to those of all the royal family." It was two o'clock by the time they returned to the Hall of the Ambassadors, where a table had been set for fifty in the Council Chamber for everyone who had been presented that day. "Everyone was served by the gentlemen of the bedchamber, and one stood behind each chair. Swiss in uniform passed the platters about,"[7] wrote Cheverny. At the head of the table was the Comte de Vergennes. The entire diplomatic corps took its places—Humphreys described them: "ambassadors, ministers or other strangers of the first fashion from all the countries of Europe," all conversing in French as the Swiss Guards came and went with the courses. Jefferson had already formed an opinion of Vergennes, "a great minister in European affairs but [he] has very imperfect ideas of our institutions and no confidence in them."[8]

Among the glittering ambassadorial corps, Thomas Jefferson was unquestionably the least adorned, his clothes and residence the least imposing. When a young American later accompanied him to a private audience with the king, he noted how different Jefferson's comportment was from the rest. As Thomas Lee Shippen wrote home,

> When we were introduced to the King, he was just pulling on his coat, a servant was tying his hair, in which there was no powder, while one of his attendants was arranging his sword belt. . . . The file of ambassadors, envoys, ministers, etc. in full dress were prostrating themselves before him. He hitched on his sword and hobbled from one side of the room to the other, spoke eight words to a few of the ambassadors and two to a German prince, who was presented with me, and left the room. Mr. Jefferson was the plainest man in the

> room and the most destitute of ribbons, crosses and other insignia of rank [but] he was the most courted and most attended to, even by the courtiers themselves, of the whole diplomatic corps.[9]

What Jefferson thought of all this folderol, whether he was flattered by it or not, he was a skillful diplomat. He was much too discreet to consign to paper—or to whisper to anyone who might—his views of the French royal family. In his report to Congress, he said simply that he had been put through the "ceremonies usual on such occasions."[10] He could not even risk a letter home to friends in Virginia. He had to assume his mail was being read. Franklin doubtless told him that everything was searched by the Sureté, the royal secret police, which had once proclaimed Franklin's underwear the whitest they had ever seen. But a few years later, when it was safer, he wrote Madison his candid opinions of Louis XVI and Marie Antoinette. "The King," he told Madison, "loves business, economy, order and justice. He wishes sincerely the good of his people. He is irascible, rude and very limited in his understanding [and] religious bordering on idolatry. He had no mistress, loves his queen and is too much governed by her."[11] He could find very little good to say about the "gaudily painted" Marie Antoinette. At the safe distance of thirty years, when Jefferson wrote his *Autobiography,* he was kinder to Louis. "He had not a wish but for the good of the nation, and for that object, no personal sacrifice would ever have cost him a moment's regret." But Louis's mind was "weakness itself, his constitution timid, his judgment null" and he was "without sufficient firmness even to stand by the faith of his word."[12] If Jefferson could forgive Louis, he put the blame for the French Revolution on the haughty, manipulative Marie Antoinette, who had "led herself to the guillotine and drew the King on with her, and plunged the world into crimes and calamities which will forever stain the pages of modern history."[13] Jefferson obviously kept these views from the queen and she actually took a liking to him, inviting him to come back and watch her play whist with her courtiers at her elegant green card table.

The formalities were to drag on for another week, with every diplomat from every nation represented at Versailles rolling up into the cul-de-sac Taitbout to call on Jefferson, sign the guest register, leave a calling card. When all the wearisome courtesies had been observed, Jefferson threw a party for his American and French friends. It was a bittersweet affair. He had drawn close to the Adamses, especially Abigail, who considered him "one of the choice ones of the earth."[14] But now they were leaving for London. After working for many months with Adams, Jefferson found he had "a degree of vanity" and was too "attentive to ceremony . . . irritable and a bad calculator of the force and probable effect of the motives which govern men," but, as he told Madison, that was "all the ill will which can possibly be said of him."[15] Jefferson would miss him.

In July, Franklin left, too. He had come to Paris on a windy, cold December day, at the darkest hour of the American Revolution. He left on July 12, 1785, a gorgeous summer day, in the queen's own litter, pulled gently by Spanish mules to ease the suffering of Franklin's journey to Le Havre and a ship home. Jefferson was there with most of the silent people of Passy, and he noted that it looked as if Passy had lost its patriarch. As a going-away present, the king had presented Franklin with a miniature portrait of himself surrounded by 416 diamonds. Some 128 crates of luggage were already on a barge floating down the Seine. Madame Helvétius, who had refused to marry him, spoke for all his friends in a last message sent after him: "I picture you in the litter, farther from us at every step, already lost to me and to those who loved you so much and regret you so. I fear you are in pain. If you are, come back, *mon cher ami,* come back to us!"[16] But Franklin did not turn back. Jefferson considered him "the greatest man and ornament of the age and country in which he lived."[17] It would be impossible to replace him, he said often. "No one can replace him, Sir, I am only his successor."[18]

The new American ambassador to France was bristling with ideas. Adams had scarcely departed for London before Jefferson began peppering him with suggestions. Jefferson missed the Adamses. Their departure, Jefferson said, left him "in the dumps." He was already fuming, however, at the way family and friends in America were neglecting him. "Monroe, I am afraid, is dead," he wrote their friend William Short on May 2. "Three packets have now come without bringing me a line from him or concerning him."[19] (He did not know that Monroe was busy courting Elizabeth Kortright those three months.) "Pray write to me, and write me long letters,"[20] he pleaded with Elizabeth Eppes, desperate for word of Polly. To Dr. Currie, he complained,

> Of political correspondence I can find enough. But I can persuade nobody to believe that the small facts which they see passing daily under their eyes are precious to me at this distance: much more interesting to the heart than events of higher rank. Fancy to yourself a being who is withdrawn from his connections of blood, of marriage, of friendship, of acquaintance in all their gradations, who for years should hear nothing of what has passed among them, who returns again to see them and finds one half dead. This strikes him

> like a pestilence sweeping off the half of mankind. . . . Continue then to give me facts, little facts, such as you think every one imagines beneath notice, and your letters will be most precious to me.[21]

As he wrote to Virginia that "I must have Polly" and waited for his daughter's arrival, Jefferson moved into his new and larger house. Now he seemed to have no desire to leave Paris. He had every reason to believe he would remain in Paris nine years, like Franklin. When Abigail wrote him in June a taunting letter from London, the beginning of a long and affectionate correspondence, Jefferson replied,

> I consider your boasts of the splendour of your city and of its superb hackney coaches as a flout, and declaring that I would not give the polite, self-denying, feeling, hospitable, good-humoured people of this country and their amiability in every point of view, (tho' it must be confessed our streets are somewhat dirty and our [carriages] rather indifferent), for ten such races of rich, proud, hectoring, swearing, squibbling, carnivorous animals as those among whom you are; and that I do love this *people* with all my heart.[22]

Now that he was the sole American minister to France, Jefferson decided he needed a more appropriate residence, where he could entertain dignitaries and feed and house visitors. After months more of house hunting, he wrote Abigail Adams on September 4, "I have at length procured a house in a situation much more pleasing then my present." The handsome villa, built for a mistress of a duke, was at the grillwork gates to Paris at the top of the Champs-Elysées, just inside the new city walls, at rue de Berri. "It suits me in every circumstance but the price, being dearer than the one I am now in." Jefferson told Abigail little more about it than that "it has a clever garden to it."[23] Named the Hôtel de Langéac after its former owner, it was the first house inside the city walls across from the tollhouse where every peasant bringing in rabbits, every wagoneer hauling salt for sale in Paris had to stop and pay taxes. Across the Champs-Elysées, Jefferson could look out at the fashionable English gardens of the Comtesse de Marbeuf.

Jefferson had to pay his litigious landlord an extra six-month's rent to get out of the cul-de-sac. His new rent was twenty percent higher than his already steep former rent and, once again, the place was unfurnished, but Jefferson was taken by the ingenious layout of the rooms, especially the oval salon with its rising-sun ceiling, and he thought that with "an almost womanly attention to the details of the household,"[24] he could manage. The handsome three-story villa had the latest gadgets, including piped water pumped from the Seine by a "fire engine better called the steam engine."[25] It also was a splendid establishment fit for the nobleman who had gone broke building it. There were stables. Jefferson could indulge his love of horses now and he bought horses and a carriage and hired a coachman. He also hired a full-time gardener for the extensive grounds, the greatest attraction of the place for Jefferson. By the next summer, he was writing home proudly to an Albemarle County relative. "I cultivate in my own garden here Indian corn for the use of my own table, to eat green in our manner. But the species I am able to get here for seed is hard, with a thick skin, and dry." Jefferson implored Col. Nicholas Lewis to send him an ear of the "small ripe corn we call hominy corn" as well as seeds of the sweet potato, watermelon, cantaloupe, some acorns, and "a dozen or two bacon hams."[26] It wasn't long before Jefferson was able to serve southern-style dinners of corn on the cob, ham, and yams, with French wine to such French noblemen as Lafayette and Chastellux. He could even begin to grow his own grapes to press his own wines, a fact he kept in mind as he went on a diplomatic mission through the Rhine Valley.

For four years, the Hôtel de Langéac, now remembered only by a plaque on a commercial building, was Jefferson's home and the American Legation in France, where Jefferson transacted all official business. Here, Jefferson welcomed not only noblemen and emissaries, but stranded American seamen, impecunious American travelers, French veterans of the American Revolution seeking their back pay, starry-eyed European adventurers in need of letters of introduction for America. His guests included illustrious American artists such as John Trumbull, who came to paint Jefferson's portrait for his epic canvas of the signing of the Declaration of Independence. And there were inventors such as Thomas Paine, who came to Paris to peddle his plans for a bridge, and to take "family soup," the informal, intimate late-afternoon meal served when Jefferson received most of his guests.

It was at these *petits soupers* that Jefferson may have begun to form some of his most important ideas as he listened to the plans, opinions, and theories of his guests. One of his visitors early in 1786 was John Ledyard of Connecticut, who had sailed with Captain Cook on his third Pacific voyage and had come to Paris with a scheme for a company to trade furs between the American Pacific Northwest and China. Ledyard had made no headway with his venture until Jefferson suggested that he explore the American northwest by journeying eastward from Paris through Siberia to Kamchatka and then on to North America and to the Atlantic coast. Ledyard became wildly enthusiastic about becoming "the first circumambulator of the globe."[27] Jefferson enlisted the support of Lafayette, and together they sought the aid of his friend, Baron de Grimm, unofficial representative of Catherine the Great, at Versailles, to obtain Ledyard free passage across Russia. When Catherine refused, Ledyard went ahead anyway and made it across almost all of Siberia to within several hundred miles of Kamchatka before Catherine's agents overtook him and turned him back. Ledyard felt he had scored a scientific triumph, nonetheless, asserting that he had proven that the Tartars and the American Indians "are the same people," he wrote Jefferson from Siberia, "the most ancient, the most numerous of any other and, had not a small sea divided them, they would have all been known by the same name."[28] Ledyard passed through Paris again in 1788 and visited Jefferson at Langéac. This time he was off on an expedition to explore Africa. Jefferson made him promise that he would next return to America and explore the continent west from Kentucky to the Pacific. Ledyard never returned, but Jefferson tucked away the idea of an expedition to the American west.

He was unfailingly helpful to American businessmen trying to launch enterprises in Europe. When Thomas Paine, author of *Common Sense,* came to Paris to promote his revolutionary iron bridge, Jefferson helped him make contacts he needed to win funding and patents, which he eventually found in England. The two became lasting friends. James Rumsey, whom Jefferson called "the greatest mechanical genius"[29] he knew, asked his help to seek exclusive rights to steam navigation in France fully twenty years before Robert Fulton's *Clermont* steamed up the Hudson. Rumsey found Jefferson "the most popular ambassador at the French Court."[30] He reported he had often enjoyed the simple *soupers* with Jefferson at the Hôtel de Langéac. "I have been frequently at Mr. Jefferson's," Rumsey wrote. "He has got all that ease, affability and goodness about him that distinguishes him as a good as well as a great man."[31]

There also were "great dinners" at Jefferson's villa, attended by the woman he would fall in love with, Maria Cosway, and her artistic friends, including the Princess Lubomirska and the Polish writer, Julian Niemcewicz. Many of the high-ranking French officers in the American Revolution, including Lafayette, Rochambeau, Chastellux, the Duc de Lauzun and Admiral de Grasse, came to Hôtel de Langéac to have Trumbull sketch their portraits. Later, Trumbull, who became Jefferson's intimate friend, painted them into "The Surrender of Cornwallis at Yorktown." Few houses in Paris could have richer associations for lovers of American history than "Mr. Jefferson's house" halfway up the present-day Champs-Elysées.

As soon as Jefferson moved into the Hôtel de Langéac with his private secretary, William Short, his old friends began to appear. Philip Mazzei, his wine-growing Albemarle neighbor, was among the first. Jefferson, he decided, "lived in a beautiful *villetta* with a charming garden" within "gunshot distance" of the gate that swung one way to Paris, the other to Versailles. Mazzei walked up the Champs-Elysées from his hotel, "arriving about an hour before dinner." He had come from Virginia hoping Jefferson could employ his services as a diplomat, and had written from Nantes to announce his arrival. Jefferson "expected me daily. Nevertheless, our meeting was moving to both."[32] They dined alone, talking for hours (or Mazzei talking for hours, usually the case). The next day, a Monday, Jefferson took Mazzei down the rue St. Honoré to visit Marmontel; the Abbé Morellet came in before they left. Then Jefferson took Mazzei on the rest of his typical Monday rounds. They visited the scientist Lavoisier; Condorcet, the *philosophe* and director of the royal mint; the Duc de La Rochefoucauld at his *château* with its elaborate gardens on the quai de Seine. And they were home in time for dinner with William Short, who gradually was becoming like a son to Jefferson. David Humphreys had gone home.

The first test of Jefferson's new diplomacy came almost at once. He had documents authorizing him to open negotiations with Morocco, Algiers, Tunisia, and Tripoli, whose corsairs had begun to seize American vessels as soon as the British Navy stopped protecting American shipping. The Barbary states along the North African coast exacted tribute from most nations whose ships sailed the Mediterranean, enforcing their demands by capturing ships and enslaving their crews in workhouses where they were tortured, starved, and held for ransom. European governments made heavy tribute payments. The now-indepedent United States could expect to pay substantial tribute, too, or forgo its substantial trade in the Mediterranean. Jefferson estimated American exports to the Mediterranean had amounted to fully one-sixth of all American wheat and flour and one-fourth of the dried fish and rice exported at the outbreak of the Revolution, and that Mediterranean trade employed some twelve hundred sailors on one hundred ships. But Jefferson opposed the very idea of paying tribute to pirates. "When this idea comes across my mind," he wrote, "my faculties are absolutely suspended between indignation and impotence."[33]

Jefferson was among the first to favor building a strong American navy to protect American commerce. Adams advocated paying the tribute and getting on with business, which he thought was far more lucrative. He estimated a $200,000 payment to Algeria would be repaid with $1 million in trade. Later Jefferson opposed navies, but when he was minister to

France, and Moroccan pirates seized the American brig *Betsy* in the Atlantic off Portugal, he declared that "we ought to begin a naval power if we mean to carry on our own commerce. Can we begin it on a more honorable occasion, or with a weaker foe? I am of opinion Paul Jones with a half dozen frigates would totally destroy their commerce . . . by constant cruising and cutting them to pieces."[34]

The crisis had reached the point where Lloyd's of London was charging twenty-five percent to insure American cargoes bound for Cadiz or Lisbon—the usual rate for these Atlantic ports was one and one-half percent. Britain stood to benefit almost as much as the Barbary states by eliminating American competition. Jefferson appealed to Versailles for reliable information on tribute payments paid by other nations while he sought to enlist French naval aid under the 1778 Franco-American treaty. The king had pledged to intervene in the Mediterranean on behalf of the Americans, but French negotiations were at a delicate stage. No European state, Jefferson learned, was interested in removing America's trade barriers. Besides, if the Barbary pirates did not have defenseless Americans to seize, they might prey on some other country.

Finally, Lafayette was able to use his influence at court to get reliable information for Jefferson. Different governments gave different gifts—gold, cloth, jewelry, carriages, food, the weapons and lumber that Christian captives were forced to use to build and arm the corsairs. It appeared that the $300,000 original estimate for an American treaty bribe would be far too low. To make matters worse, Congress had only allocated $80,000 while pressing Jefferson and his agents in Madrid and Algiers to negotiate peace "immediately, earnestly and vigorously."[35] Congress was sending out John Lamb of Connecticut, a new agent to Algiers, to assist Jefferson. Lamb, an artillerist who had marched to Quebec with Benedict Arnold in 1775 and had an eye shot away in the assault on the walled city, was tough and honest, but it took him six months to get to Paris. Jefferson did not wait for him.

Through the good offices of the Spanish ambassador, Jefferson was able to arrange the release of the crew of the *Betsy*. He quickly sent Thomas Barclay to take advantage of the good mood of the emperor of Morocco and purchase peace. Barclay carried modest gifts: a pair of gold inlaid pistols, an enameled snuffbox, a silk umbrella, perfume, a sword, phosphorous matches, a clock. In 1786, Barclay was able to negotiate with Morocco the most advanced treaty of any Christian nation with a Barbary state. The treaty cost $30,000, with no annual tribute and no ransom. American ships and cargoes were to receive the treatment of a most favored nation. Lamb, no diplomat, fared less well when he finally arrived, and Jefferson dispatched him to Algiers. Algeria, hunting Americans east of Gibraltar, had just captured two ships and was demanding ransom of $1,200 for each captive, more than Jefferson had left to pay if he wanted to. A peace treaty with Algeria was not available at any price. Lamb advised Jefferson it would "cost a tower of Constantinople."[36] Lamb failed to win either a treaty or the captives' release.

Jefferson saw the Barbary crisis as one more reason why the American Confederation must have a stronger central government. To build a navy would "arm the federal head with the safest of all instruments of coercion over their delinquent members."[37] American diplomacy would only be as effective as American power allowed it to be. To Monroe in Congress, he wrote emphatically "the necessity that the United States should have some marine force":

> It will be said that there is no money in the Treasury. There never will be money in the Treasury till the Confederacy shows its teeth. The [Barbary] states must see the rod; perhaps it must be felt by some of them. . . . Every national citizen must wish to see an effective instrument of coercion, and should fear to see it on any other element but water. A naval force can never endanger our liberties nor occasion bloodshed. A land force can do both.[38]

But Congress had already made up its collective mind, as Adams pessimistically predicted. "Your plan of fighting will no more be adopted than mine of negotiating."[39] Because of Adams's objections, a plan calling for an antipiracy naval confederacy was never forwarded to Congress. Not until Jefferson became president fifteen years later was the Barbary crisis solved. And then it was by an American navy, as Jefferson, made over into a nationalist by European diplomacy, had wanted in the first place. Meanwhile, with the signing of the Moroccan treaty, Jefferson had achieved another diplomatic coup.

Thomas Jefferson the diplomat, as the historian Bernard Bailyn has pointed out, was "unconventional"; he was also "imaginative, resourceful and tough as the best or worst of the Old World politicians and more adroit than most."[40] In his autobiography, Jefferson accurately if modestly summed up "my duties at Paris" in the order of their importance. His mission was to advance American political and economic independence and prosperity, forging diplomatic ties that assured the "receipt of our whale-oils, salted fish and salted meats on favorable terms, the admission of our rice on equal terms with that of the Piedmont [region of Italy], Egypt and the Levant, a mitigation of the monopolies of our tobacco by the Farmers-General, and a free admission of our productions into their islands [in the Caribbean]."[41] But he went far beyond this already ambitious agenda. Jefferson successfully concluded a consular convention with several European powers. After he researched and wrote a detailed study on whale fishing, he was able to open up French ports to the import of whale oil, vital to the New England economy. Deftly using his social connections, especially with the influential Lafayette, he was able to get the French Farmers-General, who paid the Crown for the exclusive franchise to collect taxes, to revise its monopoly over the American tobacco trade, which insisted that all American tobacco pass through the warehouses and the books of Philadelphian Robert Morris, and which had created, in the eyes of southern planters like himself, a double monopoly.

After long, involved negotiations, Jefferson was able to have the French exempt some fifteen thousand hogsheads (fifteen million pounds) of tobacco from Morris's monopoly, appeasing the south. By skillful personal negotiations in the Netherlands with Dutch bankers for timely loan extensions, he was able to help keep the international credit of his young nation not only sound, but, on most days, second only to England's and, on the day when there was an attempt to assassinate George III, number one in all the world on the Amsterdam bond market. Most of all, he worked persistently to improve the diplomatic status of the United States and foreign opinion of America, which he considered crucial to foreign investment. "Nothing can equal the dearth of American intelligence in which we live here," he wrote. "I had formed no conception of it. We might as well be in the moon."[42] Sometimes he took almost perverse pleasure in defending America and in forcing Europeans to revise upward their estimations of America and Americans.

He was quick to defend America against perceived slights or slurs. When Abbé Reynal stated that America had never produced "one good poet, one able mathematician, one man of genius in a single art or a single science," Jefferson wrote that America had produced a Washington, a Franklin, and the mathematician Rittenhouse:

> When we shall have existed as a people as long as the Greeks did before they produced a Homer, the Roman a Virgil, the French a Racine and Voltaire, the English a Shakespeare and Milton; should this reproach be still true, we will enquire from what unfriendly causes it has proceeded . . .

"Though but a child of yesterday," America "has already given hopeful proofs of genius."[43]

Between "squalls of work," Jefferson enjoyed considerable free time. His first winter in Paris he had rebuilt his strength by taking four- and five-mile walks around the city, renting sedan chairs in this garden or that to break up his strolls. He soon found his way to the eastern edge of the city, to the Jardins du Roi, the magnificent royal gardens. Created by Louis XIII in 1635 as medicinal herb gardens near the royal medical school, they had recently been extended all the way to the Seine by Buffon, the famous intendant of the King's Gardens. Jefferson's frequent visits enabled him to study in detail the most elaborate gardens in France and greatly whetted his appetite for botanical gardens at Monticello. His strolls helped form his ideas of gardening, even if he would reject much of what he saw as he constantly changed and expanded the gardens at Monticello. On these walks, Jefferson was inventing the American notion of a public garden.

Buffon had run dead-straight broad linden-lined *allées* where visitors could amble and converse. In the flanking *parterres,* he had planted specimens of exotic flowers, herbs, and trees brought from scores of foreign countries. At the southern end of the gardens, a man-made mountain was covered with lawns, something Jefferson had never seen before and which he would introduce to America at Monticello. From a slender, open gazebolike cage at its summit, he could gaze down at labyrinths that stretched north as far as he could see. Entering one of these labyrinths, he found a pistachio tree brought by French explorers from the Middle East. Its seeds had produced thousands of pistachio trees now growing in sunny Provence to the south. Jefferson would study the pistachio carefully on a tour a year later, recommend its cultivation to friends in South Carolina, and pay a gardener in Marseilles to send specimens to friends in Charleston.

Jefferson and Chastellux's first afternoon promenade together in the royal gardens had been timed for a visit to the sprawling residence of the famous Buffon, whom Jefferson, despite his lingering anger at the illustrious Frenchman's ignorant comments on America, acknowledged knew more about botany than anyone else alive. After turning in their cards to Buffon's steward and being invited by him to return in time for dinner, they had crossed present-day rue de Buffon and entered the gardens, passing the false acacia that was the oldest tree in the gardens, planted in 1635. They were careful not to disturb Buffon, who was walking and meditating nearby. There was much Jefferson wanted to say to Buffon, and Chastellux had brought him to introduce the two and to allow Jefferson to present a copy of *Notes on Virginia,* which was to a great extent the result of Buffon's odd theories about the inferiority of species in America. In *Notes,* Jefferson had constructed elaborate tables showing, among other things, that there were only "18 quadrupeds peculiar to Europe," while he had counted seventy-four in America. "Of 26 quadrupeds common to both countries, 7 are said to be larger in America, 7 of equal size."[44] The rest he had not personally examined. Jefferson had already sent Buffon an example: an enormous panther hide he had purchased in Boston and brought along to France. He was presently engaged at his own expense in a complicated enterprise that he had set in motion when he had gone out of his way to visit New Hampshire en route to Boston and the ship to Europe. He was determined to show Buffon how large American wildlife could be, and he had arranged with the governor of New Hampshire to hire the best hunters to go deep into the Green Mountains of Vermont, to kill and drag out an immense moose to stuff and send to Buffon, whether he wanted it or not.

There was little Jefferson could do to disprove another set of Buffon's unresearched assertions, that American Indians "lack ardor for females," "love their parents and children but little," have "small organs of generation" and "little sexual capacity."[45] Jefferson defended all Americans, including Indians, against Buffon's slights. He had known Indians personally and admired them. In his book Jefferson not only countered with the noble speech of Logan, but replied to Buffon's criticisms point for point. So much of what Jefferson wrote about Indians had echoes of himself in it. The Indian "is neither more defective in ardor, nor more impotent with his female, than the white reduced to the same diet and exercise. . . . He is brave when an enterprise depends on bravery . . . the point of honor [consisting] in the destruction of an enemy by stratagem and in the preservation of his own person free from injury. His sensibility is keen, even the warriors weeping most bitterly on the loss of their children."[46]

When Jefferson was ushered into the presence of the great Buffon, he later recalled, "I was introduced to him as Mr. Jefferson who, in some notes on Virginia, had combatted some of his opinions. Instead of entering into an argument, he took down his last work, presented it to me, and said, 'When Mr. Jefferson shall have read this, he will be perfectly satisfied that I am right.' " Unfazed, Jefferson told Buffon that he was already familiar with the volume, that Buffon had confused the red American elk with the red deer of Europe and the moose with the reindeer. "I told him that our deer had horns two feet long. He replied, with warmth, that if I could produce a single specimen, with horns one foot long, he would give up the question. I told him the reindeer could walk under the belly of our moose; but he entirely scouted the idea." Jefferson became obsessed with disproving Buffon. He continued to press Governor John Sullivan of New Hampshire to send him the horns, skin, and skeleton of a moose. Sullivan, a former Revolutionary War general, dispatched an expedition to Vermont. "The troops he employed sallied forth in the month of March" amid "much snow," attacked a herd deep in the Vermont wilderness, killing one immense moose, then cut a twenty-mile road through the forest and back to the nearest house."[47] Sullivan's bill for a staggering forty-five pounds arrived before the remains of the moose, which, after all that, had disappointingly small horns, for which a red-faced Jefferson apologized to Buffon.

He returned to the naturalist's house many times to sit through the great man's long soliloquies. (He persisted for years, until the dying Buffon finally promised to delete his theory of the diminutive American species from the next edition of his multivolume work.) Each time he visited, Jefferson could study the royal gardens, whet his appetite for gardens of his own. He found most of them too symmetrical, too cultivated and stylized for his own taste, but, in a corner, he found something closer to his preference, the controlled wildness of an English garden.

In March 1786, John Adams urgently sent for Jefferson to come to London to join him in secret negotiations with Tripoli. Adams had already met three times with a Libyan diplomat in London. Adams sent his secretary and future son-in-law, Col. Samuel Smith, to Paris early in March to explain the situation:

> There is nothing to be done in Europe of half the importance of this, and I dare not communicate to Congress what has passed without your concurrence. What has already been done and expended will be absolutely thrown away and we shall be involved in a universal and horrible war with these Barbary States which will continue for many years unless more is done immediately. I am so impressed and distressed with this affair that I will go to [Congress at] New York or to Algiers or first to one and then to the other if you think it necessary.[48]

So sensitive was Jefferson's summons to London that Adams had put out the story, seconded by Jefferson, that Jefferson was going there to help conclude a commercial treaty with Portugal before their special two-year treaty-making commission from Congress expired on May 12, 1786. Jefferson even wrote to John Jay, American Secretary of Foreign Affairs in Congress, that he was in London to help Adams with Portugal. But Jefferson had his own motive for going along with Adams's request. He hoped to force the British by his presence to accept speedily a new trade treaty with the United States that would permit Americans to trade in the British West Indies. These islands, the heart of the American carrying trade before the Revolution, had been sealed off to U.S. citizens since the end of the war. He told Jay that he and Adams would "avail ourselves of my journey here as if made on purpose just before the expiration of our commission."[49] Jefferson's seven-week stay in the homeland of his ancestors was his longest visit there. It included thirty-five diplomatic meetings and solid weeks of shopping and sightseeing.

Jefferson was ill-prepared for the cold of March in London. Paris was cool enough for him: even in July he kept the fireplaces burning at the Hôtel de Langéac, this southerner suffered so when he was out of the hot sun. But London greeted him with, he noted, "briskish wind," there was still "drifted snow"[50] in the parks, ice under the carriage wheels and underfoot. Jefferson's mood was already chilly. He had not gotten over his severe financial loss and personal humiliation during the British invasions of Virginia, and living among the anti-British French had only sharpened his criticisms, major and minor. To Abigail Adams, he had written in September 1785, "I fancy it must be the quantity of animal food eaten by the English which renders their character insusceptible of civilization. I suspect it is in their kitchens and not their churches that their reformation must be worked."[51] For an eight-week period now, he must forgo James Hemings's French cooking and think how accurate his foretaste had been.

The diplomatic side of his visit, at least, was probably doomed before he arrived. In December 1785, a desperate Adams had sent the Marquis of Carmarthen, the English Secretary of State for Foreign Affairs, a memorandum requesting the British surrender their forts in the Old Northwest in compliance with the peace treaty of 1783. But the British were using their unsettling military presence inside U.S. territory as leverage to coerce the Americans to honor prewar debts to English merchants and compel the restoration of confiscated Loyalist property, as agreed in the 1783 treaty. Two days after Jefferson arrived, on March 11, Adams requested a meeting with Lord Carmarthan, which took place on Wednesday, March 15. In the few days before the meeting, Adams and Jefferson extensively revised Adams's proposed treaty. Jefferson suggested "an exchange of citizenship for our citizens, our ships and our production generally."[52] Englishmen would have American citizenship rights on U.S. soil and vice versa. Each country would grant most-favored-nation status for the other's trade. In his earlier draft, Adams, at the instruction of Congress, had added half a dozen political demands to his request for a trade treaty, including British cession of the northwest posts, financial settlement for slaves and other property carried away by the British at the end of the war, and a proposition that Britain join with the United States to induce Spain to open the Mississippi River to free navigation.

None of these issues was germane to a commercial treaty, as Lord Carmarthan acidly pointed out at their meeting at Whitehall on March 15. With such a jumbling of political and commercial requests, Adams and Jefferson could hardly expect their negotiations to go smoothly. According to Jefferson, their first conference with the Secretary of State was a waste of time: "The vagueness and evasions of his answers to us confirmed me in the belief of their aversion to have anything to do with us."[53] In their report of that day's talks to John Jay, in New York, the two envoys added, however, that "His Lordship, after harping a little on the old string, the insufficiency of the powers of Congress to [negotiate] and to compel compliance with treaties, said he would lay the matter before the Ministry and the King."[54] Two days after their first interview with Carmarthan, Adams took Jefferson for the requisite visit to the Court of St. James. As an accredited diplomat, Jefferson was required to pay his respects at one of the king's twice-weekly levees at the palace, and it was Adams's duty to present him. It was here, on March 17, 1786, according to Jefferson, that he received a rude shock.

From the resumption of diplomatic relations and the presentation of Adams to George III in July 1785, the king had maintained his usual civility. It was even rumored in gossip-ridden diplomatic circles that, after Adams had presented his credentials, there had been a tear in the king's eye and a lump in his throat as he made Adams welcome. Adams evidently believed relations were smooth enough to present at court the man who had written the Declaration of Independence with its catalogue of the crimes of the king. But, according to Adams family history, the king took Adams's presentation of Jefferson as a particular piece of arrogance. Whatever actually happened that day, Jefferson never got over it. In his autobiography, he reported thirty-five years later,

> On my presentation, as usual, to the King and Queen, at their levees, it was impossible for anything to be more ungracious, than their notice of Mr. Adams and myself. I saw at once that the ulcerations in the narrow mind of that mulish being left nothing to be expected on the subject of my attendance.[55]

The Adams version of what happened was passed down by the envoy's grandson, historian Charles Francis Adams, who related that, when John Adams brought forward Jefferson to present him and the two bowed, suddenly Adams and Jefferson were looking at the king's and queen's backs. In other words, George III and Queen Charlotte stood up, turned around, and deliberately snubbed the American ambassadors in front of the assembled diplomatic corps. But if this were true, certainly someone, especially the hyperpunctilious John Adams, would have recorded the event. But no one did until Jefferson wrote about it in his memoirs at age seventy-seven.

There are some problems with both versions. What is certain is that the queen never attended the king's levees, where Adams and Jefferson presented themselves and where, according to Jefferson's meticulous expense records, he distributed the customary tips that day. His records show no such expenditures at either of Queen Charlotte's drawing-room receptions that week. Moreover, the customary drill for the king's levees would have made it physically impossible for the two envoys to come forward and bow or the king to rise and turn his back.

Arriving at St. James Palace between 11 A.M. and noon on March 17, Adams and Jefferson mingled with the other diplomats and visiting dignitaries, distributing the usual tips to doormen as they gradually made their way to the crimson and gold presence chamber dominated by the chair of state at the far end. At noon, the king, in his special "levee clothes" flanked by two members of his household, entered. All conversation ceased. The gentlemen in attendance arranged themselves in a large circle around the walls of the chamber. The king began his walkabout, always speaking to the gentleman on his right, who, after a brief exchange, was free to depart. As one British historian has recently pointed out, "public rudeness played very little part in George III's concept" of how a king should comport himself: "his affability, courtesy, ability to put any man at his ease, unending store of small talk, pleasantries and jokes were celebrated."[56] But Jefferson may have been put off, after Versailles, by the unexpected informality of the scene and the king just may have had very little to say to this man who had addressed him last in a declaration of his tyranny to all his subjects. Jefferson and Adams may very quickly have seen the king's back as he hurried on to the next man on his right.

Whatever did happen that day, Jefferson was still sputtering about the English a year later, when he wrote Colonel Smith, "Of all nations on earth, they require to be treated with the most *hauteur.* They require to be kicked into common good manners."[57]

Nothing went right with the Adams-Jefferson negotiations. The treaty of commerce and amity with the Chevalier de Pinto, the Portuguese ambassador, was never ratified in Lisbon. The talks with Tripoli's ambassador also aborted. Representatives of British merchants who were still owed prewar debts also balked over wartime interest they still were demanding and over American efforts to link these debts with a new trade treaty.

Jefferson's interest in the debts of Americans was personal as well as diplomatic, embarrassingly so. His losses in the war and severe inflation had made it increasingly difficult for him to pay off the debts he had inherited with his father-in-law's estate. He tried to pay off all of the debt soon after John Wayles's death by selling lands, for £4,200, but he had taken notes for the purchase prices, which had been paid off by his creditors in almost-worthless Virginia wartime currency. "I did not receive the money till it was not worth oak leaves,"[58] he explained to one creditor, Alexander McCaul. "Besides this, Lord Cornwallis's army took off thirty of my slaves, burnt one year's crop of tobacco in my [barns] and destroyed another in the fields with other damages to the amount of £3,000 or £4,000."[59] Jefferson had recently been informed by his brother-in-law Eppes that there was not enough income available for him to pay off his British debts and that British merchants were now citing the Treaty of Paris to demand payment in gold or sterling, ignoring the payments in full in Virginia Land Office currency he had made during the war. Unless Jefferson paid up, his credit would be ruined. He had tried to pay off his debts from his private income while living in France on his salary, but his payments by Congress were inadequate or tardy or both. He paid off one minor debt in London, making two separate payments to the firm of Carey, Morey and Welch totaling £168.13.1. He also met with a representative of his major creditor, who came up from Bristol, explaining to him that he had intended to pay off all his debt "in the year 1776 before there was a shilling of paper money issued,"[60] but because friends and relatives had paid him in worthless currency, he now found himself deeply in debt and unable to pay. Years of financial worries were beginning.

There had been a monetary glimmer of hope for a treaty that would allow Americans back into the British West Indies. Lord Carmarthen, probably at the same March 17 royal levee, mumbled something to Adams "in his own name and that of Mr. Pitt" that, since Adams's and Jefferson's earlier proposals had contained political conditions, they could not be accepted but, if "we should prepare a project of a treaty merely commercial," there was some hope of an accord. Adams and Jefferson called at the Foreign Office the very next day to seek details from Lord Carmarthen, but the foreign secretary was not available and an undersecretary received them. "Until we know which articles were objected to," Jefferson and Adams wrote Jay, "it would be vain for us to attempt a new draft."[61] Their attempt to elicit a counterproposal from Carmarthen failed utterly. On April 3, the undersecretary sent word that "his Lordship would wish to receive from Mr. Adams and Mr. Jefferson the project of a treaty of commerce containing only such points as are necessary for that purpose."[62] Two weeks had passed. The Americans' commission was running out. The April 3 message only put in writing what Carmarthen had said March 17. Still, returning from a tour of country houses, Adams and Jefferson overnight rewrote the treaty they had already submitted, inserting Jefferson's pet proposal for reciprocal freedom for American citizens and British subjects to trade in each nation's territories and enjoy the same rights and privileges in both and deleting the objectionable political conditions. If the treaty had been accepted, it would have marked a victory for the Americans by allowing their reentry into the lucrative West Indies trade. But Lord Carmarthen never responded.

By the time the negotiations were over, Jefferson, who had been living in a crowded boardinghouse for two months, was convinced that "that nation hates us, their ministers hate us, and their king more than all other men."[63] Whatever rosy pictures Adams had been sending back, Jefferson now insisted on darkening an extremely candid, jointly written blast to John Jay that made known to Congress for the first time that Adams and Jefferson were pulling apart on foreign policy:

> With this country, nothing is done, and that nothing is intended to be done on their part admits not the smallest doubt. The nation is against any change of measures. The ministers are against it, some from principle, others from subserviency, and the king more than all men is against it.
>
> If we take a retrospect to the beginning of the present reign, we observe that, amidst all the changes of ministry, no change of measures with respect to America ever took place, excepting only at the moment of peace, and the minister of that moment was immediately removed. Judging the future by the past, I do not expect a change of disposition during the present reign, which bids fair to be a long one, as the king is healthy and temperate. That he is persevering, we know. If he ever changes his plan, it will be in consequence of events which neither himself nor his ministers at present place among those which are probable. Even the opposition dares not open their lips in favor of a connection with us, so unpopular would be the topic.
>
> It is not that they think our commerce unimportant to them. I

> find that the merchants here set sufficient value on it. But they are sure of keeping it on their own terms. No better proof can be shown of the security in which the ministers think themselves . . . than that they have not thought it worthwhile to give us a conference on the subject, though on my arrival we exhibited to them our commission, observed to them that it would expire on the 12th of the next month, and that I had come over on purpose to see if any arrangements could be made before that time. Of the two months which then remained, six weeks have elapsed without one scrip of a pen or one word from a minister.[64]

To William Temple Franklin in Paris, he was even less diplomatic. "I found the king, ministers and nation of England hostile and averse to all arrangements. They think their commerce indispensable to us. They think we cannot unite to retaliate . . . I hope we can."[65]

Jefferson was determined not to waste any more of his time. To William Carmichael, American envoy to Spain, he wrote, "I consider the British our natural enemies and as the only nation on earth who wish us ill from the bottom of their soul."[66]

Between fits and false starts of negotiations, he went shopping for new ideas. The Adamses had welcomed Jefferson to visits at their home on Grosvenor Square as if he were a long-lost relative, just so he went back to his rented rooms on Golden Square at night. John squired him around London from the Tower to the British Museum to Drury Lane and Covent Garden, where Jefferson attended six plays, two by Shakespeare—*The Merchant of Venice* and *Macbeth*—and a new opera, Salieri's *School of Jealously*. Abigail invited many of the pro-Americans in London to dinner parties. Jefferson had to concede to Abigail that London was cleaner, roomier, handsomer than Paris, but he thought its architecture abominable, "the most wretched style I ever saw."[67] Abigail was full of advice on where to shop and what to see. Together, Jefferson and the Adamses went to theaters and to see the latest wonders of science. With Adams he went to see the great new Albion Mill at Blackfriars on the Thames, where Boulton and Watt's steam engines were launching the Industrial Revolution. They were barred admittance, however, due to England's laws protecting industrial secrets. He commissioned an invention of his own, the crafting of his first copying machine at the Cavallo works: now he could make simultaneous copies of his enormous correspondence.

They visited Broadwood's, London's leading maker of harpsichords, where Jefferson unsuccessfully plugged a technical innovation developed by his friend in Philadelphia, Francis Hopkinson. Undaunted, Jefferson ordered a harpsichord for Patsy to play on Sundays at the Hôtel de Langéac. Patsy was persevering in her music lessons as well as her French, and he wanted to surprise her. They also went to order a new carriage and leather whips, bridles, saddles, traveling trunks, boots, and shoes to ship to Paris.

In a letter to Madame de Corny, a friend in Paris, Jefferson wrote, "The splendor of their shops is all that is worth looking at."[68] But to Charles Thomson in Congress, he said, "I could write volumes on the improvements which I find made and making here in the [industrial] arts."[69] He met with Sir John Sinclair, president of the Board of Agriculture and an enlightened Scottish liberal, and learned from him the latest developments in scientific farming: both men also must have talked about their favorite poems of Ossian. He studied steam-powered innovations and visited factories, predicting that factory workers would one day rock governments. Everywhere he went, he studied the conditions of workers. French laborers, he found, were worse off than English. "They pay about one half their produce in rent; the English, in general, about a third," but English workers lived under "abject oppression." His assessment of the class structure in England was succinct and scathing. He was still horrified by it nearly thirty years later, during the second war with England, when he wrote a disciple,

> The population of England is composed of three descriptions of persons . . . 1.) The aristocracy, comprehending the nobility, the wealthy commoners, the high grades of priesthood and the officers of government. 2.) The laboring class. 3.) The eleemosynary class, or paupers, who are about one-fifth the whole. The aristocracy, which have the laws and government in their hands, have so managed them as to reduce the third description below the means of supporting life, even by labor. . . . To force the second, whether employed in agriculture or [industry], to the maximum of labor which the construction of the human body can endure and to the minimum of food, and of the meanest kind which will preserve life . . . in strength sufficient to perform its functions. To obtain food enough, not only their whole strength must be unremittingly exerted, but the utmost dexterity. . . . Those of great dexterity only can keep their ground while those of less must sink into the class of paupers. . . . The less dexterous individuals, falling into the eleemosynary ranks,

> furnish materials for armies and navies to defend the domination and vicious happiness of the aristocracy. . . . Such is the happiness of scientific England."[70]

In the brawling, class-ridden England he found, he foresaw many of the changes that were to be wrought by scientific inventions during the Industrial Revolution. He predicted that steam would be used to propel boats. He met with Samuel Romilly, the law reformer, and he began to buy scientific instruments as if they were books: a solar microscope, a globe, an expensive telescope, a protractor, a thermometer, a camp theodolite, an "air pump and apparatus,"[71] for twelve guineas, a botanical microscope, a box of tools. He ordered a fascinating portable copying press that could make two hundred copies of an invitation in forty-five minutes. He bought several upright gas lights. He had a new suit made and sat for his first portrait. The artist, a Boston Loyalist, was Mather Brown. He shopped for relatives and friends and took orders from friends in England for shopping when he got back to Paris. He met a Greek-born scholar named John Paradise who would pester him for years for money and offer in return to teach him modern Greek by mail, and he met his beautiful, Virginia-born wife, the unstable and extremely beautiful Lucy Ludwell, who flirted and flattered him as "the first person in North America."[72]

There were, between nerve-racking days in diplomatic anterooms, happy hours with the Adamses, and he was after all, in England in April: while they waited for an answer from Carmarthan, Adams and Jefferson decided to make a garden tour of southern England. Fortunately Jefferson had come prepared for just such an event, one he had dreamed of for nearly a quarter century. He had brought with him his precious copy of Thomas Whateley's *Observations on Modern Gardening*, which he had bought fifteen years earlier when he started to lay out Monticello. He turned to its thumbworn pages as he set off alone in a postchaise for the twenty-three-mile drive to Windsor Castle. "I always walked over the gardens with his book in my hand, examined with attention the particular spots he described,"[73] Jefferson wrote later in a memorandum.

Jefferson's first garden visit was marred somewhat by the fact that there was still snow in the Great Park at Windsor and by the arrival of the king and queen, who were followed by a throng of visitors. He returned to London to fetch Adams, and together they set out on April 2 "into the country to some of the most celebrated gardens."[74] What they saw stunned these two provincial Americans and strongly influenced Thomas Jefferson's years of working to transform his private mountaintop at Monticello into one of America's premier gardens.

Adams and Jefferson's tour came at a time when country house visiting was popular in England and when tourists were, as garden scholar John Murray puts it, "being bombarded with new theories of aesthetics and standards for appreciating architecture and landscape."[75] Gardens could arouse ideas in the mind of the beholder, as Addison had pointed out in his *Pleasures of the Imagination* (1712), which Jefferson certainly had read. How to approach the often-thematic garden was important, and guidebooks such as Whately's were recommended to direct the visitor on his first experience of a garden. The garden was manipulated to produce a reaction and Jefferson could see it as part of the Enlightenment, a way of presenting improved ideas on agriculture with the help of a careful display of the arts. Jefferson had long read both classical and contemporary poems in praise of gardens; he had studied the new language of aesthetics in Burke's *Philosophical Enquiry Into the Origin of Our Ideas of the Sublime and the Beautiful* (1757), published shortly before he went to study at William and Mary. Whately's *Observations* (1770) had been the first gardening book to describe individual gardens in detail. Jefferson followed it scrupulously, later pronouncing it completely accurate. There was a new word Adams and Jefferson encountered as they made their tour: picturesque. Tourists were supposed to be viewing the country with the eyes of persons accustomed to drawing and to decide whether a garden could be formed into pictures.

Dividing their tour into two segments, gardens south and north of London, they first headed west along the Thames in a postchaise. Except for their valets, Petit and John, Jefferson, Adams, and Col. Samuel Smith, his secretary, were prototypical American tourists, visiting six country estates the first day. They started their tour with the great domed villa of Chiswick, a Palladian Revival house with eight communicating rooms for entertainment under a massive rotunda. "The octagonal dome has an ill effect both within and without," Jefferson wrote, immediately differing from the popular London guidebook that hailed its "elegance of taste" that supposedly surpassed "everything of its kind in England."[76] Despite Jefferson's initial reaction, the domed rotunda on a country house he first saw that day at Chiswick made a powerful impression and was echoed closely in his final design for the façade of Monticello. His mountaintop home on the American frontier also may have benefited from Chiswick's superb gardens, which were adorned with statuary in a contrived wilderness featuring a Palladian wooden bridge. Alexander Pope had once called Chiswick "the finest thing this glorious sun has shined on"; but Jefferson was not taken by the overall attempt to evoke classical antiquity. He not only found the large rotunda on the small center of the house overpowering but he declared one garden obelisk "of very ill effect," a second, in the middle of the pond, "useless."[77]

The American envoys hurried on to Pope's garden at Twickenham, where the poet had mused over his influential theories of gardening seventy-five years earlier and had put them into effect in the pleasant village of Middlesex. Jefferson made notes on the size of the garden, the location of the house, Pope's famous grotto, mound, and rookery. He noted especially the obelisk at the bottom of the garden with its inscription to Pope's mother.

Then they were off to massive red-brick Hampton Court, built by Cardinal Wolsey in 1515 and modernized by Sir Christopher Wren in the seventeenth century, who was under William III's order to remodel it after Versailles. Jefferson gave short shrift to Hampton Court's gardens: they were not in the latest naturalistic style, which he so admired. He noted simply that he had paid four shillings, sixpence to see them and had found them "oldfashioned—clipt yews grown wild."[78]

Four miles to the south, "in a bottom near the river [Mole]," Jefferson responded more enthusiastically to Esher Place, another of Wolsey's creations, its distinctive gardens on "heights rising one beyond and above another, with clumps of trees." He admired the templelike "fine summer house built upon a hill on the left as you enter, which commands the view of the house, park, and country road on both sides of the Thames for many miles." Did he have a pang for the South Pavilion at Monticello, where he and Patty had spent their honeymoon and watched their home being built? He commented only on the gardens around him. The clumps of trees "on each hand balance finely—a most lovely mixture of concave and convex."[79]

Virtually next door to Esher Place, they next rolled through the grounds of Claremont, with its semiformal gardens, built by the architect John Vanbrugh for himself and then sold to the powerful duke of Newcastle, whose policy of "salutary neglect" in ruling the American colonies had kept the peace for so many years. On the hill which gave the estate its name, he built a belvedere, painted white, to look down on one side over a bowling green and, on the other, out over a lake and natural-looking amphitheater. He also had built an enormous kitchen garden hidden behind massive fortified walls, and installed serpentine paths to wind from the woods up to his belvedere. From its tower, Jefferson and Adams could get a fine view of the Thames and surrounding villas. But Jefferson was not impressed. He gave Claremont two words: "Nothing remarkable."[80]

Then they rattled on to Painshill, which architectural historian Marie Kimball thought would have had "almost everything that would appeal to

Jefferson, from the manner of its planting to an ingenious water wheel." While the Whately guide devoted fully ten pages to its "boldness of design and happiness of execution," Jefferson considered the house "ill-situated" and its architecture "incorrect." But he did find a Doric temple in the garden "beautiful."[81]

Spending the first night at Weybridge, a village in Surrey four miles southwest of Hampton Court, the travelers set out early the next morning to explore three great country seats clustered nearby. Jefferson was unimpressed by Oatlands, the home of the earl of Lincoln, which his guidebook said possessed majestic grandeur. He had no comment on the earl of Portmore's Ham Farm with its fine command of the junction of the Wye and the Thames. But he gave all his attention to Woburn Farm, the country seat of Lord Peters, eulogized by Whately as the kind of simple farm fit for the model for Poplar Forest in Bedford County, Virginia, the secluded retreat where he had gone when Patty died and where he would flee with increasing frequency as his political career flourished. There might have been traces, too, of Woburn Farm at Monticello, where ex-President Jefferson would install a waving walkway—bordered by small groups of shrubs, firs, small trees, and flower beds—in his western garden along a hilltop. With Whately's copious descriptions to refresh his memory, Jefferson had to jot down only the practical details about how many people it would take to install and maintain such gardens at Monticello. "Four people to the farm, four to the pleasure garden, four to the kitchen garden," he wrote. "All are intermixed; the pleasure garden being merely a highly ornamented walk through and round the divisions of the farm and kitchen garden."[82]

After returning to London for one night to rewrite the trade treaty, Jefferson evidently prevailed on Adams to return to Woburn Farm the next day. Then they went on to Caversham, which Adams pronounced "beautiful" but which Jefferson criticized in detail. Despite its "fine prospect," he found its broad, dead-straight gravel walkway "has an ill effect." Yet he liked "the lawn in front, which is pasture, well disposed with clumps of trees."[83] Spending the night at the ancient market town of Reading on the Thames, the tourists went on the next day, April 5, visiting Wotton, the marquis of Buckingham's "great and elegant though neglected" estate, as Adams phrased it. So carefully did Jefferson follow Whately's guidance through the grounds at Wotton that he could make a note, "There is a Palladian bridge, of which, I think, Whately does not speak."[84] He also studied the extensive use of water: seventy-two acres of it, Jefferson calculated. At every step, Jefferson, while he makes no comment on it, must have been plying gardeners with questions.

On April 6, Jefferson's party visited the extremes of their two-week tour, the palatial estate of Lord Cobham at Stowe and the rundown birthplace of Shakespeare at Stratford-upon-Avon. At Stowe, they followed a road two miles long into the grounds from the village of Buckingham to "a large Corinthian arch or gateway, 60 feet high and 60 feet wide, decorated on each side with a large military column." From the gateway, they could see the garden front of His Lordship's house on the summit of a hill surrounded by gardens and parks. Jefferson found "the straight approach" through wooded hills, valleys, lawns, over a river, and past an assortment of buildings "very ill," thought that the Corinthian arch "has a very useless appearance inasmuch as it has no pretension to any destination . . . Instead of being an object from the house, it is an obstacle to a very pleasing distant prospect."[85]

He made no comment on the twenty-three architectural fragments scattered throughout the park, including circular Temples of Ancient and Modern Virtue, the modern version deliberately ruined by the whimsical Lord Cobham, who had been the "boy patriot" leader of the opposition in Parliament in the mid-eighteenth century. At a time when architecture was becoming a profession by doing the bidding of competing country gentlemen, a succession of landscape architects had expensively remodeled Stowe periodically for nearly a century. Vanbrugh, Richard Kent, and Lancelot Brown, better known as "Capability" because of his answer whenever a rich landowner called him in ("This definitely has capabilities"), had all contributed to Stowe's opulence. The end result rivaled Versailles in scope, three avenues converging at a rotunda overlooking a grand canal. The Rotunda, centerpiece of the Garden of Venus, contained a gilded Venus de Milo. At the other end of the canal was a statue of Queen Caroline on fluted Ionic pillars surrounded by statues of dancing shepherds and shepherdesses. Jefferson probably agreed with a Swedish artist studying English gardens. He criticized the "unusual array of ornamental structures of a scale, size and variety that reveal a desire to win fame by outdoing every other place in respect of expense and numbers."[86] John Adams found all "the temples to Bacchus and Venus are quite unnecessary, as mankind have no need of artificial incitement to such amusements." Worse, he found, was the colossal expense: "A national debt of £274 millions accumulated by jobs, contracts, salaries and pensions in the course of a century might easily produce all this magnificence."[87]

His teeth obviously grating by this time, Jefferson kept track of every penny he laid out at their second stop that day, Shakespeare's birthplace: "for seeing house where Shakespeare was born, 1 shilling; seeing his tomb, 1 shilling; entertainment, 4 shillings, 2 pence; servants, 2 shillings."[88] Beyond that, he recorded nothing of his reactions to the shrine of his favorite playwright. Adams, on the other hand, found the house "as small and mean as you can conceive." The tourists "cut off a chip" of "an old chair in the chimney corner where he [Shakespeare] sat," which, Adams wrote, was "according to custom."[89]

It was to the Leasowes, the country house of William Shenstone, which Jefferson expected to be the highlight of their trip to the north, that Jefferson went to pay homage to the poet whose work was quoted on sculpture scattered through the "pastoral farm." For twenty years, Jefferson had worshipped the Shenstones' Leasowes from afar; now he found it rundown and expressed his disappointment at length. "The waters [were] small." This was "not even an ornamental farm—it is only a grazing farm with a path around it, here and there a seat of board, rarely anything better. Architecture has contributed nothing. The obelisk is of brick. Shenstone had but £300 a year and ruined himself by what he did to his farm."[90] But once again Jefferson found the walkway through the woods "pleasing."

As the American party headed back toward London, they visited Hagley Park, the magnificent Worcestershire estate that had contained the first Greek temple in an English garden. Disregarding the temple, Jefferson noted that "the ponds yield a great deal of trout." The next day, he found the famous palace of the Churchills at Blenheim less interesting than its gardens. He wrote down the acreage of the gardens, the water, the parks, the number of staff—two hundred. His reactions to the rest were mixed. "The water here is very beautiful, and very grand. The cascade from the lake is a fine one; except this, the garden has no great beauties. It is not laid out in fine lawns and woods, but the trees are scattered thinly over the ground, and every here and there small thickets of shrubs, in oval raised beds, cultivated, and flowers among the shrubs." Jefferson was to borrow this last idea for the gardens at Monticello. While he appreciated the broad graveled walkways, he found that "art appears too much."[91]

Jefferson and Adams made two more stops on their way back to London. Jefferson admired the use of water at Enfield Chase, country house of Lord Chatham, hero to all American revolutionaries. He especially admired Moor Park, a Palladian mansion in Hertfordshire, designed by a Venetian, Leoni. Jefferson considered the building "superb," everything he admired, but he felt Capability Brown's lawns, the broad walkway leading from house to rising terraces and then to a pool with Ionic temple "wants water."[92]

The two American ambassadors made a few stops that had nothing to do with garden architecture but everything to do with their years of struggling in the cause of American independence. At Edgehill and Worcester, sites of pivotal battles in the English civil wars, they stopped for brief visits. The battlefields "were curious and interesting to us," wrote Adams, "as scenes where freemen had fought for their rights. The people in the neighborhood appeared so ignorant and careless at Worcester that I was provoked, and asked, 'Why do Englishmen so soon forget the ground where liberty was fought for? Tell your neighbors and your children that this is holy ground, much holier than that on which your churches stand. All England should come in pilgrimage to this hill once a year.' "[93]

Adams was put off completely by what they had seen on their tour, the longest time the two men were at such close quarters. "The gentlemen's seats were the highest entertainment we met with," he complained. "Architecture, painting, statuary, poetry are all employed in the embellishment of these residence of greatness and luxury. It will be long, I hope, before riding parks, pleasure grounds, gardens and ornamental farms grow so much in fashion in America."[94] Adams must have been blind or Jefferson mute. Jefferson, ignoring the overwrought ornamentation around him, had been concentrating on the practical side of each country estate, his study the main reason for his making the tour at all. "My enquiries," he wrote to John Page in Virginia as soon as he returned to Paris, "were directed chiefly to such practical things as might enable me to estimate the expense of making and maintaining a garden in that style. The gardening in that country is the article in which its surpasses all the earth. I mean their pleasure gardening. This, indeed, went far beyond my ideas."[95]

Jefferson's rebuffs, real or imagined, in London, confirmed him in his American republicanism, especially his loathing of courtiers and royalty. In writing traveling instructions for the sons of two southern friends two years later, Jefferson, when he came to list items to be studied, sarcastically listed "courts" dead last on a list of eight headings led off by agriculture.

> Courts.—to be seen as you would see the Tower of London or menagerie of Versailles with their lions, tigers, hyenas and other beasts of prey, standing in the same relation to their fellows: a slight acquaintance with them will suffice to show you that, under the most imposing exterior, they are the weakest and worst part of mankind. Their manners, could you ape them, would not make you beloved in your own country, nor would they improve it, could you introduce them there to the exclusion of that honest simplicity now prevailing in America, and worthy of being cherished.[96]

Jefferson did not submit his final verdict on the English until 1816, when he wrote to Adams in a letter that must have reminded both men of their long-ago junket together: "Were the English people under a government which would treat us with justice and equity, I should myself feel with great strength the ties of origin, language, laws and manners which bind us together; and I am persuaded the two people would become in future, as it was with the ancient Greeks, among whom it was reproachful for Greek to be fighting against Greek in a foreign army."[97]

17

"My Head and My Heart"

I feel more fit for death than life. But when I look back on the pleasures of which it is the consequence, I am conscious they were worth the price I am paying.

—Thomas Jefferson to Maria Cosway,
October 12, 1786

Thomas Jefferson returned to Paris in mid-1786 to find himself more and more a celebrity, with *entrée* to the most fashionable salons because of three almost-simultaneous publications of his work: his pamphlet on the recently enacted Virginia Statute of Religious Freedom, Morellet's edition of Jefferson's *Notes on Virginia,* and Chastellux's *Travels in North America,* which lionized Jefferson as an American who was "at once a musician, skilled in drawing, a geometrician, an astronomer, a natural philosopher, legislator and statesman."[1] Chastellux's published praise for Jefferson came at a time when he was, in fact, immersing himself in the arts. He had just been asked by Madison, Wythe, and John Page to help plan the new state capital at Richmond. In England, he had seen Greek temples in gardens: now he urged a temple as the new State House. It was Jefferson who, after sounding out Parisian friends knowledgeable in the arts, had decided to retain the services of Jean Antoine Houdon to travel to America (he sailed with Franklin to make a life-size statue of George Washington to grace the new state capital). Houdon made a terra-cotta bust of Washington at Mount Vernon and a life mask that he carried back to his

Paris atelier, where Jefferson often broke his long strolls to visit as the work progressed.

Jefferson the diplomat intervened in a raging controversy in the Paris art world when he decided to ask Washington himself in what "particular dress" he wished to be portrayed. Houdon wanted Washington to stand eternally wrapped in flowing toga. Washington saw Jefferson's tactful diplomacy at work and was grateful Jefferson had asked. He put it as delicately as he could that "perhaps a servile adherence to the garb of antiquity might not be altogether so expedient as some little deviation in favor of the modern costume."[2] Jefferson was delighted at Washington's decision. He had talked the question over with leading American artists in London—Benjamin West, John Trumbull, and Mather Brown. They all agreed with him, he now told Washington, on "the modern dress." A "modern in antique dress [was] as just an object of ridicule as a Hercules," Jefferson wrote, "with a periwig." But Jefferson's visits to Houdon were far from over: now Houdon cast Washington as a farmer in sandals and cloak rather than a soldier, one hand resting on a walking stick, the other on "the republican fasces crowned by a liberty Cap."[3] In the end, Jefferson persuaded Houdon to throw off the cape, drape it over the fasces, and add a sword. Jefferson then had to mediate a quarrel between the Virginia Assembly, which wanted a sixty-seven-word-long inscription written by Madison, and Houdon, who complained bitterly about the length. Ten years later, when Houdon finally shipped the statue to Virginia, it was inscribed simply, GEORGE WASHINGTON.

As wartime governor of Virginia, Jefferson had instigated the removal of the state capital to Richmond. He wanted to make a clean break with the colonial past and, to symbolize the revolution, replace the Georgian English-style architecture of Williamsburg with classical Roman, which he considered more dignified. In the very first week of existence of the Virginia House of Delegates in 1776, he had introduced bills providing for construction of the new Capitol building and a hall of justice in Richmond. His plans in 1776, which called for buildings with porticos and pillars, had seemed a bit grandiose for a town of two thousand. But Jefferson had persisted. He considered the capital project his own. As governor, he had proposed separate buildings for legislative, judicial, and executive branches. It was a revolutionary plan, architecturally symbolizing the separation of powers. No other capital in the world had yet been built in this way. But as soon as Jefferson left for France, the Virginia legislature scotched the plan for want of funds, and it would not be carried out in full for another forty years. But Jefferson taught his pupil, Madison, his desire to cultivate the tastes of his countrymen for a "beautiful art of architecture" to form a "national good taste" instead of allowing "a monument of our barbarism."[4] He had made drawings of the Capitol, Hall of Justice, and Governor's Mansion that he had proposed before he left for Europe. As early as 1780, he had formed the idea of introducing to America a new kind of government building, "an example of architecture in the classical style of antiquity,"[5] a Roman temple intended for everyday government use on the edge of the American frontier. No sooner did Jefferson leave than the Assembly had hired a contractor and begun work on the new Capitol at Richmond—but according to the Williamsburg model, Georgian red brick. William Short, still in Richmond at the time, wrote Jefferson, "I do not think the directors believe it possible to build a more magnificent house than the Williamsburg Capitol. It seems impossible to extend their ideas of architecture beyond it." Short, a member of the Governor's Council of State, had argued with the directors that they should send someone "to some part of Italy for a design and workmen," and he had finally prevailed. The Capitol directors commissioned Jefferson to find an able architect in Europe and to assist him. There could only be one building, and the legislature was willing to agree to the site Jefferson had proposed, but the design was left to Jefferson, just so that it united "economy with elegance and dignity"[6] and was done quickly; a bricklayer was about to begin work.

It took Jefferson "considerable time,"[7] but he found Charles Louis Clérisseau, a French architect who had studied in Rome for twenty years and had drawn the remains of almost every ancient building there. Recently the French Academy had recommended him to design a new palace for Catherine the Great of Russia. Jefferson retained Clérisseau as his consultant and draftsman and handed over to him his own detailed drawings and list of requirements. The temple, for example, must have Ionic columns on its exterior. Jefferson himself was the architect. On his tour of English gardens, he found a replica of the Maison Carrée at Nîmes, in the south of France, the finest surviving example of a Roman temple. He commissioned Clérisseau to travel to Nîmes and draw the temple, then make a plaster model to send to Virginia for the masons to copy. When Jefferson sent on his plans and the model in 1786, he did so emphatically but without revealing the full extent of his own involvement:

> I send by this conveyance designs for the Capitol. They are simple and sublime, more cannot be said. They are not the brat of a whimsical conception never before brought to light, but copied from the most precious, the most perfect model of ancient architecture re-

> maining on earth, one which has received the approbation of nearly 2,000 years.[8]

Finally out of Franklin's shadow, now a celebrity in his own right, Jefferson moved with ease in the best salons in Paris. Lafayette, Chastellux, Marmontal, Morellet, the Duc de La Rochefoucauld, the *philosophe* Condorcet, Jefferson visited them all regularly at salons or at their country houses. Now he became the influential American who introduced young American artists and literati. He took his friend Philip Mazzei, who had written a "history of America," to Rocheguiyon, "the magnificent *château* of the Duchesse d'Anville, on the border of Normandy," Mazzei recalled. William Short, who was having an affair with the duchess's daughter-in-law, wife of Jefferson's friend, the Duc de La Rochefoucauld, went along for the ride. "The Duke," Mazzei reported, "and his friends, the most intimate of whom was the Marquis de Condorcet, were especially interested."[9] Jefferson was always pleased for an excuse to visit the duke's mother, the tall, lean, eighty-year-old Duchesse d'Anville, who had devised a "system of ethics and of government"[10] that Jefferson much admired. Franklin had described her once as "a lady of uncommon intelligence and merit."[11] Even Adams praised her "bold, masculine and original sense, [which] was translated to me."[12]

During his Tuesday afternoon visits to the duchess's salon at her Parisian *château* along the Seine at the foot of present-day rue Bonaparte, Jefferson, as Abigail Adams described the scene, found the old lady sitting in an easy chair. "Around her sat a circle of Academicians [Condorcet, Crevecoeur, and other members of the Académie Française], and by her side a young lady [her beautiful young daughter-in-law, the Duchesse de La Rochefoucauld]. "The old lady rose and, as usual, gave us a salute. As she [wore] no paint, I could put up with it, but when she approached . . . I could think of nothing but Death taking hold of Hebe." The duchess wore no makeup, no cap "but a little black gauze bonnet,"[13] a silk chemise with voluminous sleeves coming halfway down her arms, rich lace trimmings around her neck and sleeves, and a large cape. The lace, according to Abigail, was not enough to disguise her wrinkled neck and she made no attempt to hide her white hair. The duchess, Jefferson later told her, was "the foremost of your nation in every virtue," and she had only ever made one error in all the years he knew her—"treating me with a degree of favor which I did not merit."[14]

If Jefferson became somewhat like a son to the venerable old duchess, she kindly brought him together with another man his age who also virtually lived in her family, the famous mathematician Condorcet, whose office was next door on the Quai des Monnaies, where he was inspector of the royal mint. A studious man with a dead-white visage from rarely leaving his books to go out in the Paris sunshine, he had appeared at first to be cold and reserved, but his friend D'Alembert had long ago dismissed this as a façade that masked his exuberance like snow covering a volcano. Jefferson and Condorcet chatted long hours about education, their great common interest, and coinage, which intrigued Jefferson. Condorcet, like Jefferson, believed in the continuous progress of the human race from barbarism to ultimate civilization through education. It was not long before Jefferson had added the Hôtel des Monnaies to his Monday itinerary to mix with the leading literati, and they often walked and talked in the duke's famous Anglo-Chinese gardens between the mint and La Rochefoucauld's mansion (at the corner of present-day rue de Seine and the *quai*).

By late 1786, Jefferson, Condorcet, Chastellux, and a fourth man, the Marquis de Lafayette, made up an intimate circle of friends. Nearly thirty, Lafayette had matured considerably since his glory-thirsting days as Washington's surrogate son at Valley Forge and his first fitful attempts at command in the American Revolution. Married at sixteen to the thirteen-year-old Marie Adrienne de Noailles, granddaughter of the powerful Duc de Noailles, he now lived in splendor in a new *château* on the site of the present-day Musée d'Orsay on the Left Bank opposite the Tuileries Palace. His house, gardens, furnishings, and food were famous, but even if the two were not close friends, his influence at court would have made him indispensable to Jefferson. Some of the most intelligent men and women in France, including six field marshals, several bishops, and a handful of dukes, made up his distinguished lineage, which accounted for his influence at court. His amiable young English-speaking wife had helped to place Patsy Jefferson in the Panthemont convent school. She took Patsy under her wing, introducing her to Paris society. According to Abigail Adams, the Marquise de Lafayette was "fond of her children and very attentive to them, which is not the general character of ladies of high rank in Europe."[15] So pro-American were the Lafayettes that they named all their children after Revolutionary heroes.

Despite the friendship of Lafayette, Chastellux, Condorcet, and others known as friends of America, Jefferson was under a considerable strain as he tried to steer a sensible course among them. In the influence-ridden French court and society, he was utterly dependent on their goodwill no matter how well known he became. It would have been easy for him to fall under the spell of those well-connected *salonistes* or their ideas. But few of his circle were republicans, even fewer democrats like himself. Jefferson's reform-minded French friends hoped only to give their country a more balanced economy and responsible form of government to secure for themselves and their countrymen the happiness, if not all of the freedoms, of Americans. Jefferson was able to exchange ideas freely with the French and he appreciated the refined, civilized society of the upper middle class and aristocratic French he knew, but there was also much in the airy spinnings of the salon philosophers that he considered dangerous, even to the American interests he worked so hard to advance.

Among his closest friends were some of the leading physiocrats, whose beliefs formed part of a general revulsion among French liberals for mercantilism. Physiocracy appealed to natural law and naturally drew Franklin and Jefferson. The physiocrats, led by Quesnay and the elder Mirabeau, devised a system in which all moral and economic values were based on land and agriculture. Jefferson approved of much in this system, including its veneration of the farmer and dislike of excessive government interference in economic activity. But what held his physiocrat-friends together was their unbounded admiration for the United States. The ideals and goals of the physiocrats were not his own. While he was essentially a pragmatist, they were doctrinaires, as he pointed out to Madison. He knew America's needs, and while he was grateful that his circle of wealthy and powerful physiocratic friends helped him to fight a winning battle with the tobacco monopolists of the Farmers-General, he took a dim view of the low-tariff Anglo-French treaty Adams finally achieved in 1786, even if it conformed with the economic tenets of the physiocrats. He cringed at being identified with the physiocrats' "farmers." When the French talked about the freedoms of the physiocratic farmer, they meant the aristocratic values inherent in a nobleman's estate.

Jefferson idealized the small farm and the independent yeoman farmer with democratic values and American liberties that did not accord with a class-ridden society such as France's. In *Notes on Virginia,* Jefferson had eulogized the American farmer as one of the "chosen people of the earth," living by his own industry, uncorrupted by commercialized city life. He believed that a self-sufficient agrarian economy would keep America free from the clutches of British merchants, who still controlled eighty-five percent of all American trade. He knew even better than the British that Americans were economically dependent on England because of their habits of commerce and consumption. His chief goal as American minister was to make possible the transfer of American trade from England to France and other European and Middle Eastern countries. He was a politician, a statesman, a philosopher, but not abstracted from everyday life. He knew of the frailties of his countrymen, that his government was bound to follow the wishes of the majority, even if it meant following the Pied Piper of renewed trade with England immediately after the Revolution. Even though the British were undercutting America's infant manufacturing, he knew that Congress must nonetheless listen to the habits of the people. He was well aware that the United States was not the dream of the last generation of French *philosophes,* such as the physiocrats du Pont and Lafayette, made real. He feared that the French would one day wake up from the dream and find that American virtues had become tarnished by political and commercial success.

While he pressed his French friends to help him at court, he had, at the same time, to convince them that the United States was worthy of their idealistic support. His own work, his framing not only of the Declaration of Independence but of the Virginia Statute of Religious Freedom, were examples of American loyalty to these principles of freedom. He also publicized by his writings some enviable American institutions—equal opportunities for all, a representative system of government, the absence of a hereditary aristocracy. But there were problems in the American paradise. Economic and social unrest were threatening to spoil the fruits of revolution. Publicly Jefferson could not point out these dangers or contradict his aristocratic French friends for fear of seeming to betray their beliefs in America. But whenever one of them went to America, Jefferson was careful to take him aside and whisper the truth to him.

One of Jefferson's naïve commissions from the Virginia Assembly was for a bust of Lafayette, a rare honor for a foreigner in virtually artless America. Jefferson had prompted the commission as governor of Virginia; now he was once again enmeshed in a transatlantic controversy. Lafayette, always panting after glory, became annoyed after two years passed with still no bust. The Virginia Assembly, complicating matters with off-and-on instructions, finally decided that the bust should be given, not to Lafayette personally, but to the city of Paris. A second bust was to be placed at Richmond. Lafayette, astonished, could hardly complain. At least his bust would be near Washington's.

The next problem was to keep Lafayette in Paris when Houdon was there; here, too, a patient Jefferson succeeded. By late spring 1786, the bust for Paris was ready, Lafayette in uniform, bold, imperious, proud. But now Jefferson had to persuade Paris. "No instance of similar proposition from a foreign power had occurred in their history,"[16] moaned Jefferson to Governor Harrison in Virginia. He was told to petition the provost of Paris. But he would also have to take up the matter with the foreign minister and the king, who might wonder why the bust wasn't of *him.* After intricate negotiations at Versailles, Jefferson finally obtained permission "to do honor to the Marquis de Lafayette."[17] Finally, on September 27, 1786, a ceremony was arranged in the grand hall of the Hôtel de Ville, where the bust was to remain. The mayor of Paris was there. Lafayette was there, with all his family and friends. Resolutions from Virginia and a letter from Jefferson were read. Marie-Dominique Ethis de Corny, the king's attorney general and a revolutionary soldier at Lafayette's side during the American war, responded with a speech. Musicians played as Lafayette's bust was put in its place. Only Jefferson was not there. He had sent in his place his secretary, William Short. Jefferson could not possibly have attended. He was in acute pain, confined to his bed in the Hôtel de Langéac after an unceremonious interruption to a lingering love affair.

"The Halle aux Bleds might have rotted down before I should have gone to see it,"[18] Thomas Jefferson wrote to Maria Cosway on October 12, 1786, trying to explain the happy chain of accidents that had brought them together. First, there was his meeting with young Connecticut-born artist John Trumbull in London, where Trumbull was sketching Adams for his painting of the signing of the Declaration of Independence. Trumbull wanted to paint Jefferson next. He arrived at the Hôtel de Langéac early in August, where Jefferson made rooms available for him to design the layout for his vast historical canvas. Jefferson was happy to have Trumbull join "our charming coterie in Paris."[19]He made the rounds introducing Trumbull, arranging for him to sketch. On one of their outings together, they visited the *abbés* Chalut and Arnauld at Passy. Trumbull sketched *vignettes* of the house, of its terraces, of long views of Paris. One trip led to Suresnes, four miles west of the Grille de Chaillot, where they witnessed the crowning of a Rosière, chosen as the most amiable, virtuous and industrious poor girl in the parish. They came back over Jefferson's favorite bridge, the Pont de Neuilly, and stopped off for tea at Madame de Corny's *hôtel* on rue Chausseé d'Antin (near the Opéra). Lovely, delicate Madame de Corny was one of the many married French noblewomen who fussed over the handsome widower Jefferson. But it was August, and she was away in the country.

Jefferson had no idea of meeting anyone the day he went with his young protégé Trumbull, who was also interested in architecture, to see the Halle aux Bleds near St. Eustache Church. One of the public buildings Jefferson was planning for the new capital at Richmond was a covered public market. He had heard that the Halle aux Bleds had a "noble dome" and he went with Trumbull to examine it. The stone walls of the building were pierced with numerous windows and its vast dome—130 feet across—was spanned by lightweight wooden ribs supporting wide bands of windows. The overall effect was to flood with daylight the noisy, dirty transactions that provided the meal and flour to make bread and cakes for the half million Parisians. The year after the market's building opened in 1782, it had been the scene of a great public festival that had marked the signing of the Peace Treaty of 1783, guaranteeing American independence. Jefferson wanted to visit the building—"this wonderful piece of architecture"[20]—with Trumbull because he had heard that its design demonstrated the theories of Renaissance architect Philibert Delormé. He had planned to dine that evening with the Duchesse de La Rochefoucauld. The visit at the market was to be brief. But Trumbull had invited two English painter-friends, Richard and Maria Cosway, who were staying on rue Coq-Héron nearby, to meet them.

Richard Cosway, forty-six, himself a diminutive "macaroni," probably was England's best miniaturist painter. He had come to Paris to paint the children of the Duc d'Orléans, bringing along his twenty-seven-year-old wife, Maria, herself a leading miniaturist. Maria Louisa Catherine Cecilia Hadfield had been born in Florence, Italy, the daughter of the owner of a resort that catered to English travelers. Her parents were English Protestants, but Maria learned better Italian than English and became a devout Catholic. She had been raised in a convent after a lunatic nurse murdered four of her siblings. When her father died, Maria, seventeen, wanted to become a nun. She even managed to find a convent that would take her without a dowry. Her Protestant mother, horrified at the idea and worried about a livelihood, took her home to England, where she placed Maria under the guidance of the miniaturist Angelica Kaufman, who sponsored her introduction to a London social circle that included Sir Joshua Reynolds, Lord Erskine, James Boswell, and Richard Cosway. A Botticelli-like beauty with blond curling hair, brilliant blue eyes, and exquisite skin, Maria captivated a number of men. But Angelica, who saw in her a real artistic talent, and her mother, keen on a lifetime settlement for herself as well as a whopping £2,800 dowry for her daughter, sold her at age twenty to Cosway, who had made a fortune painting pornographic miniatures on snuffboxes for noblemen, including the prince of Wales. Short, shorter even than his petite wife, and nearly twice her age, Cosway was vain and invariably overdressed in the latest foppish fancy. One critic called him "an absurd little coxcomb"; another "a preposterous little Dresden china mannikin." James Northcote said Richard Cosway was "one of those butterfly characters that nobody minded, so that his opinion went for nothing."[21] But George III knew of him even if he disapproved: "among *my* painters, there are no fops."[22] Depicted by some critics as a monkey-faced little man in a mulberry silk coat ornamented with strawberries, in his numerous self-portraits he was a handsome, virile, if playful man with no trace in his face of the cruelty he was said to practice on his beautiful young wife. It was an open secret in London that Maria had been unhappy since shortly after her marriage to Richard. Northcote insisted that she had been forced to marry against her will and "she always despised him."[23]

As their London abode, Richard Cosway rented the majestic palace built by the duke of Schomberg and filled it with the cream of England's art world and the bohemian *haut monde* that the Cosways attracted to an endless round of lavish entertainments. The Cosways sent out "bushels of little Italian notes of invitation,"[24] commented Horace Walpole. Maria was a gifted hostess who presided over one of London's leading salons in Schomberg House, which was lined with European paintings and exotic furniture, including Japanese screens, ebony *escritoires* inlaid with mother of pearl, enameled and bejeweled boxes and mosaic-topped tables covered with ornaments of jasper, bloodstone, and lapis-lazuli. Tortoiseshell musical clocks chimed to punctuate Maria's playing on the piano while she sang Italian songs at their Sunday-night entertainments. In addition to painting, Maria had studied composing with William Parsons. While his wife enchanted on the harp or the pianoforte, Richard worked the crowd, "always smiling, always gay,"[25] according to his wife, selling the Oriental rugs out from under them, selling his snuffboxes, and arranging sittings for his clients. The prince of Wales commissioned him to paint portraits of at least two of his mistresses, Perdita Robinson and, later, Mrs. Fitzherbert (according to palace gossip, he was clutching Richard Cosway's portrait of Mrs. Fitzherbert when he died). Perdita and Maria Cosway talked of doing a book together, with Maria as illustrator. When the prince of Wales abandoned Perdita Robinson for Mrs. Fitzherbert, she captured the dashing little colonel Banastre Tarleton, the bloody-handed dragoon leader who had almost captured Jefferson at Monticello. Perdita had come to Paris with Tarleton that summer of 1786 when Maria met Jefferson. As an artist, Maria had progressed so well as Angelica Kaufman's student that her husband, jealous, tried to hide her work from critics and insisted she paint only their landscapes and friends. Horace Walpole, who knew everything about London art, fashion, and society, mixed praise with his derision for her work in a poem, "Verses on Seeing Mrs. Cosway's Pallet," which he composed for the London *Morning Chronicle:*

Behold this strange chaotic mass,
Where colours in confusion lie,

Where rival tints commix'd appear,
Here, *tints for water,* there *for sky.*
Kept in imagination's glow,
See now the lovely artist stand!
Grand visions beaming on her mind,
The magic pencil in her hand.[26]

Neither Richard's jealousy nor Walpole's bitchiness could, however, detract from the fact that her fine portrait of the duchess of Devonshire led to her invitations to exhibit her work in the Royal Academy along with the work of Mary Moser, Richard's mistress.

Maria Cosway's painting solaced her somewhat for the ever-worsening stories that came from her husband's libertine circle. There was a persistent rumor that a secret passageway connected Schomberg House with the residence of the prince of Wales on Bedford Square, but the stories did not make clear whether the prince had tried to seduce Maria or Richard more frequently. Jefferson may have learned of these stories. To John Jay, he wrote after an attempt on the life of George III, "the Prince of Wales on the throne and we are undone."[27] He described the prince's entourage as "the lowest, the most illiterate and profligate persons of the kingdom, without choice of rank or merit, and with whom the subjects of the conversation are only horses, drinking-matches, bawdy-houses, and in terms the most vulgar." Many young nobles had gone through the prince's circle, Jefferson wrote; they "soon leave him, disgusted with the insupportable profligacy of his society."[28] After a brief period of happiness, Maria had evidently learned that Cosway had only set her up as an ornament in his rented mansion to entice his concentric circles of friends, clients, and lovers, male and female. One bitter letter she wrote years later reminded Cosway of "how many years" they were happy "until you began to divide your thoughts, first with occupations in Bedford Square and a Miss P. Afterwards, with Hammersmith and then L. . . . Lastly with the Udneys, and this ended our happiness."[29]

To a friend, Mrs. Chambers, Maria wrote that "the moment he gave himself to Hammersmith began to lead him from me and from his home. . . . Hammersmith was dropped for another acquaintance, which kept him farther from me. Peevish, cross, we were not happy."[30] Unhappy, Maria received more and more release from her work and the fame it had begun to bring her. By 1786, as many carriages called at Schomberg House for her as for her husband. Would-be paramours were repulsed, however. She treated men like dogs, complained James Boswell. "Nothing was talked of but the great youth and talent of Mrs. Cosway,"[31] wrote Allan

Cunningham in his *Lives of the Most Eminent British Painters.* But by 1786, the years of constant strain had weakened Maria's health, and she had come to Paris that summer with her husband to recover.

Thomas Jefferson fell in love with Maria Cosway from the moment he met her. For four years, he had been faithful to the deathbed vow he had made to his wife. There is no hint that he had even the briefest liaison with any of the many Frenchwomen he had met. But no sooner did Trumbull introduce the Cosways than Jefferson began to think how he could spend every possible moment with this lively, languid, beautiful Maria, with her musical, slightly Italian accent. As he wrote weeks later, soon he was "dilating" with his "new acquaintances and contriving how to prevent a separation from them." Maria, for her part, at first seemed less smitten with the tall, celebrated Virginian, but she was obviously flattered by his attentiveness, and his brilliant talk fascinated her.

All four sightseers had other appointments later that day, but, according to Jefferson, they sent "lying messengers" off around the city "with apologies for [breaches] of engagement." Jefferson later confessed that he had told the Duchesse de La Rochefoucauld that "on the moment we were setting out to dine with her despatches came to hand which required immediate attention." Off they went in Jefferson's carriage to dinner at Saint-Cloud, a popular resort midway between Paris and Versailles where Louis XVI had recently bought an ancient *château* to add to Marie Antoinette's collection. Jefferson later exulted at the scene, "How beautiful was every object! the Pont de Neuilly, the hills along the Seine, the château, the gardens"[32] of the 965-acre Parc de Saint-Cloud with its statuary-lined fountains and Grande Cascade.

As the late summer light began to fade, the party decided to go on, this time toward Paris and the Faubourg Montmartre where, off rue St. Lazare, an Italian family, the Ruggieris, ran a Vauxhall-like pleasure garden famous for its fireworks displays. The foursome found a table in the crowded gardens and watched the Ruggieris' spectacular show. The "Salamandre" lighted up the sky in sparks and flame as its mechanized snake pursued a butterfly, the fiery serpent appearing ready to overtake its prey but never quite succeeding. In "The Forge of Vulcan," sprays of fire depicted cascades, showers of shooting stars, pyramids, revolving cones, sunbursts. "The Combat of Mars" was outlined in slender sprays of fire. The show ended with a great cluster of rockets, shot from the center of a machine, bursting high into the sky in a fan-shaped spray. From the Ruggieris', Jefferson, Trumbull, and the Cosways rolled on to the house of Johann Baptist Krumfoltz, a celebrated Viennese harpist and composer who had studied with Haydn before coming to Paris. Krumfoltz had recently designed a new harp that was the talk of the salons. Maria, a talented harpist, probably suggested they go and strike up an impromptu concert. Krumfoltz's wife, Julia, an even finer player than her husband, welcomed the party and treated them to her sadly sweet playing.

Before the August evening was over, Jefferson was hopelessly dazzled, he wrote later, by Maria Cosway's "qualities and accomplishments," by her "music, modesty, beauty and that softness of disposition which is the ornament of her sex."[33] Where they had gone together was of little consequence to Jefferson. He wrote that, "if the day had been as long as a Lapland summer day," he would have "contrived means" to keep filling it with her company. "When I came home at night and looked back to the morning, it seemed to have been a month gone."[34]

The two were rarely long apart in the weeks that followed. At first, every other day Jefferson and Trumbull called at the Cosways and they all went off together. Then Trumbull, busy with his own work, began to beg off and Richard Cosway excused himself and went back to whatever he was doing. Jefferson handed Maria up into his phaeton and they rolled off alone together to the Louvre, the Palais Royal, the Bibliothèque du Roi, to Versailles, to the salons of Houdon and David. Jefferson, lonely for so long, was enchanted with Maria's effervescence, her enthusiasm about art, her languid beauty, her easy, bright conversation. On September 9, Trumbull, with detailed instructions and introductions from Jefferson, was packed off to Germany. William Short was also away. With Trumbull gone, the artistic pilgrimages Jefferson and Maria had been making came to an end. Maria began to make unannounced morning and afternoon visits to the Hôtel de Langéac and then they would "hie away, after intervals impossible to deduce, to the Bois de Boulogne, to the Jardin du Roi, the Parc de Saint-Cloud, the cascades and ruins at Marly, the terraced gardens at Saint-Germain-en-Laye. They visited every playground of the nobility. They walked in every public garden and along the banks of the Seine, talking, picnicking, "every moment," Jefferson wrote later in a letter to Maria, "filled with something agreeable."[35]

Jefferson could not contain his happiness. To Abigail Adams, he wrote, in August 9, when he had known Maria for only a week, "Here we have singing, dancing, laugh and merriment. No assassinations, no treasons, rebellions, nor other dark deeds. When our king goes out, they fall down and kiss the earth . . . then they go to kissing one another. This is the truest wisdom. They have as much happiness in one year as one Englishman in ten."[36] The next day, he wrote William Smith that "beauty is ever leading us astray."[37] Jefferson, enthralled by a beautiful young married women, ignored the risks. When he reproached himself, weeks later, in writing, Jefferson did so in a mock dialogue between his head and his heart, and he made sure that his heart got the better of the debate: "Thou are the most incorrigible of all the beings that ever sinned!" his head told his heart:

> I reminded you of the follies of the first day, intending to deduce from thence some useful lessons for you. But, instead of listening to them, you kindle at the recollection, you retrace the whole series with a fondness which shows you want nothing but the opportunity to act it all over again. I often told you, during its course, that you were imprudently engaging your affections under circumstances that must have cost you a great deal of pain."[38]

While Richard Cosway was busy with his miniatures, the lovers went on long carriage rides usually culminating in remote gardens. On September 5, Jefferson and his "vastly pleasant" companion, as Gouverneur Morris, a notorious womanizer, described Maria, made one of their typical excursions together, driving west to the Bois de Boulogne and stopping first at the Bagatelle, an elegant little casino, its art-filled pavilion ringed by gardens sweet with magnolias, coiffed in a rotunda as if it wore an oversized bonnet. Jefferson's expense account now contained no harsh words of description. The picturesque-style gardens seem to have totally captivated him. Driving on to the Madrid, an odd little *château* built by Francis I after he returned from imprisonment in Spain, they strolled along the royal esplanade, its varnished terra cotta shimmering in the sunlight, following the semicircle of ancient trees. From the Château de Madrid, they could gaze at the Seine and a tableau of rich surrounding countryside, to Jefferson's favorite Pont de Neuilly.

It was on September 16, nearly six weeks after they had met, that Jefferson and his beloved Maria enjoyed the day that would stand out most clearly in Jefferson's recollections. Be "a kind comforter," his heart would ask his head, "and paint to me the day we went to Saint-Germain. How beautiful was every object!"[39] There was a new urgency to the day. Husband Richard had nearly finished his work for the Duc d'Orléans and was already planning to return to London. But for today, there was an unusual English-style garden at Saint Germain they had not seen. Flicking the reins, Jefferson ordered the horses west through the city gates, over the Pont de Neuilly and along the Seine Valley. At the Machine of Marly, Jefferson the man of science had once been fascinated by the great blades that raised water from the Seine to the waterfalls at Versailles. Today, it was the "rainbows of the machine" that charmed him. On they drove happily to Louveciennes and the charming pavilion of Madame Du Barry: there, they dined at a rustic inn where Petit had ordered supper for them. From here, they could see the Château de Marly, Louis XIV's nearby retreat. Its central pavilion, facing a grand basin, was flanked by smaller pavilions. Jefferson's later design for the University of Virginia bore a striking resemblance to it. From Marly, the carriage followed the highway into Saint-Germain-en-Laye. From the elaborate flower beds of the terrace, they could see the skyline of Paris. Understandably they did not stop to visit Jefferson's friends living there. Thomas Barclay was away on a mission to Algiers, but his wife, a close friend of Abigail Adams, was there. William Short was then living at Saint-Germain, studying French and in love with the daughter of his landlady.

Recollecting the glories of that day, Jefferson lingered longest over the Desert de Retz, the elaborate country estate of Racine de Monville on the edge of the Forest of Marly near Chambourcy. Monville's carefully designed wilderness was arranged around an enormous column, sixty-five feet in diameter, built to look ruined. His living quarters were wrapped around a central spiral staircase. Jefferson strolled with Maria through grounds that included a Chinese *orangerie,* a temple dedicated to Pan, an obelisk, a pyramid-shaped ice house, Gothic ruins, a small "ruined" altar. "How grand the idea excited by the remains of such a column," Jefferson reminded Maria weeks later. "The spiral staircase, too, was beautiful." He later adapted it to Monticello.

The day was memorable, too, for the passions it aroused. "The wheels of time moved on with a rapidity of which those of the carriage gave but a faint idea, and yet, in the evening, when one took a retrospect of the day, what a mass of happiness had we travelled over."[40] In these enchanted weeks, when he was away from Maria, Jefferson was often reading the poetry of love. In the third week of September, he wrote "Thoughts on English Prosody,"[41] reading again favorites that included Milton, Gray, Shenstone, Pope. "I chose," he later wrote, "the most pregnant passages, those wherein every word teems with latent meaning." His selections resonated with his new love:

And I loved her the more when I heard
Such tenderness fall from her tongue.[42]

There was a new note of sadness entering his thoughts as the day of Maria's departure loomed. He introduced the past tense:

With her how I stray'd amid fountains and bowers,
Or loiter'd behind, and collected the flowers!
Then breathless with ardor my fair one pursued,
And to think with what kindness my garland she viewed!
But be still, my fond heart! this emotion give over;
Fain woulds't thou forget thou must love her no more.

By now, Maria's return to London was imminent. Jefferson copied out melancholy lines:

I mourn
I sigh
I burn
I die
Let us part
Let us part
Will you break
My poor heart?[43]

On September 18, 1786, Jefferson planned to meet Maria for what might have been their last day together in Paris. What plans they had made beyond that can only be deduced from later letters and actions, but, soon after Maria returned to London, she wrote Jefferson, "Are you to be painted in future ages sitting solitary and sad, on the beautiful Monticello, tormented by the shadow of a woman who will present you a deformed rod, twisted and broken, instead of the emblematical instrument belonging to the Muses, held by genius, inspired by wit, from which all that is pleasing, beautiful and happy can be described to entertain." He had given her a copy of his *Notes on Virginia.* She responded passionately "Oh how I wish myself in those delightful places! Those enchanted grottoes! Those magnificent mountains, rivers. . . . Why am I not a man that I could set out immediately and satisfy my curiosity, indulge my sight with wonders!"[44] No wonder that Jefferson saved all her letters, and copies of all his letters to Maria, even if he had destroyed his correspondence with his wife. One day Jefferson's children and grandchildren would know how much he had loved Maria Cosway.

Did his daughters later blush at his letters: Had they ever known him to be like this? When he had tried to set down in a long letter the mixture of joy and turmoil she had brought him, he became something of a poet himself:

> The sun shone brightly! How gay did the face of nature appear! Hills, valleys, chateaux, gardens, rivers, every object wore its loveliest hue! Whence did they borrow it? From the presence of our charming companion. They were pleasing because she seemed pleased. Alone, the scene would have been dull and insipid: the participation of it with her gave it relish. Let the gloomy monk, sequestered from the world, seek unsocial pleasures in the bottom of his cell! Let the sublimated philosopher grasp visionary happiness while pursuing phantoms dressed in the garb of truth! Their wisdom is supreme folly, and they mistake for happiness the mere absence of pain. Had they ever felt the solid pleasure of one generous spasm of the heart, they would exchange for it all the frigid speculations of their lives.[45]

It is strange, then, that he felt he had to maintain absolute discretion about just what happened to him on September 18. He evidently was distracted as he prepared to meet Maria that day, and as he dashed to his carriage in the courtyard at the Hôtel de Langéac to go and meet Maria, tried, as if all his youth were restored, to vault over a small fountain, and tripped. He came down hard. Reflexively he tried to break his fall with his right hand, but he fell with his full weight and shattered his right wrist. It was a compound fracture, the bone piercing the skin, and must have pained him terribly. But Jefferson got up and wrapped his wrist and tried to conceal his injury and went ahead to meet Maria. For hours they rattled around Paris, Jefferson in agony, biting back the pain against the constant hard jostling, but never letting on to Maria how much his wrist, now terribly swollen, was hurting him. Finally, he could stand it no more and said he had to go home and fetch a doctor. The next item he listed in his account book after "seeing Desert [de Ritz] 6f." was "pd. two surgeons 12f." There were no more notations of expenditures for three weeks. That Jefferson had been seriously injured got around Paris and over to Passy immediately, where the malicious old Abbé Chalut gave it his own spin, putting out the story that Jefferson, strolling along the Cours de Reine embankment of the Seine near the Tuileries Gardens with Maria, had forgotten his age and, attempting to vault a fence before helping her over, had fallen. Perhaps the thought of Jefferson's taking a pratfall in front of his lady seemed more delicious to the old priest, but if Jefferson had done what Chalut had imputed to him, he might have broken more than his wrist plunging down the steep riverbank, which dropped off sharply at the Cours de Reine. In any event, Chalut's story leaped the ocean to America first. Chalut told his Passy neighbor, Veillard, who wrote to Franklin two days later that Jefferson had dislocated his wrist jumping over a fence.

Jefferson did little himself to render the incident less mysterious. To his young friend, William Smith, in London, he wrote when he could, "How the right hand became disabled would be a long story for the left to tell. It was by one of those follies from which good cannot come but ill may."[46] While he made it clear to his daughter Patsy that he had broken his wrist and not merely dislocated it, as the inept French doctors insisted, Patsy got from Jefferson the impression that he had been out for a walk with a man when he fell:

> He frequently walked as far as seven miles in the country. Returning from one of those rambles, he was joined by some friend and, being earnestly engaged in conversation, he fell and fractured his wrist. He said nothing at the moment but holding his suffering limb with the other hand, he continued the conversation till he arrived near to his own house when, informing his companion of the accident, he left him to send for the surgeon. The fracture was a compound one and probably much swollen before the arrival of the surgeon. It was not *set,* and remained ever after weak and stiff.[47]

By the next morning, Maria was fairly desperate for word from Jefferson, but she was having trouble getting away from her husband. She had, as she wrote the next day, planned a morning visit but Richard "killed my project I had proposed to him by burying himself among pictures and forgetting the hours . . . I meant to have had the pleasure of seeing you *twice,* and I have appeared a monster for not having sent to know how you was *[sic]* the *whole day."* She finally managed to get all the way across Paris to the Hôtel de Langéac, but it was too late to do anything but prove, "a disturbance to your neighbors."[48] Jefferson received this hand-delivered note on the 20th.

Maria apparently did come to see him the next day, and she came often to visit him over the next two weeks. Jefferson could not leave his room, even to attend the presentation of Lafayette's bust. Each time Maria called, he begged to go out with her in the carriage. "You repeatedly said it would do no harm,"[49] she later recalled, but it was not until October 4, the sixteenth day after the accident, that she agreed. And that day, as they rattled around the streets of Paris, Jefferson's arm in a sling, Maria had to give him the news they had been dreading. The Cosways were leaving Paris the next day. Richard, a man of so many illicit affairs, had decided to break up what he must have seen was turning into a serious relationship.

The news shattered Jefferson. The night of October 4 he could not sleep. Later he wrote Maria a letter about that night, as if he were talking to her, being careful how he worded it in case she opened it when Richard was there.

> Remember the last night. You knew your friends were to leave Paris today. This was enough to throw you into agonies. All night, you tossed . . . from one side of the bed to the other. No sleep, no rest. The poor, crippled wrist, too, never left one moment in the same position, now up, now down, now here, now there. Was it to be wondered that all its pain returned?[50]

Unable to go to say good-bye to Maria the next morning as they had planned, Jefferson struggled to write a short note with his left hand.

> I have passed the night in so much pain that I have not closed my eyes. It is with infinite regret therefore that I must relinquish your charming company for that of the surgeon whom I have sent for to examine into the cause of this change. I am in hopes it is only the having rattled *[sic]* a little too freely over the pavement yesterday. If you do not go today, I shall still have the pleasure of seeing you again. If you do, god bless you, wherever you go. Present me in the most friendly terms to Mr. Cosway, and let me hear of your safe arrival in England. Addio, Addio. Let me know if you do not go today.[51]

Jefferson called in the surgeon again, but he could not "divine the cause of this extraordinary change."[52] It is quite probable that Jefferson's fracture had started to mend and that, jostling in a hard-springed carriage over the rough *pavés* of Paris streets half the day before, he had broken it again. Maria answered his distressed note immediately.

> I am very, very sorry indeed. . . . We shall go, I believe, this morning. Nothing seems ready, but Mr. Cosway seems more disposed than I have seen him all this time.[53]

Despite the fact that he was in pain, Jefferson's agony at the thought of not seeing off Maria was greater. He dismissed the surgeon and called for his carriage and raced off over the rutted streets to rue Coq-Héron. He found the Cosways' house in chaos, their packing far from completed, and he did not stay long. They were putting off leaving until the next day. Jefferson insisted on meeting them the next morning to perform "the last sad office" of handing Maria up into her carriage.

The next morning Jefferson recruited Pierre François d'Hancarville, a puppy dog of a man who followed Maria everywhere and had often acted as their courier, as the fourth for the carriage once they had joined the Cosways. They drove to rue Coq-Héron for the last time, then rode out with the Cosways several miles to the Pavilion de St. Denis, where they stopped for breakfast. Handing up Maria to her husband one last time, Jefferson turned and walked away, as he later wrote her, "more dead than alive."

Riding back with d'Hancarville in a crowded public coach, Jefferson felt as if they were both "like recruits for the Bastille, not having soul enough to give orders to the coachman." Both in love with someone else's wife, the two men consoled each other with "a mutual confession of distress."[54]

That night, in front of a roaring fire, the pain still excruciating, Jefferson, struggling to write with his left hand, ignored the pain and poured out his thoughts and emotions in a typical eighteenth-century device, a dialogue between "My Head and My Heart," intended for Maria. It is Jefferson's single most important surviving personal letter. He apparently interrupted his writing several times. It rained hard for days. He finished the letter five days after Maria had left. Recounting the details of their farewell, he wrote Maria that he was "seated by my fireside, solitary and sad."[55] There, reason battled emotion.

> Head. Well, friend, you seem to be a pretty trim.
>
> Heart. I am indeed the most wretched of all earthly beings. Overwhelmed by grief . . . I would willingly meet whatever catastrophe should leave me no more to feel or to fear.
>
> Head. These are the eternal consequences of your warmth and precipitation. This is one of the scrapes into which you are ever leading us. You confess your follies indeed, but, still, you hug and cherish them, and no reformation can be hoped where there is no repentance.
>
> Heart. Oh my friend! This is no moment to upbraid my foibles. I am rent into fragments by the force of my grief! If you have any balm, pour it into my wounds: if none, do not harrow them by new torments. Spare me in this awful moment! At any other I will attend with patience to your admonitions.
>
> Head. On the contrary I never found that the moment of triumph with you was the moment of attention to my admonitions. While suffering under your follies you may perhaps be made sensible of them, but, the paroxysm over, you fancy it can never return. . . .

Heart. . . . Sir, this acquaintance was not the consequence of my doings. It was one of your projects which threw us in the way of it.[56] . . . You, then, Sir, and not I, have been the cause of the present distress.

Head. It would have been happy for you if my diagrams and crotchets had gotten you to sleep on that day, as you are pleased to say they eternally do [but] you were dilating with your new acquaintances, and contriving how to prevent a separation from them. Every soul of you had an engagement for the day. Yet all these were to be sacrificed, that you might dine together. . . .

Heart. Oh! my dear friend, how you have revived me by recalling to my mind the transactions of that day! How well I remember them all, that when I came home at night and looked back to the morning, it seemed to have been a month gone. Go on then, like a kind comforter, and paint to me the day. . . .[57]

Reason refuted passion point by point. Reason argued that Maria and her husband were of "the greatest merit, possessing good sense, good humor, honest hearts, honest manners and eminence in a lovely art"—once again, Jefferson had to bow to decorum to include Richard Cosway, whom he no doubt detested, in this letter to get it past the censor. Yet reason knew "that all these considerations would increase the pang of separation." That separation would, in this instance, "be the more severe as you would probably never see them again."

Heart. . . . But they told me they would come back again the next year.

Head. . . . In the meantime, see what you suffer . . . . Upon the whole, it is improbable and therefore you should abandon the idea of ever seeing them again.

Heart. May heaven abandon me if I do!

Head. Very well. Suppose then they come back. . . . What is to follow? Perhaps you flatter yourself they may come to America?[58]

Heart. God only knows what is to happen. . . .[59]

In the end, Jefferson gave Heart the last word.

I am indeed the most wretched of all earthly beings. Overwhelmed with grief, every fibre of my frame distended beyond its natural powers to bear, I would willingly meet whatever catastrophe should leave me no more to feel or to fear. . . .

> I feel more fit for death than life. But when I look back on the pleasures of which it is the consequence, I am conscious they were worth the price I am paying. . . . Hope is sweeter than despair, and they were too good to mean to deceive me. In the summer, said the gentleman; but in the spring, said the lady: and I should love her forever, were it only for that![60]

Jefferson's injury led him to enlist two of his young protégés, William Short and John Trumbull, in his continued efforts to stay in touch with Maria Cosway. Short, who was something of a playboy, became Jefferson's strong right hand, writing his letters. Trumbull, who had left America after a scandal involving an illegitimate child, became his trusted courier. In one cover letter, Jefferson urged Trumbull to "lay all my affections at her feet,"[61] in another, "Kneel to Mrs. Cosway for me and lay my soul in her lap."[62] Trumbull was on his way back to Paris from Germany when he ran into the Cosways in a tavern in Antwerp, waiting to cross the Channel. Trumbull, learning of Jefferson's accident, dashed off a long letter to Jefferson and Maria added a short postscript and signed it. When Trumbull passed through Paris on his way back to London, Jefferson asked him to carry the "Head and Heart" letter to Maria. Not only was he concerned about Richard's intercepting it, but he knew that all his mail was read both by French and English postal censors. Jefferson sent Maria instructions to send letters to him in care of his French banker.

He soon heard from her. She answered his twelve-page "Head and Heart" letter in English and Italian. She had grown melancholy without Jefferson. "My heart is," she wrote, "ready to burst." London had become unbearable. "Amid the fog and smoke, sadness seems to reign in every heart. . . . Everything is tranquil, quiet and gloomy, there are no bells ringing. . . . Your letters will never be long enough."[63] When Maria had not heard from Jefferson again by November 17, she sent him a volume of songs she had composed for him, adding, "What does this silence mean. . . . I have awaited the post with so much anxiety. . . . One lives in continual anxiety. . . . Night thoughts, before the fire, and when imagination is well warmed up, one could go cool off in a river."[64] A few days later, Jefferson was able to write with his right hand for the first time in two months. "I write with pain and must be short."[65] Ten days later, he wrote again, "I am determined, when you come next, not to admit the idea that we are ever to part again. But are you to come again? I dread the answer. . . . Say many kind things, and say them without reserve. They will be food for my soul."[66]

But Jefferson, who expected Maria to come back in spring, made no effort to go to England, to which he had vowed never to return. Alone on Christmas Eve, he wrote to Maria, "It is time therefore you should be making your arrangements, packing your baggage." He missed her terribly. "I am always thinking of you. . . . I will believe you intend to go to America, to draw the Natural Bridge . . . that I shall meet you there. . . . I had rather be deceived than live without hope."[67] But Richard kept postponing the spring trip back to Paris. When Maria finally wrote that they would not come to Paris until summer, Jefferson decided to go alone on the trip they had planned to make together, to Maria's birthplace in Italy.

18

"A Master of My Own Secret"

There never was a poet north of the Alps. . . . A poet is as much a creature of climate as an orange or a palm tree.

—THOMAS JEFFERSON TO MARIA COSWAY, 1787

SINCE HIS COLLEGE days under the enlightened tutelage of Dr. Small, Jefferson had talked and written of visiting Italy, fountainhead of the classical literature and architecture he most admired and which he considered, more than England, the proper model for the new American republic. After years of reading and conversations with his Florentine neighbor, Philip Mazzei, he had used the first possible opportunity to send Mazzei to seek the help of Italian city states for Virginia during the Revolution. His love for Maria Cosway had been set against this backdrop. They undoubtedly had spent long hours discussing the reminders of Roman and Renaissance art and architecture they found around Paris in their weeks of blissful touring.

But Jefferson always had a pragmatic side. Especially after a long series of diplomatic checks in London and at Versailles, he had become determined to break United States economic dependence on England and France by forging new trade ties with Italy. Especially interested in diversifying plantation agriculture and improving the lot of laborers, black and white, in his native region, Jefferson wrote to Governor John Rutledge of South Carolina in 1788, shortly after his tour to the Mediterranean, "Italy is a field where the inhabitants of the Southern States may see much to copy in agriculture and a country with which we shall carry on considerable trade."[1] There also, of course, was a scholarly side to his voyage. To his friend George Wythe in Virginia, he confided that he was trying to verify Hannibal's invasion route over the Alps. But he told his friends in Paris and London and even his own daughter Patsy none of this. The reason he gave for leaving his diplomatic post for three and a half months was the advice of his doctors to seek a cure for his injured wrist in the hot mineral baths at Aix-en-Provence.

Despite the physical pain that had lingered all winter, Jefferson was finally making some headway diplomatically. Before the end of 1786, he was able to propose formally to the ambassadors of Russia and Portugal a naval confederation against the Barbary states to patrol the Mediterranean and purge it of pirates. After nearly three years of studying American trade with Europe at close range, Jefferson had become convinced that Europe could not be expected to change its ways to suit Americans, but that Americans, especially farmers and plantation owners—"the mass of our countrymen"—would have "to adapt their productions to the market," which he thought should increasingly be Continental Europe. He had worked since arriving in Europe, he wrote John Jay midway through his southern odyssey, to discover and create new markets for American goods and to "endeavor to obtain favorable terms of reception."[2] Along the way, he hoped to find seeds and plants suitable for American soil. "The greatest service which can be rendered to any country," he later wrote, "is to add a useful plant to its culture."[3]

Before he left Paris, Jefferson wrote Madison in January, 1787, four months after his injury, that his wrist was scarcely better:

> I can make not the least use of it except for the single article of writing. I have great anxieties lest I should never recover any considerable use of it. I shall, by the advice of my surgeons, set out in a fortnight for the waters of Aix, in Provence. I chose these out of several they proposed to me because, if they fail to be effectual, my journey will not be useless altogether. It will give me an opportunity to examine the canal at Languedoc and of acquiring knowledge of that species of navigation. . . . It will enable me to make the tour of the ports concerned in commerce with us, to examine on the spot the defects of the late regulations respecting our commerce, to learn the further improvements which may be made . . . and on my return to get this [treaty] business finished. I shall be absent between two and three months.[4]

Jefferson gathered up a trunkful of travel guides and every map he could find of the towns and cities on his route: Lyons, Marseilles, Nice, Turin, Milan.

On February 28, 1787, "pelted with rain, hail and snow," Jefferson set out alone and "absolutely incognito" in his Monticello phaeton, its glass on four sides frosted over, his rare glimpse of a view at first only the enormous boots of the hired man riding postilion. Jefferson took his own carriage but hired post horses along the way. Jefferson left behind all of his household staff. He did not trust most of them, and he needed his faithful Petit to watch over Patsy and safeguard his mail. There were elements of his trip which he did not want reported to the French Foreign Ministry. Just before he left, he arranged to meet secretly with Brazilian colonists along the road. His negotiations with Portugal were at a sensitive stage and he already knew that the Spanish, allies of the Portuguese, were spying on him. He trusted James Hemings, but he valued his cooking more and had left him to work and study with the chef of the prince of Condé. Jefferson decided to hire servants along the way for a few days at a time. "I was quite determined to be a master of my own secret," he later wrote William Short, "and therefore to take a servant who should not know me." He concealed his identity on the road and his mission even from people from whom he had carried letters of introduction. The *abbés* Chalut and Arnauld, Lafayette and his aunt, Madame de Tessé and other *salonistes* had given him letters to get him special treatment along his route. He may have gone in disguise: the next year he urged sons of his American friends to wear large cockades in their hats so they would be considered officers.

Leaving Paris on the last day of February, as Jefferson drove southeast through the melon fields of Melun, there appeared "a few gleamings of sunshine to cheer" him.[5] Passing through the great forest of Fontainebleau, he came to the town toward evening and spent two days there. He was methodical about traveling. As he advised other American travelers, he made it his policy immediately to "walk around the ramparts when there are any" or up to the top of a steeple or the top of the nearest hill, and then allow for a long rest afterward. On arriving at a town, he wrote, "the first thing is to buy the plan of the town and the book noting its curiosities":

> When you are doubting whether a thing is worth the trouble of going to see, recollect that you will never again be so near it. . . . You may repent not having seen it, but can never repent having seen it. But there is an opposite extreme, too, seeing too much. Take care

> not to let the porters of churches, [museums], etc., lead you through all the little details in their possession, which will load the memory with trifles, fatigue the attention and . . . waste your time. They wish for your money and suppose you give it more willingly the more they detail to you.[6]

"In the great cities, I go to see what travelers think alone worthy of being seen, but I make a job of it and generally gulp it all down in a day," Jefferson told Lafayette. "On the other hand, I am never satiated with rambling through the fields and farms, examining the culture and the cultivators with a degree of curiosity which makes some take me to be a fool and others to be much wiser than I am."[7]

How much he learned from the experiences of this, his first and longest tour, can be conjectured from his long letter to Shippen and Rutledge shortly after he returned to Paris. The letter, which he probably intended for publication, is a timeless piece of warning to inexperienced travelers. He gave advice on buying wine in restaurants along the road:

> When one calls in taverns for the *vin du pays,* they give what is natural, unadulterated and cheap: when [imported] wine is called for, it only gives a pretext for charging an extravagant price for an unwholesome stuff, very often of their own brewing.

He warned against judging a country by the people travelers had to deal with along the route:

> The people you will naturally see the most of will be tavernkeepers, *valets de place* and postilions [the equivalent of cab-drivers]. These are the hackneyed rascals of every country. Of course they must never be considered when we calculate the national character.[8]

Entering Champagne, where Burgundy begins, Jefferson wrote in his log, "People are so well dressed, but then it is Sunday." On March 2, Jefferson drove on to the ancient cathedral city of Sens, where he visited the twelfth-century cathedral of Saint-Etienne. His only written comment—"seeing church, 2 francs 4"[9]—was for his *Account Book,* which he kept on rough spreadsheets that he would copy and recopy when there was less pain to writing. Leaving Sens, he made a typical agricultural notation: "The face of the country is in large hills not too steep for the plow." He scientifically arranged his notes in columns of description including soil conditions, crops, condition of the people, their diets. Later he expanded these rough notes into a long, narrative memorandum. "There are a few apple trees . . . no inclosures, no cattle, sheep or swine, fine mules. Few chateaux, no farm houses, all the people being gathered in villages. The people are illy clothed. Perhaps they have put on their worst as it is raining."[10]

As he followed the languid river Yonne through lovely farm country, Jefferson was struck by seeing women and children hard at work in the fields:

> I observe women and men carrying heavy burdens and laboring with the hoe. This is an unequivocal indication of extreme poverty. Men in a civilized country never expose their wives and children to labor above their force and sex as long as their own labor can protect them from it.[11]

Jefferson often stopped along the road to talk with the people he saw working. He blamed their suffering on the nobility. When he returned to Paris, he urged his reform-minded friend, Lafayette, to go "absolutely incognito" and see under what conditions his people live:

> You must ferret the people out of their hovels as I have done, look into their kettles, eat their bread, loll on their beds under pretense of resting yourself, but in fact to find if they are soft. You will find a sublime pleasure in the course of this investigation and a sublimer one hereafter when you shall be able to apply your knowledge to the softening of their beds, or the throwing of a morsel of meat into the kettles of vegetables.[12]

Jefferson's journey through Burgundy led him on March 4 to the Hôtel de Condé at Dijon. The weather had cleared, but there were still streaks of snow in the rows of vineyards. He "passed through about one hundred eighty miles of Burgundy," which reminded [him] of home. "The hills become mountains, larger than those of Champagne, more abrupt, more red and stony [resembling] extremely our red, mountainous country . . . all in corn and wine." Jefferson mounted a mule and, guided by a peasant, "rambled through their most celebrated vineyard."[13] He sketched a small map showing the red and white wines of the region. He was struck by the fact that the most prestigious wines, Chambertin and Montrachet, were located at opposite ends of the region, but he could not find any explanation. "I am told that the other wines are just as good but have not yet made a name for themselves,"[14] he noted. He was intrigued by everything: the capacity of the barrels, the way the cellars looked, the peasants' complexions. The owners of the more celebrated vineyards, he noted, made an excellent living. The owner of the Volnay vineyards gave him a careful tour, and he became a lifelong customer. Montrachet was to become one of his favorites.

Day after day in Burgundy, through Meursault and into the Beaujolais region to Dijon, the cleanest city he found in France, Jefferson was enthralled by the plunging and rolling highly cultivated countryside. "Nature has spread its richest gifts in profusion," he later wrote in his memorandum.

> On the right we had fine mountains lying in easy slopes in corn and vine and, on the left, the rich, extensive plains of the Saône in corn and pasture. This is the richest country I have ever beheld.[15]

In Beaune, at the medieval seat of the dukes of Burgundy with its colorful tile roofs and half-timbered façades, Jefferson made straight for a master merchant named Parent who had a wine shop in the Faubourg Bretonnière and was to become his principal wine broker. Jefferson was his ideal customer, instructing him, "Quality first, then price." Here he bought his favorite Meursault, Goutte d'Or, but it had to be the right vintage. "I want the 1784 vintage. That is the one I prefer above all others."[16] In the years to follow, Parent was to send Jefferson in Paris, in Philadelphia, at Monticello and in the White House not only this wine but casks, baskets, crates of bottles of Chambertin, Clos-Vougeot, Romancé, Latour-Montrachet, Volnay, Clos de la Commaraine. Jefferson, who deserves to be called the Father of American Wines, was also among its earliest connoisseurs.

On the road south, Jefferson stopped off in the hills not far from Villefranche at the Château de Laye Epinaye, surrounded by "a seignory of about 15,000 acres of vineyards, corn, pasture and wood, a rich and beautiful scene." Jefferson had been introduced to the Laye Epinayes by the *abbés* Chalut and Arnauld in Passy. Madame Epinaye, whose husband was in Paris, entertained Jefferson "with a hospitality, a goodness and an ease which were charming."[17] The special object of his visit, however, was Michael Angelo Slodty's statue of Diana and Endymion: a "delicious morsel of sculpture,"[18] wrote Jefferson, which "carries the perfection of the chisel to a degree of which I had no conception." When Jefferson compiled his notes later, he said it was "the only thing in sculpture which I have seen in my journey worthy of notice."[19] Prying himself away after two days, he drove on to Lyons, where he checked into the Hôtel du Palais Royal for three nights while he explored the ruins of the first Roman capital of Gaul. He was somewhat disappointed to find that "there are some feeble remains here of an amphitheater of two hundred feet diameter and of an aqueduct of brick." He was more impressed by the ancient Pont d'Ainay, its piers six feet in diameter, its arches "forty feet from center to center."[20]

Writing to Short, whom he had left at the Hôtel de Langéac as his chargé d'affaires, Jefferson wrote that he had to "hasten my movements." He would ignore the silk factories and other industries of Lyons because this interior city had no potential trade ties with the United States. "A knowledge of them would be useless," he told Short, "and would exclude from memory other things. Architecture, painting, sculpture, antiquities, agriculture, the condition of the laboring poor fill all my moments."[21]

Setting out after three days along the stark, windswept Rhône, Jefferson was moved to romantic prose by the rugged terrain of Dauphiné:

> Nature never formed a country of more savage aspect. A huge torrent rushes like an arrow between high precipices, often of massive rock, at others of loose stone with but little earth. Yet the hand of man [has] subdued this savage scene by planting corn where there is little fertility, trees where there is still less and vines where there is none. . . . It assumes a romantic, picturesque and pleasing air. The hills on the (west) side of the river [are] high, steep and laid up in terraces. The high mountains of Dauphiné and Languedoc [the Massif Centrale] are now covered with snow. The almond is in bloom and the willow putting out its leaf.[22]

Twenty miles south of Lyon, Jefferson reached Vienne, a Roman garrison town ringed by mountains. He was shocked to find that a praetorian palace had been converted into a Gothic church. To his friend Madame de Tessé he wrote,

> At Vienne, I thought of you. But I am glad you were not there, for you would have seen me more angry than I hope you will ever see me. The Praetorian Palace, as it is called, had been defaced by the barbarians who have converted it to its present purpose, its beautiful Corinthian columns cut out, in part, to make space for Gothic win-

> dows, and hewed down, in the residue, to the plane of the buildings.[23]

Awed by the scenery, Jefferson noted that "the hills come in precipices to the river," reminding him of the Susquehanna River valley of central Pennsylvania. Here Jefferson found in profusion the remains of a Roman empire he had only read about, including a temple of Augustus built by the Emperor Claudius.

He mixed wine with archaeology. The next day, he hired a guide to take him across the river to the famous vineyards of Côte-Rôti. In a "string of broken hills," he noted white soil "tinged a little sometimes with yellow, sometimes with red," piled up into terraces. He observed that "those parts of the hills only which look to the sun at midday or the earlier hours of the afternoon produce wines of the first quality." On through the "delicious"[24] country of the Rhône Valley, whose banks were flanked by stacked-up vineyards and orchards bursting into bloom, Jefferson drove to Tain, nestled at the foot of Ermitage Mountain. There he stopped to study the "wine called Hermitage, so justly celebrated." But he was barred admittance and, after a run-in at "the tavern of the post house, the master of which is a most unconscionable rascal,"[25] he drove on through Valence to Orange, where he was nourished with the remains of Roman grandeur. He admired a triumphal arch and bemoaned that the walls of an amphitheater were being pilfered to build a road. He noted that this was the farthest north that he saw olive trees. He hurtled southwest off the main Rhône road to Nîmes on the path of "the remains of Roman grandeur"—and a secret rendezvous.

From Nîmes, he wrote one of his longest, most romantic letters to his friends back in the Paris salon of his old friend Madame de Tessé, Lafayette's aunt, who was by birth a Grimaldi of Monaco, who had made the same tour and had done so much to pave his way on this trip. He found Nîmes all that he had expected, its Roman temple, the Maison Carrée, incomparable. Already he had selected it as the model for the Virginia Capitol. He traced the splendid arches of the Pont du Gard on the back of one of his ledger sheets. He studied the vast arena, the ancient engineering formula of one seat for each four Roman inhabitants suggesting just how vast their numbers had once been in France. To Madame de Tessé, he wrote:

> Here I am, Madame, gazing whole hours at the Maison Carrée like a lover at his mistress. Here, Roman taste, genius and magnificence excite ideas analogous to yours at every step . . . . I am immersed in

antiquities from morning to night. For me, the city of Rome is actually existing in all the splendor of its empire.[26]

Jefferson spent five days, his longest single stop, in Nîmes, but he was not quite immersed in the Roman era all day every day. He was surprised to find fine bookstores and skilled medal casters. But he was also keeping an appointment which, had he told anyone in Paris and word got back to Versailles, might have endangered intricate American diplomatic relations at court.

In a letter he wrote from Marseilles more than a month later, which was the earliest he could arrange to have it smuggled by Short out of the country by fast ship to John Jay in New York, Jefferson revealed the real reason for his unscheduled side trip to Nîmes. The letter to Madame de Tessé was a careful cover. He could be sure she would show it to her gossipy nephew Lafayette, which assured that everyone they both knew at court would hear about it. Writing secretly on May 4 and sending off his dispatch by an elaborate route to a ship at Le Havre, he asked Jay to lay before Congress the intelligence he had gathered over the past seven months of Brazilian plans for a rebellion against the Portuguese empire. He also passed on to Congress intelligence he had received of unstable political conditions in Mexico and the possibility of a revolution there. But his primary business at Nîmes in late March 1787, was to meet with Dr. José Barbalho Da Maia, a Brazilian from Rio de Janeiro who was studying medicine at Montpelier. As soon as Jefferson checked into Le Petit Louvre in Nîmes, he wrote to him in French to arrange a clandestine meeting.

The previous October, Barbalho had written to Jefferson that he had something of great consequence to tell him and left it to him how to arrange it. Jefferson, already planning to go south, got word to him that "I would go off my road as far as Nîmes under the pretext of seeing the antiquities of that place if he would meet me there." Their rendezvous arranged, Jefferson and Barbalho met. It was the first Jefferson knew that the subject of their talk would be a rebellion in Brazil, even though he must have suspected it. It was also the first time an American envoy was approached for aid in a revolution, and the first time a Latin American people looked north for the support of the American revolutionaries. Barbalho summed up conditions in Brazil and said that the insurgents already had the substantial war chest of $26 million promised from secret backers. To Jay, Jefferson reported:

> In short, as to the question of revolution, there is but one mind in that country. But there appears no person capable of conducting a

> revolution or willing to venture himself at its head without the aid of some powerful nation, as the people of their own might fail them. They consider the North American Revolution as a precedent for theirs. They look to the United States as most likely to give them honest support. [They] have the strongest prejudices in our favor. They would want cannon, ammunition, ships, sailors, soldiers and officers, for which they are disposed to look to the United States. . . . I took care to impress on him through the whole of our conversation that I had neither instructions nor authority to say a word to anybody on this subject and that I could only give him my own ideas, as a single individual, which were that we were not in a condition at present to meddle nationally in any war, yet a successful revolution in Brazil could not be uninteresting to us.

Jefferson had said enough to encourage Barbalho, however, and to reveal that he believed in supporting a Latin American uprising against a European power, and he made this quite clear to Congress:

> However distant we may be both in conditions and dispositions, from taking an active part in any commotions in that country, nature has placed it too near us to make its movements altogether indifferent to our interests or to our curiosity.[27]

Amazingly prescient, Jefferson was planting the first seed of his foreign policy toward Latin America, one that would bear fruit nearly forty years later when one of his disciples, as president, enunciated the Monroe Doctrine, still the credo of American dealings in the Western Hemisphere.

With a clock now ticking that dictated the exact moment he must make his rendezvous with a courier to carry his secret report to Congress, Jefferson left Nîmes on March 24 on the road to Arles, where he resumed his studies of Roman antiquity. Once again, he was dismayed at the disrespect shown for ancient artifacts by the French. At a once-grand amphitheater, he found only the portico "tolerably complete." Just how many porticos there had once been, he could not be sure, "but at one of the principal gates, there are still five. The whole of the upper entablature is gone." The attic had also been stripped bare. "Not a single seat" of the amphitheater was visible. "The whole of the inside and nearly the whole of the outside is masked by buildings. . . . There are one thousand inhabitants within the amphitheater."[28]

The evening of March 25, Jefferson arrived in the old university city of Aix-en-Provence, famous for its baths, clear air, and golden light. "I am now in the land of corn, vine, oil and sunshine. What more can a man ask of heaven?" exulted Jefferson to William Short. Jefferson was finally able to get warm after nearly a month on the road. "If I should happen to die in Paris, I will beg of you to send me here and have me exposed to the sun. I am sure it will bring me to life again." In an ebullient mood, Jefferson poured out his admiration for the Provençal city, "one of the cleanest and neatest I have seen in any country:

> The streets are straight, from 20 to 100 feet wide and as clean as a parlor floor. They have 1, 2 or 4 rows of elms from 100 to 150 years old which make delicious walks . . . . The buildings show themselves advantageously on the streets. It is in a valley where it begins to open towards the mouth of the Rhône, forming a boundless plain which is an entire grove of olive trees.[29]

Jefferson became fascinated by olive trees. He was convinced that their cultivation in America could raise the standard of living for poor farmers, improve nutrition, and provide a cash or barter crop. The olive tree, Jefferson believed, could be cultivated in Georgia and the Carolinas. "Having been myself an eyewitness to the blessings which this tree sheds on the poor," he wrote to one of his Charleston correspondents, "I never had my wishes so kindled for the introduction of any article of new culture into our country."[30] After studying the proper soil conditions on this trip, Jefferson sent off the first of five thousand cuttings to the United States where, unfortunately, his friends were not so impressed and used the few that survived merely for ornamental trees on their plantations. Jefferson also admired the olive tree because "it takes well in every soil, but best where it is poorest, or where there is none."[31] So precious was the topsoil in Provence, he observed to Short, that it was not wasted on planting grass —which, he felt, led to a salubrious chain of events:

> The want of dung prevents the progress of luxury in Aix. The poverty of the soil makes its streets clean. The preciousness of the soil prevents its being employed in grass. Hence, the dung gatherers (a numerous calling here) hunt it as eagerly in the streets as they would diamonds. Everyone, therefore, can walk cleanly and commodiously. [Because walking was possible, there were] few carriages, few assembly [balls] and other occasions for the display of dress.[32]

But, after noting that the waters at Aix were ninety degrees Fahrenheit, he found the baths an utter waste of time and money. When Short asked if they had any effect on his wrist, Jefferson replied, "None at all. But time is doing slowly what they cannot do." Between "douches," Jefferson went sightseeing and enjoyed evenings at the theater, where he loved to hear an actress from Marseilles who "has ear, voice, taste and action: she is, moreover, young and handsome."[33] The unidentified actress also had the advantage over "the celebrated ones of Paris in being clear of that dreadful wheeze, or rather whistle in respiration, which resembles the agonizing struggles for breath of a dying person."

Provençal speech fascinated him. "Every letter is pronounced, the articulation is distinct, no nasal sounds disfigure it and on the whole it stands close to Italian and Spanish in point of beauty."[34]

Staying in Aix only four nights, "long enough to prove the inefficacy of the waters,"[35] the restless Jefferson hurried on to the great Mediterranean seaport city of Marseilles, where he remained for an entire week. He had come to study the port's commerce, but he was charmed by the dramatic situation of the city in a vast natural amphitheater jammed against the sea by jagged "high mountains of naked rock,"[36] the steep banks of calcaire pocked with caverns.

Staying at the "tolerably good" Hôtel des Princes, he found "the commerce of this place most remarkable."[37] He went down to the docks and hunted up American merchant ships. He interviewed Marseillaise merchants and tried to find out exactly how many American ships had called in four years of peace. From them, he learned about the sources of an important trade in rice funneled through Marseilles to Paris and the north. Two varieties of rice were being sold, small amounts of Carolina rice from the United States, preferred only for serving with milk and sugar, and Piedmont rice, from the high northern plains of Italy, preferred for almost all cooking with meat or oils because fewer grains had been broken. He had recently been elected honorary vice president of the South Carolina Society for Promoting and Improving Agriculture: he wanted to report back to its rice-growing members. Jefferson now learned that the Italians had developed a special machine for husking the rice without breaking it. He also studied the "cultivation and species of the dried raisin," which he was sure would succeed in South Carolina. He spent whole days talking to gardeners:

> From that class of men I have derived the most satisfactory information in the course of my journey and have sought their acquaintance with as much industry as I have avoided that of others who would have made me waste my time [in] a good society.

From gardeners, he learned of new varieties of figs from Smyrna, of olives, capers, pistachio nuts, almonds. "All of these articles may succeed on, or southward of, the Chesapeake."[38] He toured the notable wine cellar of a merchant named Bergasse, took excursions to the Château Borely on a hill commanding the harbor, and a boat trip to the medieval fortress, the Château d'If, where the legendary Count of Monte Cristo had been imprisoned. Jefferson, fleeing all-too-familiar scenes where he had gone with his Maria, most of all finally was managing to relax in noisy, lively, anonymous Marseilles. He wrote back to *salonistes* in Paris about a profound transformation that had overtaken him. "The plan of my journey as well as my life," he explained to his artist-friend, Madame de Tott, "being to take things by the smooth handle, few things occur which have not something tolerable to offer me":

> [The inn] in which I am obliged to take refuge at night presents in the first moment nothing but noise, dirt and disorder. But the *auberge* is not to be too much abased. True: it has not the charming gardens of Chaville without, nor its decorations nor its charming society within. I do not seek, therefore, for the good things which it has not, but those which it has. A traveler, says I, retired at night to his chamber in an inn, all his effects contained in a single trunk, all his cares circumscribed by the walls of his apartment, unknown to all, unheeded and undisturbed, writes, reads, thinks, sleeps, just in the moments when nature and the movements of his body and mind require. Charmed with the tranquility of his little cell, he finds how few are his real wants, how cheap a thing is happiness, how expensive a one, pride.

Far from Monticello, Paris, and Versailles, Jefferson viewed "with pity the wretched rich whom the laws of the world have submitted to the cumbrous trappings of wealth." He saw the rich man "laboring through the journey of life like an ass oppressed under ingots of gold, little of which goes to feed, to clothe or to cover himself, the rest gobbled up by harpies of various descriptions with which he has surrounded himself. These, and not himself, are its real masters. Jefferson was alone and lonely, he admitted, but he could find "no middle ground" between "the society of real friends and the tranquility of solitude."

That Jefferson managed to think at all between "the cruel whip of the postilion by day" and the noise around him was, he thought, worth having Madame de Tott relate to her salon:

> I should go on, Madam, detailing to you my dreams and speculations, but my present situation is most unfriendly to speculation. 4,350 market women [I have counted them] brawling, squabbling, jabbering *patois;* 300 asses braying and bewailing to each other and to the world their cruel oppressions, 4 files of mule carts passing in constant succession, with as many bells to every mule as can be hung about him, all this in the street under my window, and the weather too hot to shut it.[39]

At last, Jefferson was in a place where he felt warm enough not only to speculate but to complain. He also delved seriously into "the subject of rice, as [Marseilles] is a great emporium for most of the Levant and of Italy," he wrote to John Jay. He was convinced that the Italians possessed a superior machine "for cleaning which brought European rice to market less broken than ours, as had been represented to me by those who deal in that article in Paris." In Marseilles, "I found several persons who had passed through the rice country of Italy, but not one who could explain to me the nature of the machine." But he could "see it myself on entering Piedmont," if he took a three-week detour from his itinerary. He had planned to go to the Palladio country around Venice and then south to Rome. But, if his trade mission succeeded in the rice country, he could quite possibly place our rice above all rivalship in quality as it is in color "by introducing a better machine."[40]

Jefferson was not just indulging a spur-of-the-moment scheme by dropping everything and going off to Italy. He was deeply concerned that the manner of cultivating rice and the type grown in the Carolinas was one reason for the stubborn adherence of the planters to slavery there. He had a long-standing interest in changing the type of rice from the Carolina variety that only grew in fiercely hot, malarial swamps that took a high annual death toll of slave laborers to an upland variety that could be grown by yeomen farmers without slaves. He had tried growing some in his meadows at Monticello as early as 1774. In *Notes on Virginia,* he had written, "The climate suits rice well enough wherever the lands do."[41] He later wrote that "I cultivated it two or three years at Monticello and had good crops, as did my neighbors, but not having conveniences for husking it, we declined it."[42] Jefferson probably experimented with upland rice seeds supplied him by Philip Mazzei. Rice had been introduced in Tidewater Virginia in 1647 and had reached the Carolinas by 1685, then spread into Georgia. Jefferson understood the dependence of its planters on the crop, but he also was aware of the widespread loss of health and life resulting from cultivation in flooded swampland. To solve the mystery of producing a less-dangerous type of rice that was at the same time acceptable to the lucrative northern European market, he decided to go over the Alps into Italy and explore the rice country. He told no one that to do so involved great physical danger on the journey through the mountains. There also was the matter of a ban on bringing out any Piedmont rice unmilled. The penalty was death.

Eating a hearty breakfast, settling his bill at the Hôtel des Princes and generously tipping the servants, Jefferson set off from Marseilles for Toulon, his next stopover, where he spent two days questioning the peasants about the cultivation of capers, which he later sent back to America. He went to the theater at night, then drove on to Hyères, the oldest Mediterranean winter resort. He noted the colorful cherry trees, the hedges of pomegranate, sweetbriar, and broom, the thyme growing wild. He was entering the French Riviera, its dark wooded massive promontories reaching straight down to sugar-white beaches washed by an azure sea. At Hyères, covered with olive trees with black trunks "four feet thick" and orange trees in full fruit, he visited the gardens of Monsieur Fille, who cultivated some fifteen thousand orange trees. "They are blossoming and bearing all the year," noted Jefferson, "flowers and fruit in every stage at the same time." He found Hyères "delicious," described it carefully. A plain of two or three miles was covered not only with scented orchards but the shore was lined with "palm trees twenty to thirty feet high." The streets were steep, in stairs of steps eight feet wide. "No carriage of any kind can enter it. The wealthiest inhabitants use [sedan] chairs. But there are few wealthy." Hyères reminded him of a feudal manor. Everyone else was a "laborer of the earth."[43] Following the strand of Riviera fishing villages—St. Raphael, La Napoule, Cannes, Antibes—he hugged the sea, the rough precipices on his left forcing his carriage to stay on a narrow track over the edge of the sea. He crossed the Var on a ferry, leaving France. He had to show a special passport issued him by the French foreign minister and entered the kingdom of Sardinia and, a few miles later, the free city of Nice, a walled town named for Nike, the goddess of victory. In his log, he noted that the climate was "quite as delightful as it has been represented."

At Nice, already an English resort colony, Jefferson stayed at the Hôtel de York, in the new English section outside the city walls. He considered it "a fine English tavern very agreeably situated. . . . The mistress [is] a friendly, agreeable woman."[44] Tobias Smollett, taking a cure for tuberculosis, had been one of the first English visitors here more than twenty years earlier and had popularized Nice for his English audience in his *Travels Through France and Italy* (1766). A Scottish physician who had known Jefferson's teacher William Small, Smollett was interested in some of the same things Jefferson was, especially classical archaeology and literature. Jefferson's favorite novelist, Laurence Sterne, had also made the same voyage and had written *Sentimental Journey*, excoriating his rival Smollett. Jefferson undoubtedly was familiar with Smollett's descriptions of Nice and its lawless, sharp-trading Niçoise merchants. The tax-free city, where any fugitive could come, live, and open a business, had become a crossroads of Mediterranean trade. A walled city once dominated by a powerful *château* on the summit of a towering rock formation, it had held out for centuries against raids from North Africa, only to be stripped of its ornaments, its castle, and its cathedral, which were deliberately blown up by French royal engineers so that it could no longer pose a threat to the authority of Louis XIV. Its ancient walls now formed a walkway where ladies and gentlemen of fashion as well as the local prostitutes took promenades along the sea without struggling to walk on the rocky beaches. Jefferson joined Niçoise gentry and, to his horror, the large entourage of the duke of Gloucester, George III's brother, who was visiting the Riviera on his own grand tour of France and Italy.

Jefferson adored Nice. He spent more time in the crowded town of eighteen thousand than anywhere else along his route—"you may pass many days here very agreeably"[45]—and he advised Short and his other young Virginia friends. He declared the Côte de Provence wine of the Bellet vineyard "the most elegant" and "remarkably good."[46] With a letter of introduction from Abbé Chalut, he paid a call on the owner of Château Bellet, the wine merchant Sasterno, at his villa across the river Paillon from the city walls in a new *quartier* of palm-shaded villas. Sasterno told him how he could find his way up to the heights of Cemenelum and the ruins of vast Roman baths and a three-thousand-seat arena at Cimiez overlooking Nice. Here, in the shade of olive groves and orange trees, he found ruts in the stone road between the sandy arena and the Via Julia, scored by the chariots of legionnaires who had come to soak away the dust of travel between Rome and the Gallic war front. On the ruins of the Roman Empire, a monastery had been built: he could visit the exquisite wooden Brea triptyches hanging above its altar. Strolling through the monastery gardens, the highest ground in Nice, he could look south over the azure crescent of the Bay of Angels at the old Greek colony on its hummock above the walled city or, to the north, to the Maritime Alps, still snow-covered in mid-April, as they started to climb toward Italy.

From the merchant Sasserno, Jefferson confirmed what he had heard at Marseilles. Trains of up to one thousand mules loaded with sacks of rice and other goods were guided each week over the Alps from Turin to Nice. Only two years earlier, a Scottish engineer had supervised the blasting of a new tunnel road through the Alps to make the voyage passable even for the Niçoise ladies, who were carried all the way to the court city of Turin in carriages and sedan chairs to go shopping each summer.

Despite the fact that there were still snowstorms and drifts on the 120-mile road through the Alps, which made a carriage ride very dangerous so early in the year, Jefferson decided to make the journey anyway. He hired a string of bell-bedecked mules and joined the muleteers on their perilous trek north. But first, after leaving the arched Turin Gate, there was the obligatory stop along the Paillon at the little baroque chapel of Notre Dame de Bon Voyage for prayers, candlelighting, incense, and donations. On virtually every voyage, a muleteer followed his pack animal over the edge of a precipice down a thousand meters or more to his death. They would not leave without the special blessings of the priest at Notre Dame. Climbing back into his phaeton, Jefferson began his four-day journey over the towering Italian Alps. As the road climbed past tilled sedimentary hills and wound and switched back and forth like an agitated snake hour after hour, he noted that there were "speckled trout from Nice to Turin" in the cold mountain streams. He wrote down at what elevation the olive trees gave out and the thick, fog-shrouded spruce forests began. He took scientific notes as he passed his forty-fourth birthday one mile into the Alps. He urged Shippen and Lee to observe the "many curious and enchanting objects on this passage. . . . Watch where you lose and where you recover the olive tree in rising and descending the several successive mountains." On three ever-steeper mountains, he lost sight of the olive tree as he climbed. As he passed through the first Alpine tunnel at the Col de Tende seven thousand feet high on the Italian frontier, this sure horseman who had never been higher than the two-thousand-foot Blue Ridge of the Appalachians breathed the thin mountain air with exultation and admired the handiwork of the road builders. "This road is probably the greatest work of this kind which ever was executed either in ancient or modern times. It did not cost as much as one year's war." At the end of the third day, Jefferson had to send his carriage back to Nice and mount a mule, his feet and greatcoat dangling near the road. Even surefooted mules had to pick their way along at a mile an hour from here on up. Sometimes, Jefferson had to dismount and lead the animal so it could rest. But he found the scene breathtakingly romantic. This man who loved his mountaintop home could occasionally make out a castle high on a peak above him. To Shippen and Lee, he wrote, "Fall down and worship the site of the Chateau di Saorgio; you never saw nor will see such another."[47]

A few weeks later, he wrote of the scene in his most romantic prose in a letter smuggled to Maria Cosway in London:

> Imagine to yourself, madam, a castle and village hanging to a cloud in front, on one hand a mountain cloven through to let pass a gurgling stream; on the other, a river over which is thrown a magnificent bridge, the whole formed into a basin, its sides shagged with rocks, olive trees, vines, herds, etc. I insist on your painting it.[48]

All along the road, Jefferson passed narrow rock terraces where oats to feed livestock grew and were cultivated by peasants whose day began at 3 A.M. and continued until after dark. At night he read classical and modern books he had brought in his saddlebags, trying to resolve the scholarly controversy then raging over Hannibal's route. He wrote George Wythe that there was insufficient evidence to be sure.

Jefferson arrived in Turin, the capital of the vast kingdom of Sardinia, which extended from Geneva, Switzerland, to the boot of Italy and included Nice and Sicily, on April 16, 1787. Presenting his credentials was out of the question: he still was traveling incognito. Passing up the opportunity to visit the resplendent palaces around the Piazza San Carlos, he stayed at the Hôtel d'Angleterre and was free to attend the theater two out of his three nights in Turin and visited the museum of antiquities, which he found "worth seeing." He advised his protégés to "ask for Nebuile wine" and said the Montferrat wines were also "worth tasting." From Turin, he drove out into the rice country, quickly making two discoveries. First, the rice region was not the Piedmont; it was farther north, in Lombardy, and he struck out on excursions to Moncaglieri, Stuponigi, Superga to study the rice fields, which lay on either side of the road. He also made a "fine excursion to Lake Como." He then headed toward Milan, studying the great rice fields between Vercelli and Pavia, stopping to talk to peasants in the fields and interviewing owners. He drove on to Novara, where there were "fields of rice all along the road." Permitted to see the husking machinery, he later sketched it from memory to illustrate a long memorandum describing rice processing. He made excursions to Rozzano, saw how Parmesan cheese was made and studied their icehouse to learn how the cheese was stored all year in snow. He toured the dairies that produced the cheese. Here he learned how to make and store ice cream at Monticello. He discovered mascarponi, "a rich and excellent kind of curd'' and, from the Count Francisco Del Verme, learned how it was made. From Rozzana, he went on to the ''celebrated church'' of Certosa near Pavia, ''the richest thing I ever saw''; he copied its cloister in his original design for the University of Virginia. His curiosity never let up: he noted in his log an ''excursion to hear the echo of Simonetta.''[49]

His second and most important discovery in Italy was that it was not only the husking machine that accounted for the superiority of Piedmont rice. Piedmont was a different species. To Edward Rutledge in rice-growing South Carolina he wrote that this was the ''one conclusion to be drawn.'' Despite the fact that to take any of the unhusked rice out of Italy to a place where it could be used to plant a crop in competition to the Piedmont was a crime punishable by death, Jefferson, as he told Rutledge, ''determined to take enough to put you in seed.'' He told Rutledge he knew this was illegal. ''They informed me that its exportation in the husk was prohibited.'' Bribing the man enough to make the risk worth his while, he ''took measures with a muleteer to run a couple of sacks across the Apennines to Genoa,'' where he could take it by boat to Nice. But just in case the muleteer was caught or decided to pocket the money, Jefferson decided to ''bring off as much as my coat and surtout pockets would hold.'' Weeks later, he shipped the contraband rice off to Rutledge and urged him to use it sparingly. ''The little I brought myself must be relied upon for fear we should get no more.''[50]

His pockets stretched as he carried out daring agricultural espionage, Jefferson hurried on as fast as his carriage would take him over the terrible mountain roads of the Appenines to Milan, where he admired the salon of the Casa Belgioiosa—''superior to anything I have ever seen''—and Casa Candiani by Appiani with its small museum ''the ceiling of which is in small hexagons within which are small cameos and heads painted no two the same.'' But he found the famous cathedral of Milan revoltingly opulent:

> The cathedral of Milan [is] a worthy object of philosophical contemplation, to be placed among the rarest instances of the misuse of money. On viewing the churches of Italy, it is evident without calculation that the same expense would have sufficed to throw the Appenines into the Adriatic and thereby render it *terra firma* from Leghorn to Constantinople.[51]

Jefferson's impromptu trade mission to the rice country had made his original objective impossible; he could not go on to Rome and see the ruins of antiquity. The rich banking-center city of Genoa was to be ''the spot at which I turned my back on Rome and Naples," he later wrote apologetically to Maria. "It was a moment of conflict between duty which urged me to return and inclination urging me forward."[52] Rice, tobacco, foreign loans, and the impending arrival of his nine-year-old daughter, Polly, all pulled at him to abandon his romantic adventure.

In Genoa, home of Columbus, its streets crammed with the palaces of the merchants and bankers who had financed the Crusades and bankrolled the exploration of the Americas, the Middle East, Africa, and the East Indies, Jefferson gulped down "the abundance to be seen here" in four last days of imbibing Italian culture. He attended plays, gaped at the harlequin marble floors and exquisitely decorated ceilings, strolled through the inner courtyard gardens overhanging the half-moon harbor. On April 27, 1787, he made a final thirty-mile excursion along the magnificent Italian Riviera. He found that "the Prince Lamellino's gardens at Sestri are the finest I ever saw out[side] of England."[53]

The journey back to Nice was to prove the most dangerous part of his tour. Deciding to make up time by chartering a small sailing vessel, he barely escaped a raid on the harbor of Genoa by Algerian warships. Then the small, crowded vessel became becalmed in the hot sun for two full days. "Mortally sick," he asked the captain to put him ashore as soon as possible. They landed at Noli after making only fifty miles of headway. His mind functioning better than his stomach, he was able to notice "the aloe on a precipice hanging over the sea." Since Alexander the Great had ordered his soldiers to carry aloe on their marches of conquest, the plant had been associated with healing under adverse conditions. At Noli, a small fishing village, Jefferson found only "a miserable tavern," but one that provided "good fish, namely sardines, fresh anchovies,"[54] and fresh strawberries. After a good dinner and a bad night at Noli, Jefferson spent the next two days, as he wrote to his daughter Patsy, "clambering the cliffs of the Apennines, sometimes on foot, sometimes on a mule, according as the path"—barely four feet wide here—"was more or less difficult, and two [days] travelling through the night as well as day without sleep."[55] Yet, as Jefferson walked with the bright-blue sea at his left, the steep, snow-covered mountains close on his right, he mused,

> If a person wished to retire from his acquaintance, to live absolutely unknown and yet in the midst of physical enjoyments, it should be in some of the little villages of this coast, where air, water and earth concur to offer what each has most precious. Here are nightingales, beccaficas, ortolans, pheasants, partridges, quails, a superb climate, and the power of changing it at any moment from summer to

> winter by ascending the mountains. The earth furnishes wine, oil, figs, oranges and every production of the garden in every season.[56]

After a night at Albenga in the "most detestable" tavern he had ever seen "in any part of the earth and dearest, too,"[57] an exhausted Jefferson and his tired mules went on slowly toward Nice, where he rested four days and then climbed into his carriage for the drive to Marseilles, where he safely packed off letters and packages of rice toward Paris and continued on his tour of French ports.

What Jefferson considered the high point of his fourteen-week odyssey was a nine-day excursion along the two-hundred-mile Canal du Midi. One of the great engineering achievements of the reign of Louis XIV, the canal was built to allow shipping between the Mediterranean and the Atlantic without exposure to pirates or enemy navies. Jefferson hired a boat and a horse and driver and entered the canal at Cette. He had his carriage detached from its wheels and sat in it, writing down his observations. He had to do without fresh butter for breakfast and the bread was "rarely eatable." "I chose to go leisurely," he wrote Shippen and Lee. He could have trimmed five days by taking a mail boat instead. He passed through the Frontignan wine country around Beziers, saw the last of the olive trees slide by at Carcassone. Giving the boat horse a day of rest, he hired horses and a guide and rode up into the hills to examine "the sources of water which supply the canal of Languedoc." Back on the canal, most of the time, Jefferson walked along beside the boat:

> Of all the methods of travel I have ever tried, this is the pleasantest. I walk the greater part of the way along the banks of the canal, level and lined with a double row of trees which furnish shade. When fatigued, I take seat in my carriage where, as much at ease as if it were my study, I read, write and observe. . . . I have had some days of superb weather, enjoying two parts of the Indian's wish, cloudless skies and limpid waters. I have had another luxury which he could not wish, since we have driven him from the country of mockingbirds—a double row of mockingbirds along the banks of the canal in full song.

Arriving at Vaucluse "little fatigued," Jefferson sat down at a fountain in a secluded hollow of a mountain, two hundred feet above the gushing fountain, "the ruins of Petrarch's chateau on a rock." To his delight, "every tree and bush" was "filled with nightingales in full chorus. Their song is more varied, their tone fuller and stronger than on the banks of the Seine:

> It explains to me another circumstance, why there never was a poet north of the Alps, and why there never will be one. A poet is as much a creature of climate as an orange or a palm tree.[58]

Spending two days at Toulouse, the western end of the canal, Jefferson floated down the Garonne through the Bordeaux region and some of the best vineyards on earth. He stopped at Sauterne, noting it was remarkable for its Château Margaux wine. He promptly sent off six cases to his brother-in-law, Eppes, in Virginia; he ordered 250 bottles sent to the Hôtel de Langéac at Paris! At Bordeaux, while straightening out the twisted financial affairs of the American consul, Thomas Barclay, Jefferson toured the brick remains of a Roman circus, "enquired into" a wide variety of Médoc red and white Graves Bordeaux wines, and attended plays at the magnificent Théâtre de Bordeaux. Forty years later, he would adapt its rotunda to his University of Virginia.

Returning to Paris by way of Nantes and L'Orient, the Atlantic ports trading with the United States, Jefferson studied the poverty-stricken people of Brittany, the "hilly country with poor grey soil," the "ragged" people living on rye bread. "The villages announce a general poverty as does every other appearance. Women smite on the anvil and work with the hoe. There are but few chateaux here."[59] As if for sheer contrast of condition, he hurried on to Amboise and toured the famous pagoda and gardens of the Duc de Choiseuil, Madame de Pompadour's château at Menars, and the beautiful gardens along the Loire, strolling on a plateau of espaliered roses, heliotropes, and fuchsias. The last night of his voyage, he stopped in Orléans, only long enough to have his phaeton repaired and visit a barber for a shave. On June 10, 1787, he drove into Paris after an absence of 104 days. "I never passed three and a half months more delightfully,"[60] was his verdict.

Pulling into the courtyard of the Hôtel de Langéac, he was soon flipping through the basket of letters for a familiar handwriting. Maria Cosway was cross with him: had he "gone to the other world"? Before answering any other letter, he answered hers. "I am born to lose everything I love. Why were you not with me? So many enchanting scenes which only wanted your pencil to consecrate them to fame." Urgently, Jefferson demanded, "When are you coming here? If not at all, what did you ever come for? Only to make people miserable at losing you. Consider that you are only four days from Paris. Come then, dear madam, and we will breakfast every day à l'Angloise, hie away to the Desert, dine under the bowers of Marly and forget that we are ever to part again."[61] It was more than two months before Maria took the bold step of coming to Paris without her husband, staying for four months with the Princess Lubomirska, a cousin of the king of Poland who entertained at one of the great Paris salons. Although they renewed their round of impromptu rendezvous at the Hôtel de Langéac and strolled through gardens, by November 13, Jefferson was writing to Trumbull in London,

> Mrs. Cosway is well but her friends are not so. They are in continual agitation between the hopes of her stay and the fear of her recall. A fatality has attended my wishes and [our] endeavors to see one another since she has been here. From the mere effect of chance, she has been away from home several times when I called on her and I when she has called on me. I hope for better luck.[62]

Jefferson was busier than ever with affairs of state. By December 1, Maria was complaining that she wanted to see him more. "I have felt the *loss* with displeasure."[63] She did not believe how long she had been in Paris, how little she had seen him. On December 7, she invited him to breakfast the next day. When he arrived, she was already gone to England. She had left a note. To bid him "adieu once is sufficiently painful. . . . I leave you with very melancholy ideas."[64] Two days later, she wrote again, making it evident that they had met so little because he did not want to share her company with a roomful of even the most brilliant.

As early as February 1787, three weeks before he had set out on his southern odyssey, Jefferson had hinted at his dissatisfaction at Maria's habit of surrounding herself with an impenetrable circle of artistic friends and admirers. He made no allowance for the fact that she was a celebrity in her own right. He had grown tired of the incessant round of parties and salons, and of the "empty bustle" of "gay and thoughtless Paris." Complaining to the beautiful Ann Willing Bingham of Philadelphia of "the absence of conjugal love and domestic happiness" in France, he drew "a contrast between the simple innocence of Americans and the worldly sophistication of Europeans":

> Thus the days of life are consumed, one by one, without an object beyond the present moment. . . . In America, on the other hand, the society of your husband, the fond care for the children, the arrangements of the house, the improvements of the grounds, fill

> every moment with a healthy and useful activity. They are contented to soothe and calm the minds of their husbands returning ruffled from political debate. . . . They have the good sense to value domestic happiness above all other and the art to cultivate it above all others.[65]

Maria envied Jefferson's freedom as a man to come and go as he pleased. "I do not know how to go about it."[66] By March 1788, after he had written less and less that winter, she wrote to him angrily, "Your long silence is unpardonable."[67] She had thought of sending him only a blank sheet of paper, she said. But Jefferson was not in Paris when her diatribe arrived. The pace of American politics at home and the diplomatic consequences in Europe were filling up more and more of his time.

The failure of Congress to collect taxes from individual states had triggered a financial crisis that required Jefferson to rush off to the Netherlands to confer with Adams, before he headed home to take his post as vice president of the new federal government. After a tortuous five-day journey, the two met at The Hague. Their different ideas had never been more evident. Adams thought that the Dutch bankers had immeasurable avarice, were bluffing when they threatened the near-bankrupt United States. Jefferson was not so sure. America needed the Dutch more than the bankers needed such a feeble customer. Jefferson wanted to pay the $1 million in interest due so they could negotiate a new two-year loan. During his travels in France and Italy, a special constitutional convention had been called in Philadelphia. Under its strengthened central powers, which Jefferson had urged five years earlier as a member of Congress, the new government could expect to be able, by that time, to collect taxes and meet its foreign obligations. The two envoys argued over a new loan, but it was Adams who finally backed down and agreed to initiate negotiations with the Dutch bankers. Adams then left, annoyed at Jefferson, who stayed on for nine days of successful negotiations.

On his way back to Paris, Jefferson rewarded himself with a trip through the Rhine wine country of Germany, which he found "sublime."[68] At Frankfort, he had a warm reunion with Baron de Giusmar, once his ward as a prisoner-of-war in Virginia. But they could not renew their old habit of playing violin duets. Jefferson's bow hand had been permanently crippled: he would never play well again.

Jefferson returned in late April 1788 to a life in Paris that gradually was growing more domestic. His nine-year-old daughter Polly had at last arrived from Virginia. For three years since the death of little Lucy, Jefferson had been trying to bring Polly to Paris to be with him and with Patsy. He desperately wanted what was left of his family reunited. But Polly had been only four years old when her mother died and she clung to the aunt who had raised her while resisting going to the father she scarcely knew. In the spring of 1786, when she was seven, Jefferson finally insisted that the Ep-peses send her to him. "I am sorry you have sent for me," the child wrote back. "I don't want to go to France. I want to stay with Aunt Eppes."[69] Finally Eppes resorted to trickery to pull her away from her aunt, taking her aboard a ship with her body servant, thirteen-year-old Sally Hemings, and playing with her until she fell asleep. When she awoke, the ship had sailed. The slave Sally was to prove worthless as a nanny—only a child herself, sick and frightened from the sea voyage—but the ship's captain was kind to Polly and helped her arrive in London safe, sound—and clinging to *him*. Again Polly transferred her affection, this time to Abigail Adams. "I show her your picture," wrote Abigail. It was probably the miniature Trumbull had done for Abigail about the time he had introduced Jefferson to Maria Cosway. "She says she cannot know it, how should she when she should not know you."[70]

Abigail was annoyed at Jefferson, pleading a pile of work in Paris, when he did not come to London to meet Polly but instead sent his servant Petit. After several more days of tears and coaxing, the distressed child, always being torn from anyone close to her, went off with Petit to meet her father. Abigail was again distressed when Jefferson placed her in the convent school at Panthemont with her sister, but she quickly adjusted and the two joined him for weekend visits, bringing with them another American schoolmate, Kitty Church, the daughter of Angelica Schuyler Church, who was the daughter of Gen. Philip Schuyler of Albany and the sister-in-law of Alexander Hamilton. Mrs. Church, married to an Englishman, was a close friend of Maria Cosway.

The Hôtel de Langéac was filled on weekends with the happy noise of children. The two sons of Hector St. John de Crèvecoeur and the sons of General Nathanael Greene often left L'Ecole pour Jeune Noblesse, their boarding school across rue de Berri, to join the American girls at Jefferson's villa. Jefferson had less need to disturb the peace of his reunited family with his reinvolvement with Maria. For the moment, he was enchanted by his daughter Polly, who had his late wife's beauty, was slight and high-strung as her mother had been, bewitching if less studious than her serious, tall, redheaded older sister. Polly was his favorite now and he wrote gratefully to his in-laws, "Her reading, her writing, her manner in general show what everlasting obligation we are under to you."[71] Music once again soothed Jefferson as Patsy played her new harpsichord for her father. And they all went with him in the phaeton to concerts at the Tuileries Palace. On Monday, Jefferson dropped them off in their red smocks at school and turned his full attention to the growing political crises in France and the United States.

With Polly from America had come Sally Hemings, the thirteen-year-old sister of Jefferson's chef, James. Abigail Adams warned Jefferson that "the girl [Polly] has with her wants more care than the child and is wholly incapable of looking properly after her without some superior to direct her."[72] Jefferson could turn her over to James to help in the kitchen, but Polly needed her at school. Some writers have tried to paint her as the concubine of a lonely, lovesick Jefferson. In *Thomas Jefferson: An Intimate History,* the late Fawn M. Brodie suggested that, when Jefferson traveled through France and Germany and eight times described soil as mulatto in his twenty-five sheets of notes, he was not referring, as he labeled the appropriate column of his charts, to yellowish soil in the hills and valleys he traveled through but was really thinking of the contours of Sally's body. And when he was taking notes on a new kind of mold-board plow that he invented shortly after the journey, he was really thinking of plowing the fertile Sally as soon as he returned to Paris. But *mulatto* is a precise term describing yellowish-brown soil. And when Jefferson used the term *mulatto* to describe soil during his French travels, Sally was still on a ship with Polly, accompanying her to France. If he had ever noticed her or remembered her at all, Sally had been only ten years old when Jefferson last visited Monticello hurriedly in 1784 to pack James Hemings off to France with him. She was only eight when Jefferson had last resided at Monticello and was mourning his wife's death. Unless Brodie was suggesting that Jefferson consoled himself by having an affair with an eight-year-old child, the whole chain of suppositions is preposterous.

Jefferson had enemies, however, who, in his lifetime, broadcast stories of his cavorting through the slave quarters on Mulberry Row. Chief among them was a disgruntled former employee, the worst sort of witness. James Callender had worked as a political writer in Philadelphia for years, selling his services free lance. During Jefferson's third campaign for the presidency, Jefferson's enemies hired Callender and apparently leaked to him a highly charged account of his flirtation in 1768 with Betsy Walker and added for good measure lurid descriptions of Sally Hemings's arrival from Paris pregnant with Jefferson's child, the first of many he supposedly sired by her as the master of Monticello made her the mistress of his wifeless household. Historians dismissed the Callender charges for nearly two centuries until Fawn Brodie dusted off a highly inaccurate and uncorroborated memoir by a man who described himself as Madison Jefferson, son of Jefferson and Sally Hemings. The account, published by an abolitionist journalist only in the *Pike County* (Ohio) *Republican* in 1873, where the aged former slave was homesteading, resembles many uncorroborated slave narratives and cannot be credited. It is full of hearsay about events that the would-be former house slave could not have seen or known first-hand, if only because of his age, and must be put down as mere gossip about a great man published in the absence of journalistic standards, much less historical ones. Also offered as evidence by Brodie and other writers since her book was published in 1974 are descriptions of a host of mulattoes at Monticello who resembled Jefferson, and a declaration that one even played the fiddle!

President Jefferson refused to dignify Callender's charges publicly, but privately he wrote to subordinates to confront the charges in general, admitting he had once made a fumbling pass at Mrs. Walker when he was very young but saying that this was the only incident they should credit in the slightest. Thus, he denied the charges of having a slave family with Sally Hemings. And when Henry Randall journeyed to Monticello half a century later to interview Jefferson's family, Jefferson's granddaughter, Ellen Randolph, suggested that the father of Sally Hemings's children who so resembled Jefferson actually was Jefferson's nephew, Samuel, "the most good-natured Turk that ever was master of a black seraglio kept at other men's expense."[73] Sally Hemings's lover was, in other words, a son of Dabney Carr and Jefferson's sister Martha. It is impossible to believe that Jefferson abandoned his love for Maria Cosway to force his affections on even the most beautiful adolescent slave girl, just as it is beyond belief that Jefferson would pay the blackmail demanded fifteen years later by a hireling political scrivener to keep the affair out of print.

Shortly before Jefferson had left Paris in February 1787 for his grand tour, he had attended the opening session of the Assembly of Notables at Versailles. Jefferson could not foresee the rapid onset of revolution in France. As he rode south into the sunshine, the Assembly had listened to the king's finance minister, Calonne, and, rejecting his bold plan for reforms, instead had insisted that he be dismissed. Along with Lafayette and his reform-minded friends, Jefferson agreed on the need for structural changes. France was debt-ridden and the middle classes, asked to shoulder an ever-increasing tax burden, would not be placated without solid reforms. Jefferson favored Calonne's call for creating provincial assemblies as the first stop to bringing an experienced *bourgeoisie* into the government. But there was drought in the wheat fields that summer, and doubling prices for bread accompanied doubling unemployment in Paris. As mobs began to riot in Paris, every journalist denounced the royal government. When Jefferson, worried about the stability of France–U.S. relations, returned to his post, he took the opportunity to press Versailles to put into effect Calonne's agreement to allow the United States to sell France fifteen thousand tons of tobacco above and beyond Robert Morris's twenty-thousand-ton monopoly and to refuse to renew Morris's three-year franchise agreement when it expired. The decree finally was signed at Versailles in December 1787.

Jefferson reported this diplomatic victory to Congress along with his candid assessment of the deteriorating state of affairs at Versailles in the face of a "revolution of public opinion." "The king," he reported to John Jay, "long in the habit of drowning his cares in wine, plunges deeper and deeper, the queen cries, but sins on."[74] But this victory was to prove fleeting.

Even as France opened her doors to American goods a little wider, Americans used their newfound profits to increase their orders for more British manufactured goods instead of spending their money in France, confirming the suspicions of French merchants who could only think in mercantilist terms and were worried that French support of the American Revolution had drained the last drop out of the French treasury. In fact, nothing remained of Louis XIV's mercantile riches, and a corrupt palace guard of *noblesse* could not serve the interests of France, let alone her allies. All that remained of a Franco-American policy after the death of the foreign minister Vergennes was that the French government wanted the United States strong enough to foil British interests but weak enough to remain dependent on French economic and political support. As the confusion of purpose of French policymakers trickled down to French merchants with no other channel for their anger, the merchants, who would have been the first to profit from a free exchange of American tobacco, whale oil, rice, and the products of French colonies in the Caribbean, resented the free-trade movement, fearing foreign competition. Confirmed in Louis XIV's mercantilism, they saw free trade with the United States as only a step on the way to the British inundating France with their cheap manufactured goods that undersold French products. They could not condone either the free-trade spirit of the physiocrats or the enlightened imperialism of Jefferson's friend Lafayette, who had been successfully advocating at Versailles for enhanced commercial privileges for the U.S., the same privileges that Jefferson had been seeking as a check on British interests.

Enlightened French leaders like Lafayette understood that if America was to be won over permanently to France's sphere of influence, it was essential she be permitted to trade in the part of the French empire where American ships most needed access, the Caribbean. From the merchants of Le Havre, Bordeaux, Amièns, and Nantes had come a flood of anti-American protests: to admit American ships and products into the West Indies would deprive French sailors of jobs, hurt native business, loosen ties between France and her colonies, and destroy the mercantilist policy under which the colonies had been formed. After five years of steady diplomatic pressure supported by Lafayette and physiocrats such as Chastellux, Jefferson's efforts at winning more trade concessions kept seesawing. The hard-won whale-oil concession so vital to New England was closed off. As the French government became more paralyzed, Jefferson was able to exploit the situation by winning the reopening of French ports to American whale oil, rice, and almost twice as much tobacco—even as he won further extensions on payment of American war debts. He expected that time would bring still greater dividends. Famine and French bankruptcy would force Versailles to allow Americans to trade with Santo Domingo and even to supply starving Paris with salted meats, fish, and American wheat.

At the same time, war loomed with America's other strong treaty partner, the Dutch, whose dominant democratic Patriot Party, linked by treaty with France, was being challenged by royalists backed by the English and Prussia. On a pretext, Prussia sent in twenty thousand troops and Britain armed to intervene if France honored her alliance with the Dutch patriots. Jefferson foresaw great commercial benefits from a neutral America free to trade with both sides. But he did not want to see war break out in Europe, he wrote Washington, because

> in the first place, no war can be safe for us which threatens France with an unfavorable issue. And in the next, it will probably embark us into the ocean of speculation, engage us to overtrade ourselves, convert us into sea-ravens under French and Dutch colors, divert us from agriculture which is our wisest pursuit because, in the end, it contributes most to real wealth, good morals and happiness.[75]

When the British ambassador sounded out Jefferson on America's policy in the event of war, he promptly replied that his country would be neutral unless England foolishly attacked her ally, France's territories in the Caribbean. "Then it will be war." Fortunately for an unprepared America, war in Europe was averted. Ominously, France reneged on her treaty of support for the Dutch patriots. To John Jay, Jefferson wrote that the episode had given him an "important lesson, that no circumstances of morality, honor, interest, or engagement are sufficient to authorize a secure reliance on any nation at all times and in all positions."[76]

Events in America also had been moving faster than the transatlantic mails could keep Jefferson posted. The lockout of American trade by both England and France from its historic markets in the Caribbean had helped to trigger a depression in New England that had touched off Shays's Rebellion in western Massachusetts. Farmers, many of them Revolutionary War veterans, were unable to pay their debts to British factors and had no markets for their goods. Also unable to pay their taxes, they faced widespread foreclosures. Armed and organized by their former officers, they had closed the courts to prevent further debt collections and marched on the federal arsenal in Springfield before federalized militia under Gen. Henry Knox divided and routed them. Fear of spreading rebellion led first to an emergency convention in Annapolis and then to a special constitutional convention in Philadelphia to reform the Articles of Confederation. Jefferson had been one of the first to press for more centralized powers for Congress. He sent his protégé, Madison, some three hundred books on politics and government to help him formulate a written critique of the Confederation's weaknesses and the abuses in the individual states. Jefferson was not, at first, alarmed by Shays's Rebellion when Adams informed him of it, but as more and more of the old revolutionary leaders clamored for a crackdown on Shaysites and called for a far more powerful national government, Jefferson became isolated among all the national leaders in minimizing the dangers of the shortlived tax revolt. He began to fire off letters to influential Americans that show how disturbed he had been during his travels in Europe to see the growth of despotism. To Edward Carrington in Virginia he wrote,

> Experience declares that man is the only animal which devours its own kind, for I can apply no milder term to the governments of Europe and to the general prey of the rich on the poor.

Jefferson was afraid that a counterrevolution in America would concentrate power in the hands of a despotic minority, as in Europe. Most of all, he feared repression of any kind, and he was not afraid of popular resistance to governmental measures.

To Carrington, a conservative Virginian, he voiced his concern about counterrevolution:

> The people are the only censors of their governors, and even their errors will tend to keep these to the true principles of their institutions. To punish these errors too severely would be to suppress the only safeguard of the public liberties.[77]

To Abigail Adams, on learning that the Shaysites had surrendered, he wrote that he hoped the Massachusetts government would pardon them:

> The spirit of resistance to government is so valuable on certain occasions that I wish it to be always kept alive. It will often be exercised when wrong, but better so than not to be exercised at all. I like a little rebellion now and then. It is like a storm in the atmosphere.[78]

To Carrington, and others closer to the incipient revolt, Jefferson sounded more radical than ever after his travels and reflections in Europe:

> I am persuaded myself that the good sense of the people will always be found to be the best army. They may be led astray for a moment, but will soon correct themselves.

He worried that the insurrection would be used as an excuse to abridge freedom of the press:

> Were it left to me to decide whether we should have a government without newspapers or newspapers without a government, I should not hesitate a moment to prefer the latter. . . . Every man should receive those papers and be capable of reading them.[79]

When he read the new Constitution drafted at the four-month-long convention in Philadelphia, he was stunned. For days he poured out his fears. He wrote to Adams, who had sent him a copy, that there were "things in it which stagger all my dispositions to subscribe" to it. He would have shocked Madison, who had labored so long on the draft constitution, when he wrote to Adams that "all the good" of the new frame of government "might have been couched in three or four new articles" added to the Articles of Confederation, a "good, old and venerable fabric."[80] To Adams's son-in-law, William S. Smith, he wrote that he was opposed to the notion of a powerful president as an antidote to the "anarchy" of Shays's Rebellion:

> . . . Where does this anarchy consist? Where did it ever exist except in the single instance of Massachusetts? And can history produce an instance of rebellion so honorable conducted? God forbid that we should ever be 20 years without such a rebellion. What signify a few lives lost? The tree of liberty must be refreshed from time to time with the blood of patriots and tyrants.[81]

Writing to Adams, he worried about the creation of a presidency. The president envisioned in Philadelphia "seems a bad edition of a Polish king."[82] To Washington, he bluntly wrote that he feared a trend away from civil liberties in America toward despotism could result from the provision for an unlimited number of presidential terms. Would he not be, in effect, a king?

> I was much an enemy to monarchy before I came to Europe. I am ten thousand times more so since I have seen what they are. There is scarcely an evil known in these countries which may not be traced to their king as its source, nor a good which is not derived from the small fibres of republicanism among them. I can further say with safety there is not a crowned head in Europe whose talents or merits would entitle him to be elected vestryman by the people of any parish in America.[83]

As the fight over ratification of the new Constitution raged in America, Jefferson publicly declared himself a neutral, as his diplomatic position required. He apparently thought the document would be rebuffed at first and then would have to be revised by a second convention to provide a declaration of the rights of citizens. "Were I in America," he wrote young Smith, "I would advocate it warmly till nine [states] should have adopted [it] and then as warmly take the other side to convince the remaining four that they ought not to come into it till the declaration of rights is annexed to it."[84] But as the bloodless battle proceeded from state convention to state convention, Jefferson became convinced that it was Washington's looming if silent figure that was carrying the Constitution through to ratification. He had "boundless confidence" in Washington, but he worried about the men around Washington and what would happen *after* Washington. Yet he was becoming proud of this peaceful reform movement. To John Rutledge, Jr., he sent his congratulations at the "beautiful example of a government reformed by reason alone without bloodshed."[85] As he waited for Virginia to ratify, he wrote diplomatically to Madison, his protégé, "It is a good canvas on which some strokes only want retouching."[86] And to his old protégé, David Humphreys, he wrote, "The example of changing a constitution by assembling the wise men of the state instead of assembling armies will be worth as much to the world as the former examples we had given them."[87]

When Jefferson returned to Paris after seeing off Adams in May 1788, he found that the "gay and thoughtless Paris is now become a furnace of politics." The king had been compelled to call the Estates General for the first time in a century and a half. Jefferson wrote to Jay that some leaders were maneuvering secretly to get a declaration of rights passed by the Estates General and reported uncertainty about the loyalty of the army to the king. "There is neither head nor body in the nation to promise a successful opposition to 200,000 troops." He forecast that "whatever turn this crisis takes, a revolution in their constitution seems inevitable."[88]

By July 1788, there was rioting; by early September the deaths of several protesters in Paris who had been killed in a bayonet charge by the Swiss Guards. After the king's government consented to call the Estates General early in 1789, calm was restored, partially by martial law. Jefferson was optimistic there would be a successful and peaceful reform of the French government. The king needed money and would cede a larger role to the Third Estate, which was made up of the taxpaying people, at the expense of the nobles and the clergy, who controlled half of the wealth of the nation but paid no taxes. Jefferson thought that the Estates General would share power, like Congress, in the reform government. He favored an orderly transition. Periodic meetings must first be provided, then the exclusive right of the Estates General to set taxes; finally, the authority to codify laws and amend them. In time, the people would gain the right to originate laws.

Jefferson, after viewing the poor and passive majority of Frenchmen during his travels, did not believe they were yet ready for republican government. "They are not yet ripe for receiving the blessings to which they are entitled," he wrote Madison. But he was sure there would be a successful movement toward self-government. "The nation has been awakened by our revolution. They feel their strength, they are enlightened, their lights are spreading and they will not retrograde."

So confident was Jefferson of gradual and orderly change that in November 1788 he requested six months' home leave. He had carried out all his major objectives, including the conclusion of a consular treaty with France. His affairs in Virginia needed his attention. And he wanted to take his daughters back to Virginia where Patsy could meet young American boys. He was, moreover, worried that Patsy, who for a year had talked of converting to Catholicism, was considering becoming a nun. Jefferson had told her to wait to make a decision until she was eighteen, still two years away, but he was eager to get her away from the convent and back to his idea of a normal life in Virginia. But his application for home leave was stalled for five months by the change in government. If Congress had approved his leave promptly, Jefferson would have missed the onset of the French Revolution and his opportunity to play a major role in it.

On May 5, 1789, Thomas Jefferson, nearing the end of five years as an American envoy in France, joined the entire diplomatic corps, the court, and twelve hundred delegates to the Estates General at its opening ceremonies at Versailles. Gouverneur Morris of New York, a former colleague from Congress, obtained a ticket from Jefferson and rode out with him to the opening of the historic two-week meeting of the national assembly, which had not been convened since 1648. "There was displayed everything of noble and of royal [rank] in this titled country," reported Morris, "a great number of fine women and a very great number of fine dresses" in the latest hues of shimmering satin and silk laden with gold and silver. Bejeweled coiffures a foot high contrasted with faces daubed dead-white with flour.

Above them all, his corpulence wrapped in ermine, Louis XVI sat enthroned on a high stage against a backdrop of "gold-fringed violet velvet embroidered with fleurs-de-lis of gold." Marie Antoinette was seated a little below him. Abundant princes and princesses flanked them, the royal couple and "a cluster of guards of the largest size dressed in ancient costume taken from the times of chivalry" stood behind the throne. Just below them, a dazzling array of bemedaled marshals of France faced the beribboned ministers of state. In front of them, on the second tier, were three hundred "priests of all colors, scarlet, crimson, black, white and gray"[89] opposite (in front of the marshals) an equal number of the *noblesse* dressed in black robes, gold waistcoats, and gold cowls widening to the waist. On the third tier down from the king, on long rows of benches facing him, sat the representatives of the Third Estate, the people.

"The king's speech was what it should have been, and very well delivered," Jefferson reported to John Jay. But he was dismayed, after all his years in the company of La Rochefoucauld, Chastellux, and Lafayette, to find that "the Noblesse, on coming together, show that they are not as much reformed in their principles as we had hoped they would be." Like many reformers, Jefferson had hoped that the lower orders of clergy would join the Patriot Party behind Lafayette, but the reform-minded clergy evidently feared that the upper clergy and nobility might withdraw from the Estates General if the reformers of all three estates united. Jefferson was still optimistic, he reported to Jay, that nothing would happen to "render things desperate. If the king will do business with the [Third Estate], which constitute the nation, it may well be done without priests or nobles."

By the end of two weeks, however, the three estates were deadlocked, sitting in separate chambers. They could not agree whether voting should be *en bloc* by orders of nobility, clergy, and commoners or by individual members, the upper clergy and a majority of *noblesse* favoring voting by orders, the Third Estate unanimous in insisting on individual voting. "All the world are conjecturing how they are to get over the difficulty," wrote Jefferson, attending daily. "Abundance are affrighted and think all is lost," he told Jay, calling this attitude "rank cowardice." Finally, the Third Estate, announcing it would go on without the other orders if they went on without the people's representatives, declared themselves a National Assembly on June 17.

Jefferson, hearing the debate, realized how serious the crisis had become. To Jay: "The fate of the nation depends on the conduct of the King and his ministers. Were they to side openly with the commons, the revolution would be completely without a convulsion. A constitution would result," Jefferson wrote excitedly, "totally free and in which the distinction of noble and commoner would be suppressed." But Jefferson was pessimistic. He predicted that the queen would join the nobles and higher clergy if the king sided with the people, and a civil war would result. "The king is honest and wishes the good of his people but the expediency [of retaining] an hereditary aristocracy is too difficult a question for him," he wrote Jay. "On the contrary, his prejudices, his habits and his connections decide him in his heart to support it." Jefferson thought the only possible solution was creation of a two-house parliament similar to England's. This was what Jefferson's friend, finance minister Jacques Necker, had already suggested to the king. Jefferson applauded the newly formed National Assembly as "cool, temperate and sagacious" when, locked out of its assembly hall at Versailles, it boldly reconvened in a royal tennis court and swore an oath not to disperse until they had created a constitution. When some clergy and nobles joined the National Assembly, Jefferson told Jay that the "great crisis"[90] was over. Until this moment, Jefferson's analysis had been shrewdly accurate, but even this canny veteran of revolution could not see into the future.

Jefferson could not remain a spectator any longer. When the Comte de Mirabeau, a shrewd economist who had been rejected by the *noblesse* and elected as a commoner, falsely accused Necker of refusing to buy corn and flour from the Americans while the people of Paris went hungry, Jefferson was dragged into the fray, resenting that the U.S. was being injected into internal French politics. When Lafayette demanded clarification, Jefferson replied anxiously that he "never in my life made any proposition to Mr. Necker on the subject"[91] and asked Lafayette to convey the true facts to the National Assembly. When Lafayette obliged him, he reported that Necker and Montmorin, the foreign minister, were extremely angry at Lafayette. Lafayette, the leader of the patriots, was so worried that he suggested, only half joking, that he might have to become a citizen of the United States.

Secretly Jefferson and Lafayette were already working during long evenings at the Hôtel de Langéac to frame a declaration of rights based on Virginia's declaration of 1776 with its antiroyalist preamble by Jefferson. Jefferson had already widely circulated the Virginia declaration in France and Lafayette had sought his help. Jefferson acted as Lafayette's editor on early drafts of the now-famous French constitution. Initially he suggested that Lafayette remove such aristocratic-sounding words as the right to "property" and "the care of his honor." He said they were too inflammatory to the majority of Frenchmen. His major contribution was to the concluding article of Lafayette's draft, which gave to coming generations the right "to examine and, if necessary, to modify the form of government," a radical objective creating a republican form of representative government.

After working closely with Lafayette for months, on September 6, 1789, Jefferson for the first time expounded in writing this explosive doctrine of perpetual revolution more fully in a letter to Madison. "I set out on this ground, which I suppose to be self-evident, that the earth belongs in usufruct to the living: that the dead have neither powers nor rights over it. . . . What is true of every member of the society individually is true of them all collectively, since the rights of the whole can be no more than the sum of the rights of the individuals. . . . No society can make a perpetual constitution or even a perpetual law. . . . The earth belongs to the living and not to the dead." His head screaming with the pain of a migraine, as it so often did when he forged a new doctrine, this man who had lost so much to death hammered out the doctrine that was to renew his own country time and time again. And when, after numerous revisions, the French Declaration of the Rights of Man and Citizens was enacted by the National Assembly in late October 1789, it followed closely the Virginia Declaration of Rights and the American Declaration of Independence. "Our proceedings," wrote Jefferson to Madison, "have been viewed as a model for them on every occasion and, though in the heat of debate, men are generally disposed to contradict every authority urged by their opponents, ours has been treated like that of the Bible, open to explanation but not to question."[92] It was one of the great moments of Jefferson's life as he sat among the leaders of the new France, one of the mightiest nations on earth, while his ideas were debated, adapted, and adopted.

On June 29, 1789, Jefferson was so confident that a bloodless revolution had transformed France that he wrote, "At this moment, the triumph of the Third Estate is complete. Tomorrow, they will recommence business, voting by persons on all questions. All danger of civil commotion is at an end."[93] On July 4, Lafayette and his friends gathered with Jefferson and many other Americans in Paris to celebrate the thirteenth "anniversary of our independence" and to offer a congratulatory address written by the poet Joel Barlow of Connecticut to Jefferson "who sustained so conspicuous a part in the immortal transactions of that day."[94] But by July 8, Jefferson felt he had to inform Montmorin, the foreign minister, that the Hôtel de Langéac had been robbed three times in recent weeks and that the entire *quartier* needed protection by troops.

On July 12 the Customs House, a hated symbol of taxation, which was outside Jefferson's front window, was attacked and burned by rioters after the king dismissed Necker, the popular finance minister. That same day, Jefferson was driving his phaeton through the Place Louis XV (the present-day Place de la Concorde) when he encountered a crowd gathered at the entrance to the Tuileries Gardens. As he drew closer, he could see that the crowd was menacing the Prince de Lambesc's royal cavalry. The crowd had armed itself with stones piled up for the construction of the Pont Louis XVI. "I passed through the lane they had formed without interruption," Jefferson recounted. "But the moment after I had passed, the people attacked the cavalry, and the showers of stones obliged the horse to retire."[95]

That night, wrote Jefferson, the people of Paris "now armed themselves with such weapons as they could find in armorers' shops and private houses and were roaming all night through all parts of the city without any decided object." The next morning, July 13, in an attempt to restore order and protect private property, Lafayette formed a citizen's militia and took command, arming his men from the vast Garde Meublé, the warehouse of royal furniture and weapons, most of them antique, at Place Louis XV, where the draft Franco-American treaty had been signed in 1778. The militiamen broke into milliners' shops and seized red, white, and blue ribbon to make cockades for their hats.

The next afternoon, July 14, Jefferson was visiting his friend Madame de Corny near St. Eustache Church, when her husband burst in, terribly agitated. Ethis de Corny was *procurateur* of Paris and was in charge of all weapons. He was also a member of the Paris Commune that had requested weapons for the National Guard from the officers of the Garde Meublé, only to learn that Lafayette's militia had already seized them and then marched on the Bastille, the hated old prison where royal prisoners had been sent for centuries. When de Corny and his citizens' committee met with Monsieur de Lounay, Swiss governor of the Bastille, an inconclusive parley was interrupted, Jefferson wrote to John Jay. "The people rushed forward and almost in an instant were in possession of a fortification of infinite strength. They took all the arms and discharged the prisoners [they found only seven sick or insane inmates], and such of the garrison as were not killed in the first moment of fury, carried the governor and lieutenant governor to the Place de Grève (the place of public execution), cut off their heads and sent them through the city to the Palais Royal."[96]

Over the next few days, Jefferson and William Short rode through the city almost constantly, "in order to be sure of what was passing," wrote Jefferson. He paid six livres to see the Bastille, already being torn down, and contributed sixty livres, according to his account book, "for widows of those who were killed in taking the Bastille." On July 17, Jefferson stood on his balcony at the Hôtel de Langéac on the Champs-Elysées as Lafayette rode out to Versailles at the head of his militia and then rode back, his armed guards in triumph marching before and behind the royal carriage bearing Louis XVI "through the streets of his capital" as a crowd of "about 60,000 citizens of all forms and conditions armed with the muskets of the Bastille and the Invalides [and] pistols, swords, pikes, pruning hooks, scythes, etc., lined the streets. . . . People in the streets, doors and windows saluted them everywhere with cries of 'vive la nation'."[97]

By July 22, as Jefferson wrote to Madison that the "tumults of the city had pretty well subsided," they flared again with the capture of Foulon, a fugitive minister of the king. "He was forced from the hands of [Lafayette's] guards by the mob, was hung and, after severing his head, the body was dragged by the enraged populace through the principal streets of Paris,"[98] presumably past the Hôtel de Langéac. Six days later, after Louis XVI recalled Necker, the Estates General adopted the Declaration of Rights. On the night of August 4, Jefferson reported to Jay, the National Assembly "mowed down a whole legion of abuses—abolished all titles of rank, all the abusive privileges of feudalism": titles, provincial privileges, "the feudal regimen generally."[99]

Through the weeks of tumult, Jefferson kept up a pretense of diplomatic neutrality—until Lafayette put him on the spot on August 25: "I beg for liberty's sake you will break every engagement to give us a dinner tomorrow, Wednesday. We shall be some members of the National Assembly." Eight of them, Lafayette added, whom he needed to bring together from deadlocked factions in a coalition "as being the only means to prevent a total dissolution and a civil war. . . . These gentlemen wish to consult you and me. . . . I depend on you to receive us."[100] On August 26, they came. Lafayette, du Pont, Darnave, Lameth, Blacons, Mounier, La Tour, Maubourg, d'Agoult. In his memoirs, Jefferson described the scene:

> The cloth being removed and wine set on the table . . . the Marquis introduced the objects of the conference. . . . The discussions began at the hour of four and continued till ten. . . . I was a silent witness to the coolness and candor of arguments. . . . The result was [the] concordats [which] decided the fate of the constitution.

The next morning, Jefferson felt obliged to hurry out to Versailles to explain to the foreign minister "how it had happened that my house had been made the scene of conferences of such a character."[101] But Montmorin already knew everything that had transpired. For the next six days, an anxious Jefferson was bedridden with a severe migraine, and he sent off Trumbull with a message for Maria.

When Maria Cosway learned that Jefferson was going home on leave, she did not know that he would never return. She implored him to visit her in London en route: "Don't go to America without coming to England."[102] Jefferson, whose hatred of England made the request impossible, suggested that Maria and her friend Angelica Church come to America on the same ship with him. When she did not reply at once, he wrote her that he might make the crossing with Angelica, without Maria. "We shall talk a great deal of you."[103] Early in September, Jefferson suffered a migraine that lasted six days. His headaches seemed often triggered by loss: the death of his mother, his wife. This time, he was bedridden until he could wait no longer to pack. As the day to leave Paris, September 26, 1789, arrived, Jefferson wrote, "Adieu, my dear friend. Be our affections unchangeable, and if our little history is to last beyond the grave, be the longest chapter in it that which shall record their purity, warmth and duration."

As Jefferson's ship waited on the English coast in mid-October for a favorable wind to take him to America, he had time to dash off two personal letters on October 14, 1789. To Paris, he wrote to his old friend Madame de Corny that he feared events in France had moved "to the brink of a civil war." He had just seen the London newspapers: the situation in France was "much worse than when I left you." Jefferson had come to love the people of France and he dreaded the prospect of civil war. He knew from his own experience in America how awful the prospect was. "From this, heaven preserve your country and countrymen, whom I love with all my soul."[104] There was still time for one last note to Maria, who had already written that she was too ill to see him off:

> As we think last of those we love most, I profit of the latest moment to bid you a short but affectionate adieu. . . . My daughters are with me and in good health. We have left a turbulent scene, and I wish it may be tranquilized on my return, which I count will be in the month of April. . . . Spring might give us a meeting at Paris with the first swallow. . . . Remember me and love me.[105]

As he wrote his last letters from Europe, Jefferson could not know that, only the day before, George Washington had written to him from New York, asking him to be his Secretary of State. He had considerately left the door ajar for Jefferson to return to Europe if he wished by waiting to name a successor as ambassador to France. But Washington clearly expected Jefferson to accept the post and was already suggesting who should be Jefferson's assistant. Washington's appointment went off toward Virginia and was waiting for him when he landed.

Expecting to return in the spring, unaware that he was leaving France forever, Jefferson wrote few farewells. It would be years before he could write to many of his closest friends in Paris. Soon they would be scattered by the French Revolution, and many would be killed by the guillotine or, like his close friend La Rochefoucauld, chased and stoned in the streets in full view of his family. But Jefferson, once so critical of France and the French, enjoyed no period of happiness quite so much as his five years in Enlightenment France. There was, Jefferson wrote in his autobiography thirty years later, no "more benevolent nation" than France, and he always hoped to go back, especially if he could have gone back to the prerevolutionary Paris of the salons. Paris was, in the years just before 1789, when Jefferson was there, the seat of the most amiable and polished society of the world, as he wrote to his old friend, Madame de Staël, in 1813. In old age, in retrospect, Jefferson wrote in his autobiography, "So ask the travelled inhabitant of any nation in what country on earth would you rather live? Certainly in my own, where are all my friends, my relations, and the earliest and sweetest affections and recollections of my life. Which would be your second choice? France."[106]

19

"The Blessings of Self-Government"

To take a single step beyond the boundaries specially drawn around the powers of Congress is to take possession of a boundless field of power, no longer susceptible to definition.

—Jefferson to Washington, February 1791

That Jefferson had every intention of returning to France appears certain from his last scribbled note to Maria Cosway and from the possessions he took home to Virginia with him—everything that belonged to Patsy and Polly, crates of books for Franklin, Washington, Madison, his old Virginia phaeton, sapling trees for Monticello—but none of his furniture, his new English carriage, artworks, or the hundreds of handsome leatherbound books he had so carefully acquired. Jefferson had only come home for a visit, hoping to find a suitable husband for Patsy and a school for Polly. He had made no arrangements to leave Paris permanently and fully planned to return to the scene of a revolution in which he was playing such an important part. On his arrival aboard the *Clermont* in Norfolk on November 23 after a fast, storm-wrecked passage, however, he learned from the newspapers that Washington had appointed him Secretary of State, despite the fact that Jefferson had rebuffed a feeler from Madison some six months earlier. A formal letter from Washington overtook him two weeks later at Eppington where Jefferson and his daughters were visiting the Eppeses. Seven years after he had left behind Monticello, Jefferson returned to Virginia in culture shock at the rural society he had outgrown.

He wrote to William Short in Paris that he was about "to plunge into the forests of Albemarle."[1] He conveyed to Washington his "gloomy forebodings" of "criticisms and censures" if he accepted the post. Unless Washington insisted, Jefferson preferred to return to Paris. "It is not for an individual to choose his post. You are to marshal us as may be best for the public good."[2] But a gala welcome by both houses of the Virginia Assembly and a tour of the new capital at Richmond, which he had designed personally, filled Jefferson with a civic pride that made him waver in his decision to return in the spring to Paris. On December 15, Jefferson wrote Washington a conditional letter of acceptance.

Two days before Christmas of 1789, despite his written orders that he wanted only a skeleton staff of slaves on hand for his anticipated two-month visit, Jefferson rode up his little mountain to the cheers of scores of his slaves, who had permission to come to Monticello from all his plantations. Unhitching his carriage, they pushed and pulled Jefferson the last yards of his long journey home. "Such a scene I never witnessed in my life," Patsy wrote years later. "When the door of the carriage was opened, they received him in their arms and bore him to the house, crowding round and kissing his hands and feet, some blubbering and crying, others laughing."[3] Over the next few years, Jefferson was to begin to emancipate these devoted slaves, one at a time: Sally Hemings's brother Robert first, then James Hemings. He evidently had decided that the time was not ripe to openly defy the slave system all around him and reopen the debate over emancipation at a time when the new government was so unstable.

When, after a whirlwind two-month courtship, his seventeen-year-old daughter, Patsy, married Thomas Mann Randolph, Jr., her third cousin and the son of Jefferson's old grammar-school mate, "Tuckahoe Tom" Randolph, Jefferson settled twenty-five slaves and one thousand acres of land on her, thus perpetuating the system of slavery within his own family.

Enthusiastic public receptions for him at Norfolk, Richmond, and in Albemarle only increased the pressure on Jefferson. Madison, who called Jefferson's association with Washington a matter of "infinite importance,"[4] informed him in a private meeting at Monticello that the south and west were depending on him. Washington sought Jefferson's opinions on many thorny questions, even if he did not always follow them. When Baron von Steuben organized the Order of the Cincinnati as a powerful elite group of veteran officers with Washington at its head, Jefferson had warned that the people would not tolerate an aristocracy. Yet Washington accepted the post. He usually deferred to Jefferson as the resident expert on Europe and, as a southerner, knew how much he had done to help southern trade. Washington nevertheless was annoyed at first at Jefferson's reluctance to serve with him but he wrote to Jefferson that he would have to decide for himself whether to accept. He tried to reassure Jefferson that, despite the fact that home and foreign affairs had been combined in one post, his duties would be lighter than Jefferson might imagine. If domestic duties became troublesome, as Jefferson feared, Washington would ask Congress to create a new post to cover them. Still, the decision oppressed Jefferson "greatly."[5]

At the same time, his Virginia neighbors were urging him to take up his old seat in the Assembly. He delayed answering Washington for two months, until only a few weeks before Washington's inauguration. Not until February 12, 1790, did he accept Washington's offer to become a leader of the new nation, against his own will. But he must have made up his mind even earlier, as a speech to his Albemarle neighbors, one of his most important and moving and an accurate reflection of his experiences in Europe, makes clear. Addressing them at Charlottesville as "fellow-laborers and fellow-sufferers" in the "holy cause of freedom," he spelled out his credo:

> It rests now with ourselves alone to enjoy in peace and concord the blessings of self-government so long denied to mankind, to show by example the sufficiency of human reason for the care of human affairs. . . . The will of the majority, the natural law of every society, is the only sure guardian of the rights of man. Perhaps even this may sometimes err. But its errors are honest, solitary and short-lived. Let us then, my dear friends, forever bow down to the general reason of the society. We are safe with that, even in its deviations.[6]

The speech, printed in a Richmond newspaper, was soon reprinted in the *Gazette of the United States* and widely circulated throughout the new country as Jefferson's acceptance of his appointment as the first American Secretary of State.

Jefferson waited to travel to the capital in New York City until after Patsy's wedding. Described by a shipmate on the *Clermont* as "amiable," tall and genteel, "with a winning simplicity and good-humored reserve,"[7] Patsy had scarcely begun to readjust to the home she had not seen since age twelve when a former childhood playmate, her cousin, Thomas Mann Randolph, Jr., twenty-one, proposed marriage. Jefferson wrote to Madame de Corny in Paris that although young Randolph was "the son of a bosom friend of mine" whose "talents, dispositions, connections and fortune were such as would have made him my own first choice," he had been careful to say nothing to influence his daughter. "It ended in their marriage."[8] Jefferson was too shrewd to do anything but say nothing against the match, Patsy too sensitive to her father's wishes to do anything but accept the proposal. To the groom's father, his cousin and lifelong friend, Jefferson was less circumspect: "the marriage of your son with my daughter cannot be more pleasing to you than to me."[9]

On March 1, 1790, one week after Patsy's wedding at Monticello, Jefferson left for his new post. The winter journey took three weeks, made longer by the sadness of visiting the dying Benjamin Franklin in Philadelphia. By the time he reached New York City, Jefferson was already homesick for Monticello and his daughters. To Patsy, he wrote that after "having had yourself and dear Poll to live with me so long, to exercise my affections and cheer me in the intervals of business, I feel heavily the separation from you."[10] There were to be few intervals from business. Finding a house, one he did not like, to rent at 57 Maiden Lane, he spent far too much to add a windowed library in the back to house his books, papers, and plants. Then he plunged into his new duties. As the ranking department head of the Washington administration, Jefferson found that the entire State Department staff consisted of two clerks, two assistant clerks, and a translator. Excluding the overseas diplomatic establishment, its entire budget was under $8,000, including his $3,500 salary.

Jefferson knew all the other department heads and had no reason to expect trouble. In one capacity or another, he had worked with Alexander Hamilton at Treasury, Henry Knox at the War Department, and his young cousin, Edmund Randolph, now the attorney general, as well as Vice President Adams and Chief Justice John Jay, the former foreign secretary. Washington did not consider them powerful ministers, as are today's cabinet members. All of the department heads served Washington as his assistants. The president expected all policy questions to be brought to him for his approval. From the outset, Washington ran his administration as he had the army. As Jefferson soon discovered, to his horror, there was no idea that department heads should keep to their own affairs. While Washington strictly controlled the War Department as commander-in-chief, he deferred to Hamilton on finance and to Jefferson on foreign affairs. Because he was least sure about finance, Washington allowed Hamilton more freedom than all the others.

Never before an official within a bureaucracy, Jefferson hated his new office job. Within six weeks he was stricken with a severe migraine that kept him in bed for a week and out of the office for most of a month. At his rented home, he was able to think and write; he drafted his report to Congress on weights and measures there. Proposing the rod as the standard unit of measure, he divided everything from it decimally, as he had American currency, "bringing the calculation of the principal affairs of life within the arithmetic of every man who can multiply and divide plain numbers."[11] But Congress had trouble with numbers, couldn't grasp Jefferson's revolutionary system, and did nothing about his report, allowing France to introduce the revolutionary metric system first. Jefferson was still suffering from his migraine when he first appeared before a Senate committee at Congress Hall to explain his proposal, which may in part explain this description by the rough-hewn William Maclay of Pennsylvania:

> Jefferson is a slender man. Has rather the air of stiffness in his manner. His clothes seem too small for him. He sits in a lounging manner on one hip commonly and with one of his shoulders elevated much above the other. His face has a scrawny aspect. His whole figure has a loose, shackling air. He had a rambling vacant look and nothing of that firm collected deportment which I expected would dignify the presence of a secretary or a minister. I looked for gravity, but a laxity of manner seemed shed about him. He spoke almost without ceasing, but even his discourse partook of his personal demeanor. It was loose and rambling and yet he scattered information wherever he went, and some even brilliant sentiments sparkled from him.[12]

Maclay was undoubtedly one of the causes of Jefferson's discomfort during his first year in his new post. The Scots-Irish senator from backwoods Pennsylvania had set ideas about government. He did not believe there should be any foreign service. Money spent abroad was wasted. "I know not a single thing that we have for a minister to do in a single court in Europe," he said. "Indeed, the less we have to do with them, the better."[13] As a result, Jefferson had to scale down the foreign service even further. Only France kept a minister plenipotentiary; Madrid and Lisbon, chargés d'affaires; London and The Hague, each an agent; and Morocco a consul, all at a cost of only $40,000 a year. Washington made one modification. If and when Britain sent a minister to New York, an American minister would again be sent to London.

But the importance of diplomacy almost immediately became clear, even to Congress, in a complicated crisis involving the threat of a European war in America. The Spanish navy seized British ships off Vancouver Island. When British Prime Minister William Pitt made the crisis public in May 1790, Gouverneur Morris, now the American agent in England, reported that, if war came, England and Spain would both be willing to pay for American neutrality. Jefferson was in accord with Morris's analysis. On July 4, he wrote to Edward Rutledge that, in the event of a European war, he now thought it advisable to "become the carriers for all parties as far as we can raise vessels,"[14] in effect, to allow America to fatten on the folly of European war. But to Washington, he admitted that there were dangers to this course and he understood them. Jefferson was "decidedly of the opinion we should take no part in European quarrels but cultivate peace and commerce with all." Yet he saw the source of such wars "in the tyranny of those nations who deprive us of the natural right of trading with our neighbors."[15] Jefferson was clearly enunciating for Washington a diplomacy based on the revolutionary natural-rights doctrine. This held that any nation had the right to remain neutral and trade with any other nation. He was denying the mercantilist system still followed by European colonial nations that bound their own colonies to trade only with them and forbade all trade with other countries, effectively locking out the United States if it remained neutral or dragging it into war if it sided with any combatant.

The British prime minister, meanwhile, had instructed the governor general of Canada, Lord Dorchester, to find out the American position in the event of war. Dorchester sent his aide, Maj. George Beckwith, as a secret agent to New York, where he had made many intelligence-gathering missions since his days as a Loyalist officer on the staff of British spymaster John André during the American Revolution. On earlier visits he had talked with Hamilton, who had not reported these conversations either to Washington or Jefferson, who should have been informed as Secretary of State. After Beckwith and Hamilton met in July 1790, Hamilton reported it in a meeting with Washington that Jefferson attended. That the Secretary of the Treasury had intervened so intimately in the business of the Secretary of State was only the first sign of a growing rift between the two cabinet officers.

More important, the pro-English Hamilton was working at cross purposes with Jefferson, who considered Hamilton's predilection for an alliance with Britain a clear step toward once again subjugating the United States to England. After conferring with Madison, who was then a key member of Congress and a trusted adviser to Washington, Jefferson drew up a memo to the president on July 12 proposing an American policy aimed at keeping open the Mississippi with its access for westerners to the Caribbean. If England seized Louisiana and Florida from Spain, the United States would be completely surrounded by the British army and navy. "Instead of two neighbors [England and Spain] balancing each other, we shall have one, with more than the strength of both,"[16] draining off American settlers and trade from the west.

Even had Jefferson seen the minutes of the British committee of the King's Privy Council for Trade for April 17, 1790, he could not have been more accurate in his predictions. The Privy Council report declared that "it will be for the benefit of the country to prevent Vermont and Kentucky and all the other settlements now forming in the interior . . . from becoming dependent on the government of the U.S. . . . and to preserve them on the contrary in a state of independence and to induce them to form treaties of commerce and friendship with Great Britain."[17] Already alarmed at Hamilton's meddling, Jefferson nevertheless coolly urged President Washington to use his contact with the spy Beckwith and inform the British through him that the United States, while still remaining a treaty ally of France, favored a commercial treaty with "perfect reciprocity" with England, something Jefferson had been seeking since his visit to England in 1786. In case of war, he advised being "strictly neutral"[18] with England and Spain, but he urged speedy secret negotiations with Spain to urge her to grant independence to Louisiana and Florida. But Washington overruled Jefferson, instead instructing Hamilton to meet privately with Beckwith and attempt to extract as much information as possible from the British agent without committing the United States.

According to Beckwith's report to Lord Dorchester, Hamilton went far beyond Washington's instructions, assuring him that the United States wanted to clear up "all matters unsettled between us and Great Britain in order to effect a perfect understanding between the two countries, and to lay the foundation for future amity." Further, Hamilton had assured Beckwith that the United States had no ties with Spain, even though he must have known that Jefferson had already dispatched an envoy to Spain to open talks. Worse still, Hamilton had warned Beckwith that, although Washington was "perfectly dispassionate," Jefferson could be a problem "from some opinions he has given respecting your government." Beckwith said Hamilton had also told him:

> I shall certainly know the progress of negotiation from day to day. . . . In case any such difficulties should occur, I should wish to know them in order that I may be sure they are clearly understood and candidly examined.[19]

In other words, the American Secretary of the Treasury had offered to subvert the efforts of his own Secretary of State and to assist the agents of a foreign power secretly. Fortunately war did not materialize. Revolution-racked France once again abrogated a treaty alliance, and her ally Spain had to back down and make peace with England. In his first months in office Jefferson had learned what he could expect from the scheming, insinuating Hamilton, but not fast enough to extricate himself from another Hamiltonian scheme that he would come to deplore for the rest of his life.

In January 1790, before Jefferson was sworn in, Hamilton, in charge of creating a financial system for the new government, had presented his first report on the public credit to Congress, which divided sharply over his proposals to fund a national debt and have the federal government assume the unpaid debts incurred by the states during the Revolution. Southern delegates, including Madison, opposed allowing speculators from New York and New Jersey who had bought up depreciated securities to make huge windfall profits by being paid off at par value with accrued interest. Madison's proposal to distinguish between original bondholders and speculators was voted down, but Hamilton's proposal to assume the northern states' debts had been defeated by the southern states, which had paid their war debts already. With Congress deadlocked, Jefferson feared that, if there were no compromise, the funding bill would die and the national credit would "burst and vanish and the states separate to take care everyone of itself."[20]

One day Jefferson ran into Hamilton outside Washington's door, looking "somber, haggard and dejected beyond description."[21] Hamilton urged him to use his good offices with southerners in Congress. It was part of Jefferson's job to handle domestic as well as foreign diplomacy, but he refused to appear directly involved in Hamilton's affairs. He did agree to invite Hamilton, Madison, and delegates from Maryland and Pennsylvania to his house for dinner the next day to discuss the impasse. That night and the next day before the dinner meeting, he worked out a compromise that led to congressional passage of Hamilton's debt-assumption act and location of the new national capital on the Potomac after a ten-year stopover in Philadelphia.

In July 1790, when the debt-funding, debt-assumption, and capital-location bills came before Congress, all three sailed through to passage. In exchange for the siting of the capital in the south, two Maryland and two Virginia members changed their votes on funding and assumption of debts to support Hamilton and the northern financial interests in Congress. Within two years, Jefferson, who played the key role in the compromise, was claiming that Hamilton had "duped" him and made him a "tool for forwarding his schemes, not then sufficiently understood by me."[22] But in 1790 Jefferson thought that compromise with Hamilton was urgently needed to save the Union.

In mid-August, Jefferson sailed to Rhode Island to welcome that state, the last to ratify the Constitution, into the union. Returning to New York, he drove south with Madison on September 1. In Philadelphia, they shopped for office space and living quarters for the temporary transfer of capitals. In Georgetown, they inspected lands for permanent capital buildings: Jefferson was moving quickly because he feared that Congress, once it had moved to Philadelphia, would settle in there and decide not to move again unless capital buildings were already under construction. Reluctant to go before a Congress that could turn him down for funds, he brilliantly devised a plan to encourage Georgetown landowners to contribute lands that could be sold to raise construction money in exchange for the profits that would accrue from the increased value of their adjacent lands. He urged Washington to move promptly to appoint a commission to design the capital and begin construction. "If the present occasion of securing the Federal seat on the Potomac should be lost, it could never be regained."[23]

When the new congressional session began in late November 1790, Jefferson moved into a large, half-completed brick house on poplar-lined Market Street at Eighth Street, which he had rented and had extensively renovated to include a large library and a carriage house for three carriages and five horses. Then he began unpacking seventy-eight crates shipped to Philadelphia from Paris, containing furniture and books, fine writing paper, a dozen cases of wine, two cases of macaroni, a favorite food he had discovered in Italy, even a fountain. As he settled in, he entertained at meals cooked according to his own handwritten recipes. Adrien Petit, his faithful *maître d'hôtel* from the Hôtel de Langéac, whom Jefferson had lured to Philadelphia, was to take charge of his house. Jefferson had brought James Hemings from Monticello to serve up French food.

That fall, Jefferson brought thirteen-year-old Polly to Philadelphia to live with him in his elegant new town house across the street from the State Department and two blocks from the president's house. He was also within easy walking distance of Peale's Natural History Museum in Independence Hall, Congress Hall, the College of Philadelphia, and the American Philosophical Society, which elected him vice president in January 1791, and included in its membership the semblance of a scientific salon. There, in 1792, a family of plants, *Jeffersonia diphylla,* was named after him to celebrate his preeminence in natural history. There were also a theater, botanical gardens, assembly halls, and the soirées of society ladies. Of all

American towns, Jefferson liked cosmopolitan Philadelphia best, but he was too busy for its pleasures through most of the winter of 1790–91, completing a major report virtually every week. For all of his apparent extravagance in private, Jefferson was doing what he always did when he was away from Monticello—creating a comfortable atmosphere where he could live and work surrounded by his books. He shunned public display and he came to criticize openly Washington's "court," his personal aloofness, his formal weekly levees, and his use of a carriage drawn by six horses as he rode out on official occasions with his "family" following in a parade of coaches.

Jefferson's first order of business was to study relations with England. He concluded that the British had no intention of evacuating the northwestern forts or of negotiating a commercial treaty. He recommended to Washington closing the American mission in London and recalling Gouverneur Morris, to which Washington agreed. He also advocated increasing trade with France by diverting large quantities of fish from sales to England to any country maintaining friendly relations with the United States. When Jefferson presented his report on fisheries to Congress, Hamilton worked against it within the administration. In a cabinet meeting in December 1791, Jefferson recommended retaliating against British trade restrictions. Hamilton argued hotly against retaliation, saying that Jefferson's policy would endanger negotiations over the western forts with the newly arrived British envoy, George Hammond. Jefferson held off reporting to Congress until the session ended. He was angered when he finally concluded that the British envoy had no instructions to negotiate and that Hamilton had been leaking the contents of their meetings to Hammond all along, making possible further British delays.

But Jefferson expected little from the British. His chief concern was to develop a policy for dealing with Algeria and what to do about American captives there. Because of secret negotiations with the Algerians, it was December 30, 1790, before he submitted a full report to Congress. Jefferson still favored joining naval forces with other countries trading in the Mediterranean. "Upon the whole, it rests with Congress to decide between war, tribute and ransom," he told Congress. Senator Maclay saw through Jefferson's careful neutrality. The papers Jefferson presented, he wrote, "seemed to breathe resentment and abounded with martial estimates in a naval way."[24] Britain's policy of paying tribute encouraged Algerian preying on vulnerable American ships, but Hamilton, eager to appease the British, refused to join Jefferson's effort to get Congress to protect the Mediterranean trade. But a Senate committee heavy with Hamilton supporters declared that the "trade of the U.S. to the Mediterranean cannot be protected but by a naval force" adding that "the state of public finances" made that impossible.[25] The Senate, overriding the Secretary of State's recommendations, finally resolved to appropriate $40,000 to ransom U.S. citizens.

Sharp differences between Jefferson and Hamilton were becoming evident in dealings with not only the Barbary pirates but England and France, in a word, in all foreign policy, because the two cabinet ministers had diametrically opposed views. By early 1792, President Washington was so alarmed at the gulf in his cabinet and the resulting division in the nation that he asked the two men to account for their rivalry. Their answers showed that their policies were irreconcilable. Jefferson, in a rare outburst, wrote Washington on September 9, 1792, that Hamilton's history at Treasury "from the moment at which history can stoop to notice him, is a tissue of machinations against the liberty of the country which has not only received and given him bread but heaped its honors on his head."[26] Jefferson's scathing attack was an allusion to Hamilton's arrival in New York as the illegitimate son of an English admiral and a West Indies Frenchwoman, his acceptance in Manhattan society, and his service as Washington's favorite aide-de-camp during the Revolution. By late 1792, Jefferson was deeply alarmed both by Hamilton's program, now fully revealed, and by his palace-guard politics.

For his part, Hamilton's Federalist financial system rested on the funding and assumption of the Revolutionary War debt, a program of external and internal taxes—customs duties and excise taxes—as well as creation of a national bank and the encouragement of domestic manufacturing. In a series of four reports to Congress between 1790 and 1792, Hamilton contended that each of these four elements was indispensable to establish national power. By establishing the credit of the United States on a firm foundation, the federal government would, in Hamilton's words, restore the confidence of foreign and domestic lenders, which would enable the United States to borrow immense sums. Hamilton apparently had not consulted the facts that Jefferson could point out to him about the already high credit ratings of the United States in Amsterdam and Paris. Hamilton further argued that it was essential to create a pool of capital to provide a circulating medium of currency. He meant to eliminate the shortage of hard money in the country as a further means of encouraging investment in large-scale economic enterprises. Federal taxes levied to support Hamilton's program would give power to the central government while binding its creditors to support it politically. Hamilton believed that, to make the United States independent of European mercantilism, it was essential to develop "an extensive domestic market" based on American manufacturing that could absorb the profits of American farm surpluses. As long as the economy depended mainly on agriculture, the United States would not be able to "exchange with Europe on equal terms." Hamilton told Congress that " 'tis for the United States to consider by what means they can render themselves least dependent on the combinations, right or wrong, of foreign policy."[27]

Hamilton's isolationism could not have been more antithetical to Jefferson who, with Madison, was working to create a republican system based on an entirely different set of assumptions. While Hamilton preached adjusting to "the general policy of nations,"[28] Jefferson sought to revolutionize it. Since before the Revolution, he had believed that Anglo-American commercial ties were inherently unequal and exploitative. He contended that England actually depended on American commerce, which gave the United States a weapon vital in reforming foreign-trade policy. As Madison put it succinctly, England supplied Americans "chiefly with superfluities."[29] Trade with America employed a large part of the British workforce. England depended on the Americans, among other imports, for the grain to make her bread, the whale oil to light her lamps, the wood to build her navy. If the United States played its card of nonimportation, Jefferson believed, the great mass of English workers would be thrown out of work, West Indies planters and their slaves would go hungry. But it was only necessary to threaten these two pressure points, Jefferson argued, in peacetime to persuade the British to open all their domestic and colonial ports to American goods and shipping, and, in wartime, this would force the British to respect American neutral trading rights. Jefferson was relying on his memory of the effects of revolutionary nonimportation in 1773–75 to gauge the impact of his theory. At that time, fully one-third of British trade had been with her American colonies. By 1792, trade with the United States was only about one-sixth of total British imports and exports, while fully one-half the value of all American trade was with England and her possessions. This worsening imbalance worried Hamilton and the Federalists, who pointed to it whenever the hint of commercial war came up. Hamilton argued that Americans would feel the effects of a commercial war long before England. Jefferson countered that, in a contest of self-denial, Americans would, as they had in the 1770s, surely make the sacrifices necessary to prevail. The threat of peaceable coercion conceived by Jefferson and pursued by Madison became the centerpiece of their Republican foreign policy.

Foreign policy differences between Jefferson and Hamilton also became evident when the French protested Congress's failure to exempt France from Hamilton's higher tonnage duties levied on foreign vessels.

France protested that the new duties contradicted the Franco-American treaty of amity and commerce of 1778. While rejecting the French argument, the pragmatic idealist Jefferson recommended to Washington that, considering the importance of Franco-American relations and the small amount of customs duties involved, a concession be made to France. Hamilton openly opposed Jefferson's proposal in the cabinet and it failed to win support in Congress, which had earlier rejected Madison's efforts to impose higher duties on ships of countries which did not have commercial treaties with the United States. Just as Hamilton sought to weaken the American links with France, Jefferson was intent on strengthening them. In his September 1792, report to President Washington, Jefferson wrote,

> In the case of two nations with which we have the most intimate connections, France and England, my system was to give some satisfactory distinctions to the former, of little cost to us, in return for the solid advantages yielded us by them, and to have met the English with some restrictions which might induce them to abate their severities against our commerce. Yet the secretary of the Treasury, by his cabals with members of the legislature and by high-toned declamation on other occasions, has forced his own system, which was exactly the reverse.[30]

Jefferson and Hamilton were also clashing bitterly on domestic politics. Secretary of the Treasury Hamilton increasingly saw himself as something of an American prime minister. The Treasury Lord in England had long served as the prime minister, controlling Parliament and cabinet for the king. When Hamilton tried to impose the English model on the American cabinet, Jefferson stopped cooperating with him. And when Hamilton proposed to Congress a national bank similar to the Bank of England, Jefferson privately began to voice his concerns about Hamilton's policies in general while Madison, his lieutenant, led open opposition in Congress to the bank bill.

After the national bank bill passed the House and the Senate, Washington, before signing it, asked for opinions on its constitutionality from his attorney general first, and then from Jefferson and Hamilton, all three of whom were lawyers. The confidential memoranda of response, kept from the public, revealed to Washington the extent of the split in his government. Attorney General Randolph was the first to declare the national bank unconstitutional. Jefferson had not seen Randolph's memo when he wrote his own. One of the most important documents in American history, it carried Jefferson's arguments for strict construction of the Constitution that he was to follow all the rest of his political life, the very arguments that laid the foundation for the doctrine of states' rights.

Jefferson's brief was concise. He argued that to charter a national bank was unconstitutional because the

> still-unratified Tenth Amendment held that "the powers not delegated to the United States by the Constitution, nor prohibited by it to the States, are reserved to the States respectively, or to the people. . . . To take a single step beyond the boundaries thus specially drawn around the powers of Congress is to take possession of a boundless field of power.

The Constitution had not delegated incorporation of a bank to Congress, not under its powers to tax, to borrow, or to regulate commerce. Congress could not *"do anything they please."* Jefferson feared that free interpretation of any aspect of congressional power would reduce the Constitution to "a single phrase, that of instituting a Congress with power to do whatever would be for the good of the United States. . . . As [Congressmen] would be the sole judges of good or evil, it would be also a power to do whatever evil they pleased." Jefferson also anticipated Hamilton's argument of implied powers granted by the Constitution to Congress to make all laws necessary and proper to carry into execution the powers specifically enumerated in the Constitution. Congress could only do what was necessary. To form a national bank was not necessary, only convenient. Without his insistence on limiting Congress to enacting necessary laws, there was nothing "which ingenuity may not torture into a convenience, in some way or other, to someone." With his opinion on the bank bill, Jefferson also gave the president his reasoning on using the presidential veto. Despite Jefferson's arguments, unless Washington was convinced that the bank was unconstitutional, "A just respect for the wisdom of the legislature would naturally decide the balance in favor of their opinion."[31]

With two opinions against the bank in his hands, Washington turned them over to Hamilton, who refuted Randolph and Jefferson point by point. He directed his main attack at Jefferson's interpretation of the "necessary and proper" clause. Jefferson insisted on using the word *necessary* as if it were prefixed by "absolutely" or "indispensably." What is constitutional or not, Hamilton countered,

> is the *end* to which the measure relates as a *mean*. If the end be clearly comprehended within any of the specified powers, and if the measure have an obvious relation to that end and is not forbidden by any

> particular provision of the constitution—it may safely be deemed to come within the compass of national authority.[32]

When Washington was persuaded by Hamilton's end-justifies-the-means argument, Jefferson was truly alarmed. He now considered the door open to the incessant extension of what he considered a heresy concocted to serve the special moneyed interests, what he called the "stockjobbers," without the consent of the majority of people, who were yeomen farmers.

It was by accident that it became known in public that there was a sharp breach in Washington's government. Jefferson, returning a borrowed copy of Paine's revolutionary new pamphlet, *The Rights of Man*, to a Philadelphia printer planning to publish it, enclosed a brief note that he was pleased the radical work was to be reprinted in the capital and that "something is at length to be publicly said against the political heresies which have sprung up among us. I have no doubt our citizens will rally a second time round the standard of Common Sense."[33] He was shocked when the letter was abstracted atop Paine's pamphlet when it was printed in May 1791. Adams, whose recent political writings Jefferson had in mind, was greatly offended. "I tell the writer freely that he is a heretic," Jefferson confessed privately to Madison, "but certainly never meant to step into a public newspaper with that in my mouth."[34] Jefferson tried to placate his old friend Adams. But he did not try to soothe Hamilton who, he said, was more and more "open-mouthed" against him. All up and down the country, newspaper editors eagerly reported the controversy. Jefferson was becoming overnight the voice of republicanism, pitted against Hamilton and the special interests. When Jefferson and Madison made a trip through New York and New England in May and June 1791—sailing along Lake George and Lake Champlain and driving back through Bennington, Vermont; western Massachusetts; Connecticut; and Long Island—editors and other political experts read in partisan motives for the tour. One Hamilton ally reported "every appearance of a partisan courtship"[35] with Robert R. Livingston and Aaron Burr, with Hamilton as their target. In the cauldron of intensifying party politics, few were willing to believe that two old friends were on vacation, studying flora and fauna.

The first unconscious step Jefferson and Madison took toward organizing a political party to oppose Hamilton and the Federalists came when they persuaded the radical New Jersey poet and journalist Philip Freneau to set up a newspaper in Philadelphia. Jefferson and Madison wanted a fairer hearing for their views, "a whig vehicle of intelligence" to counteract Federalist editor John Fenno's *Gazette of the United States*, what Jefferson called "a paper of pure Toryism, disseminating the doctrines of monarchy, aristocracy and the exclusion of the influence of the people."[36] Philip Freneau's Princeton classmate, Henry Lee, had told Madison, another Princetonian, of the poet's impecunious state and of his desire to leave a poorly paid journalist's job in New York City and move to the capital. Jefferson offered Freneau a $250-a-year part-time translator's job that "gave him so little to do as not to interfere with any other calling,"[37] but Freneau held out until Madison negotiated with a New York printing firm to take Freneau into partnership to establish the *National Gazette* in Philadelphia with Freneau as editor. Jefferson actively, if secretly, participated in the venture, insisted on paying Freneau the translator's salary for a no-show job, and promised him "the perusal of all my letters of foreign intelligence and all foreign newspapers, the publication of all proclamations and other public notices within my department and the printing of the laws." Just the printing business was enough to ensure the newspaper's survival. Jefferson and Madison then set to work drumming up subscriptions, especially in Virginia. They were bent on making Freneau's paper truly national. All that Jefferson asked in return for this was that Freneau "would give free place to pieces written against the aristocratical and monarchical principles."[38]

Freneau needed no coaxing. He promptly attacked the Hamiltonians and their policies. Disingenuously, Jefferson insisted he had not expected Freneau to go so far. "My expectations looked only to the chastisement of the aristocratical and monarchical writers and not to any criticisms of the proceedings of the government,"[39] he told the president, defending his role as merely support for an artist. Washington eventually asked Jefferson to do what he could to tone down Freneau's paper. But by that time, a full-fledged newspaper war had helped to bring on, in the Second Congress, the formation of opposing blocs, which eventually became the Federalist and Republican parties. James Madison had emerged as the leader of the anti-Hamilton bloc by the end of the First Congress in March 1791.

Nine months after Freneau broke into print and while Washington, Jefferson, and Madison were out of town, Hamilton personally wrote attacks on Jefferson and placed them in Fenno's paper, assuming Jefferson was writing for Freneau's. Hamilton first exposed Freneau's salary at State and then charged that his paper had been "instituted with the public money," its editor "regularly pensioned with the public money in the disposal of that officer."[40] The war escalated rapidly, Hamilton writing under a variety of pen names, assuming Jefferson did likewise. Freneau called Fenno, who received the Treasury's printing business, "a vile sychophant" with "emoluments from government" who was "disseminating principles and sentiments utterly subversive of the true republican interests of this country."[41] Madison contributed eighteen unsigned articles in the two years before Freneau's paper folded. Jefferson flatly denied to Washington that he ever wrote anything for Freneau's paper under any name, but he encouraged others to write for Freneau and he gave him invaluable access to State Department papers he personally selected, careful to protect the confidentiality of reports and equally careful not to let him see what he didn't want him to see. When Washington asked Jefferson to find a way to curtail Freneau, Jefferson refused. Freneau's paper had

> saved our constitution which was galloping fast into monarchy . . . the President, not sensible of the designs of the party, has not with his usual good sense and sangfroid looked on the efforts and effects of this free press and seen that, though some bad things had passed through it to the public, yet the good have preponderated immensely.[42]

By 1792, both factions were accusing each other of contributing to what Jefferson termed "the heats and tumults of conflicting parties."[43] Founded in unanimity, only five years after the federal government was formed at Philadelphia, a two-party system had emerged in the United States, with Hamilton at the head of the Federalists and, in Hamilton's words to President Washington, "Mr. Madison cooperating with Mr. Jefferson . . . at the head of a faction decidedly hostile to me and my administration. Jefferson and Madison were in my judgment subversive of the principles of good government and dangerous to the union, peace and happiness of the country."[44]

In the late summer of 1792, Washington wrote confidentially to Jefferson and Hamilton, urging reconciliation and regretting that "internal dissensions should be harrowing and tearing our vitals."[45] His letters to the two rivals differed. He urged Jefferson to have "more charity for the opinions and acts of one another."[46] He cautioned Hamilton about his temper and mentioned "irritating charges"[47] in the gazettes. He urged both to make "allowances, mutual forebearances and temporizing yieldings on *all sides.*" But Jefferson was implacable, placing all the blame on Hamilton, calling his policies "adverse to liberty" and predicting they would "undermine and demolish the republic" by giving his Treasury Department too much influence over Congress. But he steadfastly denied that "I have ever intrigued"[48] in Congress or the state legislatures to thwart the government's policies.

But Hamilton kept up the attack in the press. Jefferson's friends anonymously took up the cudgels against Hamilton. Hamilton, sarcastically referring to Jefferson as "the quiet, modest retiring philosopher," said it was time he should be revealed "the intriguing incendiary."[49] Under the pseudonym Catullus, Hamilton kept up the attack through six articles. Madison and Monroe openly defended Jefferson in a series of six essays in David Claypoole's New York *American Daily Advertiser,* Monroe doing the bulk of the writing. Both Hamilton's attacks and their "Vindication of Mr. Jefferson" appeared in print during the presidential election of 1792. Washington wanted to retire, but Madison and Jefferson pleaded with him to make "one more sacrifice,"[50] as Madison put it. Jefferson told Washington he feared a "corrupt squadron"[51] in Congress would take unlimited power and create an English-style monarchy. Hamilton, warning against republicanism, also asked Washington to run again.

As congressional campaigns unfolded, Jefferson stood aside but allowed Madison and Monroe to organize support in Pennsylvania and New York for Governor George Clinton of New York to oppose John Adams for the vice presidency. Madison, Monroe, and Aaron Burr carried out most of the behind-the-scenes campaigning, but Jefferson clearly knew what was going on.

On a trip to New York in September 1792, John Beckley, clerk of the House of Representatives and an able party organizer, carried a letter from a close friend of Jefferson's, Dr. Benjamin Rush of Philadelphia, stating that Beckley was trusted by Jefferson and Madison and urging Burr "to take an active part in removing the monarchical rubbish of our government."[52] The Republicans succeeded remarkably, garnering unanimous electoral blocs for Clinton in Virginia, New York, North Carolina, and Georgia and giving him fifty electoral votes to Adams's seventy-seven. Washington, unchallenged, was unanimously elected, but as the party around Jefferson made abundantly clear, his succession would certainly be contested. Jefferson saw the results of the first contested election as "generally in favor of republican and against the aristocratical candidates." Hamilton had lost much of his support and Jefferson's party would certainly have "a decided majority in favor of the republican interest"[53] in Congress. By the time Congress reconvened in December 1793, Jefferson had walked into Washington's office and handed him his letter of resignation.

Jefferson had planned to resign at the end of Washington's term. He had given up his lease on his town house effective March 1793, and had begun packing. He looked forward to retiring to Monticello. When Washington decided to run again, Jefferson first rejected a second term at State, then declined Washington's request that he return to Paris as minister plenipotentiary. He yearned for Monticello "with the fondness of a sailor who has land in view."[54] Under Hamilton's continuing attack, however, Jefferson grew reluctant to appear to be running from the field of battle. To his daughter Patsy he confided his fear that his resignation would injure him in the public eye as if he were retreating "from investigation" and was too weak "to meet slander."[55]

He could not have chosen a more difficult year to linger at State. The Reign of Terror swept France as war spread across Europe and the oceanic possessions of England, France, and their allies. In Philadelphia, a yellow fever epidemic killed six thousand people, fully one-fifth of the population, forcing the government to take refuge in Trenton, New Jersey. Worldwide war and local pestilence eclipsed the Jefferson-Hamilton feud.

Jefferson had found it increasingly difficult to follow the fast-moving events in France. Many of his friends were dead or had scattered to find refuge at their country *châteaux* or in exile outside France. He relied almost entirely on letters from his protégé, William Short. By the summer of 1792, Jefferson's attitude toward the French Revolution was changing with the spirit of the revolution. He had expected rational reforms to produce a constitutional limited monarchy on the British model. But as Hamilton and his cohort, whom Jefferson called "monocrats," kept up their attack and royalists in Europe formed a coalition backed by the armies of Prussia and Austria to crush the French Revolution, Jefferson abandoned his dispassionate neutrality toward France and became an ardent exponent of French republicanism. He no longer relied on a stable French government that he could exploit to advance American interests. With the French Revolution threatened, he worried that American republicanism was also in danger. A royalist victory in Europe would inspire Hamilton's "monocrats" to institute monarchy in the United States on the British model.

From Philadelphia, Jefferson could not distinguish at first between the moderate Girondists friendly to America and the far more radical Jacobins when they both joined the Paris Commune in deposing and imprisoning Louis XVI after he tried to flee France to assemble a royalist army against the revolutionaries. When a radical National Convention swept away the National Assembly as well as the monarchy, Jefferson's early fears that reforms were moving too fast were realized. Worse news came with each ship. Lafayette had gone into exile and then been imprisoned. La Rochefoucauld had been overtaken by a mob near his *château* and stoned to death in front of his family.

Jefferson reacted by embracing the Girondists, who had brought about the republic that led to war. When, during the American presidential campaign, news arrived in America that the revolutionary armies had defeated the royalists, Jefferson joined in the celebrations. The French militia victory surely helped his party nearly unseat the Hamiltonians in Congress. Remembering personal friendships in Paris with physiocrats, such as Brissot de Warville, who had become leaders of the Girondists, he tried to ignore the first reports of the Reign of Terror. When William Short wrote of his revulsion and disenchantment, Jefferson tried to persuade him to keep faith in the revolutionary movement. "The liberty of the whole earth was depending on the issue of the contest, and was ever such a prize won with so little innocent blood?"[56] And when Washington appointed the Hamiltonian Gouverneur Morris as the new minister to France, Jefferson gave him specific instructions in a series of letters to keep cordial relations with any French regime.

> It accords with our principles to acknowledge any government to be rightful which is formed by the will of the nation substantially declared. . . . We surely cannot deny any nation that right whereon our own government is founded, that every one may govern itself under whatever form it pleases, and change these forms at its will. It may transact its business with foreign nations through whatever organ it thinks proper, whether king, convention, assembly, committee, president or whatever else it may choose. The will of the nation is the only essential thing to be regarded."[57]

In private, Jefferson linked the French Revolution with his struggle against the Hamiltonians at home. The revolution in France was absolutely essential to the spread of liberty not only in Europe but also in the United States. "A check there," he wrote to Short, "would retard the revival of liberty in other countries. I consider the establishment and success of their government as necessary to stay up our own and to prevent it from falling back to that kind of half-way house, the English constitution."[58]

The news from Paris lagged six to twelve weeks behind events. Jefferson learned in December 1792 that French victories over Prussia had been followed in September by establishment of a French republic. He reported "universal feasts and rejoicings"[59] to Short. There were more celebrations when the first ambassador of the new French republic arrived in April 1793, but, one day before he landed, news reached America that Louis XVI and Marie Antoinette had been guillotined. Jefferson had to inform Washington that the royal executions had been followed almost immediately by a French declaration of war against England and the Netherlands.

Many Americans were saddened by the death of the man who had generously aided their own revolution.

In reporting to Washington that war had erupted between France and England, Jefferson advised taking "every justifiable measure for preserving our neutrality." He had foreseen the event and had warned American envoys in Europe "not to meddle with the internal affairs of any country. Peace with all nations and the right which that gives us with respect to all nations are our object."[60] Despite Jefferson's personal belief in the French Revolution, he never wavered from pursuing a policy of strict neutrality in foreign affairs.

Not so in domestic politics: in early 1793, he still was keeping up the struggle with Hamilton. He was intimately involved in the demands of his Virginia lieutenants in Congress for an investigation into the affairs of Hamilton and the Treasury Department. The drive, led by Representative William Branch Giles, culminated in a motion to censure Hamilton that was then defeated in Congress. Jefferson denied that he had any contact with Congress in the affair, but his interest in this matter can be deduced from his own handwritten copy of the Giles resolution of censure, found in the Library of Congress.

Upon hearing from Jefferson that the war in Europe had widened with a French declaration of war on Spain, President Washington cut short a vacation at Mt. Vernon to return to Philadelphia in mid-April of 1793. But Washington did not confide in Jefferson by seeking his advice on the appropriate American diplomatic response. Jefferson received a list of questions, ostensibly from the president, which was sent to all cabinet members, along with a call for a cabinet meeting at Washington's office in Germantown, eight miles outside the city, at nine the next morning. The memo asked whether a neutrality proclamation should be issued. It also raised the question of whether to recognize the French Republic by asking whether an envoy should be received, and whether the 1778 Franco-American treaty was still valid. Though Jefferson recognized Washington's handwriting, he concluded that it was Hamilton's language and "the doubts his alone."[61]

At the April 18 cabinet meeting, Jefferson argued vigorously that the president had no authority to issue any declaration of neutrality and could do nothing more than declare peace. "It would be better to hold back the declaration of neutrality as a thing worth something to the powers at war, that they should bid for it. . . . We might reasonably ask a price, the *broadest privileges* of neutral nations."[62] But Jefferson was overruled. Only the word *neutrality* was omitted when Washington issued a proclamation warning Americans against involvement in the hostilities.

Jefferson agreed to make the cabinet motion unanimous, but Washington asked Attorney General Randolph, not Jefferson, to draw up the proclamation. The cabinet also voted to receive the envoy from the French Republic, but only over the objection of Hamilton. Jefferson noted Hamilton's great regret that the United States was obliged to recognize the French Republic. Hamilton also provoked a long debate over the 1778 treaty with France, insisting that, because it had been signed when France was a monarchy, it should now be suspended. Jefferson countered that treaties were made between nations and that the people were the source of all authority and had the right to change their agents of government at any time without affecting the acts of a nation. He also pointed out that both nations had changed their forms of government since 1778 without abrogating the treaty because the Constitution of the United States replaced the confederation with a republic in 1789.

Hamilton and Jefferson sparred over receiving Ambassador Edmond Genet without qualifications. But the provision of the 1778 treaty triggering the greatest debate was a clause pledging the United States to help France defend her West Indian possessions against the British. Jefferson said he doubted this obligated the United States actually to go to war. The cabinet meeting was deadlocked and was adjourned without deciding the question. Asking Jefferson, Hamilton, and Randolph to give him written opinions, Washington privately told Jefferson that he had believed all along that the Franco-American treaties were still valid. Later, Washington upheld Jefferson's opinion on the treaties and ordered the new French minister received but "not with too much warmth or cordiality."[63]

Despite the cool civility of Washington's reception of Genet, he had been treated like a conquering hero from Charleston to Philadelphia. Jefferson's reaction to Genet was more enthusiastic than Washington's. He found the Frenchman "magnanimous,"[64] especially after Genet told him the United States would not be called upon to protect the French West Indies. But their honeymoon soon cooled when a French frigate towed a British prize into Philadelphia and thousands of farmers crowded the Delaware waterfront, bursting into "peals of exultation." Jefferson wished "we may be able to repress the spirit of the people within the limits of a fair neutrality."[65]

Misgauging Jefferson's warm reception, the French minister began to ignore strict American neutrality. As soon as he had landed at Charleston, he began fitting out privateers and raising American recruits for a sea and land campaign against Spanish-held Florida. Genet gave the French consul at Charleston authority to sell prize merchant ships taken from the British. When Jefferson protested that this violated American sovereignty, Genet rejected Jefferson's interpretation of the 1778 Franco-American treaty. Genet insisted that it provided that the French could fit out privateers or sell captured ships in American ports. When Jefferson upheld one of Genet's arguments that French prizes were French property that could be sold by a French consul, President Washington overruled Jefferson and ordered all French privateers out of American ports.

When Jefferson warned Genet to avoid any further violations of American neutrality, Genet rejected his definition of neutrality and continued to challenge him. On June 17, Jefferson again warned him to stop outfitting vessels in American ports, but Genet apparently believed that the administration was divided on the issue and that Jefferson was only reluctantly passing along Washington's orders. When Genet confided to Jefferson that he was planning to recruit an army on the Kentucky frontier to invade Spanish-held Louisiana, "I told him that his enticing officers and soldiers from Kentucky to go against Spain was really putting a halter about their necks, for they assuredly would be hung if they commenced hostilities against a nation at peace with the United States." But Jefferson, who believed the Spanish were also picking a fight with the United States, added that "leaving out that article, I did not care what insurrections were incited in Louisiana."[66]

For four months, Genet continued his machinations and argued with Jefferson, who found it was because of the Frenchman's intrigues that he was "worn down with labors from morning to night and day to day."[67] Adding to his burden was renewed party strife over the declaration of neutrality. Jeffersonians found neutrality pro-British because it disregarded American treaty obligations to France. Madison went so far as to call it unconstitutional "by making the Executive," not Congress, "the organ of the nation in relation to war and peace."[68] But Jefferson, caught between his own followers and Washington, upheld the administration because he believed it necessary to maintain "even a sneaking neutrality."[69]

The war in Europe had another effect on Americans, Jefferson soon noted: it "kindled and brought forward the two parties with an ardor which our own interests merely could never incite."[70] The old epistolary warfare flared anew in the Philadelphia press in June 1793. Hamilton, as "Pacificus," launched a series of defenses of neutrality exactly echoing his arguments in the closed cabinet meeting. After a second Hamilton salvo, an exasperated Jefferson wrote Madison, "For God's sake, my dear sir, take up your pen, select the most striking heresies and cut him to pieces."[71] The grating result was a series of five Madison letters under the pen name "Helvidius" which underscored the differences between Hamilton and Jefferson in ideas of executive power. Few people doubted who the warring scriveners were—two former collaborators of the Federalist Papers.

The Genet affair finally came to a head in July 1793, when a French ship towed the *Little Sarah,* a captured British merchantman, into Philadelphia to have it more heavily armed. Jefferson angrily demanded that Genet detain the ship until Washington arrived. Genet refused to do so. While Hamilton and Knox clamored to set up a battery of artillery on the Delaware to impede the ship's departure, Jefferson feared such a belligerent act would draw the United States into open warfare with their French ally. Besides, he knew a large French fleet was approaching Philadelphia. When the cabinet met before Washington could arrive, Hamilton demanded decisive action. Jefferson was outvoted. As Knox rushed work on batteries on Mud Island, Genet defiantly ordered the French ship to drop downriver beyond their range and said it would sail when it was ready.

When Washington arrived on July 11, he found Jefferson's packet of letters, marked "instant attention," on his desk. Washington, unaware that Jefferson was ill, vented his rage at Genet on his Secretary of State. He sent a steaming message off to Jefferson at his temporary quarters on the Schuylkill, accusing him of "submitting"[72] to Genet, then called a cabinet meeting for the next morning. A calmer Washington, finding his cabinet divided on appropriate action, accepted Jefferson's advice to ask the Supreme Court for an opinion on guidelines for American neutrality. When the Court refused to consider the question, the cabinet formulated an unprecedented set of "Rules Governing Belligerents." On August 2, the cabinet also voted unanimously to demand Genet's recall. After a forty-five-minute harangue in which Hamilton demanded that Genet's letters to Jefferson be made public, Jefferson and Randolph blocked Hamilton but voted to have the correspondence sent to France to ensure Genet's recall.

It was Jefferson's last battle as Secretary of State. On July 31, he had informed Washington he would retire September 30. When Washington once again pleaded with him to stay on, Jefferson assured him that the Republicans would support him through the remainder of his presidency. He also said he was tired of a post that forced him "to move exactly in the circle which I know to bear me peculiar hatred, that is to say the wealthy aristocrats, the merchants connected closely with England, the new created paper fortunes."[73] Finally Jefferson agreed to stay on until the end of 1793, close to the time when Hamilton had announced he would resign.

That summer, in swampy, mosquito-ridden Philadelphia, foreign and domestic intrigues had to be set aside temporarily as the worst yellow fever epidemic in American history combined with the bleedings and purgings of weakened patients by unenlightened doctors to kill thousands. Hamilton was stricken and went home to New York to recover; Washington went to the safety of Mt. Vernon. Jefferson insisted on going into his office, but the government offices had been decimated by the disease and he, too, finally went home, taking Polly to Monticello with him.

When the government reassembled in September, it was in the safer but crowded village of Germantown, where Jefferson took up residence in a bed in the corner of a public room in a tavern. Washington and his cabinet did not move back into the city until December 1, and then Jefferson only had time to clean out his desk and report to Congress on restrictions on American commerce before going home. His final report gave Madison the ammunition to counterattack British trade restrictions in Congress. He met with Washington one last time on December 21, but the president's pleadings fell on unsympathetic ears. Jefferson, he told a friend, had made up his mind "to be liberated from the hated occupations of politics."[74] Writing to Patsy at Monticello on December 15, he said he was coming home after eleven years away from Monticello in the national service, "no more to leave you."[75]

In nearly four years in Philadelphia, Jefferson had heard very little about Maria Cosway, but no one had taken her place in his affection. He must have missed their long talks, their relaxing rides and strolls. Now he was all work. Almost as soon as he took charge at State and immersed himself in official business, Jefferson began to suffer from chronic migraine headaches. The "severe attack of periodical headache," he wrote Thomas Cooper of South Carolina, "came on every day at sunrise and never left me till sunset."[76] In April 1790, he learned that Maria Cosway was pregnant. Lucy Paradise wrote from London, "This lady is with child for the first time." Maria was thirty. "She has been extremely ill but is now perfectly recovered and expects in a few months to lie in."[77] The impact of the news that Richard Cosway had found a way to keep Maria at home and put a crimp not only in her travel plans but in her career as an artist hit Jefferson just as his struggles with Hamilton were beginning. Four days after learning of Maria's pregnancy, Jefferson's migraine returned, this time for six weeks. Still, he apparently did not write to Maria.

On June 11, he received a reproachful letter from her. "I fear my dear friend has forgot me: not one line ever since your departure from this part of the world!"[78] She did not mention being pregnant, but Jefferson wrote at once to congratulate her, adding,

> You may make children there, but this is the country to transplant them to. . . . There is no comparison between the sum of happiness enjoyed here and there. All the distractions of your great cities are but feathers in the scale against the domestic enjoyments and rural occupations and the neighborly societies we live amidst here. I summon you, then, as a mother to come and join us. You must tell me you will, whether you mean it, or no. En attendant je vous aimerai toujours [while waiting, I will always love you]. Adieu, my dear Maria.[79]

Just as Jefferson had kept his pledge to his dying wife never to remarry, he evidently kept his promise to love Maria all the rest of his life, even as the opportunities of corresponding with her became fewer and fewer. She was unable to tolerate her increasingly psychotic husband's continued illicit *amours*. Moreover, he was probably suffering the early symptoms of syphilis. Maria left him and their daughter three months after the child was born in 1790. She fled to Paris with a celebrated Italian opera singer, the castrato Luigi Marchesi, and tried unsuccessfully to revive her art career at a time when it was difficult to find aristocratic clients to sit still to pose for miniatures. Her flight was duly noted by diarist Horace Walpole: "Surely it is odd to drop a child and her husband and country all in a breath."[80]

Maria did not write Jefferson another letter for three years, but he had news of her from her brother, a young architect who came to America and obtained his sponsorship. By 1793, Maria was again trying to enter a convent. To her husband, who still sent her money, she wrote, pleading with him to raise their daughter as a Catholic and remonstrating with him over his libertine life. "May it please the Almighty to enlighten you and show you the blind errors in which you are not only unhappily fallen into but, more unfortunately, persist in them."[81] When Cosway became seriously ill in 1795, she went back to London, nursed him, took care of her daughter. Writing to Jefferson about her husband's recovery, she reported, "I have found a pretty little girl. She shows natural talent and a soft disposition."[82] Maria took the child on a visit to Italy, but the youngster was never well and died at age six. Richard Cosway had a friend decorate a marble sarcophagus and kept his daughter's embalmed remains in their living room while he pursued his penchant for animal magnetism and believed that he could raise the dead. Finally Maria had the child's body interred, and then she left for Paris, where she made a last futile attempt to earn a living as a painter. After a brief stay in a convent at Lyons, she went to Lodi, Italy, and raised money to open a convent school for girls. Eventually the emperor of Austria made her Countess Maria of Lodi in honor of her labors in his dependency of northern Italy.

In 1795, shortly after Jefferson retired to Monticello, he received two letters from Maria. He did not reply for six months, but then he composed a long letter to her: "I am eating the peaches, grapes and figs of my own garden and I only wish I could eat them in your native country, gathered on the spot and in your good company." After suggesting a trip to Italy, he returned "to what is real":

> I am become, for instance, a real farmer, measuring fields, following my ploughs, helping the haymakers. How better this than to be shut up in the four walls of an office, the sun ever excluded . . . the morning opening with the fable repeated of the Aegean stable, a new load of labors in place of the old. . . . From such a life, good Lord, deliver me! I am permitted from the innocence of the scenes around me, to learn to practice innocence toward all, hurt to none, help to as many as I am able."[83]

Jefferson's idyllic descriptions mask a long period of depression and reclusiveness intensified by his retirement at age fifty from politics. On April 27, 1795, he wrote to Madison, "My health is entirely broken down within the last eight months."[84] Seven years later, in advising his daughter Polly to avoid withdrawal from society after her husband became mentally ill, he could have been talking to himself:

> I think I discover in you a willingness to withdraw from society more than is prudent. I am convinced our own happiness requires that we should continue to mix with the world, and to keep pace with it as it goes; and that every person who retires from free communication with it is severely punished afterwards by the state of mind into which they get, and which can only be prevented by feeding our sociable principles.
>
> I can speak from experience on this subject. From 1793 to 1797, I remained closely at home, saw none but those who came there, and at length became very sensible of the ill effect it had upon my own mind, and of its direct and irresistible tendency to render me unfit for society, and uneasy when necessarily engaged in it. I felt enough of the effect of withdrawing from the world then, to see that it led to an antisocial and misanthropic state of mind, which severely pun-

> ishes him who gives in to it: and it will be a lesson I never shall forget as to myself.[85]

During his thirty-seven-month retirement from politics, Jefferson never went more than seven miles from Monticello.

But what a hive of new construction and experiments was Monticello during those years! After ten years away when overseers farmed its twelve hundred acres of fields, the lands were played out. Jefferson immediately devised a seven-field system of crop rotation, dividing the tillable acreage into wheat, peas and potatoes, corn and potatoes, peas and potatoes, rye, and clover. Jefferson, like Washington, preferred to grow wheat and not tobacco because wheat preserved the soil and gave greater benefits to society, providing food and requiring less labor. To produce fruit and beauty, he bordered his fields with nine hundred peach trees. Most of his harvests went to feed the 105 slaves Jefferson had left on his Albemarle farms in 1795 and his own extended family. Another forty-nine slaves lived in Bedford County.

Jefferson had lost money steadily on his government posts for twenty-five years, making up the shortfall by selling slave-cultivated tobacco and, when his debtors grew shrill, families of slaves. By 1795, he was £7,500 in debt and indefinitely dependent on slave labor. Yet he continued to make fine distinctions on the subject. He never used slave labor to grow rich. His only cash crop, wheat, used few slaves. An unusual proportion of his slaves were household servants or, more and more, skilled artisans he trained and supervised. Rather than perpetuate the cycle of newly purchased field hands producing more soil-depleting tobacco and cotton which needed more lands heavily mortgaged to buy more lands, he tried to modernize his farms and make them more efficient. To supplement his cash and retire some of his debts, he began to produce iron nails, bringing a few slave artisans to Philadelphia to learn the craft of making them. Soon, he was training a dozen slave boys from ten to sixteen years old, personally supervising them. "With my farming and my nail manufacturing, I have my hands full," he told Adams. "I am on horseback half the day and counting and measuring nails the other half."[86] The horseback rides had a fringe benefit, helping his migraines to abate. He produced a ton of nails a month in 1796, but he learned a harsh lesson in the meantime. Local merchants customarily bought nails along with other goods from English distributors and were reluctant to alienate them. Jefferson found the British factors were guilty of "a principle of suppressing every effort towards domestic manufacture."[87] Jefferson responded by capitalizing his own distribution network for dealers willing to sell on consignment.

He also worked to modernize farming implements. For years, he had thought of ways to improve the mold-board plow. His eventual design, successfully tested at Monticello, increased his fame as a scientist and agriculturalist and won him a gold medal from the Societé d'Agriculture du Département de la Seine at Paris in 1805. He also built a new compact horse-powered threshing machine to streamline wheat harvesting, introducing it in 1796. But for all of his efforts, he confessed to Washington in late 1795, "it will be the work of years before the eye will find satisfaction in my fields."[88]

The eye would find only tumult atop Monticello during this period as Jefferson tore apart his house and doubled it according to plans he had been formulating for ten years. "Architecture is my delight," he confessed, "and putting up and pulling down is one of my favorite amusements."[89] In France, he had observed that the latest style was for one-story houses. Day after day in Paris, he had gone to the Cours de Reine along the Seine and stared transfixed at the Hôtel de Salm (now the Palais de la Légion d'Honneur) on the Left Bank near Lafayette's *château*. He told Madame de Tessé he was "violently smitten"[90] with the one-story palace that concealed two stories of private rooms behind entertainment rooms eighteen feet high. Small private staircases avoided the wasted space of great staircases, "which are expensive and occupy a space which would make a good room in every story."[91] Also, preferring the discreet elegance of concealed second-story windows tacked away in a façade that appeared one story high, Jefferson topped off the new Monticello with a rotunda inspired by the Halle aux Bleds, where he had first met Maria Cosway.

With plans he had drawn while Secretary of State, Jefferson launched a great rebuilding program that doubled the exterior of Monticello and eventually, in the words of the visiting Duc de La Rochefoucauld-Liancourt, assured that "his house will certainly deserve to be ranked with the most pleasant mansions in France and England."[92]

But politics always interrupted his plans for Monticello. The massive renovation all but stopped when Jefferson decided he could no longer ignore the call of his followers to resist the growing power of what Madison called "the British party."[93]

It is safe to say that Thomas Jefferson actually thought he was finished with politics when he resigned from Washington's cabinet in 1794 and only aspired to be "the most ardent and active farmer" in Virginia. "I cherish tranquility too much to suffer political things to enter my mind at all,"[94] he told Washington in 1794. He swore to his cousin Edmund Randolph, who succeeded him at State, that "no circumstance will ever more tempt me to engage in anything public."[95] But by 1796, Madison, who managed the Republicans, was touting Jefferson as a candidate for the presidency. By this time, Jefferson's letters show, he was once again becoming interested in national events. He denounced the Jay Treaty with England as an "infamous act," a hidden treaty of alliance "between England and the Anglomen of this country against the legislature and people of the U.S."[96] Among other provisions, the Jay Treaty allowed British merchants to repudiate paper currency payments and sue in American courts to collect their full pre–Revolutionary War debts plus accumulated interest for twenty years. Jefferson had accepted paper payments for lands he sold to clear his father-in-law's debts but now he was sued twice and lost. He owed another £1,000! When Hamilton used troops to enforce a new excise tax in Pennsylvania, Jefferson protested this excessive use of the military. Jefferson supported Republican efforts in Congress to block the implementation of the Jay Treaty, and he pressured Madison to take on Hamilton in the press again. When Jefferson's disciples talked about the person Republicans should support to succeed Washington, Madison considered Jefferson the natural candidate, but Jefferson told him that "the little spice of ambition which I had in my younger days has long since evaporated."[97] By early 1796, Jefferson asked Madison to send him a weekly report of what was happening but to keep it "behind the curtain."[98] By late February 1796, Madison told Monroe that Washington would not seek a third term. Adams was expected to be the Federalist candidate, according to Madison. "The Republicans, knowing that Jefferson alone can be started with hope of success, mean to push him."[99] But Madison apparently did not consult Jefferson in his plans that summer and the campaign did not get under way until September, when Washington released his farewell address to the newspapers. The Republicans protested this late announcement was a scheme by Hamilton "to prevent a fair election and the consequent choice of Mr. Jefferson."[100]

The first political party contest for the presidency was a strange affair. John Adams, who considered himself Washington's heir-apparent, did not campaign, thinking it undignified. And Jefferson never left Monticello, never lifted a finger, but he didn't ask his lieutenants to stop pushing his candidacy either. It was a hard-fought campaign and a close election nonetheless. John Beckley, House clerk, organized an especially aggressive campaign in Pennsylvania, where Republicans won fourteen of fifteen electoral votes. The pivotal issue was the Jay Treaty, but it was really a campaign of personality cults, with Jefferson hailed by Republicans as a "steadfast friend of the rights of the people" and Adams called the "advocate for hereditary power."[101] The election, according to a typical handbill, was over "whether the Republican Jefferson or the Royalist Adams shall be President."[102] Federalists attacked Jefferson for mismanaging the war governorship of Virginia, for opposing the Constitution, for quitting his post as Secretary of State under fire and for leading a pro-French party bent on overthrowing the government. Presidential campaigns have had a vicious tone from this time onward.

The charge that probably hurt Jefferson most was that he was pro-French. It became damaging when the new French minister to the United States, Pierre Adet, announced that he was being recalled to France and that there was to be no replacement until the U.S. government "returns to sentiments and to measures more conformable to the interests of the alliance."[103] By fostering American fears of a confrontation with France that only one candidate, Jefferson, could avert, Adet was interfering deliberately in the election. His stratagem backfired. Jefferson lost key support in three states. When the electoral votes were counted, Adams won by a thin three-vote margin of 71–68, the Federalist vice presidential candidate, Thomas Pinckney, trailing with forty-eight votes and the Republican, Aaron Burr, with thirty votes. Adams's victory margin was based on single votes lost by Jefferson in three populous states, Virginia, Pennsylvania, and North Carolina. Jefferson carried all the votes in South Carolina, Kentucky, Georgia, and Tennessee; Adams all of New England, New York, New Jersey, and Delaware. In essence, Adams and the Federalists had carried the mercantile and New England states, Jefferson the south and the farming west.

Under the 1787 Constitution, as the second-highest vote getter, Jefferson had been elected vice president. To the surprise of many, including Madison, Jefferson was willing to accept the second-highest office: The "second office is honorable and easy," Jefferson wrote. "The first is but splendid misery. If Mr. Adams can be induced to administer the government in its true principles and relinquish his bias to the English constitution," Jefferson was willing to work with him. "He is perhaps the only sure barrier against Hamilton's getting in."[104] On March 2, 1797, Vice President-elect Jefferson returned to Philadelphia in a public stagecoach after an absence of three years. Word had leaked out that he was coming. As a crowd lined the streets and cheered him, his stagecoach passed under a banner, JEFFERSON THE FRIEND OF THE PEOPLE.

There were to be few moments of peace in Philadelphia for Vice President Jefferson. On March 3, 1797, a highpoint of his life came when he was installed as president of the American Philosophical Society, succeeding his friend the late David Rittenhouse in the nation's preeminent scientific society, founded by Benjamin Franklin. The next day, Jefferson, a month short of fifty-four years old, was sworn in as vice president in the Senate chamber at Congress Hall, where he was to preside for the next four years. The vice presidency consisted chiefly of presiding over the Senate, where there were no rules, so Jefferson invented them, writing the parliamentary handbook still used today. Congress Hall stood next to the old Pennsylvania State House, where Adams and Jefferson had decided between them, twenty-one years earlier, just who should write the Declaration of Independence. The two old friends began their administration together by marching into the House chamber where Adams, wearing a handsome new sword, took the oath of office and gave a conciliatory inaugural address.

There were to be few such peaceful moments between them, however. Two months later, Adams called a special session of Congress to consider what to do about the rejection of the new American minister to France, Charles Cotesworth Pinckney, by Napoleon's French government. At the opening of the congressional session, Adams accused the French of treating her old American ally "neither as allies or friends nor as a sovereign state" and urged Congress to show the French that the Americans were not "the miserable instruments of foreign influence" who had no regard for "natural honor, character and interest."[105] Jefferson was shocked at Adams's speech and later said he was convinced that it alone almost brought on war with France.

But any remaining chances for cooperation between the two old treaty friends dimmed almost immediately. A private letter Jefferson had sent to Philip Mazzei more than a year earlier during the Jay Treaty controversy now appeared in print in New York City in Noah Webster's Federalist *Minerva.* Mazzei had released the letter to a newspaper in Florence, which printed it in Italian. Translated into French, it was next published in the Paris *Moniteur.* Then it was translated back into English and printed by Webster with a few errors and one major addition. Jefferson had written his old friend and neighbor Mazzei an acid portrayal of American politics where liberty and republican government had been displaced by "an Anglican monarchical and aristocratical party . . . whose avowed object is to draw over us the substance, as they have already done the forms, of the British government." While Jefferson argued that most Americans were still republican, he said that all government officials, merchants trading with England, speculators, bankers, and others preferring "the calm of despotism to the boisterous sea of liberty" were working hard to turn the American republic into an English-style system. "It would give you a fever were I to name to you the apostates who have gone over to these heresies, men who were Samsons in the field and Solomons in the council but who have had their heads shorn by the harlot England."[106] The simple change by Webster of the word *forms* to *form* of British government infuriated Adams and the Federalists. Jefferson had meant to criticize Washington's trappings of royalty; his state carriage, his weekly levees, the official celebration of the president's birthday, and the "inauguration pomposities." But, even worse, someone had added a long sentence that had Jefferson charging Americans with ingratitude and injustice toward the French while accusing the pro-British Hamiltonians of trying to alienate Americans from the French to bring the United States even further under British influence.

Whatever shadings had been added, however, Jefferson had obviously meant to count Adams and Washington among the apostates of the Federalist party. As the letter was reprinted and the controversy spread, Jefferson never publicly acknowledged the letter nor responded in the press. But he warned his correspondents not to let his letters get out of their hands, and he asked Madison and Monroe if they thought he should answer in print. Monroe said he should. But Madison, his senior adviser, thought it a "ticklish" proposition that might backfire. Jefferson kept silent, but Republican editors, principally Benjamin Franklin Bache, editor of the Philadelphia *Aurora*, strongly supported Jefferson's critique of the pro-British party, no matter who its author was.

With Madison out of office and back in Virginia, Jefferson now personally assumed Republican party leadership. Worried about eroding Republican strength in Congress, he sought the help of Aaron Burr, who had only feebly supported his candidacy. He wrote Burr to ask if New York would "ever awake to the true state of things. . . . Can the middle, Southern and Western states hold on till they awake? These are painful and doubtful questions."[107] Jefferson was openly asking for Burr's help; Burr was flattered by Jefferson's confidence, but he was worried that their letter might get into print and proposed to come to Philadelphia and meet secretly with the vice president. On a Sunday morning, Jefferson, Burr, Albert Gallatin (the Republican leader in Congress), and James Monroe, just arrived from France, met on the Delaware River in the ship that had brought Monroe to report to Adams. Their two-hour "greeting" of Monroe was followed a few days later by a public dinner billed as a demonstration of Republican support for Monroe, but it was actually a clear signal that Vice President Jefferson was vigorously leading a well-organized and experienced opposition party.

The next bitter clash of parties came in Jefferson's home congressional district where a federal grand jury with Supreme Court Associate Justice James Iredill, a Federalist, presiding made an official court presentment of the circular letters of several Republican members of Congress, especially Rep. Samuel J. Cabell. Accusing him of spreading "unfounded calumnies" that were "a real evil" against the government, the grand jury presentment asserted that the Republicans had been alienating the people and encouraging "a foreign influence ruinous to the peace, happiness and independence of these United States."[108] Jefferson had been increasingly alarmed by the political usage of the federal judiciary. "The charges of the federal judges have for a considerable time been inviting the grand juries to be inquisitors on the freedom of speech, writing and of principle of their fellow citizens."[109] He anonymously drafted a petition for his Virginia neighbors protesting the grand jury's presentment as a clear violation of the natural right of free communication among citizens and from Monticello, asked the Virginia House of Delegates to impeach and punish the members of the grand jury for committing "a great crime."[110] He waited to have the petition submitted in the Virginia capital until Congress reconvened and he got his supporters in the Virginia House to arrange the printing and distribution of one thousand petitions. The Virginia House voted to denounce the grand jury but ordered no arrests when the grand jury did not press its charges against Cabell. Jefferson's vigorous attack on the Federalists in Virginia showed not only his spirited party leadership but his belief in a states'-rights doctrine. He opposed petitioning Congress, he told Monroe. "The system of the [federal] government is to seize all doubtful ground. We must join in the scramble or get nothing. . . . It is of immense consequence that the States retain as complete authority as possible over their own citizens."[111]

By the time Jefferson returned from the fray in Virginia to Philadelphia, the crisis with France had deepened with the news from Paris that three American envoys sent to protest Monroe's eviction had also been rebuffed in the XYZ affair. Jefferson was totally unprepared for this latest crisis. On March 4, 1798, Secretary of State Timothy Pickering received a batch of dispatches from the American commissioners to France, Charles Cotesworth Pinckney, John Marshall, and Elbridge Gerry. Before the messages were even decoded, Adams went before Congress to report that there was no hope of having envoys received by the French and, further, that the French Directory had ordered the seizure of all neutral ships carrying goods made in England or its possessions, a devastating blow to the American merchant fleet. French agents, their names encoded as X, Y, and Z, had demanded that President Adams apologize for his message to Congress of May 1797, and had asked for a substantial loan to their government and a $50,000 bribe, or *douceur,* before they would even open negotiations. Adams initially wanted a declaration of war but, unless he was willing to spread the entire correspondence before Congress, this was impossible. Instead, he urged Congress to protect American shipping by building a navy, to defend the coasts, restore arsenals, manufacture arms, and levy a special tax to pay for defense. He further announced he was rescinding President Washington's ban on arming American ships. While Republicans demanded full disclosure of the dispatches from Paris, Jefferson called Adams's new policy "insane" and his party spokesmen said the president's message was "a declaration of war."[112] On April 2, 1798, the Republicans, bolstered by alarmed crossover Federalists, voted 65–27 to demand the XYZ papers. The next day, as Congress read transcripts behind closed doors, Jefferson and his Republicans sat stunned.

When Congress voted to publish the documents, an anti-French fever spread across the land. Republicans tried to characterize the crisis as a "nefarious scheme."[113] Jefferson, dispatching a set of the papers to Peter Carr, said, "You will perceive they [the American envoys] have been assailed by swindlers, whether with or without the participation of Talleyrand [the French foreign minister] is not very apparent."[114] But Jefferson, who had known Talleyrand during his exile in Philadelphia, where twenty-five thousand Frenchmen had taken refuge from the Reign of Terror, believed it possible and that this still did not mean that Napoleon's Directory had any knowledge of the affair. Instead, Jefferson blamed Adams for stirring up the crisis with his intemperate remarks about the French and placed the blame squarely on the president for provoking the French into refusing to negotiate or receive Pinckney, a pro-British fixture, as American minister. Jefferson believed that Adams's speech was the only obstacle to reopening negotiations. But Jefferson could see that the war publication of the documents had "carried over to the war-party most of the waverers"[115] in Congress, and he could not have been surprised when Congress approved the creation of a Navy Department, the building of a fleet authorized to capture armed French ships off American shores, and the rapid buildup of coastal fortifications. Congress next expanded the regular Army even as it voted Adams emergency power to raise a ten-thousand-man volunteer army and buy large quantities of arms and munitions. To pay for these warlike measures, Congress voted a direct federal tax on land, houses, and slaves.

At the same time, Jefferson began to hear rumors that the Federalists planned a series of harsh measures for internal security, even though war had not been declared. To Madison in Virginia, a thoroughly alarmed Jefferson wrote, "They have brought into the lower house a sedition bill which, among other enormities, undertakes to make printing certain matters criminal, though [the First Amendment] to the Constitution has so expressly taken religion, printing presses, etc., out of their coercion."[116] He was sure that the Republican newspapers, especially Bache's *Aurora,* were the target. The war fever did not abate when envoy John Marshall arrived in New York from Paris and reported that the French had no intention of making war on America. When Marshall reached Philadelphia, enormous crowds thronged the streets. At a banquet welcoming him, Federalists raised their glasses in the toast, "Millions for defense but not one cent for tribute."

Still the eternal optimist, Jefferson wrote before leaving the capital on June 27, "A little patience and we shall see the reign of witches pass over, their spells dissolved, and the people recovering their true sight, restoring their government to its true principles."[117] Two days before he left Philadelphia, Congress passed the first Alien Act, a temporary peacetime measure. The president could deport any alien he considered "dangerous to the peace and safety" of the United States. Only after receiving a deportation order could an alien present evidence in his defense. Congress passed a second and permanent Alien Act on July 6, 1798, allowing the president to imprison or deport all aliens of an enemy power in wartime. On July 14, Congress passed a Sedition Act, to be enforced in peacetime or war, making it unlawful for American citizens to combine or conspire to oppose any government measure, to prevent government officers from doing their duties or to aid "any insurrection, riot, unlawful assembly or combination." But it also targeted any person writing, speaking, or publishing "any false, scandalous and malicious writing" against the president, Congress or the government. The act was to expire the last day of Adams's term.

Jefferson denounced the Alien and Sedition Acts as a blatant political attempt to silence Republicans and their newspapers, calling it "an experiment on the American mind to see how far it will bear an avowed violation of the Constitution."[118] As vice president and the specific target of the Sedition Act, he had to work in secret to appease the suppressive measures, but by October 1798, he was casually speaking with members of the state legislatures of Virginia and North Carolina who dropped by to visit him that he was sure their legislatures would react strongly. In October, he finished drafting a set of resolutions attacking the unconstitutionality of the acts and dispatched them to Wilson Cary Nicholas, a Republican in the Virginia Assembly. He also arranged for them to be introduced in the North Carolina legislature and when a former neighbor, John Breckenridge, chanced by Nicholas's house, Nicholas gave Breckenridge a copy to carry to Kentucky, where he was to sponsor them in the House of Representatives. Nicholas revealed Jefferson's authorship to Breckenridge, who swore secrecy and decided not to visit Jefferson at Monticello to protect the vice president from criminal prosecution. But Jefferson personally approved the arrangements and encouraged Nicholas to confer with Madison, who visited Monticello in October to discuss a similar set of resolutions Madison had secretly drawn up for John Taylor to present to the Virginia Assembly.

In the opening sentence of his Kentucky Resolutions, Jefferson affirmed that the "several states" were not united on "the principle of unlimited submission to their general government." The states had only delegated "certain definite powers." When the national government "assumes undelegated powers, its acts are unauthoritative, void and of no force." In November 1798, the Kentucky legislature adopted Jefferson's resolutions, declared the Alien and Sedition Acts null and void, and called on other states to express their outrage. In late November, Virginia followed suit. The two states had planted the seed, borrowed from Jefferson, of the doctrine of the state's right to nullify federal law, and had fired the first shots in the presidential election campaign of 1800. A reluctant candidate in 1796, Jefferson by early 1799 wrote to Madison that it was high time now, "the season for systematic energies and sacrifices. The engine is the press. Every man must lay his purse and his pen under contribution."[119]

20

"The Empire of Liberty"

I steer my bark with hope in the head, leaving fear astern. My hopes indeed sometimes fail, but not oftener than the forebodings of the gloomy.

—Thomas Jefferson to John Adams, April 8, 1816

After four years of powerlessly presiding over the Senate while the Federalists created a standing army and navy, abolished the Franco-American alliance, and virtually outlawed open opposition by the Republicans, Vice President Jefferson was ready for an all-out run for the presidency—even if the tradition of the time forbade campaigning openly for the highest office. By 1798, halfway through Adams's term, Jefferson began to plan his campaign. Without making a single speech, he would write hundreds of letters and political pamphlets that would circulate among Republican newspapers and supporters to become an underground campaign within a year. Early in 1799 Jefferson in Philadelphia wrote Madison in Virginia that the "season for systematic energies and sacrifices" had come and that Madison must devote a portion of every day to writing political letters for the papers in the capital and elsewhere. "When I go away, I will let you know to whom you may send so that your name shall be sacredly secret."[1] Jefferson gave similar instructions to other lieutenants while himself keeping out of print. But in fact, the campaign of 1800 had begun almost as soon as Jefferson became vice president in 1797.

The emergence of a second national political party came as the immediate result of the enactment of the Alien and Sedition Acts by the Adams administration. Vice President Jefferson, remaining at Monticello for much of the time the administration was in the capital city of Philadelphia, grew more and more outraged at the "Consolidators," as he called the Federalists. It was Jefferson who had the idea of a nationwide legislative protest against the Alien and Sedition Laws, passed by Congress in July 1798, at the peak of the war fever against France. The genesis of the protest came in a conversation Jefferson later had with John Breckinridge of Kentucky and Wilson Cary Nicholas, one of Jefferson's closest friends and a leader in Virginia politics. The two men, according to Jefferson, were visiting Monticello, where they pressed Jefferson to draft resolutions to protest the new laws as unconstitutional encroachments on the rights of the states. All three must have realized that the second highest officer in the U.S. government could be charged under the new law with sedition if he openly associated himself with a protest movement to declare two laws of the current Congress unconstitutional. But Jefferson was by this time convinced that the laws were directed at him and his supporters, especially the newspapers that had helped him come so close to winning in 1796 and were even more vital now.

But Jefferson was eighty and his memory dim when, more than twenty years later, he recounted how the protest—and the Democratic Party—had come about. He was the originator of both. Jefferson had begun to be alarmed at the Federalist attack on his Republican adherents in May 1797, when the Federalist-appointed grand jury of the Richmond federal circuit court "presented" Republican Congressman Samuel Jordan Cabell because he had sent circular letters to his constituents violently attacking the Adams administration. Jefferson labeled the presentment a "perversion" of a legal institution into a political one.

Returning to Monticello that summer, Vice President Jefferson drafted a petition denouncing the presentment and sent it to Madison for revisions. He also consulted Monroe. When the petition was submitted to the Virginia state legislature, its authors remained unknown. The petition charged that the presentment obstructed the proper relations between a representative and his constituents and reaffirmed the Republican principle that there could be no intrusion of the judicial power into the legislative branch. The House of Delegates passed a resolution calling "said presentment a violation of the fundamental principles of representation . . . a usurpation of power . . . a subjection of a natural right of speaking and writing freely."[2]

Shortly before the Sedition Act was passed, an angry Jefferson wrote his reaction to Madison:

> Among other enormities, [it] undertakes to make printing certain matters criminal, tho' one of the amendments to the Constitution has so expressly taken religion, printing presses, etc. out of their coercion. Indeed this bill and the Alien bill are so palpably in the teeth of the Constitution as to show they mean to show no respect to it.[3]

In these two bills, Jefferson and the Republicans saw clearly the threat to their existence: the Federalists were closing off the avenues of political propaganda to their rivals before the 1800 presidential election. Silenced, the Republicans could not win. Moreover, they feared a Federalist effort to alter basically the principles of the federal government. Republican Senator Henry Tazewell of Virginia said the laws "indulge that appetite for tyranny that alone could have occasioned the introduction of the principle [of monarchy]." The Sedition Act negated one of Jefferson's sacred beliefs, he wrote to a student at William and Mary in June 1799:

> I join you in branding as cowardly the idea that the human mind is incapable of further advances. This is precisely the doctrine which the present despots of the earth are inculcating and their friends here re-echoing, and applying especially to religion and politics: "that it is not probable that anything better will be discovered than what was known to our fathers. We are to look backwards then and not forwards for the improvement of science, and to find it amidst feudal barbarisms. . . . But thank heaven the American mind is already too much opened to listen. . . . While the art of printing is left to us, science can never be retrograde. . . . As long as we may think as we will, and speak as we think the condition of men will proceed in improvement."[4]

The enactment of the Sedition Act signaled one of the most secretive periods in Thomas Jefferson's long life. His letter book reveals that, while they usually exchanged several letters a month, he did not exchange a single letter with his closest ally, Madison, for more than three months in 1798 even as they launched the Republican counterattack against the Alien and Sedition Laws. Yet it is clear that the two men consulted each other regularly. With two close confederates, Breckinridge and Nicholas, each giving "solemn assurances" to shield Jefferson's identity and apparently using their land speculations as a cover for meetings, Jefferson and Madison launched a series of clandestine legislative initiatives.

Originally Jefferson planned to have a resolution introduced in the

North Carolina legislature calling for a legislative protest against the Federalists' odious acts. This device could then be emulated by other states, as had been the method in the long crisis of the Revolution of 1776. Breckinridge had come to Virginia to meet Jefferson and carry the resolutions to North Carolina: Vice President Jefferson had stopped using the mails, too easily monitored by Federalist officials. But Breckinridge was warned it would be too dangerous for Jefferson to meet him. Apparently receiving the documents from Nicholas, he changed his plan and went directly back to Kentucky, introducing Jefferson's resolutions there.[5] That summer of 1798, there were mass Republican rallies all over Virginia and Kentucky calling on the state legislatures to demand that Congress repeal the Alien and Sedition Acts. In November, when the Kentucky house reconvened, Governor James Gerrard in his opening message urged the need for a "protest against all unconstitutional laws."[6] When Breckinridge introduced Jefferson's resolution, it sailed through with only three dissenting votes; the Senate then passed it unanimously.

In this first states'-rights protest against federal power, Jefferson invoked the constitutional theory that the federal government was made up of a compact of states. He forthrightly held that where the national government exercised powers not specifically delegated to it, each state "has an equal right to judge for itself as well of infractions as of the mode and measure of redress." The federal Congress had exercised a power not specifically granted when it passed the Alien and Sedition Acts. The Sedition Act especially contradicted the First Amendment protection of freedom of the press. Jefferson's resolutions pointed out that, in such cases, the states "have the right and are in duty bound to interpose" and arrest "the progress of the evil." Declaring that the state governments, under the compact called the Constitution, had assigned only limited powers to the federal government, Jefferson held that it was only logical to conclude that each state retained the right to judge for itself if there had been such an infraction. It was the first time any legislature had put forth the proposition that a law could be repealed by some body other than Congress. What that redress should be had been toned down by Kentucky. While, in his draft, Jefferson had used the word *nullification* for the first time in a state's-rights protest, the Kentucky resolutions only called for the support of other states.

When Madison introduced his version, the Virginia Resolution, that autumn, it only called for support from sister legislatures, stopping short of declaring the Alien and Sedition Acts null and void. Dismayed, Jefferson called unsuccessfully for more radical wording. Kentucky did follow one recommendation of Jefferson's without tempering it: a committee of correspondence was empowered to confer with other states, a protorevolutionary step borrowed from the revolutionary protests of the 1770s.

In October 1798, Madison visited Jefferson at Monticello. The two men differed on their theories of representation. While Jefferson saw the state legislatures as the proper venue for protest, Madison favored a special convention, arguing that just such a convention had drafted the Constitution. Jefferson and Madison, as a letter written soon afterward makes clear, disagreed about the power of either the state, acting through a popular convention, or the state legislature, to react to laws contrary to the federal Constitution. Madison accepted Jefferson's argument that the Union was based on a compact among states but rejected Jefferson's theory that the state was ever justified in declaring federal laws null and void.

The Kentucky and Virginia resolutions were rejected by every other state as the Federalists accused the Republicans of threatening disunion. Rumors flew wildly by the spring of 1799 that Virginia Republicans were arming and planning to use violence. By March 1779, Jeffersonians learned that the aged Patrick Henry, long Jefferson's bitter enemy, was planning to reenter the legislature to lead the fight to repeal the resolutions. Republican lawmaker John Taylor worried to Madison that "if Mr. Henry prevails in removing [Virginia's] resistance to monarchical [Federalist] measures, the whole [nation] will be dispirited and fall a sudden and easy prey to the enemies of liberty." Taylor pleaded with Madison to run again for the House of Delegates to oppose Henry. "If you will not save yourself or your friend—yet save your country."[7] As historian Adrienne Koch has put it, "Political tempers had risen in the year following the first Virginia and Kentucky resolutions, earning for the Republicans the sincere fear and genuine hatred of the Federalists on an intenser scale than ever before."[8] The Federalist governor of North Carolina was so alarmed after a visit to Virginia that he wrote Supreme Court Justice James Iredell that he feared the "leaders there are determined upon the overthrow of the general government. . . . They would risk upon it the chance of war. . . . Some of them talked of 'seceding from the Union' while others boldly asserted . . . the practicability of 'severing the Union.' "[9]

On August 23, 1799, Jefferson launched his presidential campaign, breaking his silence with a long letter to Madison. "Perhaps you could take a ride." Jefferson suggested they meet with Nicholas the following Thursday to work out "a concert of action" in Virginia, Kentucky, and other states "as have ventured in the field of reason." Jefferson spelled out a detailed platform for Republican opposition to the Federalists that ended on a chilling note. He was, he assured Madison, "confident in the good sense of the American people," but if they did not rally round "the true principles of our federal compact," he was "determined were we to be disappointed in this, to sever ourselves from that union we so much value rather than give up the rights of self-government . . . in which alone we see liberty, safety and happiness."[10]

In probably the most extreme statement he ever made on this or any other question, Jefferson, after mulling over the Alien and Sedition Laws for more than a year, showed that he valued liberty over the Union. Destruction of the liberties guaranteed by the compact of the states was a far worse thing than secession from such a ruthless union. By late 1799, with the Kentucky and Virginia resolutions as their battle flags, Jefferson and Madison were ready to lead the Republican resistance in what was perhaps the dirtiest presidential campaign in American history.

At their Monticello meeting and in meetings over the next three months, Jefferson and Madison laid the groundwork for their campaign. Further meetings with Madison were considered unwise, so Jefferson took the risk of writing, on November 26, 1799, the Republican platform. "Our objects," he wrote Madison, "should be these." He called for peace, "even with Great Britain." "A sincere cultivation of the Union" was to be the *public* stance. The army was to be disbanded "on principles of economy and safety," but while protesting unconstitutional laws, nothing was "to be said or done which shall look or lead to force and give any pretext for keeping up the army." The 1798 Resolutions had become the fundamental planks of the Democratic-Republicans in domestic affairs. The explicit pledge to avoid the use of force was to guarantee a bloodless revolution, one far too long unrecognized as such by historians.

Building on Jefferson's list of objectives, Madison, as chairman appointed by the Virginia legislature to study the answers of other states to the 1798 resolutions, wrote his "Report of 1800." The report, approved by the House of Delegates on January 7, denounced the centralizing measures of the federal government—the Alien and Sedition Acts, a tax on carriages, the bank law—restating as a dangerous tendency the Federalist trend toward granting new powers to Congress. Vindicating Virginia's declaration against federal consolidation of power, Madison argued the importance of interposing state governments between the national government and the people. But Jefferson pulled the Democrats back from the brink of threatening nullification or secession. Acquiescing in their great collaboration, he quickly reprinted Madison's report and saw that it was spread throughout the country. His mind was now on a sweeping new declaration against the Alien and Sedition Acts that all the states would embrace. "As soon as it can be depended upon, we must have a Declaration of the principles of the Constitution." Was it to be like the document he had labored over long ago with Lafayette in Paris of 1789? All he said was that it was to be a "declaration of rights, in all the points in which the Constitution has been violated."[11]

During the eighteen months that the Sedition Act was in force, the federal government prosecuted twenty-five writers, editors, and printers in fourteen separate cases: ten were convicted and imprisoned.

Jefferson arranged financial aid and comfort to Republican newspaper writers and editors suffering under the Sedition Act, including James Thomson Callender, a fugitive Scottish journalist who had attacked President Adams in print and was sentenced to nine months in prison. While Callender's writings hurt the Adams administration, Jefferson later said the fifty-dollar payments—sizable at the time—he twice gave the man were "mere charities" and not fees for a hired gun. But during the 1800 campaign, when Callender sent him the proof sheets of his anti-Federalist book *The Prospect Before Us,* Jefferson, despite warnings from Virginia friends not to trust the man, wrote to encourage Callender that "such papers cannot fail to produce the best effect; they inform the thinking part of the nation [and] set the people to rights."[12]

Personally distributing campaign propaganda throughout the country, Jefferson typically sent one batch of a dozen to Monroe in February 1799, to spread among "the most influential characters among our countrymen," fence-sitters who could bring in other influential new voters to Republican ranks. "It would be useless to give them to persons already sound." But Monroe was also admonished to keep secret Jefferson's involvement. "Do not let my name be connected with the business."[13] He sent eight dozen offprints of Dr. Thomas Cooper's *Political Arithmetic* (Cooper was to serve six months in jail for writing it), which argued that it cost far more to build and maintain an American navy than the trade it protected, to the chairman of the newly created Virginia Republican Committee with a note: "Though I know that this is not the immediate object of your institution, yet I consider it as a most valuable object to which the institution may most usefully be applied." Jefferson warned the chairman that disclosure of the vice president's active role in the distribution of political propaganda could prove explosive. "You will readily see what a handle would be made of my advocating their contents. I must leave to yourself therefore to say how they come to you." More actively involved in a presidential campaign than any candidate to date, Jefferson hammered home the theme in the press that American merchants were getting all the benefits from an expensive government while "the consumer, the farmer, the mechanic, the laborer, they and *they alone* pay."[14]

But Jefferson's most important contribution was to develop the Republican platform in letters that explained his own political principles to party leaders around the country. To his old friend Elbridge Gerry of Massachusetts, he asserted that his views were "unquestionably the principles of the great body of our fellow citizens." The Constitution was to be kept intact "according to the true sense in which it was adopted by the States" to prevent the Federalists from "monarchising" it and to preserve "to the States the powers not yielded by them to the Union." Jefferson fervently opposed "transferring all the powers of the States to the general government and all those of that government to the Executive branch." He called "for a government rigorously frugal and simple, applying all the possible savings of the public revenue to the discharge of the national debt." He opposed a standing army in time of peace and was willing to rely on the militia for defense except in the case of actual invasion. By 1800, he had come to fear an expensive navy in peacetime because he predicted "the eternal wars in which it will implicate us, grind us with public burdens and sink us under them." The United States must most of all stay out of European politics, he insisted. "I am for free commerce with all nations, political connection with none and little or no diplomatic establishment." Defending his cherished beliefs in freedom of religion and freedom of the press against recent attacks by the Federalists on the First Amendment, Jefferson told Gerry he was "against all violations of the Constitution to silence by force and not by reason the complaints or criticisms, just or unjust, of our citizens against the conduct of their agents."[15] Jefferson made sure that his political credo circulated widely through Republican newspapers, party pamphlets, flyers, and broadsides. He gave the Republicans a clear and strong stand on the issues in the 1800 election while he eschewed a personal attack on President Adams, even if his followers did not always heed his example. Some Republicans, in fact, complained that Jefferson's campaign was too high-toned, that with all the talk about principles, not enough was being done to stir up people's feelings so they would get out and vote. Benjamin Franklin Bache, grandson and namesake of the great inventor and printer, insisted on simplifying the campaign; under his editorship, the Philadelphia-based *Aurora* proclaimed, "The friends of peace will vote for Jefferson—the friends of war will vote for Adams or for Pinckney."[16]

Jefferson's presidential campaigns, as well as his opposition role as vice president, depended on his close and secret working relationship with Republican newspaper propagandists in New York, Richmond, and the capital at Philadelphia. As early as his first run for the presidency in 1796, Jefferson had learned how to carry out a concerted newspaper campaign through a chain of command that stretched from Monticello through Madison to John Beckley, his campaign manager in Philadelphia, to a handful of trusted political writers, including James Thomson Callender. Policy formulated by Jefferson and Madison was passed down to such ardent Republicans as Bache, who in turn bought columns from free-lance journalists such as Callender.

Exactly how this chain worked can be seen in the controversy surrounding Dr. James Reynolds, a speculator in the New York financial markets and a man close to Alexander Hamilton. For five years, Jefferson's allies, including Congressman James Monroe, had known of Hamilton's long-running affair with Dr. Reynolds's wife, Maria, which they discreetly covered up. They had first learned of Hamilton's financial dealings from Reynolds when the latter was jailed on a charge of subornation to perjury arising from shady stock dealings. From jail, Reynolds threatened to make disclosures "injurious to some head of a department,"[17] unmistakably indicating Hamilton. Monroe and two other Congressmen had visited Reynolds in jail and taken affidavits from the man: that same day, all charges against Reynolds were dropped, and the man disappeared. The papers were evidently sent to Jefferson, who recorded in his personal diary a short time later that he knew all about Hamilton's admitted affair with Maria Reynolds.

In the 1796 presidential campaign, hints were placed in the *Aurora* by editor Bache and manager Beckley, possibly at Jefferson's behest, that Hamilton, while Secretary of the Treasury, was involved in "very suspicious"[18] financial dealings with the absconded Dr. Reynolds. Bache had his own strong motive for attacking Hamilton: crude reports of the late Dr. Franklin's sexual prowess were beginning to appear in the Hamiltonian Federalist press.

In the summer of 1797, as Jefferson struggled to keep his political party alive, he paid a call on the Philadelphia printshop where James Callender was seeing his scurrilous new anti-Federalist book, *History of 1796,* through the press. Callender later claimed that

> Mr. Jefferson gave me no sort of countenance upon my arrival in this country. I never spoke to him, nor, to my knowledge, did I ever see him for upwards of four years after my arrival. Even then I did

> not introduce myself to him. It was Mr. Jefferson that introduced himself to me. He called at the office of Snowden and McCorkle in Philadelphia in June or July, 1797, asking for me.[19]

A short time later, evidently after John Beckley turned over copies of the Reynolds papers to Callender, a detailed account of Hamilton's supposed shady dealings with Reynolds in the stock market appeared in bookstores in Callender's *History of 1796.* Hamilton countered with a pamphlet in which he claimed his own dealings with the Reynoldses were limited to a long-standing extramarital affair with Maria Reynolds. Maria Reynolds later denied the affair; Monroe insisted that both the affair and the stock speculations had taken place; and Callender reiterated the charges in his next annual volume of billingsgate, *Sketches of the History of America.*

While Callender had become the most effective political writer in American politics, he had no regular income and, time after time, Jefferson had to arrange help when Callender and other propagandists needed money: being out of power, he had no government sinecures in his gift. In the thick of protests over the 1798 Kentucky Resolutions, Jefferson related to Madison his worry that the feeble Republican papers in Philadelphia and New York would "totter for want of subscriptions . . . We should really exert ourselves to procure them for, if these papers fail, republicanism will be entirely brow-beaten"[20] by the well-oiled Federalist press. The Federalists had already targetted Callender, in their flagship *Gazette of the United States,* calling him the "scum of party filth and beggarly corruption."

By the time the Alien and Sedition Acts took effect in 1798, Callender had moved to Richmond, where Jeffersonians were solidifying their power base. With access to the homes and libraries of Jefferson's lieutenants if not to Monticello itself, Callender began assembling his next anti-Federalist tome, *The Prospect Before Us,* the publication of whose first volume was timed to coincide with the opening of the Presidential campaign of 1800. Callender had opened a familiar correspondence with Jefferson, who answered his queries and supplied him with papers he needed for his research. Jefferson personally kept the project alive by twice sending Callender fifty dollars as advances against orders for his book. In addition, Jefferson's well-known patronage helped Callender sell his columns to the Republican papers of Virginia, from the Richmond *Examiner* to the Staunton *Scourge of the Aristocracy.* Galley proofs of Callender's writings were sent to Jefferson, containing gossipy allegations of a host of political corruptions practiced by the Adams administration, such as the furnishing of luxurious diplomatic outfits for John Quincy Adams as he moved from one European embassy to the next, and the lavish redecoration of the

President's House by Abigail Adams. Jefferson made no attempt to censor or suppress Callender's lurid prose.

By May 1800, in the heat of the presidential campaign, U. S. Supreme Court Justice Samuel Chase had become the self-appointed Federalist inquisitor, seeking out seditious libels against the Adams government. Chase headed for Richmond after securing the imprisonment of Jeffersonian journalists in Philadelphia and Delaware. He read the first volume of Callender's *Prospect* in the coach and decided to seek a conviction for "a libel so profligate and atrocious" against President Adams. To a fellow passenger, Chase commented, "It is a pity you have not hanged the rascal."[21] The Sedition Act had made it a crime "to write, print, utter or publish . . . any false, scandalous or malicious writings" against the government, the President or Congress or to stir up hatred or disrespect for any of them. Callender was charged with defaming Adams in twenty passages of his book and with the crime of printing it. As members of the Jeffersonian gentry packed the courtroom and shook his hand, Callender received a nine-month jail sentence and a stiff four-hundred dollar fine. In Richmond jail, he was allowed to dine at the sheriff's dinner table, stroll in his garden and receive a parade of prominent Jeffersonians, including then-Governor Monroe and Chancellor George Wythe. His expenses, including payment of the fine and the support of his children, were underwritten by a pro-Jefferson subscription drive, to which Jefferson himself contributed. Callender had become a martyr of the Alien and Sedition controversy.

No presidential election since 1800 has taken place without an attempt to damage at least one candidate's reputation by innuendo, rumor, and ridicule in order to make him appear unworthy of the nation's highest office, but none has more brutally combined these tactics than the 1800 campaign, which left Jefferson stunned and the country deeply divided for years. In the first knock-down, drag-out campaign, Americans proved they preferred newspapers to pamphlets or books, and, further, that they preferred their newspapers crammed with items of scandal. It was the first modern campaign and, as one historian put it, "The assaults on the virtue and integrity of the candidates in that election have never been surpassed either in their ferocity or in their departure from the truth. His reputation suffered blows in that campaign from which it has not yet recovered." Jefferson was pilloried for his unorthodox religious views, his scholarly interests, his absolute devotion to a quite literal democracy, his cultured tastes. The attack was public and private and there is no way to calculate or document the full extent of a smear campaign of drawing-room whispers. The Federalists unleashed a savage counterattack against Jefferson's pro-

French connections and his deist religious views from the pulpit and the press. Yale University President Timothy Dwight, a Congregationalist divine, had set the tone in a fiery sermon on July 4, 1798, which was now widely reprinted. To elect Jefferson would be to put the most radical of Jacobins into power. Dwight's sermon, reprinted in, among other papers, "A Christian Federalist," reached the voters of rural Delaware:

> Can serious and reflecting men look about them and doubt that, if Jefferson is elected, and the Jacobins get into authority, that those morals which protect our lives from the knife of the assassin, which guard the chastity of our wives and daughters from seduction and violence, defend our property from plunder and devastation and shield our religion from contempt and profanation, will not be trampled upon? For what end? . . . that our churches may become temples of reason . . . the Bible cast into a bonfire . . . that we may see our wives and daughters the victims of legal prostitution? Shall our sons become the disciples of Voltaire and the dragoons of Marat?[22]

Already, antislavery innuendo was fueling a puritanical Federalist campaign against the morality of such slaveowners as Jefferson. "Mr. Jefferson's Congo Harem" became the topic of northern tea-party whispers.

Pound for pound of mud slung, the charge of atheism in a nation still dominated by its clergy was the most devastating attack on Jefferson. In fact, Jefferson's religious views, except on the point of religious toleration, differed little from those of his cynical Federalist opponents.

Jefferson's Enlightenment deism substituted a natural religion in which reason dictated social morality for adherence to any single church. The *Gazette of the United States,* the flagship Federalist paper, asked voters the question

> to be asked by every American, laying his hand on his heart. . . . Shall I continue in allegiance to God—and a Religious President, or impiously declare for Jefferson—and No God!!![23]

The Rev. John M. Mason fairly clucked that Jefferson could be convicted as an infidel from his own writings: "Happily for truth and for us, Mr. Jefferson has *written;* he has *printed.*"[24] In searching out evidences of Jefferson's heresy, the divines found he doubted the reality of the Flood. This was apostasy! And he had sinned in questioning the age of the earth. Heresy! Even questioning that there were biological differences between the white and black races was a denial of the brotherhood of man and challenged the notion of the "chosen people." And Jefferson was indicted for opposing reading of the Bible by schoolchildren. But for sheer proof that he was an atheist, there was his remark, "It does me no injury for my neighbor to say there are twenty gods, or no god. It neither picks my pocket nor breaks my leg." As the Rev. Mason summed up the clergy attack, "What *is* he, what *can he be,* but a decided, a hardened infidel?"[25]

The Rev. William Linn of New York dipped into Jefferson's *Notes* to answer Mason from the candidate's own words: "The election of any man avowing the principles of Mr. Jefferson would destroy religion, introduce immorality and loosen all the bonds of society." To vote for Jefferson was "no less than a rebellion against God."[26] Republicans countered feebly that Jefferson was being attacked "because he is not a fanatic." He was not willing to force one denomination of Christians to pay for the clergy of another and "because he does not think that a Catholic should be banished for believing in transubstantiation or a Jew for believing in the God of Abraham."[27] But still there was no way to stop increasing whisperings coming from Virginia that Jefferson had had illicit relations with his best friend's wife as well as with his beautiful mulatto concubine. Jefferson himself was called a mulatto by one Federalist who proclaimed that Jefferson was "a mean-spirited, low-lived fellow, the son of a half-breed Indian squaw sired by a mulatto father . . . raised wholly on hoe-cake made of coarse-ground Southern corn, bacon and hominy with an occasional change of fricasseed bullfrog."[28] If Jefferson were elected, warned the Hartford *Connecticut Courant,* "Murder, robbery, rape, adultery and incest will all be openly taught and practiced."[29]

A Federalist lady in a small Connecticut town was so terrified of what would happen to the family Bible if Jefferson became president that she took it to the only Jeffersonian she knew and asked him to hide it for her. The Jeffersonian tried to allay her fears about Jefferson, but she remained unconvinced.

"My good woman," he finally said, "if all the Bibles are to be destroyed, what is the use of bringing yours to me? That will not save it when it is found."

"I'm sure it will," she insisted. "It will be perfectly safe with you. They'll never think of looking in the house of a Democrat for a Bible."[30]

The Federalists' attack on Jefferson's religious views may have backfired, however, instead advertising his championing of freedom of religion, which won for him many of the leaders and followers of the Second Great Awakening, a religious revival that was sweeping New England and the frontier. The Federalists again were unlucky in raising the issue of Jefferson's Enlightenment philosophy. Republicans reminded voters, making it public knowledge for the first time, that Jefferson was the author of the Declaration of Independence—"the patriot of his country and the friend and benefactor of the whole human race."[31]

A decisive factor in the campaign may have been Jefferson's superior organization of the Republicans. He had to offset the natural advantage of the Federalist incumbent, who could claim that the "sage maxims of administration established by the immortal Washington" were being "steadily pursued by his virtuous successor." Washington had only died at the end of 1799. Why should his conservative party be cast aside to set sail on "the tempestuous sea of liberty"[32] with Jefferson? But the Republicans proved innovators in building a party organization, pushing through revisions in election laws and procedures, campaign tactics, and in turning out the vote. Jefferson and Madison skillfully delegated party chores to congressmen, to state and local party leaders. In Virginia, the most populous state, Jefferson's Republicans controlled the Assembly and were able to change from a district system of choosing presidential electors to a general election, which, because of Jefferson's popularity, assured him of the largest bloc of electoral votes in the nation. In Massachusetts, Adams's home state, the Federalists reversed the process from popular to legislative selection of electors. When all the jockeying was over, voters in only five states directly chose their own representatives to the electoral college. In the majority of states, the real power was now concentrated in the legislatures, where the fight was to take place with legislative elections scattered throughout the year 1800.

In the first major test of party strength, Aaron Burr led the Republicans in New York against Hamilton and the Federalists in a struggle for control of the state legislature. Burr's victory in New York City packed the Assembly with Republican legislators and guaranteed Jefferson's critical victory in the state. But Hamilton nearly short-circuited the process by appealing to the Federalist governor, John Jay, to call a special Assembly session to change the process by which electors were chosen *before* the terms of Federalist assemblymen expired July 1. "In times like these," wrote Hamilton, "it will not do to be overscrupulous." He urged Jay to dispense with "ordinary rules" to "prevent an atheist in religion and a fanatic in politics from getting possession of the helm of state."[33] Jay refused, assuring a Republican victory. The early New York victory gave Jefferson a great boost and led almost immediately to Burr's nomination as vice presidential candidate by a caucus of Republican congressmen. The New York vote also further crippled Adams's chances for reelection when Federalists in Congress voted to support two candidates equally, Adams and Pinckney, and let the electoral college decide the outcome. "To support Adams and Pinckney equally is the only thing that can possibly save us from the fangs of Jefferson," wrote Hamilton. Secretly, however, Federalist leaders began to push Pinckney's candidacy, not Adams's, dividing the party.

It was Hamilton who now worked within his own Federalist party to unseat the incumbent president. In May 1800, he was privately saying he would "never more be responsible for Adams—even though the consequence should be the election of Jefferson. If we must have an *enemy* at the head of the government, let it be one whom we can oppose."[34] By October, as the campaign ground to its conclusion, Hamilton was openly opposing Adams, obviously with his eyes on the 1804 elections.

But Hamilton's support or opposition did not decide the outcome of a close election that went down to the last state to vote. Adams was ahead of Jefferson by a single electoral vote going into South Carolina, where cousin opposed cousin, with Charles Pinckney leading the vigorous Republican opposition to Federalist presidential candidate Charles Cotesworth Pinckney. After some last-minute maneuvering that may have included unauthorized promises of offices, the Republicans triumphed, 8–0 in South Carolina. When Pennsylvania tied 8–8, Jefferson and Burr became clear winners, with Adams running a distant third.[35] The Republicans had won the entire electoral vote of New York, Virginia, South Carolina, Georgia, Kentucky, and Tennessee and split the votes evenly in Pennsylvania, Maryland, and North Carolina. Adams carried only his home region, New England, New Jersey, and Delaware. And in Congress, the Republicans had won a stunning upset, garnering 67 out of 106 seats in the House. "The Jig's Up!"[36] trumpeted the *Baltimore American.* But a presidential election is not over until the electoral college votes. When all the state votes were counted, Jefferson and Burr were tied, 73–73, with Adams, 65; Pinckney, 64; and Jay, 1, throwing the contest into the old Federalist-dominated House of Representatives for a final vote. The Republican tie vote inspirited the Federalists, who now gloated over Jefferson's failure to cast one electoral vote to someone else the way Adams's forces had arranged a single vote for John Jay.

Embarrassed at the impasse, Jefferson worried that, "after the most energetic efforts, crowned with success, we remain in the hands of our enemies."[37] It had been Jefferson himself who had overruled his friends, Madison and Wythe, who had doubted assurances by Burr supporters that he would relinquish one electoral vote to allow Jefferson to win. But Burr had warned his supporters that "I should not choose to be trifled with."[38] In state after state, as the Republican party triumphed, not a single vote was withheld from the Burr column. Now, the decision fell to a House divided equally. Under the Constitution, votes in a presidential electoral tie had to be by states, not individual electors. Despite a Federalist majority of congressmen, the Federalists controlled only six states, the Republicans eight. But two state delegations, those of Vermont and Maryland, were equally divided. Jefferson was in the new capital city of Washington, D.C., when he heard an erroneous early report that a vote *had* been withheld from Burr in South Carolina. He immediately wrote to Burr, congratulating him on the vice presidency. Burr replied that he could "never think of diverting a single vote from you."[39] But when Burr learned that he was tied with Jefferson, he refused to promise he would not accept election to the presidency by the House of Representatives. The revived House Federalists saw a chance to embarrass Jefferson by voting for Burr or at least forcing a new election.

In the end, Alexander Hamilton liked Aaron Burr even less than he liked Thomas Jefferson, and this was to prove decisive. With fatally accurate prefiguring, he let it be known that he thought anyone, even Jefferson, would make a better president than Burr:

> There is no doubt that, upon every virtuous and prudent calculation, Jefferson is to be preferred. He is by far not so dangerous a man and he has pretensions to character. As to *Burr*, there is nothing in his favor. His private character is not defended by his most partial friends.[40]

Hamilton pointed out just how little character Burr had in letter after letter to Federalist leaders week after week until the House voted. Hamilton wound up defending his old rival Jefferson as a man who would uphold values and systems dear to the Federalists.

On February 9, 1801, the House agreed that, if there were a tie electoral vote, they would vote continually until a president were chosen, eating and sleeping in the new and unfinished Capitol. When the state electoral votes were opened February 11, the two men were tied. By one that afternoon, the House began balloting. Jefferson had eight votes on the first ballot, Burr six, Vermont and Maryland were evenly divided. Not a single vote changed as nineteen ballots were taken by midnight. Balloting went on hourly all night. A bed was brought in to allow one ailing congressman, Republican Joseph Nicholson of Maryland, to get some sleep. His vote was crucial. After the twenty-seventh ballot, the House voted to suspend voting until the following day. There were two ballots that day, three the next. After four inconclusive days, Congress adjourned over Sunday. But when no vote changed by Monday, the Federalists began to plan their surrender. Finally it came down to one man, James A. Bayard, the sole congressman from Delaware and a staunch Burrite who had for months refused to cast his vote for Jefferson despite persistent pleading by Hamilton to do so. Now Bayard forced the Federalists' hand by saying he intended to switch to Jefferson. Delaware's vote would prove decisive. On February 17, on the thirty-sixth ballot, Jefferson won. Bayard put in a blank piece of paper, as did several Federalist delegates from Vermont and Maryland, whose states now counted in the Republican column. South Carolina also cast a blank ballot. No Federalist-dominated state voted for Jefferson, who ended up with ten states to Burr's four. According to Jefferson, the Federalist surrender prevented "the certainty that a legislative usurpation would be resisted by arms." Had the Federalists not backed down, the Republicans were ready to call a convention to reorganize the government. "The very word convention gives them the horrors," he wrote. "In the present democratical spirit of America, they fear they should lose some of the morsels of the constitution."[41] The Federalist Party had been crushed and was never to return to power. The party of Washington and many of the Founding Fathers had been swept away. Jefferson had given birth to a two-party system in the United States and it had stood firm against the Federalists' last concerted attack. In 1800, for the first time, American voters had brought about a bloodless transfer of power, which Jefferson called "the revolution of 1800." The momentous change of government was, said Jefferson, "as real a revolution in the principles of our government as that of 1776 was in its form; not effected indeed by the sword . . . but by the rational and peaceful instrument of reform, the suffrage of the people."[42]

At four o'clock the morning of March 4, 1801, a humiliated President Adams left Washington, D.C., to avoid seeing his successor's inauguration. Shortly before noon, President-elect Thomas Jefferson, a month shy of fifty-eight years old, stepped out of Conrad and McMunn's boardinghouse on New Jersey Avenue at C Street and joined an escort of officers of the Alexandria, Virginia, militia and stepped off briskly through the muddy streets of the still-unfinished federal city escorted by District of Columbia marshals and congressmen. Walking to the Capitol, he shunned the splendor of Washington's and Adams's inaugural parades in ceremonial carriages. He wore no elegant suit, no sword as his predecessors had. The Alexandria *Times* reported that "his dress was, as usual, that of a plain citizen without any distinctive badge of office."[43] The first president to be inaugurated in the capital city he had made his brainchild, Jefferson had closed his ears to the criticisms of even his closest friends who despised the raw federal town. One cabinet officer said Washington, D.C., was "hated by every member of Congress without exception."[44] Jefferson hiked up Capitol Hill, where he had insisted the Capitol be built on higher ground than the President's House to symbolize the preeminence of the people. He acknowledged the cheers of Republicans lining his route. Arriving at the unfinished Capitol, Jefferson strode confidently between ranks of Alexandria riflemen who presented arms as he entered the only finished room, the Senate Chamber.

One of the crowd packing the Senate Chamber to see Chief Justice John Marshall swear in Jefferson was Mrs. Samuel Harrison Smith, wife of the editor of the Washington *National Intelligencer.* She was thrilled by "one of the most interesting scenes a free people can ever witness. The changes of administration, which in every government and in every age have most generally been epochs of confusion, villainy and bloodshed, in this our happy country take place without any species of distraction or disorder."[45] Mrs. Smith strained to hear as Jefferson read from his carefully written inaugural address, his voice barely filling the chamber. In one of the immortal inaugural speeches, Jefferson began by stressing the "sacred principle that, though the will of the majority is in all cases to prevail, that will, to be rightful, must be reasonable." In a pointed reference to the Alien and Sedition Laws, he reminded his listeners that "the minority possess their equal rights, which equal laws must protect, and to violate would be oppression." He called for reconciliation with the defeated Federalists:

> Every difference of opinion is not a difference of principle. We have called by different names brethren of the same principle. We are all republicans: we are all federalists. If there be any among us who would wish to dissolve this Union, or to change its republican form, let them stand undisturbed as monuments of the safety with which error of opinion may be tolerated where reason is left free to combat it.

Reiterating the political principles that had framed his election campaign, President Jefferson pledged "equal and exact justice to all men of whatever state or persuasion, religious or political." In a single sentence, he declared his program as president: "Peace, commerce and honest friendship with all nations, entangling alliances with none." He spelled out his policies in broad political and philosophical terms: commitment to states' rights as well as the powers of the central government; dismantling of the standing army and navy in favor of a well-trained militia and merchant marine, as part of his belief in civilian control of the military; frugality in government spending and payment of the national debt; the promotion of agriculture as the basis of trade surpluses. The "bright constellation" of political beliefs—he enumerated freedom of religion, of the press, habeas corpus, and jury trial—which have "gone before us and guided our steps through an age of revolution and reformation"—he pledged as the personal "creed of our political faith."[46]

The moderate tone of Jefferson's speech displeased the Republicans around him, but they need not have worried. Jefferson wanted to convert Federalist voters to Republican ranks, but he fully planned to purge Federalist officeholders. Three days later, he wrote Monroe he would not try to mend fences with Federalist leaders, "whom I abandon as incurables and will never turn an inch out of my way to reconcile."[47] As Republican newspapers clamored that "the board should be swept" and key offices "filled by men of republican principles,"[48] Jefferson shrewdly appointed his cabinet, balancing able old friends and men who could win Republican support in areas where the party was weakest. Madison was Secretary of State; Albert Gallatin of Pennsylvania, Secretary of the Treasury; Revolutionary War veteran Henry Dearborn, who had fought at Quebec and Saratoga, was Secretary of War; Levi Lincoln, Attorney General; Robert Smith of Baltimore, Secretary of the Navy. Four days after his inauguration, Jefferson ordered a halt to all prosecutions under the Sedition Act and ordered all fines refunded.

At the first full cabinet meeting on May 15, President Jefferson confronted his first foreign policy crisis, one he had tackled first as minister to France fifteen years earlier. Tripoli had attacked American ships in the Mediterranean. Putting into effect his long-held views on the subject, Jefferson had already assembled an American naval squadron at Norfolk that was ready to sail. An American navy sailing off Tripoli, he told his cabinet, "might lead to war."[49] He wanted his cabinet's opinions and approval. All five cabinet members agreed on sending the squadron but disagreed over Jefferson's authority to act while Congress was adjourned. Navy Secretary Smith and Treasury Secretary Gallatin backed Jefferson's position that the president could use military force to defend the United States, but Attorney General Lincoln argued that without a formal declaration of war by Congress, American warships could only repel attacks on individual vessels and could not "destroy the enemy's vessels generally."[50] But Lincoln was outvoted. American naval ships could destroy the North African pirates wherever they could be found. Refusing to pay any further tribute to the marauders, Jefferson argued that Congress, when it reconvened seven months later, must decide "whether to abandon the Mediterranean or to keep up a cruise in it."[51] Meanwhile he ordered the squadron to sail. By conferring with his entire cabinet on the question, Jefferson had reversed early practices, the president personally taking charge of foreign affairs and consulting his entire cabinet instead of delegating this crucial responsibility to his Secretary of State. The cabinet was to be the principal policymaker of Jefferson's presidency.

The lull before Congress returned enabled Jefferson to tackle an urgent domestic policy problem—patronage. He said that he hated the chore of removal and appointment: "All others compared with these are as nothing."[52] He was furious at Adams's midnight appointments of federal judges, his "indecent conduct in crowding nominations after he knew they were not for him,"[53] and he vowed to dismiss them, but they were, he found to his chagrin, protected by the Judiciary Act of 1801—passed by the Federalists *after* Jefferson's election. Since he could not remove the judges, he replaced all Federalist marshals and district attorneys as well as all officers found guilty of misconduct in office. He spared many incumbent Federalist officeholders, appointing Republicans only to vacant offices and defending his leniency to his Republican critics. "Good men, to whom there is no objection but a difference of political principle," he said, "are not proper subjects of removal."[54] But there was mounting disapproval in Republican ranks at the idea that Jefferson's doctrine would not apply at the state level. From Connecticut to Virginia, there were yowls of protest. The New York *American Citizen* was fairly typical:

> It is rational to suppose that those who removed John Adams from office would naturally expect the removal of the lesser culprits in office. If this should not be the case, for what, in the name of God, have we been contending? Merely for the removal of John Adams that Mr. Jefferson might occupy the place which he shamefully left?[55]

In July 1801, under growing pressure, Jefferson issued a public statement on the patronage question. His inaugural address was being misinterpreted, he wrote: he had never intended to retain all Federalist officeholders, continuing the Federalist Party's monopoly on power after its defeat. "Is it political intolerance to claim a proportionate share in the direction of the public affairs?" he asked. He would not wait for Federalists to die off or resign. "This is a painful office, but it is made my duty, and I meet it as such."[56] Once there was equilibrium between Federalists and Republicans in office, he vowed, he would "return with joy to that state of things" when honesty, ability, and loyalty to the Constitution were the only questions. According to his own careful record, of 316 federal offices subject to presidential appointment and removal, within two years of his inauguration 158—exactly half—were filled by Republicans, 132 by Federalists, and 26 by independents. This was keeping faith by the numbers: in fact, most of the key offices were in Republican hands on both federal and state levels. The change of parties in power in fact ushered in the spoils system, which was to plague American politics and government throughout most of the nineteenth century.

It also led to one of the most important tests of the Constitution, the case of *Marbury v. Madison,* which established the Supreme Court's real power in the first of a series of bitter clashes between President Jefferson and his kinsmen, Chief Justice John Marshall. The last-minute appointment of Federalist magistrates by Adams infuriated Jefferson: it seemed to rekindle all his old animosity to the courts left over from his days as a young lawyer. One of Adams's last acts had been to appoint Marshall as chief justice. At the same time, Adams made the appointment of Republicans more difficult by the Judiciary Act of 1801, which reduced the number of Supreme Court seats to five while it created sixteen circuit courts with a judgeship each and increased the number of U.S. marshals, attorneys and clerks, making Adams's "midnight appointments" possible.

Jefferson made the repeal of the Judiciary Act a prime target. On January 8, 1802, he had a repeal motion introduced in the Senate, but it deadlocked and Aaron Burr was able to block the repeal temporarily. Finally, after a two-month struggle, the act was repealed on March 8. Jefferson ushered in a new Judiciary Act in April, which restored a sixth justice to the Supreme Court and created six circuit courts, each headed by a Supreme Court justice. Adams's newly created circuit court judges were stripped of their offices.

In February 1803, the contest between President Jefferson and Chief Justice Marshall came to a head when Jefferson ordered Secretary of State Madison to withhold from William Marbury his commission of appointment as justice of the peace for the District of Columbia signed and sealed by Adams on March 2, 1801, under the old Judiciary Act. When Marbury sued for a writ of mandamus to compel delivery of his commission, Chief Justice Marshall, arguing that the Court lacked jurisdiction, dismissed Marbury's suit. Marshall's strategy cleverly avoided an open battle with the executive branch, which was responsible for enforcing the writ. It was the first time the Supreme Court had taken on itself the power to hold an act of Congress unconstitutional. Marshall declared that Section 13 of the Judiciary Act of 1789, which empowered the Court to issue such a writ, was unconstitutional and therefore invalid. At a stroke, Marshall had established the doctrine of judicial review of the actions of the other branches of government.

But Jefferson and his Republicans kept up their campaign against hold-over Federalist judges. In March 1804, Judge John Pickering, federal district judge in New Hampshire, was impeached by the Senate, adjudged guilty, and removed from his post. The Senate also impeached Supreme Court Justice Samuel Chase of Maryland for his bias in the 1798 sedition trials of John Fries and James Callender. Chase was acquitted, however, and the Jeffersonian campaign to purge the federal courts came to an end, after it had dragged on throughout Jefferson's first term.

Throughout his first term, Jefferson in fact pressed his bloodless revolution, systematically eliminating the long-entrenched Federalist aristocracy from every level of federal office, inflicting permanent damage on the party of Washington, Adams, and the hated Hamilton and banishing from power many of the oldest and best-established families in America. Not only did he oust the top elective officials but he removed the second tier as well by personal intervention or by wielding his power in Congress to eliminate whole departments. He evicted eighteen of thirty federal judges, thirteen of twenty-one United States attorneys, eighteen of twenty U.S. marshals. He purged Hamilton's old power base at the Treasury, sacking fifteen of sixteen revenue supervisors, twelve of twenty-one inspectors and fifty of 146 customs officials, including all but one collector of customs in ports from Maine to Georgia. The case of Joseph Tucker, collector at York, Maine, was fairly typical. His dossier contained the allegation that he was "constantly drunk and incapable of business and a violent *federalist.*"[57] The York Town Meeting voted 106–0 refuting the charges and petitioning Jefferson for Tucker's reinstatement; instead a loyal Republican was put in his place.

More than half the ousted Federalist officeholders were the scions of aristocratic families. They included thirty-six lawyers, twenty-four merchants, numerous shipowners, physicians, bankers, planters, five United States Senators, nine members and thirteen former members of Congress. Twenty-seven percent had been officers in the Revolutionary War. No purge in nineteenth-century American politics, not even those of the Jackson years, surpassed the thoroughness of Jefferson's housecleaning. The Revolution of 1800 shattered the Federalist Party by permanently damaging the underpinnings of the party in every state, at the same time breaking the power of the Old Guard Federalist aristocracy.

The new president adapted old habits that made his tenure one of the most innovative and interesting in the history of the presidency. He worked long, hard days and exercised great personal influence rather than control. He was aware of his immense popularity and the power that it gave him, especially to propose the legislation that he wanted introduced in Congress and which so often became the law of the land. Often he drafted the laws himself, but even more often his cabinet ministers and the department heads of his government drew up the bills for Congress, supplied Congress with supporting information, and testified before congressional committees. For years, until the Capitol was finished, Congress had no offices other than each member's desk on the floor of the House or Senate, no individual staffs, only a pool of clerks. All of the staffwork had to be done in department offices—or at the President's House at the foot of Pennsylvania Avenue. There, an informal president wrote all his own letters and drew up his own state papers, using his private secretary, Army Lieut. Meriwether Lewis, his cousin, more as an aide than a secretary.

Jefferson devised "a steady and uniform course" for his hardworking presidency. "It keeps me from 10 to 12 and 13 hours a day at my writing table, giving me an interval of 4 hours for riding, dining, and a little unbending."[58] He stayed hard at work at the reins of the executive branch for all but three weeks a year in the spring during congressional adjournment and during the annual summer recess of August and September. When Federalists criticized his administration for shutting down in Washington summer heat, Jefferson was indignant. "I consider it as a trying experiment for a person from the mountains to pass the two bilious months on the tidewater. Nothing should induce me to do it. Grumble who will, I will never pass those two months on tidewater."[59] As president, Jefferson took charge of his administration and full responsibility for its conduct. At the center of his administration was his close-knit cabinet, which Jefferson hailed as "an example of harmony" which he summoned whenever it was needed. "We sometimes met under differences of opinion but scarcely ever failed, by conversing and reasoning, so to modify each other's ideas as to produce an unanimous result." The ordinary business of his government was done "by consultation between the President and the head of the department alone to which it belongs."[60] He maintained a close working relationship not only with his cabinet officers but with Republican leaders in both houses of Congress.

His rigorous self-discipline led him to organize his days carefully, a habit he had usually maintained since his college days. He still rose regularly at five each morning and did paperwork until nine, when he received cabinet officers or important visitors. Lieutenant Lewis would greet a visitor at the door and usher him into the president's office. One senator was nonplussed when he first called on President Jefferson to find "he was dressed, or rather undressed, with an old brown coat, red waistcoat, old corduroy small clothes, much soiled, woolen hose and slippers without heels."[61] Jefferson always felt the drafts, wherever he was, and the large, unfinished President's House was no exception. That he was dressed for comfort as he worked long hours eluded many visitors, especially the new British ambassador, who nearly made an international incident out of Jefferson's appearance. Jefferson had learned to hate diplomatic ritual in Europe and had abolished all diplomatic receptions, scrapping all the formal rules of diplomatic etiquette. In a memo he wrote for his administration, he said, "When brought together in society, all are perfectly equal, whether foreign or domestic, titled or untitled, in or out of office."[62] When Anthony Merry drove up to the President's House to present his credential as the new British ambassador, Jefferson welcomed him in his usual morning dress, including his heelless French slippers. Invited to dine with the president, Ambassador Merry considered himself further offended when the president offered his arm to Dolley Madison, wife of the Secretary of State, instead of to Mrs. Merry. But when Jefferson allowed his guests to find their own seats at the big oval table, it was too much for Merry. There was nothing accidental in Jefferson's apparently relaxed etiquette. All cabinet officers were to follow his democratic style, their oval dinner tables also lacking a head or a foot. And if the British ambassador could drop into the President's House, so could any member of Congress or any private citizen.

After his morning office hours, Jefferson scheduled cabinet meetings at noon or got in a little letter-writing until one o'clock, when he went for a long horseback ride. At three-thirty, he had dinner. The centerpieces not only of his social life but of his informal style of government were his thrice-weekly dinner parties for a dozen invited guests. Guests were invited by printed invitations which began "Th. Jefferson requests." He dressed elegantly for dinner. The same senator who had decried his morning attire recorded that he found the president was "well-dressed" for dinner in "a new suit of black, silk hose—shoes—clean linen, and his hair highly powdered." The dinner was equally elegant, with eight different kinds of very good French wine and the president "very social and communicative."[63] Another guest, a New York congressman, said dinner was "easy and sociable," noted there were no toasts and reporting that the French chef "understands the art of preparing and serving up food to a nicety,"[64] especially the ice-cream balls wrapped in warm pastry. Keeping the food and wine and conversation flowing, Jefferson was apt to give little after-dinner talks on the latest product or invention, domestic or foreign: silkworms raised in Virginia, cheese from Cheshire, Massachusetts. Federalists were invited at first to the same dinners as Republicans but eventually were kept separate. When one member of Congress criticized Jefferson for these dinners, Jefferson readily explained his social philosophy. "I cultivate personal intercourse with the members of the legislature that we may know one another and have opportunities of little explanations of circumstances which, not understood, might produce jealousies and suspicions injurious to the public interest."[65] The president himself paid for all entertainment out of his $25,000-a-year salary, each year running, on average, $4,000 further into debt. He nonetheless considered the expenditure of his personal funds essential to his presidency.

Yet Jefferson had not risen above criticism. The attacks on his religious views during the 1800 campaign had wounded him. By six every night, the guests had left and he wrote more letters; then, late into the night, he read and thought. As the Federalists kept up their attacks on him as irreligious and an enemy of Christianity, Jefferson studied the Gospels of Matthew, Mark, Luke, and John, marking the passages that he thought represented the simple beliefs of Jesus Christ and ignoring those he considered later corruptions. After reading Joseph Priestley's *Socrates and Jesus Compared,* Jefferson wrote a two-page summary of his own religious beliefs, entitling it "Syllabus of an Estimate of the Merit of the Doctrines of Jesus Compared with Those of Others."[66] He sent copies to his daughters, to at least two of his cabinet officers, and to Benjamin Rush in Philadelphia. Then, he spent several evenings cutting out his favorite Gospel passages and pasting them onto blank pages. "It was the work of two or three nights only at Washington after getting through the evening task of reading the letters and papers of the day,"[67] he insisted. But he had produced his little book, *The Philosophy of Jesus,* which he later expanded into *The Life and Morals of Jesus,* printed for his own use and for a few family members and friends. He continued to try to pin down his own religious beliefs. He became convinced that early Christians most closely resembled the Unitarians of the early nineteenth century and he found that his concept of God most closely resembled theirs. After stripping away the accretions of time since the first century, he found a simple, moral Jesus. After his presidency was over and he resumed his friendship with John Adams, he sent him a copy of his little book. Editing out all but what he considered the authentic words of Jesus, "there will be found remaining the most sublime and benevolent code of morals which has ever been offered to man. I have performed this operation for my own use, by cutting verse by verse out of the printed book and arranging the matter which is evidently his, and which is as easily distinguishable as diamonds in a dunghill . . . 46 pages of pure and unsophisticated doctrines such as professed and acted on by the *unlettered* apostles, the Apostolic fathers and the Christians of the first century."[68]

But there seemed to be nothing that could quell the persistent rumors of President Jefferson's sexual misconduct. The first revelation came in the Federalist *Gazette of the United States,* which reported in 1802 that President Jefferson, when he was twenty-five, had "offered love to a handsome lady," the wife of his neighbor, John Walker. In Richmond, James Thomson Callender, now the editor of the *Recorder,* was busily savaging the Virginia gentry after he had been refused a high federal office despite his repeated personal appeals to President Jefferson and Secretary of State Madison. Jefferson had ignored earlier hints that Callender would release information damaging to the president. To Monroe, Jefferson wrote that Callender was only looking for "hush money" and "knows nothing of me which I am not willing to declare before the world."[69]

For weeks, Callender had been issuing revealing glimpses at the sexual behavior of the landed aristocracy and was fast building a highly successful newspaper with his peeks at white masters frolicking with African-American mistresses in the bordellos and theaters of Richmond. When one Loyalist Jeffersonian editor suggested that there was at least one man above Callender's criticism, President Jefferson—"there is not one recorded or unrecorded truth which can in the smallest degree militate against the integrity or talents, the patriotism or wisdom of Mr. Jefferson,"[70] Callender revealed that it was Jefferson who had financially backed his political writings during the 1800 campaign.

Then, on August 25, 1802, William Duane, editor of the Philadelphia *Aurora,* Callender's former employer, charged that Callender's wife had died "overwhelmed by a created [venereal] disease on a loathesome bed . . . while Callender was having his usual pint of brandy at breakfast."[71] Three days later, a copy of the *Aurora* reached Richmond. In the next week's edition of the *Recorder,* Callender published the first account of the Jefferson–Sally Hemings affair:

> It is well known that the man, *whom it delighteth the people to honor,* keeps, and for many years past has kept, as his concubine, one of his own slaves. Her name is SALLY. The name of her eldest son is TOM. His slaves are said to bear a striking although sable resemblance to those of the president himself. The boy is ten or twelve years of age. His mother went to France in the same vessel with Mr. Jefferson and his two daughters. The delicacy of the arrangement must strike every person of common sensibilities. . . . By this wench Sally, our presi-

> dent has had several children. . . . THE AFRICAN VENUS is said to officiate as housekeeper at Monticello.[72]

Callender had never been to Monticello and relied on papers and letters supplied him by Jefferson's political enemies. But one of America's most durable myths, unproven and unprovable, burst upon the sleazy scene of yellow-rag political journalism and produced an even stronger reaction than even Callender may have expected after a rival Richmond newspaper reported that, on the very day of the first Sally Hemings story, Chief Justice John Marshall, Jefferson's cousin, Federalist nemesis, and the close personal friend of Alexander Hamilton, had been seen visiting the offices of Callender's *Recorder*.[73]

In subsequent weeks and months, Federalist and Republican newspapers reprinted and argued over the particulars of the Sally Hemings story. Its basic outlines remained unchanged as it became the sensation of President Jefferson's reelection campaign of 1804. Callender himself was dead long before the election. His body was fished out of the James River one Sunday morning in July 1803 and was buried without ceremony on the same day, after a hurried coroner's inquest which pronounced he had drowned while intoxicated.

In 1803, in the middle of the reelection campaign, Henry "Light-Horse Harry" Lee married Ann Hill Carter, the niece of Betsy Walker. Lee was Jefferson's avowed enemy in the vicious struggle between Jefferson Democrats and Federalists in Virginia. The year before President Jefferson had openly broken with his old friend, John Walker. Walker told Lee the affair between his wife and Jefferson had lasted from 1768 to 1779. As the stories of Jefferson's alleged peccadilloes gained currency during the bitter 1804–1805 elections, they appeared as far away as the *New England Palladium* on January 18, 1805, and even were debated on the floor of the Massachusetts House of Representatives.

Walker sent a statement of his allegations to his friend and kinsman General Lee, which Lee copied out and which is quoted here in part:

> . . . I was educated at William and Mary, where was also educated Mr. J.
>
> We had previously grown up together at a private school and our boys' acquaintance was strengthened at college. We loved [at least I did sincerely] each other.
>
> My father was one of his father's executors and his own guardian and advanced money for his education. . . .
>
> I took Mr. J. with me the friend of my heart to my wedding. He

> was one of my bridesmen . . . in '64. In '68, I was called to Fort Stanwyx. . . . I left my wife and infant daughter at home, relying on Mr. Jefferson as my neighbor and fast friend, having in my will, made before my departure, named him first among my executors. . . . I returned in November, having been absent more than four months.
>
> During my absence, Mr. J.'s conduct to Mrs. W. was improper, so much so as to have laid the foundation of her constant objection to my leaving Mr. J. my executor, telling me that she wondered why I could place such confidence in him.
>
> At Shadwell, his own house, in '69 or '70, on a visit common to us, being neighbors and, as I felt, true friends, he renewed his caresses. [He] placed in Mrs. W.'s gown sleeve cuff a paper tending to convince her of the innocence of promiscuous love.
>
> This Mrs. W. on the first glance tore to pieces.
>
> After this we went on a visit to Colonel Coles, a mutual acquaintance and distant neighbor. Mr. Jefferson was there. On the ladies retiring to bed, he pretended to be sick, complained of a headache and left the gentlemen, among whom I was.
>
> Instead of going to bed, as his sickness authorized a belief, he stole into my room, where my wife was undressing or in bed.
>
> He was repulsed with indignation and menaces of alarm and ran off. . . .

Walker swore that Jefferson "continued his efforts to destroy my peace until 1779." When Jefferson sailed for France in 1784, Walker wrote, his wife once again asked him why he kept Jefferson as his executor, then "related to me these base transactions, apologizing for her base silence."[74] In 1785, Walker said he had written to Jefferson in Paris. Walker's original statement was enclosed in his letter to General Lee of March 28, 1805. Callender had apparently been told some of the allegations as early as 1802. Jefferson left a copy of the letter containing the changes when he left his papers to his family. The Library of Congress later acquired them and included them in the Jefferson Papers, "with doubtful propriety,"[75] historian Dumas Malone wrote in 1948. Ultimately, in 1805, Walker was to demand satisfaction from President Jefferson, part of the process of challenging to a duel. Lee wrote the letter to Jefferson, as Walker's intermediary.

Whatever really had happened that long-ago summer in the hills of Virginia, Jefferson refused ever to write or to speak publicly in answer to the charges. To reassure his friends, however, when one of them, Secretary of the Navy Robert Smith, brought up the attacks of his enemies, he replied in writing, "You will perceive that I plead guilty to one of their charges, that when young and single, I offered love to a handsome lady. I acknowledge its incorrectness." But he went on to deny the alleged liaison with his slave, Sally Hemings. His "offering of love" to Betsy Walker was "the only one founded in truth among all their allegations against me."[76]

Jefferson's private statements to his aides leave little doubt that Thomas Jefferson, at twenty-five, tried to seduce the wife of a close friend, and it is impossible to know whether he was in love. Thirty-seven years later, as president, he privately assured Walker that Betsy Walker had been blameless, and there was no duel. As a gentleman, it was the least Jefferson could do. Years later, in love again in Paris, again with a married woman, he may have been referring to this "scrape," as he put it, when he wrote about "the eternal consequences of warmth and precipitation." Jefferson told his presidential secretary, in 1805, that "the affair had long been known." Still, Jefferson's rival, Alexander Hamilton, "had threatened me with a public disclosure." Jefferson's attempt to seduce Betsy Walker was "without premeditation and produced by an accidental visit."[77]

With a Republican majority in both houses of Congress, Jefferson could expect cooperation for sweeping changes, but he decided to tread cautiously. A month before Congress convened, he sent a draft of his annual message to Congress to all cabinet members, soliciting criticism. Several objected to one Jefferson proposal: an executive declaration that the Sedition Law was unconstitutional. Jefferson no doubt wished to bypass the Federalist-dominated Supreme Court and exercise his talents as a constitutional lawyer. "I took that act into consideration, compared it with the constitution, viewed it under every aspect of which I thought it susceptible,"[78] he wrote. But Smith and Gallatin questioned the constitutionality of such a declaration, predicting a split among Republicans over giving the president such prerogative power to nullify an act of Congress. Jefferson struck out the paragraph. Recognizing the importance of an annual State of the Union message, he also always circulated its text and revised it according to the advice of his cabinet. The man who had helped to invent the United States was inventing the modern American presidency. But he also decided to send the message to Congress, not read it in person and listen to Congress's reply, as Washington and Adams had done. Their method struck him as too closely resembling the English king's annual Speech from the Throne opening Parliament. Republicans applauded Jefferson's dispensing with "all the pomp and pageantry, which once dishonored our republican institutions" as Congressman Michael Leib of Philadelphia put it, grateful to Jefferson for eliminating the spectacle of a president "drawn to the Capitol by six horses and followed by the creatures of his nostrils and gaped at by a wondering multitude."[79] From 1801 until Woodrow Wilson stood before Congress in 1913, Jefferson's practice of sending his private secretary to Congress with his written message was the accepted practice.

Jefferson wrote Congress's agenda. He urged sweeping reforms that would virtually dismantle the Federalists from the entire fiscal and military program. Jefferson recommended repeal of all internal federal taxes all the way back to and including Hamilton's excise tax on whiskey and abolished the internal revenue service created to collect them. He reduced the residency requirement for naturalization from Adams's fourteen years back to its original five years. He slashed away at federal spending. "We shall push you to the uttermost in economizing,"[80] he told Congressman Nathaniel Macon. To enable the deepest cuts, he attacked military spending, which the Federalists had linked to the growth in American trade. The Federalists had steadily increased the peacetime military establishment from 840 men in 1789 to 5,400 in 1801: the Republicans cut it back to 3,300 men. The navy, with thirteen frigates on active duty and six state-of-the-art ships-of-the-line under construction, was sharply cut back to six frigates, most of those decommissioned and allowed to rot in drydock while construction of the men-of-war was halted. Instead, Jefferson ordered construction of four new sloops-of-war suited for inshore fighting in the Tripolitan War. Jefferson was consistent in his naval policies: he had long seen the need for warships to patrol the Mediterranean to protect American trade from North African marauders, but he always opposed a costly arms race which, he believed, could only bring about constant friction and the threat of war with much wealthier naval powers. Shifting to a strict defensive policy at home, Jefferson built up coastal fortifications, beginning an extensive building program that cost a staggering $2.8 million between 1801 and 1812, three times what the Federalists had spent in the 1790s. He also created what his critics called a "mosquito fleet" of small and inexpensive gunboats that had proved effective in Mediterranean warfare where heavier ships could not maneuver, spending $1.5 million on them in eight years.

But while Jefferson believed that the United States should be neutral and assume a defensive posture with European powers, he can hardly be considered a pacifist. Thomas Jefferson's views on war appear to be inconsistent, especially since he never wrote them down systematically, and they have to be disentangled from his written reactions to changing circumstances and his own experiences. He began as a philosophical pacifist. He abhorred violence, confrontation, and debt. He believed that standing armies and navies in peacetime caused dangerous financial burdens for society. He dreaded power, especially his own. During the long, drawn-out American Revolution, he saw idealism turn into personal greed, friendship into hostility as war subjugated everything in the struggle for survival of a new nation. His habitual optimism had faded before the lamp of experience as he metamorphosed from abstract political thinker into battered pragmatist.

A reluctant revolutionary at first, by 1774, when he had drafted instructions for Virginia's delegates to the First Continental Congress, he had argued that resisting King George III's troops was justifiable if they refused to follow colonial laws: they would then be behaving as a hostile invading force. When Jefferson learned of the British march on Lexington, he had been outraged, and had dismissed the possibility of reconciliation. Elected to Congress, he had written the American justification of a de facto war. He linked the necessity and justice of the Revolutionary War to the nature of government itself and he added his own personal determination to fight.

Chosen to draft the Declaration of Independence, he had refined the consensus position of eighteenth-century Whig republicans that a defensive war was a just and necessary step toward national survival. Free men could wage defensive war even against aggression from their own government. The king, who had "plundered our seas, ravaged our coasts, burned our towns and destroyed the lives of our people," was conducting an unlawful war and Congress, representing law-abiding people, had "full power to levy war against a tyrannical monarch."

Jefferson's views on war shifted and evolved throughout his forty-year career as a public servant. His last words based on his humiliating experiences during the Revolutionary War were written while he was preparing to become American minister to France. As a congressman in December 1783, he wrote instructions for American diplomats serving in Europe. Jefferson clung to the Enlightenment belief that war was a limited tool of governments; typical of eighteenth-century rulers, he sought to shield productive subjects from the ravages of warfare.

In *Notes on the State of Virginia,* published in Paris in 1785, Jefferson suggested that war was a learned habit, a habit to blame for making men "honor force more than finesse." But he upheld the national right to self-defense. In his last words on the subject for many years, he had written, "While an enemy is in our bowels, the first object is to expel him[81] . . . Wars, then, must sometimes be our lot; and all the wise can do will be to avoid that half of them which would be produced by our own follies and our own acts of injustice; and to make, for the other half, the best preparations we can."[82]

As president, Jefferson's attack on military spending did not mean he intended to abandon defensive preparations. He believed that, while the United States did not have to remain bellicose and respond to every insult with force, some threats could "be met by force only," and it was the president's duty to "recommend such preparations as circumstances call for."[83] He created West Point and actually increased military spending, emphasizing professional officers, fortifications, and gunboats rather than militia. He farsightedly plumped unsuccessfully for a huge drydock to house frigates kept in mothballs in peacetime to spare the "great annual expense and be an encouragement to prepare in peace the vessels we shall need in war."[84]

Even as the Napoleonic Wars ravaged Europe and terminated the Age of Reason, Jefferson clung to an Enlightenment determination to stress reason as a deterrent against war:

> My hope of preserving peace for our country is not founded on the greater principles of non-resistance under every wrong, but in the belief that a just and friendly conduct on our part will procure justice and friendship from others.[85]

During his embargo of American trade with the warring Europeans five years later, even as America tumbled again into a depression because of his halfway pacifism, he remained adamant that only reason and justice could provide bulwarks for his country. "If nations go to war for every degree of injury, there would never be peace on earth.[86]

Yet when the United States was attacked, Jefferson was quick to respond. His years as American envoy in Europe had left him frustrated with failure to bring about peace between piratical North African powers and the United States even at a price. In the spring of 1801, the pasha of Tripoli alleged that the Americans had insulted him by paying more tribute to Algiers, and declared war on the United States by having the flagpole in front of the American consulate cut down. After a brief effort to negotiate, Jefferson decided to wage war. Despite the fact that his powers as commander-in-chief were probably sufficient, Jefferson called on Congress to authorize him to conduct offensive operations. Eventually Jefferson was to send a squadron of some thirty American warships to blockade and bombard Tripoli. The ships constituted the largest American fleet to date and went far beyond anything Hamilton had proposed.

Sending a naval squadron to the Mediterranean with orders to protect American commerce, he demonstrated that he would not hesitate to use force. Jefferson carried out what would today be called a police action when Tripoli, breaking treaties concluded under Adams in 1795 and 1796, declared war on the United States. Jefferson dispatched four warships to sink or destroy the vessels of any Barbary power at war with the United States, to interdict enemy vessels in their home waters and at Gibraltar, to keep them out of the Atlantic. When possible, U.S. warships were to provide convoy escorts for American merchant vessels in the Mediterranean. By May 1801, all six American frigates were convoying ships past the Barbary pirates. By 1804, Jefferson had issued positively aggressive orders: "You will by all the means in your power annoy the enemy."[87] Jefferson's years of European diplomacy now paid dividends. Augmenting American frigates with smaller gunboats borrowed from the Kingdom of the Two Sicilies, Commodore Edward Preble fought the Tripolitan cruisers in their own harbor while the forty-four-gun *Constitution* pounded the fortress and town of Tripoli with barrages. In the opening engagement of the summer 1804 campaign, Stephen Decatur set afire the captured *Philadelphia.* While the ship's loss embarrassed Jefferson's administration, it did help to bring about peace. After ransoming the three-hundred-man crew of the scuttled frigate, Jefferson's negotiators paid for peace. The ransoms, cost of naval patrol vessels, and any future tribute were to be made from a special Mediterranean Fund made up of import duties, an ingenious innovation of Jefferson's.

But a limited show of force to cow the pasha of Tripoli was far from a military buildup and confrontation with European superpowers. Whether Jefferson's military cutbacks saved or cost money in the long run, he was determined, as were his Republican followers, to avoid entanglement in the spreading Napoleonic Wars, and he believed that his defensive policy would not only prevent war but spur commercial growth. If anything, the Republican Congress was more hostile to an expensive military establishment than was Jefferson. "Every nation which has embarked to any extent in naval establishment has been eventually crushed by them,"[88] argued Congressman James Fisk. "Show me a nation possessed of a large navy and I will show you a nation always at war,"[89] echoed Congressman Samuel McKee. Jefferson and the Republicans were willing to cut the nation's regular forces because they thought they could rely on the state militia, and their naval equivalent, privateers, which were democratic in character, posed no threat to republican institutions and were cheap. The militia wasn't paid until it was called into active service; privateers brought in revenues from taxes on prize sales.

By eliminating the United States as a military threat to either combatant in the Napoleonic Wars, Jefferson, adhering to his long-held natural-rights doctrine of trade, hoped to reap far greater profits from the neutral American merchant marine and its business of carrying American exports to both sides. Increased trade would bring increased customs duties, the main income of the federal government. By eliminating unpopular internal excise taxes, cutting forty percent of the Treasury Department workforce and shifting to reliance on external customs taxes, Treasury Secretary Gallatin gave Jefferson the ammunition to persuade Congress that the entire national debt could be paid off in sixteen years.

Jefferson's fiscal policies had their roots in a remarkable letter he had written just before he left Paris in 1789 at a time when he was helping Lafayette to reform the government of France. Writing to James Madison on September 6, 1789, Jefferson had set out a new philosophy of government spending—"the earth belongs in usufruct to the living." Jefferson maintained that no government had the right to incur debts that could not be paid off by the same generation of taxpayers: he allowed nineteen years for a generation. "This would put the lenders and the borrowers also on their guard. By reducing, too, the faculty of borrowing within its natural limits, it would bridle the spirit of war."[90]

Jefferson had every reason to expect his frugal fiscal program to succeed when he took office in 1801. American maritime commerce was booming. Exports had nearly trebled in only seven years from $33 million in 1794 to $94 million in 1801. But this was largely as a result of the 1795 Jay Treaty with England that Jefferson so detested. The British had allowed direct American trade that had brought on unparalleled American prosperity. French seizure of $20 million in American shipping in the Caribbean during the Adams administration's quasi-war with France and the costs of building a navy to protect American shipping had not begun to compare with the growth in American profits. But continued American prosperity, as Jefferson could not see at the time, was largely a function of continued war in Europe. When the English and the French signed the Peace of Amiens in late 1801, it helped Jefferson to cut military spending still further, but it also dried up European demand for American products. American exports plummeted from $94 million in 1801 to $54 million in 1803, a forty percent drop.

Thomas Jefferson's idealistic democratic principles made him president. The circumstances of his presidency made him set aside his principles at strategic moments, enabling him to make decisions that only a pragmatic realist reacting to circumstances would consider and insist upon. More than once he had to rationalize his fondest principles to achieve his greatest victories. He had decided to run for president largely because he despised John Adams's administration for its arbitrary use of presidential power, especially for setting aside the First Amendment to the Constitution. He hated the British and their mercantilist system, yet he ruthlessly played the card of threatening to side with the British in the Napoleonic Wars to win concessions from France and her Spanish ally. He decried expensive government, yet he overspent his personal fortune to remain in office and live in the style which he thought befitted even a republican chief of state. In the process he sank ever deeper into debt, relying increasingly on the income from slave-labor estates—even as he publicly opposed slavery.

There is no clearer example of his ambivalent pragmatism than his greatest presidential triumph, his acquisition of the vast Louisiana Territory. No sooner had he become president than he received reports of a secret treaty between Spain and France to turn vast Spanish Louisiana over to France. Jefferson promptly launched a diplomatic campaign in Madrid and Paris to discourage the retrocession. When that failed, he tried to buy Florida from Spain. Jefferson kept these efforts from Congress all through 1801, by the next year becoming convinced that, despite French disclaimers, the transfer to France was about to take place. Jefferson turned up the heat by writing to Robert R. Livingston, his envoy in Paris, "There is on the globe one single spot the possessor of which is our natural and habitual enemy. It is New Orleans. France placing herself in that door assumes to us the attitude of defiance." Spain was too weak to threaten American expansion and trade in the Mississippi Valley, but Jefferson found it "impossible that France and the U.S. can continue long friends when they meet in so irritable a position." If France occupied New Orleans, "from that moment we must marry ourselves to the British fleet and nation."[91] He dispatched the letter unopened to Paris with his friend Pierre Samuel du Pont de Nemours as courier, giving him permission to impress on the government of France the inevitable consequences of their taking possession of Louisiana.

The crisis deepened before Jefferson could confirm French intentions. The Spanish intendant at New Orleans, violating the 1795 treaty with the United States, suspended the American right to deposit goods at New Orleans before shipment to Europe. Westerners now numbered more than one million and made up fully twenty percent of the United States population. Jefferson's eyes had always looked to the west and he knew these people well. He did not exaggerate when he wrote to Ambassador Charles Pinckney at Madrid, "The Mississippi is to them everything. It is the Hudson, the Delaware, the Potomac and all the navigable rivers of the Atlantic States formed into one stream."[92] Madison echoed Jefferson's concern. He warned Livingston in Paris of "200,000 militia on the waters of the Mississippi, every man of whom would march at a moment's warning . . . every man of whom regards the free use of that river as a natural and indefeasible right."[93] But Jefferson reported none of this in his annual message to Congress in December 1802. Instead, he urged Congress to further cut naval expenses. On Louisiana, he only hinted at "a change in the aspect of our foreign relations";[94] he didn't even mention the suspension of the right of deposit.

But Jefferson was not only growing increasingly secretive as president, he was also making procrastination the chief virtue of his foreign policy. Napoleon had tried to crush a successful slave revolution on Santo Domingo, but his expedition to the malarial island had cost the lives of thousands of his best troops just as it appeared that war was about to break out again in Europe.

Jefferson stalled for time, overtly ignoring Federalist demands for war even as he secretly reinforced American outposts on the frontier, including the massing of infantry and artillery at Fort Adams on the Mississippi on the Spanish border. Two thousand Mississippi Territory militia bivouacked at Natchez. When the Federalists tried to force Jefferson's hand by introducing resolutions in Congress demanding the presidential papers on Louisiana and a full report on the crisis, the House, after a secret session, overwhelmingly defeated the resolutions and supported Jefferson's efforts.

Concerned that the Federalists were trying "to force us into war,"[95] Jefferson named Monroe as his special envoy to France and Spain and kept secret du Pont's latest dispatches, received ten days earlier, that France was willing to negotiate. In fact, he even had passed along a selling price—$6 million—for New Orleans and all Spanish territory east of the Mississippi, including Florida, if the United States would renounce all claims west of the Mississippi. In a secret session of Congress, Jefferson wrung a $2 million appropriation "to defray any expenses,"[96] giving no hint that Monroe carried presidential instructions authorizing him to pay up to $9 million for New Orleans and Florida. In Paris, meanwhile, Livingston had suggested to the French that they sell Louisiana in order to place an American buffer zone between the French in New Orleans and the British in Canada. While Monroe was sailing to France, Jefferson learned that the Spanish had reopened New Orleans to Americans: his policy of procrastination had averted a needless war.

On April 11, 1803, before Monroe could reach Paris, Talleyrand, the

French foreign minister, summoned Livingston and asked him if the United States might be interested in purchasing Louisiana; earlier in the day, Napoleon had told his finance minister, Jefferson's old Philadelphia friend Barbé-Marbois, that he intended to sell the huge territory. He was abandoning his plans to reconquer Santo Domingo. About to reopen the war with England, he regarded Louisiana as vulnerable to British conquest from Canada. Barbé-Marbois knew the rest: Napoleon needed the money. By April 30, Monroe and Livingston initialed an agreement ceding Louisiana to the United States for $15 million. The French signed two days later. An ecstatic Jefferson released the news in time for the Fourth of July edition of the *National Intelligencer.* It was "a proud day for the President," a day of "widespread joy of millions at an event which history will record among the most splendid in our annals."[97]

Jefferson's greatest diplomatic triumph created his trickiest constitutional dilemma. As the strict-constructionist president read the Constitution, "The general government has no powers but such as the constitution has given it; and it has not given it a power of holding foreign territory, and still less of incorporating it into the Union."[98] A constitutional amendment was needed, he believed, and he drew one up. But an amendment would take time, Jefferson argued with himself, and Napoleon might change his mind. He decided to seize "the fugitive occurrence which so much advances the good of their country," urge Congress to put behind them "metaphysical subtleties,"[99] buy Louisiana, and *then* go to the public to seek a constitutional amendment. When he opened the Eighth Congress in October 17, 1803, he called for purchase of Louisiana but did not mention the constitutional problem. The treaty, presented to the Senate the same day, was ratified only three days later.

When France officially transferred Louisiana to the United States in a ceremony in New Orleans on December 20, Jefferson in a speech before Congress hailed the doubling of territory of "the empire of liberty," of its "ample provision for our posterity and a widespread field for the blessings of freedom."[100] There was no question that these freedoms extended only to whites, who were free to double the territories where slavery could flourish. In 1804, a Connecticut senator offered an amendment to the bill, organizing the Louisiana Territory to prohibit slavery. Jefferson and the Republicans did not support it, instead taking the much weaker step of forbidding importation of slaves from foreign countries.

All of Jefferson's attention was riveted on the huge new territory. For nearly a year, he had been working quietly to gather a small expedition to explore the Missouri River and search for an overland route to the Pacific.

Indeed, for twenty years, Jefferson had bided the right time to explore the American west. As early as 1783, when he returned to Congress after the Revolution, he helped draft reports on the lands along the Ohio and Mississippi, portraying their importance for future American development. By December 1783, he first expressed his fear that England intended to colonize the trans-Mississippi country. He proposed that George Rogers Clark lead an expedition to California, but Congress could offer no funds and Clark was broke after his years of conquering and holding the Illinois country for Virginia. Later schemes by John Ledyard, Moses Marshall, and André Michaux also failed, but as early as 1793, ten years before the Louisiana Purchase, Jefferson began to draft instructions for a western exploration.

Immediately after his election, Jefferson began to plan for a secret expedition to explore as far as the Pacific. In February 1801, he wrote to Meriwether Lewis, then paymaster with the rank of captain in the Army of the West, to become his private secretary. "Your knowledge of the Western country, of the army and all of its interests" was Jefferson's reason to single out the son of an old family friend: together, Jefferson's father and Lewis's grandfather had mapped Virginia so long ago. Jefferson also hinted at other "private purposes."[101]

Jefferson's lifelong fascination with the west combined his romantic streak and his hard-headed pragmatism. He knew that the fledgling United States needed to improve her relations with the Indians, who tended more and more to side with her enemies. He worried that several European powers—Britain, France, Spain, and Russia—all had claims that conflicted with those of the United States. Jefferson also sought a navigable "river to the West" that could carry U.S. trade not only to Indians and European ports along its route but to the Pacific and Asia. And he was fascinated with the flora and fauna off the vast western wilderness: any expedition would take along naturalists to carry out systematic botanical and zoological studies. The Louisiana Purchase finally provided the impetus for his decades-old dreams to become his greatest scientific contribution.

President of the American Philosophical Society since he had become vice president, Jefferson had for thirty-five years continued his amateur scientific studies, filling volumes of ledgers with notes on weather and the change of seasons, the blooming of the first spring flowers. of minerals and medicinal springs, of migrating birds and the bones of extinct animals, of all these interests concentrating most on observing plant life. There was no distance between his private studies and his public offices. When he was away from Monticello, he urged his family to keep up his observations, writing to Maria on March 9, 1791,

> I hope you have, and will continue to note every appearance, animal and vegetable, which indicates the approach of spring, and will communicate them to me. . . . By these means, we shall be able to compare the climates of Philadelphia and Monticello.[102]

From Philadelphia, he noted that he had seen his first robin of the season that February 27—and his first blackbird. When he had arrived in Philadelphia in 1797 as vice president–elect, he carried in his baggage the bones of a prehistoric animal for the American Philosophical Society collection. At the President's House in Washington, he was often seen working with flowers and plants; a pet mockingbird entertained him as he plied his garden and carpenter's tools or, seated at his drafting board, pored over maps and charts. And he sometimes stole away on secret solitary expeditions up the Potomac and into surrounding hills and woods. Wrote Mrs. Smith in her diary, "Not a plant from the lowliest weed to the loftiest tree escaped his notice. . . . He would [get off his horse and] climb rocks or wade through swamps to obtain any plant he discovered or desired and seldom returned from these excursions without a variety of specimens."[103]

Just how often Jefferson took along Lewis, his latest surrogate son, on these forays is unclear, but Jefferson did see to it that Lewis had the money and the time to go to Philadelphia for a full year to study botany, zoology, and the rudiments of medicine to prepare him for the expedition: Meriwether Lewis was one of Jefferson's greatest Enlightenment experiments. But Jefferson's greatest scientific contribution was to coax Congress in a secret session to appropriate a scarce $2,500 to underwrite the expedition, the first scientific expedition of the United States. He sold it on the basis of competition for the rich fur trade that could be opened and the fear that European trading companies would otherwise at least economically exploit the vast new American territory. To Congress, all in secret, he described the expedition as vital "for the purpose of extending the external commerce of the U.S.";[104] to European governments, he claimed it was only a "literary pursuit."[105] To Meriwether Lewis, he wrote secret instructions, signed June 20, 1803, that he was to take careful notes on

> . . . the soil and face of the country, its growth and vegetable productions, especially those not of the United States, the animals of the country generally, and especially those not of the United States; the remains and accounts of any which [may] be deemed rare or extinct . . . the dates at which particular plants put forth or lose their flower or leaf, times of appearance of particular birds, reptiles or insects. . . .[106]

Lewis was also instructed on June 20, 1803, to study whether the furs of the northwest could be collected at the headwaters of the Missouri; ironically, the expedition's encounters with the beaver led to the wild and rapacious boom years in the American fur trade after the War of 1812. In all, Lewis and Clark brought back from their two-year-three-month-long, eight-thousand-mile trek to the Pacific journals crammed with meticulously drawn and annotated records of 122 species and subspecies of vertebrate animals and 178 plants never previously described. Only one man had died. To Lewis, Jefferson wrote "with unspeakable joy" in September 1805, when he learned of his safe return. "The length of time without hearing of you had begun to be felt awfully," Jefferson wrote, reminding Lewis of "my constant affection for you."[107]

Amid the celebrations over the peaceful doubling of American territory, Jefferson's administration was hailed as a towering success by the nation's Republican majority, but the Federalists never let up their attacks. In early March 1804, Jefferson, citing the unbounded calumnies of the Federalist party, announced that Federalist charges against him "have obliged me to throw myself on the verdict of my country for trial."[108] One week earlier, 108 congressional Republicans had caucused and unanimously renominated him, dropping Aaron Burr from the ticket with "not one single vote."[109] By 1804, Jefferson had long since rejected Burr's patronage recommendations and Republican leaders had read him out of the party. When the Republicans refused him the nomination for governor of New York, Burr ran anyway, hoping to divide the Republican vote. After his defeat, he challenged Hamilton to a duel and fatally wounded him, further alienating both parties. Jefferson's landslide reelection with George Clinton of New York as his running mate was even more of a walkaway than he expected, Republicans carrying every state but Connecticut and Delaware. The electoral vote, unanimous in fourteen states, was Jefferson, 162, Charles Cotesworth Pinckney, 14. But Jefferson's triumph was instantly marred by the childbirth-related death of his twenty-four-year-old daughter Polly. Instead of walking to his second inauguration, President Jefferson rode in a carriage, dressed in black, including black stockings. His grief was made worse by his guilt: when he had heard that Polly was not recovering after giving birth to her second child, he had put off visiting her with the excuse that Congress could not function without him. As her condition worsened, he had again procrastinated: "Nothing but impossibilities prevent my instant departure to join you, but the impossibility of Congress proceeding a single step in my absence presents an insuperable bar."[110]

Jefferson wrote Patsy of his "inexpressible anxiety" over Polly's condition, but still he did not go. He could not grasp the idea that Polly was fading away, just as her mother had. Jefferson poured out his grief to his oldest friend:

> Others may lose of their abundance but I, of my want, have lost even the half of all I had. My evening prospects now hang on the thread of a single life. The hope with which I had looked forward to the moment when, resigning public cares to younger hands, I was to retire to that domestic comfort from which the last great step is to be taken, is fearfully blighted.[111]

All through his presidency, the grim tolling continued. In 1806, Jefferson lost his law mentor and oldest friend, George Wythe. Wythe, who had served in Congress with Jefferson and had been appointed, at Jefferson's behest, the first law professor in the United States, was eighty. He left Jefferson his law library, the most valuable thing he owned, and the money to raise his freed mulatto son, Michael Brown. The son was to share Wythe's estate with the chancellor's white grandnephew—unless the freed slave boy predeceased the grandnephew. In May 1806, Richmond police arrested the grandnephew and jailed him for forgery. But the charge should have been murder, for by June 1, 1806, Wythe, his freed slave mistress Lydia Broadmax, and his mulatto son all had suffered arsenic poisoning. Only "Lyddy" survived.

Jefferson received yet another shock in October 1809, when Meriwether Lewis, the man he had made his closest aide, head of his fondest project, the Lewis and Clark Expedition, and the first governor of the vast Louisiana Territory, died only months after Jefferson left office. The circumstances of Lewis's death have left lingering doubts: was he robbed and murdered as he slept alone in a cabin in central Tennessee en route to Monticello? Shot twice, once in the head and once in the chest, with his own pistols, and slashed with his own razor, his gold watch and money were missing when he was found. Or, as more recent investigations surmise, did he commit suicide, despondent over his debts and half-crazed from weeks of heavy drinking?

Jefferson, from shock or from habit or both, recorded no reaction to the death of either friend.

After a second bitter presidential campaign in which he was pilloried by the Federalist press for his alleged relationship with Sally Hemings, Jefferson declared in his Second Inaugural Address on March 4, 1805, that he had decided that libelous publications ought to be prosecuted. During his administration he wrote, "in order to disturb it, the artillery of the press has been levelled against us, [loaded] with whatsoever its licentiousness could devise or dare." The man who had become president by campaigning against a law that controlled the press now suggested that the states invoke their laws "against falsehood and defamation." Jefferson himself was too busy to pursue his critics, he said, but it would be a public service "to public morals and public tranquility" if anyone "who has time" should reform the press's "abuses by the salutary coercions of the law."[112]

In 1806, Jefferson appointed Republican Pierpont Edwards as a U.S. district court judge for fiercely Federalist Connecticut. Almost immediately Edwards asked a Republican marshal for indictments against publishers of libels against the United States on the grounds they would "sap the foundations of our Constitution of Government [more] than any kind of treason."[113] The grand jury returned indictments for seditious libel against Judge Tapping Reeve of the Connecticut Superior Court for articles he had published in the Federalist *Litchfield Monitor,* against Joseph Collier, its publisher, and against Thaddeus Osgood, a candidate for the Congregationalist ministry. A few months later, the editors of the Federalist Hartford *Connecticut Courant* and the Rev. Azel Backus were also indicted. All had publicly attacked Jefferson's candidacy and his presidency: all were charged with seditious libel of President Jefferson.

Three years later, Jefferson was to insist in a letter to Wilson Cary Nicholas that the prosecutions "had been instituted, and had made considerable progress, without my knowledge, that they were disapproved by me *as soon as known,* and directed to be discontinued."[114] The first time that official Washington learned of the indictments was on January 2, 1807, when Federalist Congressman Samuel W. Dana of Connecticut asked the House of Representatives to support his bill for making truth a defense in federal libel trials. The bill aborted, but President Jefferson, writing to a Hartford Republican leader, said he supported the truth as a defense. It could "not lessen the useful freedom of the press." But he would "leave to others" the task of holding the press to the truth."[115]

Jefferson did not, however, criticize or stop the federal prosecutions—not until he learned the exact nature of Reverend Backus's libel against him from the pulpit. Backus demanded an immediate jury trial when he learned that Judge Edwards was issuing subpoenas for witnesses to testify. Jefferson later recalled that "I heard of subpoenas being served on General [Henry] Lee" and other Virginians as witnesses to Backus's charges that Jefferson had had an affair with Betsy Walker. John Walker had turned over his account of the incident to General Lee. That was enough for President Jefferson. "I immediately wrote to Mr. Granger [Attorney General Gideon Granger] to require an immediate dismission of the prosecution."[116] There was no further prosecution for sedition under Jefferson.

Long ago, Jefferson the philosopher of reason had written from Paris, "The basis of our governments being the opinion of the people . . . were it left to me to decide whether we have a government without newspapers or newspapers without a government, I should not hesitate a moment to prefer the latter."[117] Twenty-five years of politics and governing later, Jefferson had another view of the press. He had come to "deplore" the "putrid state into which our newspapers have passed, and the malignity, the vulgarity and mendacious spirit of those who write for them."[118]

Almost from the beginning, Jefferson's second term was clouded by events in Europe, where British victory at Trafalgar had broken French naval power, making England master of the seas. On the Continent, Napoleon's armies had crushed the allied powers at Austerlitz. "What an awful spectacle does the world exhibit at this instant," wrote Jefferson as 1806 began, "one man bestriding the continent of Europe like a Colossus and another roaming unbridled on the ocean."[119] When the Anglo-French war had resumed in 1803 shortly after the Louisiana Purchase, American exports, which had plummeted from $94 million in 1801 to $54 million in 1803 during the Peace of Amiens, once again skyrocketed to a record $108 million by 1807. Anglo-American relations had settled down after Napoleon's sale of Louisiana defused tensions in North America. From London, American ambassador Monroe reported, "The truth is that our commerce never enjoyed in any war as much freedom and indeed favor from this government as it now does."[120]

But within a year, after their victory at Trafalgar, the British were planning, according to Monroe, "to subject our commerce at present and hereafter to every restraint in their power."[121] Monroe was alarmed by a British navy crackdown ordered by the King's Privy Council on the American reexport trade. Enforcing the half-century-old Rule of 1756, the British tried to break up the use of American ships to carry French goods to French colonies when French colonies were blockaded by the British. American ships had been circumventing the rule by making a token stopover in the United States. In 1800, in the *Polly* case, the British had ruled that this did not violate their rule and American ships had since cornered most of the trade between Europe and the Caribbean, with reexports amounting to half of the export trade by 1805. In the *Essex* decision that year, the High Court of Admiralty ruled that landing goods and paying duties in the United States was no longer proof of importation and Royal Navy ships began seizing American ships, in all taking some three hundred to four hundred ships and crippling the reexport trade. A further complication came from the shortage of American sailors as her carrying trade doubled while British sailors seeking to escape terrible conditions and a high risk of death in battle deserted English men-of-war. Probably a quarter of the fifty thousand to one hundred thousand seamen on American merchant ships were British deserters. British navy officers increasingly sent boarding parties to impress suspected British deserters, seizing an estimated six thousand between 1803 and 1812. British officers sniffed at papers of "protection" issued in lieu of American passports, contending that a British subject could become American for only one dollar.

When the British Admiralty ignored American complaints, Jefferson sent Monroe and Pinckney to London to seek a new Anglo-American treaty in 1806. Among other concessions, the British agreed to allow the reexport trade as long as American ships paid a small transit duty during their United States stopover. In exchange, the United States pledged neutrality in the European war. But there was a catch. Napoleon had just retaliated diplomatically by issuing his Berlin Decree, a meaningless piece of paper declaring a naval blockade of the British Isles. The new Anglo-American agreement provided that the British could retaliate against France if the United States honored the blockade, which American merchantmen were not about to do. This proviso clouded Anglo-American relations. President Jefferson refused to submit it to the Senate for ratification. "To tell the truth," he said, "I do not wish any treaty with Great Britain."[122] Unless the British gave up imprisonment of American seamen, he argued, all other British concessions were trifles.

From that moment on, Anglo-American relations deteriorated. On June 22, 1807, H.M.S. *Leopard* approached the U.S. frigate *Chesapeake* as it sailed out of Hampton Roads, Virginia, and demanded that a British boarding party be permitted to search for deserters. The *Chesapeake* had a large number of British deserters in her crew and the British commander-in-chief at Halifax, who knew the names of four of them, took matters into his own hands and ordered his commanders to recover them, even if by force. When the *Chesapeake*'s skipper refused permission to board, *Leopard* fired three broadsides at close range, killing three sailors and wounding eighteen others. When the crippled *Chesapeake* struck her colors, the British took off four suspected deserters (three, Americans, were later returned; the sole British deserter was hanged). When the shattered *Chesapeake* reached port, outraged Americans clamored for war. "But one feeling pervades the nation," reported one congressman. "All distinctions of Federalism and Democracy are banished."[123] Refusing to fan the flames, Jefferson ordered all British warships out of American waters and demanded a British apology. It did not come for four years.

The same week that the *Chesapeake* limped back into Hampton Roads, Aaron Burr was indicted for treason, charged with plotting with Spain to overthrow Jefferson's administration and lead the secession of all American territory west of the Appalachians. Shortly after killing Hamilton and while still vice president, Burr had met secretly with British Ambassador Merry, who reported Burr's offer "to lend assistance to His Majesty's government in any manner in which they may think fit to employ him, particularly in endeavoring to effect a separation of the western part of the United States."[124] After leaving office, Burr met with Merry again about an independent Louisiana. He was also in contact with the Spanish government. Traveling through the west, he aroused suspicion after the publication, in the Federalist *United States Gazette* in July 1805 of an anonymous letter accusing him of plotting to form a separatist government for the west, seize public lands there, and invade Mexico. President Jefferson received the first anonymous letter warning him about Burr in December 1805 and simply filed it. He also filed without comment a report from the governor of the Louisiana Territory that Burr had met with the former Spanish intendant in New Orleans as well as with disgruntled westerners. He took more seriously a report from Joseph H. Daveiss, the U.S. district attorney for Kentucky, that warned Jefferson of Spanish intrigues and "traitors among us—a separation of the Union in favor of Spain is the object."[125] The attorney general implicated by name Gen. James Wilkinson, commanding general of the army and governor of Upper Louisiana, who was in the Spanish pay.

When Jefferson circulated the letter to part of his cabinet, no one could believe the involvement of Wilkinson, a former aide to Benedict Arnold and to Horatio Gates in the Revolution who had been a general since the age of twenty-one. The president decided to demand a full report from Daveiss, a Federalist, who promptly supplied the names of Burr, Wilkinson, Attorney General John Breckinridge, Henry Clay, and William Henry Harrison. Jefferson refused to credit the report, even after Daveiss struck the last two names from the list of the accused. But Jefferson also heard from Col. George Morgan of western Pennsylvania, whom Jefferson trusted, that Burr had tried to recruit his sons in a military expedition. In October 1806, when Jefferson received a confirmed report from his postmaster general, Gideon Grainger, that Burr had offered Wilkinson command of a separatist expedition, he called a cabinet meeting that decided to send a federal agent to investigate Burr and, if Burr carried out any overt plot, to arrest him. The governors of the western territories and naval officers at New Orleans were also alerted. Jefferson took no action on Wilkinson, who had already decided to abandon the conspiracy and expose Burr.

Two weeks earlier, a Burr aide had delivered a ciphered letter to Wilkinson: funds were ready, England would provide ships, the plan was being set in motion. Burr was to go west in August and five hundred to one thousand men, the first wave, would leave Louisville in November and meet Wilkinson at Natchez in December. After a two-week delay, Wilkinson sent the cipher and a packet of letters to the president. The documents reported "a numerous and powerful association"[126] from New York to the Mississippi formed to launch a ten-thousand-man invasion of Mexico at Vera Cruz. Wilkinson, in his covering letter, lied to the president that he did not know the instigator or the objectives of the plot. On November 25, as soon as the courier from Wilkinson arrived, Jefferson called an immediate cabinet meeting, then issued a proclamation warning all citizens against the plot and ordering the conspirators and their weapons seized. Without waiting for an indictment, trial, or corroborating evidence, in a special message to Congress in January 1807, Jefferson exposed the plot and denounced Burr, the man whose support had once made possible his election as president. Burr's "guilt is placed beyond question."[127] The cipher letter to Wilkinson was printed in the *National Intelligencer* the next day. By this time, Burr had already surrendered in Mississippi Territory, where he had arrived with ten boats and sixty men and had learned of Wilkinson's betrayal and Jefferson's order for his arrest. When a grand jury did not find enough evidence to indict him, Burr was released. But he was quickly arrested again and brought under military guard to Richmond in March 1807 for trial on a charge of treason.

The preliminary hearing before Jefferson's cousin, Chief Justice Marshall, only two weeks later, began in a Richmond tavern and was adjourned to the House of Delegates. It arrayed Caesar A. Rodney, the U.S. attorney general, against two prominent lawyers defending Burr, including Jefferson's cousin Edmund Randolph. The chief justice stunned the audience and infuriated Jefferson when he ruled there was no proof of Burr's treason and ordered Burr released on $10,000 bail to face only a misdemeanor charge for instigation of an expedition into Spanish territory. Jefferson had become obsessed with Burr and ruthlessly kept up the pressure for his conviction, despite the lack of evidence of treason. He had clashed repeatedly with Marshall over judicial appointments, and now, bypassing the attorney general, personally meddled while railing at "the tricks of the judges to force trials before it is possible to collect the evidence" and "their new born zeal for the liberty of those whom we would not permit to overthrow the liberties of their country."[128]

When the grand jury met at Richmond in May 1807, the court was crowded with spectators from all around the United States. Marshall appointed Jefferson's leading critic in Congress, John Randolph, as jury foreman. The government, at Jefferson's insistence, again moved to commit Burr for treason and to deny bail. Marshall refused to rule on the motion until the grand jury deliberated but he increased bail to $20,000, a fortune at the time. Since Wilkinson had not arrived, court was adjourned from day to day for three weeks. Meanwhile Burr asked the court to subpoena key documents in the case and to subpoena President Jefferson. Marshall responded by issuing a subpoena to Jefferson or whichever department secretary had the papers in question. Jefferson, again circumventing his attorney general, sent the court the papers even before he received Marshall's order. But he made it known that he was less willing to appear in person. He said he had "paramount duties to the nation"[129] and would invoke the executive privilege of the president to refuse to appear in a court outside Washington. He also made it clear that he resented Marshall's arguments that he had the authority to subpoena the president to Richmond:

> Would the executive be independent of the judiciary if he were subject to the *commands* of the latter and to imprisonment for disobedience if the several courts could bandy him from pillar to post, keep him constantly trudging from north to south and east to west and withdraw him entirely from his constitutional duties?[130]

When Wilkinson finally appeared on June 15, 1807, he narrowly escaped indictment for treason by a 9–7 vote of the grand jury. But Burr was indicted. His trial in August ended abruptly when the chief justice ruled on narrow grounds that, unless witnesses could place him in person with the sixty-man force that had assembled on Blennerhassett's Island in the Ohio River on December 10, 1806, no other collateral evidence of conspiracy could be used against him. The jury deliberated for twenty-five minutes, then found Burr "not proved to be guilty under this indictment by any evidence submitted to us."[131]

But Jefferson was implacable against what he perceived to be betrayal by his former vice president. Privately he had written earlier to du Pont that "Burr's conspiracy has been one of the most egregious of which history will ever furnish an example"; but he was not surprised by the legal outcome. "Such are the jealous provisions of our laws in favor of the accused and against the accuser that I question if he can be convicted." Yet publicly he denounced the trial at Richmond as "equivalent to a proclamation of impunity to every traitorous combination which may be formed to denounce the Union."[132] Jefferson the lawyer never read the trial record or weighed the evidence presented against Burr. He considered Marshall's ruling to be politically motivated. But by the autumn of 1807, Jefferson had little time left for Aaron Burr.

On his annual vacation atop Monticello, Jefferson concluded that the British were not about to meet his demands for "reparation for the past, security for the future"[133] against further attacks on American ships and continued impressment. He became reconciled to war with England and began to think also of fighting Napoleonic Spain. If Jefferson was not an imperialist, he definitely was an expansionist who considered it necessary to remove foreign powers from U.S. borders to provide security. He also saw that Napoleon, blockaded by the British navy, would be unable to defend his empire against a massive American volunteer force allied with revolutionaries in an American-led confederation of former Spanish territories. To Madison, his Secretary of State, he confided:

> I had rather have war against Spain than not if we go to war against England. Our southern defensive force can take the Floridas, volunteers for a Mexican army will flock to our standard and a rich pabulum will be offered to our privateers in the plunder of their commerce and coasts. Probably Cuba would add itself to our confederation.[134]

Jefferson's long-standing detestation of the British now seemed to ride the crest of American anger, at its highest point since the beginning of the Revolution. If he had been a pacifist then, now he was ready for war:

> I never expected to be under the necessity of wishing success to Bonaparte. But the English being equally tyrannical at sea as he is on land, and that tyranny bearing on us in every point of either honor or interest, I say "down with England" . . . what Bonaparte is then to do to us, let us trust to the chapter of accidents. I cannot, with Anglomen, prefer a certain present evil to a future hypothetical one.[135]

In his walks around Monticello, Jefferson realized he must keep the war spirit alive in America if he hoped his diplomacy would succeed. The news had arrived from Halifax that the four seamen seized from the *Chesapeake* had been court-martialed at Halifax and that one had been hanged. The tone of his annual message to Congress was more stridently anti-British than ever before. Gallatin, reading the draft, thought it was too much a manifesto against Great Britain that would lead to war. Navy Secretary Smith, who thought "peace is our favorite object,"[136] worried that Jefferson would incite a congressional declaration of war. Even Secretary of War Dearborn questioned whether Jefferson wanted to threaten offensive actions against the British. But when Jefferson returned to Washington, he found that the mood of Congress had once again shifted in the wake of news from Europe that, because England's allies, Russia and Prussia, had abandoned the war against France, the British might be ready to negotiate with the United States. Shifting his own ground, he wrote, "we are all pacifically inclined here."[137] No sooner had he become optimistic again than he learned from London that negotiations there had broken down. His diplomatic initiative to avoid war had failed. Disgusted, Jefferson wrote his son-in-law that it was now up to Congress whether there was to be "war, embargo or nothing."[138]

At first making no recommendation to Congress, by December 18, 1807, Jefferson, citing "the great and increasing dangers with which our vessels, our seamen and merchandise are threatened on the high seas,"[139] now proposed an embargo on the departure of all American vessels from U.S. ports. In the ten days between his first neutral stand with Congress and his strong appeal for embargo, Jefferson had learned that George III had issued a proclamation in October requiring all naval officers to enforce impressment. Less than a month later, learning that Napoleon was blockading American as well as British ships, he issued an Order-in-Council prohibiting Americans to trade with ports from which British ships were excluded and ordering that American vessels bound for European ports must stop at British ports, pay taxes, and get British clearance. News of these latest British coercions had already appeared in the Philadelphia *Aurora* when Jefferson submitted his request for an embargo. The measure sailed swiftly through both houses of Congress. What everyone hoped to accomplish by boycotting trade with the belligerent European powers was summed up by a Tennessee congressman writing to his constituents:

> We may complain because we cannot sell for a good price our surplus provisions and other productions; they will *suffer* because

> they cannot procure a sufficient quantity of those articles to subsist upon—*to support life.*[140]

But the embargo was not to bring England and France to their knees. In one year it destroyed eighty percent of all American trade and brought on the worst depression since the Revolution. At the same time, it was widely ignored by Americans, who smuggled across the northern frontier to supply British bases in Canada and carried on such a wholesale illicit carrying trade to the Caribbean and Europe that not even the seizure of nine hundred American ships by the British and French could completely stop what Jefferson thought he could do by the application of law and reason. In his last years in the President's House, Jefferson became embittered at the widespread flouting of the embargo.

At first, Jefferson's embargo flattened shipping from Atlantic coast ports: New Englanders and southerners trading with England and the Caribbean were thrown out of work. But this still left wide open the lucrative overland lake-and-river traffic between border states and British and Spanish territories. Here, the lure that had attracted many Americans to leave the more cosmopolitan east coast and settle in the raw frontier clearings and towns of the frontier—and the bedrock of Jefferson's popular agrarian philosophy—was the chance to sell surplus crops and timber products for export to British and Spanish merchants and the quartermasters of hungry garrisons in St. Louis, St. Augustine, Montreal, Quebec. A hemorrhage of exports, for example, defied the blockade as it flowed north along Lake Champlain into lower Canada: cattle and pig herds were driven through Smuggler's Notch to feed English soldiers and sailors and fetch high prices paid in gold.

On March 3, 1808, Jefferson decreed a second phase of the trade boycott, the land embargo, closing off all overland and waterborne trade across American frontiers.

In Vermont, at least one-third of all income in the Green Mountain State depended on shipping fine white pine planks and white oak spars for shipbuilding and barrels of potash made from rendering timber scraps. Potash was particularly vital to the British chemical industry, badly needed to provide chemicals for processing soap, glass, and textiles.[141] U.S. customs officials in Vermont quickly protested to Secretary of the Treasury Albert Gallatin the "impossibility of executing" the new law "without military force."[142] Gallatin informed Jefferson, who was quick to issue a proclamation on May 8, 1808, pointedly addressed at Vermonters. (Up to now, frontier Vermont was the only Jeffersonian stronghold in New England.) Jefferson warned Vermonters that further trade with the British in lower

Canada would be treated as an insurrection and treason and would be quelled by force of arms.[143] To put teeth in the president's decree, Vermont Gov. Israel Smith, a recently elected Jeffersonian, called up the Franklin County militia, many themselves suspect of being involved in the traffic from their hardscrabble border farms, posting some at Windmill Point to intercept smuggler's rafts running the frontier on the strong current at night. But in the darkness, militiamen looked the other way, and fast smuggling ships slid into Canada. The Republican governor finally relieved the frontier militia of their duty and sent in 150 volunteers from Rutland, his home district eighty-five miles south of the border. Militia and cavalry of the Second Vermont Brigade marched north on June 4, 1808, "to stop the potash and lumber rebellion on Lake Champlain."[144] Soon they were on the lookout for one smuggler's ship in particular, the *Black Snake.*

Built as a Lake Champlain ferry, this beamy, tar-blackened forty-foot sloop drew only four feet and was ideal for slipping in and out of shallow creeks and coves. It could be sailed or rowed by fourteen oarsmen and could carry one hundred barrels of potash; the black-market price of a three-hundred pound barrel was to soar from $25 to $300 during the embargo. On August 1, the capture of the *Black Snake* was turned over to the *Fly,* a sleek, red-trimmed revenue cutter, its twelve-man crew commanded by Lt. Daniel Farrington of Brandon. The next night, smuggler Truman Mudget, captain of the *Black Snake,* rowed in a small boat and picked up his crew along with a gun over nine feet long with a one-and-one-quarter-inch bore to mount in the bow of their sloop. They rowed up the Winooski River to Joy's Landing where the smugglers sipped rum and made eleven hundred bullets and loaded their muskets and waited for their cargo of potash to arrive.

A mysterious informant, meanwhile, flagged down the *Fly* and told Lieutenant Farrington the whereabouts of the smugglers and their names. At dawn, *Fly* sailed up the Winooski, soon in hot pursuit of a smuggler rowing frantically to warn his friends, who were busily trying to conceal the sloop. As the militiamen headed for the *Black Snake,* the *Fly*'s skipper, a musket at his shoulder, shouted, "I swear by God I will blow the first man's brains out who lay hands on her."[145] The lieutenant and some men boarded the *Snake* and seized her, cutting her mooring line as his men rowed the sloop toward midriver. Shots came from the shore: as Pvt. Ellis Drake took the rudder two bullets hit him in the head. As the militiamen landed and prepared to return fire, Samuel Mott opened fire with his oversize musket, discharging fifteen bullets that cut down three militiamen, killing two more men.

Less than three weeks later, eight smugglers stood accused of their murders in a showcase trial. Potential jurors were accepted or rejected according to their party affiliation. In his closing argument, State's Attorney William Harrington underscored the political nature of the trial:

> The defense will say to you that the law laying an embargo has occasioned this unhappy affair. . . . It is painful to find that party spirit . . . has already assumed an alarming attitude.

Three months passed, and no reprieve came from President Jefferson. On November 11, 1808, Cyrus B. Dean, who had urged Mott to fire, became the first Vermonter to be hanged. An estimated ten thousand Vermonters followed the death wagon, flanked by militia, to the gallows on a knoll near the present-day campus of the University of Vermont. That autumn, too, Vermonters marched to the polls, defeating the Jeffersonian governor and replacing him with a Federalist: indeed, it was a vote for smuggling, against Jefferson and his hated embargo. After the election, the other smugglers received lighter punishments. The two smugglers who had fired the fatal shots received public whippings and ten-year jail sentences, later commuted. Mudget, the gang leader, was released after a mistrial.

Appalled by such incidents, Jefferson wrote to all state governors in January 1808. He defended his embargo. "While honest men were religiously observing it, the unprincipled along our sea-coast and our frontiers [have been] fraudulently evading it."[146]

The embargo crisis came at a turning point for Jefferson, the eve of the 1808 presidential elections. By late 1806, Republicans had been holding meetings and sending resolutions to Jefferson, pleading with him not to retire. The legislatures of nine states and one territory entreated him to stay on for four more years. "The voice of the people were never so strong in your favor,"[147] wrote one Pennsylvania congressman. But Jefferson feared that his acceptance of a third term would make the presidency a lifetime office. Jefferson had long ago decided to throw his support behind his longtime protégé and friend, James Madison.

Jefferson turned his mind to making the embargo work in his last year in office. But, as he found it unenforceable, he admitted privately to Gallatin's pleas for harsher laws of enforcement that "this embargo law is certainly the most embarrassing one to execute." He grew increasingly disillusioned with the American people as they widely ignored the trade boycott. "I did not expect a crop of so sudden and rank growth of fraud and open opposition by force."[148] In his last annual message to Congress on November 8, 1808, Jefferson looked for something good to say about the embargo: at least it had saved some American lives and property and given the United States time to build up its defenses against the war he now feared was inevitable. With the embargo the main election issue and himself gone from the Republican ticket, Jefferson also had reason to worry about the survival of his party. It was with vast relief that he watched the results come in. Madison had won, 122 electoral votes to 47 for the Federalist Pinckney. The party of Jefferson lived.

A milestone of Jefferson's presidency was his signing into law the Act of Congress which banned the further importation of slaves to the United States. In the Federal Constitution of 1787, a moratorium had been imposed that enjoined Congress from banning the slave trade until twenty years had passed. By 1807, only the state of South Carolina still allowed the foreign slave trade: all other states had banned importation from foreign countries. Jefferson, of course, could not foresee that the introduction of the cotton gin would nurture the cotton economy and that the domestic sale of slaves from one state to another would burgeon as slave owners sold off their surplus laborers for cash, more than offsetting the need for further imports from Africa.

As he prepared to retire from the presidency, Jefferson showed in his writings that he had, over the years, outgrown his once deeply provincial views on slavery and had come to realize that African-Americans were as intelligent as whites if they only had the opportunities to develop their talents. Shortly after he signed the ban on foreign importation, he proudly wrote to a group of Philadelphia Quaker abolitionists, "It is honorable to the nation at large that their legislature availed themselves of the first practicable moment for arresting the progress of this great moral and political error."[149]

21

"A Fire Bell in the Night"

I am happy to find you are on good terms with your neighbors. It is almost the most important circumstance in life, since nothing is more corroding as frequently to meet persons with whom one has any difference. The ill-will of a single neighbor is an immense drawback on the happiness of life.

—Thomas Jefferson to his daughter, Patsy, May 1790

Except for working vacations, Thomas Jefferson had been away from his beloved Monticello, from his daughter Patsy, his sister, nieces, nephews, and grandchildren for most of the twenty years since he had returned from France. In all, he had served the public for forty years since his first election to the House of Burgesses. Cleaning out his desk and files in the President's House at age sixty-five, he sent a cavalcade of wagons ahead of him and left Washington after Madison's inaugural, arriving at Monticello after three days on horseback in a snowstorm. The hard winter journey buoyed him. For months, he had been brooding about his lagging vigor and a "memory not so faithful as it used to be."[1] He was eager to get back to farming. For one thing, he needed the money. He drew his last federal check—expense money owed him since 1789—and wrote one friend that he had added not a penny to his fortune from all his years in office, his "hands as clean as they are empty."[2] There was no presidential pension and Jefferson confided to his daughter that he had gone $30,000 deeper into debt as president. "My own personal wants will be almost nothing beyond those of a chum of the family,"[3] he told her.

Patsy and her husband, Thomas Mann Randolph, Jr., had built their home a few miles east of Monticello so that they could visit him easily when he came home. Now, she was waiting for him with her eight children, six under the age of thirteen. She never left him for the last eighteen years of his life, eventually taking over his household. Her husband, often depressed, at times filled with resentment, managed their farm and came and went, leaving Jefferson to be father and grandfather. "I live in the midst of my grandchildren,"[4] Jefferson wrote John Adams in 1812. In one of his last letters to Maria Cosway, he wrote in 1820 that he had "about half a dozen" great-grandchildren and was living "like a patriarch of old."[5] His grandchildren adored him. "Our grandfather read our hearts to see our invisible wishes,"[6] wrote Ellen. He bought his granddaughters their first silk dresses, and they were married in the drawing room. He walked and talked in the gardens with them, taught them games, handicapped their footraces, dropping his handkerchief at the start and giving prizes of dried fruit at the finish. He became especially close to his grandson, Thomas Jefferson Randolph, whom he called Jeff, gradually turning over the management of Monticello to him.

Honored for his botanical experiments, he introduced many of his European discoveries into daily life around him: ice cream, macaroni, capers, olives, fine white wines, all graced his table, making his dinners half-Virginian, half-French.

As he grew old, he believed more in education as the only hope to teach not only useful skills but to change attitudes, to improve morality and spread civilization. "A part of my occupation," he said two years out of office, "is the direction of studies of such young men as ask it."[7] Young would-be lawyers took rooms in nearby Charlottesville and came to Jefferson's library, where he guided their studies.

He rarely left Monticello except to make the ninety-mile trip to Poplar Forest, the octagonal house he had designed and built as his secluded retreat. He had originally intended it for his daughter, Polly, but it was not until after she died that he designed this revolutionary house of pure octagonal forms: octagonal exterior, octagonal rooms around a square dining room, even octagonal outhouses. He made the ride there until he was eighty, then turned over the property to Polly's only surviving child, his scholarly grandson, Francis Eppes. He never really did finish Monticello: that seems never to have been his intention. There were always piles of wood and treacherous catwalks and half-lighted narrow stairs; the thrill of building always satisfied him more than the result. An endless stream of visitors came to see him, many merely curious, some devoted old friends. Thousands wrote to him and he insisted on answering all of them.

Friendship mattered more to him as he left politics further behind.

With the patient intermediation of his friend Benjamin Rush, he reopened his correspondence with John Adams. Adams sent the first brief note on New Year's Day, 1812; Jefferson happily responded in a long letter, touching on their years as "fellow laborers in the same cause," on the years of struggle of the young republic and its leaders, "in your day French depredations, in mine English."[8] A remarkable literary exchange developed that flourished for their last years, each man trying to correct and comprehend the record of what had passed, commenting on passing events. In these letters, Jefferson revealed his essential self probably more clearly than at any other time. When the British burned the Library of Congress in 1814, Jefferson offered to sell his library of nearly sixty-five hundred books, his collections of fifty years, to Congress for about four dollars a volume. The last wagon to cart them down from Monticello left Jefferson feeling hollow. "I cannot live without books,"[9] he wrote to Adams. He immediately began to buy more.

He rarely allowed himself to interfere in politics: "I do not often permit myself to think of that subject,"[10] he wrote. He did write to his son-in-law, John W. Eppes, chairman of the House Ways and Means Committee, suggesting a way to finance the War of 1812 after the charter of the Bank of the United States expired and was not renewed. When his cousin, Chief Justice John Marshall, borrowing the arguments of Hamilton, upheld the constitutionality of a national bank in *McCulloch v. Maryland,* Jefferson vented his hostility to Marshall, who had also asserted sweeping powers for the national government. He was worried about the centralization of power in America and wrote privately to Judge Spencer Roane, "We find the judiciary on every occasion still driving us into consolidation."[11]

By 1821, as he worried about the survival of the Union, he was working hard to build a university that would be "the most eminent in the United States," where Virginians could learn from "characters of the first order of science from Europe as well as our own country."[12] Jefferson himself drew up the plans for the great domed building that was to serve as the university library. Using Palladio's descriptions and drawings of the Pantheon in Rome, Jefferson recreated it half-size for the Rotunda. The campus he designed was an academic village with lawns, the forerunner not only of many campuses around the country but the prototypical American suburb. When the state paid the university's construction debts and pledged $50,000 for books and equipment, Jefferson at eighty moved quickly to recruit a faculty of six professors from Europe. Several able Americans had refused to leave their northern campuses to support Jefferson's experiment. As rector, he personally drew up class schedules, student rules, and faculty bylaws and handled a host of other administrative details, all ratified by a Board of Visitors. He insisted on freedom for students "to attend the schools of his choice and no other than he chooses." Students would have the responsibility "to pay especial attention to the principles of government." Specific books were to be required, including the works of John Locke, Algernon Sidney, *The Federalist Papers,* Jefferson's Virginia Resolutions of 1799, Washington's Farewell Address. Despite this one requirement, Jefferson insisted in his last year of life that the university would be "now qualified to raise its youth to an order of science unequalled in any other state; and this superiority will be greater from the free range of mind encouraged there, and the restraint imposed at other seminaries by the shackles of a domineering hierarchy and a bigoted adhesion to ancient habits."[13]

Shortly before the university opened, Jefferson received a great surprise. Lafayette was coming to America. In November 1824, Lafayette with his military escort drove up to Monticello where, before a crowd, the two old friends, who had not seen each other since the opening days of the French Revolution, embraced. The next day, they were honored at the inaugural ceremonies for the University of Virginia. Beneath the high dome of the Rotunda, at a three-hour dinner, Jefferson, now eighty-one, seated between Lafayette and Madison, also now enfeebled, toasted Lafayette. He was himself toasted as "founder of the University of Virginia."[14] He had written a brief address, his last, but his voice was now too weak, and he asked someone to read his words of praise for Lafayette, his thanks to his neighbors and friends. The next spring, the first thirty students moved onto a campus a visiting Harvard professor described as "more beautiful" than any in New England and "more appropriate to a university than can be found, perhaps, in the world."[15]

On New Year's Day, 1826, Jefferson wrote an Albemarle neighbor that he had been confined to his house for three weeks "and indeed to my couch"[16] by recurring bouts of diarrhea (he may have been suffering from colon cancer) as well as diabetes and a urinary tract infection. "I write slowly and with difficulty," he said in one of the last of his twenty-eight thousand letters. He was "weakened in body by infirmities and in mind by age, now far gone into my 83rd year, reading one newspaper only and forgetting immediately what I read."[17] Jefferson had prized himself all his life on his regimen of physical fitness. At college, he had begun to run every day. In Paris, he walked a measured daily course; at Monticello, he laid out walks of one to five miles around his mountain. He watched his diet, preferring vegetables and fruit, eating little meat or animal fat. He drank a glass of wine each day. But for his last five years, he had suffered chronic digestive and urinary problems.

On January 2, 1826, a sick Jefferson received another rude shock. His son-in-law, Thomas Mann Randolph, the former governor of Virginia, had to sell his estate at auction to satisfy his creditors. All over Virginia, in the aftermath of the Panic of 1819, the farms of cash-poor, land-rich planters who could no longer sell their lands were going under the gavel. Jefferson himself was, on the day of his death, $107,000 in debt, his financial condition made impossibly worse when two $10,000 loans he had cosigned for his political lieutenant, Wilson Cary Nicholas, had been called at Nicholas's death and the obligation shifted over to him.

Jefferson at eighty-three was lying awake nights trying to think how he could save Monticello for his family. The night of January 19, an idea hit him "like an inspiration from the realms of bliss."[18] At dawn, he sent for his grandson, Jeff, and proposed a lottery that would sell off his nail-making mills and about a thousand acres of land to clear his debts. (He had long opposed lotteries or gambling of any kind and had recently banned cards and dice at the University of Virginia. In protest, a student had broken a window in the rector's house and lobbed in a deck of cards and a pair of dice.)

All Jeff's efforts to sell off his land had failed. An excited Jefferson said he was sure they could sell large numbers of low-cost lottery tickets with land as the prize. But any lottery would have to be approved by the state's General Assembly. All that day, Jefferson dashed off letters to his friends in the capital about a matter of "ultimate importance."[19] He drew up a paper, "Thoughts on Lotteries," to send along with Jeff to Richmond. Citing examples of worthwhile lotteries in Virginia's history, Jefferson argued that a lottery was the only way in which he could receive fair value for his lands. So desperate had he become that he pleaded that he was entitled to special consideration because of his sixty-four years of public services, which he then catalogued. Then Jefferson slumped back on his couch, surrounded by his books and his inventions, and fretted. On February 1, Hetty Carr, widow of his nephew, Peter, wrote her son that Jefferson was very ill: defeat of the lottery would kill him.

In Richmond, Jefferson's grandson was having trouble getting the lottery bill introduced. "A panic seized the timid and indecisive among your friends as to the effect it might have upon your reputation, which produced a reaction so powerful that yesterday and the day before I almost despaired of doing anything."[20] But four judges of the Virginia supreme court threw their weight behind the lottery, and on February 8 a motion was made to introduce the bill. But it was tabled by a one-vote margin. The next day, the motion to introduce carried, but only by four votes. Forty years before, Jefferson had considered himself humiliated by this same House; now the painful memory seemed to haunt him. "I see, in the failure of this hope, a deadly blast of all my peace of mind during my remaining days," he wrote Jeff. "I should not care were life to end with the line I am writing, were it not that I may yet be of some avail to the family."[21]

The shock of the news of Jefferson's financial condition led to a citizens' meeting amid a public outcry in Richmond. When the motion to introduce the lottery bill squeaked past by only a few votes, Senator William Cabell, who had worked with Jefferson to found the state university, wrote Jefferson, "I blush for my country and am humiliated to think how we shall appear on the page of history."[22] Jefferson wrote back, "I count on nothing now."[23] Two weeks later, Cabell was able to report to Jefferson that the bill had passed the House by a healthy two-to-one margin and the Senate by three-to-one. Jefferson would be able to pay all his debts. As Americans read of the voting in their newspapers, donations began to pour in to the lottery managers. Jefferson should have been heartened, but now he was hit by an aftershock. The bill, as passed, required that he sell *everything,* not just his mills and some lands but Monticello, its furnishings, all his slaves, his horses, everything. He could live in Monticello until his death while the tickets were printed and distributed, but his only daughter, Martha, would have to move out within two years of his death, and Monticello would pass out of his family. Indeed, Monticello was sold after his death and remained outside the family for half a century. Neither the Commonwealth of Virginia nor the United States attempted to buy it and preserve it as a national shrine; even today it is maintained by a family association.

Sinking back on his couch, Jefferson barely had the strength left to make out his will. He had already bequeathed his octagonal summer retreat to Francis Eppes. He left his gold-headed walking stick, which he had only recently begun to need, to his friend Madison, along with the request that he watch over affairs at the University of Virginia. He left a watch to each of his grandchildren, all his books to the University of Virginia. To his grandson Jeff, he left all his personal papers, including his farm books and account books, and his greatest treasure, his vast personal correspondence.

Unlike Washington, who had been able to manumit his slaves by his will, Jefferson now had lost the ability to grant freedom to his slaves as his dying wish. But he could only plead with his creditors to be generous and grant his deathbed request to free five of his servants: his valet, Burwell, a glazier by trade, John Hemings, a carpenter, and two of his sons, Madison and Eston, his apprentices, who were to be freed on reaching maturity, and Joe Fosset, the blacksmith. He asked that houses be built for each of them near their work and he asked the University of Virginia to employ them. He left each man his tools, and he left Burwell $300 in cash, requesting that he stay on and be employed at Monticello. In his will, he asked the legislature to exempt these five men from the Virginia law that required all freed blacks to leave the state within a year of their emancipation, one of many obstacles Jefferson had fought against for more than half a century. As a fledgling legislator as long ago as 1769, in the Virginia courts as a young lawyer, in his Declaration of Independence, and in his first draft of the Northwest Ordinance, Jefferson had worked to make his contemporaries recognize the evil of slavery and to set free all enslaved African-Americans. "Nothing is more certainly written in the book of fate than that these people are to be free,"[24] he wrote. But he considered it irresponsible, indeed cruel, to turn loose his slaves until they were self-sufficient and prepared to remain free. He had freed his favorite chef, James Hemings, who then drifted from job to job, became an alcoholic, begged to be allowed to return to Monticello, and finally committed suicide.

Jefferson's views on slavery had evolved over the years as he wrestled with America's worst problem. As a young lawyer, he had argued in court the birthright of freedom of all men that he had imbibed from his study of the classics. As governor of Virginia, he wrote in his *Notes* racist views of blacks as inferiors who would need to be treated like children once they were freed gradually. Now, an old man, he worried about the growing dissension over slavery that had flared into angry debate in Congress over the admission of Missouri as a slave state. "This momentous question, like a fire bell in the night, awakened and filled me with terror. I considered it," he wrote, "as the knell of the Union."[25] Congress had finally agreed on the Missouri Compromise of 1820, but Jefferson was still worried. "It is hushed, indeed, for the moment, but it is a reprieve only, not a final sentence." He had once hoped that the next generation would abolish slavery. "I regret that I am now to die in the belief that the useless sacrifice of themselves by the generation of 1776 to acquire self-government and happiness to their country is to be thrown away by the unwise and unworthy passions of their sons, and that my only consolation is to be that I live not to weep over it."[26] But he insisted on keeping his views private so that he would not jeopardize political support for his final public project, the establishment of the University of Virginia. He believed that all men and women had been created equal and he considered blacks as MEN—he had capitalized the word in the clause in the Declaration of Independence that had been stricken by the Second Continental Congress. Unlike many of his time, he believed that blacks did not belong in a separate and unequal category but had been endowed by the Supreme Being with the sacred rights of life and liberty that he wrote in the preamble of his Declaration were the natural endowments of all mankind. He had known few blacks who had not been degraded and intellectually stunted by the institution of slavery:

> It will be right to make great allowances for the difference of condition, of education, of conversation, of the sphere in which they move.[27]

And he continuously searched for examples of talented blacks to hold up as examples to his contemporaries, who overwhelmingly believed in the inferiority of blacks.

All his life, Jefferson believed unshakably that mankind was advancing steadily, that education and scientific progress would eliminate social evils such as slavery. But he believed that the process was gradual. At age seventy-three, he had written that "laws and institutions must go hand in hand with the progress of the human mind:"

> As that becomes more developed, more enlightened, as new discoveries are made, new truths disclosed, and manners and opinions change with the change of circumstances, institutions must advance also, and keep pace with the times. We might as well require a man to wear still the coat which fitted him when a boy as civilized society to remain ever under the regimen of their barbarous ancestors.[28]

Since he did not trust cities as civilizing institutions, he had placed all his hopes in the university, its reinvention as an "academic village" his last and most utopian act.

But there was ever the realist at the core of Jefferson's idealism. He did not believe that integration of blacks and whites could work in his lifetime, and he had grave misgivings about the possibility of integration in the future because of

> deep rooted prejudices entertained by the whites; ten thousand recollections, by the blacks, of the injuries they have sustained; new provocations; [and] the real distinctions which nature has made.[29]

Jefferson had long believed that it was the responsibility of the state and society to free all slaves. He advanced many plans: one was to free all slaves at birth and require that they be educated, raised at the expense of the state, trained in a useful skill, and then be set free at age thirty to leave for a state west of the Mississippi, in the new Louisiana Territory, in Santo Domingo or in Africa, where they could prosper and rule themselves. Until society took this responsibility, however, he considered mass emancipation impossible and believed it his burden of duty to care kindly for his slaves, freeing them individually as they became skilled enough to find jobs.

In the end, Thomas Jefferson ran out of time. It was, ironically, his Louisiana Purchase and the invention of a new machine, the cotton gin, that fueled the rapid expansion of slavery by the 1820s. Cheap land to the west drew whites from the played-out fields of the Old South. Toward the end of his life, he had given up waiting for public opinion to change quickly. He foresaw a cataclysm between north and south after the bitter congressional debates leading to the Missouri Compromise of 1820. When a young abolitionist wrote to him with yet another scheme in 1825, urging Jefferson to lead the crusade against slavery, Jefferson wearily replied,

> This, my dear sir, is like bidding old Priam to buckle the armor of Hector. No, I have outlived the generation with which mutual labors and perils begot mutual confidence and influence. This enterprise is for the young, for those who can follow it up and bear it through to its consummation. It shall have all my prayers, and these are the only weapons of an old man.[30]

In the end, the American apostle of the Age of Reason, a man who believed that education, self-discipline, and hard work could produce self-perfection and bring about social change, felt defeated. "All is circumstance," he lamented. "All is circumstance."[31] Leaving the glassed-in library of Monticello rarely now, he rode his stallion Eagle to Board of Visitors meetings at the university in April and May and, according to his family, still went for horseback rides, sometimes impatiently lashing his horse down Monticello. On May 22, he made his last entry in his *Farm Book*. He noted that "a gallon of lamp oil, costing $1.25, has lighted my chamber highly 25 nights, for 6 hours a night, which is 5 cents [a night for] 150 hours."[32] Virtually a shut-in now, this dying American apostle of Enlightenment tried to calculate the economics of his last confinement. By late June, he knew he could leave Monticello no more. He declined an invitation from the citizens of Washington, D.C., to attend the fiftieth anniversary celebration of his Declaration of Independence, but he wrote a public letter to the *National Intelligencer.* It was to be the last letter he ever wrote and in it he reaffirmed his faith in the principles of that declaration: "May it be to the world what I believe it will be," he wrote on June 24,

> to some parts sooner, to others later, but finally to all, the signal to assume the blessings and security of self-government. . . . All eyes are opened, or opening, to the rights of man. The general spread of the light of science has already laid open to every view the palpable truth, that the mass of mankind has not been born with saddles on their backs, nor a favored few booted and spurred, ready to ride them legitimately, by the grace of God. These are grounds of hope for others. For ourselves, let the annual return of this day forever refresh our recollection of these rights, and an undiminished devotion to them.[33]

After he finished writing the letter, he summoned Dr. Dunglison. One week later, on July 1, 1826, Thomas Jefferson, eighty-three, lapsed into unconsciousness. He roused himself several times to ask if it was yet the Fourth of July. He looked steadily at Burwell until Burwell came to arrange the pillows higher, the way he liked them.

On July 2, he gave his daughter Martha a small jewelry box she was not to open before his death. The morning of July 3, he had a little tea. He slept most of the day, then woke around seven and asked Dr. Dunglison, "Is it the Fourth?" The doctor woke him at nine to give him the customary laudanum (a mixture of opium and honey) to help him sleep. He refused it. "No, Doctor. Nothing more."[34] He tossed and turned. After midnight, he sat up, lifting his right hand and his elbow, as if to write, moving it slowly sideways. Then he slumped back on the pillows. At 4 A.M., he woke briefly and asked that his family and servants be brought into the room. At eleven the morning of July 4, he gazed once more at Jeff and moved his lips silently until Jeff moistened them with a little water on a sponge. He never stirred again. At 12:50 P.M., July 4, 1826, he stopped breathing and Dr. Dunglison pronounced him dead.

When Patsy opened the small jeweled box, she found her father had penned and folded up some verses for her. They ended:

> *Then farewell, my dear, my lov'd daughter, adieu!*
> *The last pang of life is in parting from you!*
> *Two seraphs await me long shrouded in death;*
> *I will bear them your love on my last parting breath.*[35]

Before sundown that same day, July 4, 1826, the man who had insisted that Thomas Jefferson write the Declaration of Independence also died. Shortly after noon, at about the time Jefferson died, John Adams moved on his deathbed and whispered, "Thomas Jefferson survives."[36] Before the day was over, he, too, was dead.

They buried Thomas Jefferson beside his wife and near his sister Jane and his dear friend Dabney Carr in the little graveyard on the hillside at Monticello. He had designed even this and he had written the script for his headstone himself:

> *Here was buried*
> *Thomas Jefferson*
> *Author of the Declaration of American Independence*
> *of the Statute of Virginia for religious freedom*
> *and Father of the University of Virginia.*[37]

He did not think it important enough to mention that he had been twice elected and served as the president of the United States.

Key to Abbreviations in Notes and Bibliography

AA = Abigail Adams
AH = Alexander Hamilton
AHM = *American Heritage Magazine*
AHR = *American Historical Review*
AHI = *American History Illustrated*
AJH = *American Jewish History*
AJLH = *American Journal of Legal History*
APS = American Philosophical Society
AS = *American Scholar*
AM = *Atlantic Monthly*
ATJ = TJ, *Autobiography*
BF = Benjamin Franklin
DAR = *Daughters of the American Revolution Magazine*
EAL = *Early American Literature*
EIHC = *Essex Institute Historical Collections*
GW = George Washington
HEH = Henry E. Huntingden Library
HSP = Historical Society of Pennsylvania
HP = *Historic Preservation*

HEQ = *History of Education Quarterly*
HT = *History Today* (Great Britain)
JM = James Madison
JA = John Adams
JJ = John Jay
JP = John Page
JH = *Journalism History*
JAH = *Journal of American History*
JSH = *Journal of Southern History*
JCC = *Journals of the Continental Congress*
JHD = *Journal of the House of Delegates* (of Virginia)
JHBV = *Journal of the Virginia House of Burgesses*
LDC = *Letters of Delegates to Congress*
LC = Library of Congress
L&B = Lipscomb & Bergh, *Writings of TJ*
LCB = TJ, *Literary Commonplace Book*
MT = Madame de Tessé
MACH = *Magazine of Albemarle County History*
MC = Maria Cosway
ML = Marquis de Lafayette
MJR = Martha Jefferson Randolph (TJ's daughter)
MWJ = Martha Wayles Jefferson (Mrs. TJ)
MJE = Mary Jefferson Eppes (TJ's daughter)
MHS = Massachusetts Historical Society
MHSB = *Massachusetts Historical Society Bulletin*
MVHR = *Mississippi Valley Historical Review*
NHB = *Negro History Bulletin*
NEQ = *New England Quarterly*
NYPL = New York Public Library
NYT = *New York Times*
PAH = *Papers of Alexander Hamilton*
PBF = Labaree et al., *Papers of Benjamin Franklin*
PTJ = Boyd, *Papers of Thomas Jefferson*
PJ = Patsy Jefferson
PMHB = *Pennsylvania Magazine of History and Biography*
PAPS = *Proceedings of the American Philosophical Society*
QJLC = *Quarterly Journal of the Library of Congress*
RAH = *Reviews in American History*
SR = Sarah Randolph, *Domestic Life of TJ*
SAQ = *South Atlantic Quarterly*
TJ = Thomas Jefferson
TJP = Thomas Jefferson Papers, Library of Congress
TJR = Thomas Jefferson Randolph
TMR = Thomas Mann Randolph, Jr.

VC	=	*Virginia Cavalcade*
VHS	=	Virginia Historical Society
VMHB	=	*Virginia Magazine of History and Biography*
VQR	=	*Virginia Quarterly Review*
VSL	=	Virginia State Library
VSP	=	*Virginia State Papers*
WMQ	=	*William and Mary Quarterly*
WS	=	William Short
WLCL	=	William L. Clements Library
WLB	=	*Wilson Library Quarterly Bulletin*
WJM	=	*Writings of James Madison*
WW	=	Fitzpatrick, *Writings of George Washington*

Notes

Preceding the numbered notes in each chapter are background subjects, each with sources that were found especially valuable and pertinent. Full particulars of each work together with other readings carefully considered are listed in the bibliography on page 641.

Introduction

1. Peterson, 456.
2. Malone, *TJ and His Times, I: Jefferson the Virginian,* 174.
3. D.L. Wilson, *LCB,* 6.
4. Chinard, ed., 3.
5. Kimball, *Jefferson: The Road to Glory,* 84–89.
6. Brodie, *TJ: An Intimate History.*
7. Barbara Chase-Ribaud, *Sally Hemings,* New York: Viking, 1979.

1. *"I Cannot Live Without Books"*

The Jefferson Family: D. Malone, *TJ and His Times,* 1:3–33; M. Peterson, *TJ and the New Nation,* 3–9; H.S. Randall, *Life of TJ,* 1:1–17; F.M. Brodie, *TJ: An Intimate History,* 19–35; C.G. Bowers, *Young Jefferson,* 1–13; S. Randolph, *Domestic Life of TJ,* 17–26; S.H. Hochman, "Thomas Jefferson: A Personal Financial Biography," 1–34; TJ, *Autobiography;* TJ, *Notes on the State of Virginia;* M. Kimball, *Jefferson: The Road to Glory,* 3–38.

TJ's Boyhood and Early Schooling: J.R. Anderson, "Tuckahoe and the Tuckahoe Randolphs," *VMHB,* 45 (1927): 55–86; H. Cripe, *TJ and Music,* 12–13; D. Malone, 1:37–42; S. Randolph, 22–26; Bowers, 10–13; J. McLaughlin, *Jefferson and Monticello,* 39–41; *Jefferson's Literary Commonplace Book,* 3–13.

1. *ATJ,* 3.
2. TJ to Thomas Adams, Feb. 20, 1771, *PTJ,* 1:62.
3. *ATJ,* 3.
4. Henrico Order Book, 1694–1701, 169.
5. Henrico County Misc. Court Records, 1650–1807, 849; will dated Mar. 13, 1725.
6. SR, 21.
7. Quoted in Malone, 1:13.
8. SR, 21.
9. Peter Collinson to John Bartram, February 17, 1737, in W. Darlington, *Memorials of John Bartram and Humphrey Marshall,* 89.
10. Bartram to Colonel Custis, November 19, 1738, ibid., 312.
11. Bartram to Collinson, December, 1738, ibid., 122.
12. SR, 22.
13. Ibid., 21–22.
14. *ATJ,* 3–4.
15. Randall, 1:13.
16. SR, 19.
17. *ATJ,* 4.
18. Goochland County Will & Deed Book, No. 5:73; will dated March 2, 1742.
19. J.W. Wayland, ed., *The Fairfax Line: Thomas Lewis's Journal of 1746,* 8.
20. TJ, *Notes,* 18.
21. Ibid., 3.
22. Ibid., xviiin.
23. Shelburne Papers, mss.; WLCL, L, 93–5.
24. TJ, *Notes,* 137.
25. SR, 20.
26. Quoted in Bowers, *Young Jefferson,* 42.
27. *ATJ,* 3–4.
28. J. McLaughlin, *Jefferson and Monticello,* 39.
29. Quoted in M. Kimball, *Jefferson: The Road to Glory,* 30.

30. *ATJ*, 4.
31. TJ, *Notes*, 210.
32. TJ to JA, June 11, 1812, *Adams-Jefferson Letters*, 307.
33. TJ to J. Priestley, Jan. 27, 1800, L&B, 10:146–7.
34. TJ to TJR, November 24, 1808, *Family Letters*, 362–3.

2. "I Am Surrounded With Enemies"

TJ's School Days: R.B. Davis, *Intellectual Life in the Colonial South*, 1:302–5, 327–30. Anne Maury, ed. *Memoirs of a Huguenot Family*, 379–424; L.B. Wright, *The First Gentlemen of Virginia*, 95–116; H.D. Bullock, ed., "A Dissertation on Education," 36–60; D. Malone, *Jefferson*, 1:37–48; M. Kimball, *Jefferson: The Road to Glory*, 26–38; D.L. Wilson, ed., *Jefferson's Literary Commonplace Book;* Wilson, "TJ's Early Notebooks," *WMQ*, 3rd ser. 42 (1985): 433–52; S.H. Hochman, "TJ: A Personal Financial Biography," 35–43; Mayer Reinhold, "Classical World," in "TJ: A Reference Biography," 135–56.

1. J. Boucher, *Reminiscences of an American Loyalist*, 61.
2. Ibid.
3. TJ to Joseph Priestley, January 27, 1800, L&B, 10:146–47.
4. J. Maury to Robert Jackson, July 17, 1762, in *PAPS*, 2:47.
5. Quoted in Malone, 1:45.
6. Rev. James Maury to his son, James, Feb. 17, 1762, mss., Maury Deposit, UVA.
7. Quoted in Kimball, 30.
8. TJ to James Maury, April 25, 1812, L&B, 13:149.
9. *LCB*, ed. D.L. Wilson, 13.
10. Ibid., No. 171, 82.
11. Ibid., No. 407, 153.
12. Ibid., No. 60, 56.
13. Ibid., No. 61, 56.
14. Ibid., No. 66, 57.
15. Ibid., No. 71, 59.
16. TJ to JP, July 15, 1763, *PTJ*, 1:10–11.
17. Wilson in *LCB*, 19.
18. Ibid., No. 303, 117.
19. Ibid., No. 314, 120.
20. Ibid., No. 77, 61.
21. Ibid., No. 79, 61.
22. TJ to John Harvie, Jan. 14, 1760, *PTJ*, 1:3.
23. Ibid.
24. Ibid.

3. *"A More Universal Acquaintance"*

At William and Mary College: L.H. Johnson III, "Sharper Than a Serpent's Tooth"; R.B. Davis, *Intellectual Life,* 1:330–49; L.G. Tyler, "Early Courses and Professors at William and Mary College," *Jefferson,* 1:49–61; D. Malone, "Jefferson Goes to School in Williamsburg"; M. Kimball, *Jefferson: The Road to Glory,* 39–72; C.G. Bowers, *Young Jefferson,* 16–28; M. Peterson, *TJ and the New Nation,* 11–15; R.P. Thomson, "The Reform of the College of William and Mary;" F. Brodie, *TJ: An Intimate History,* 55–59; "Williamsburg—The Old Colonial Capital."

1. TJ to William Wirt, August 5, 1815, Ford, *Writings,* 9:473.
2. J.F.D. Smythe, *A Tour in the United States,* 15–16.
3. Ibid.
4. Ibid., 17–18.
5. TJ, *Notes on Virginia,* 152–53.
6. J.D. Schöpf, *Travels in the Confederation,* 2:78.
7. *Virginia Gazette,* December 19, 1776.
8. Smythe, 21–22.
9. TJ to John Minor, Aug. 30, 1814, Ford, 11:420–21.
10. *ATJ,* 4–5.
11. TJ to Benjamin Rush, Aug. 17, 1811, Ford 9:328.
12. TJ to TJR, Nov. 24, 1808, Ford 9:231.
13. *ATJ,* 4.
14. TJ to William Duane, October 1, 1812, Ford, L&B, 2:420–21.
15. Malone, *Jefferson,* 1:55.
16. TJ to Dr. Vine Utley, Mar. 21, 1819, Ford, 9:126.
17. F.W. Gilmer, c. 1816, quoted in *Francis Walker Gilmer,* 350.
18. TJ to MWR, March 28, 1787, *PTJ,* 11:250.
19. JP, "Autobiography" in *Virginia Historical Register,* July, 1850, 151.
20. Quoted in G.P. Coleman, ed., *The Flat Hat Club and the Phi Beta Kappa Society.*
21. Quoted in Lyon Tyler, *Letters and Times of the Tylers,* 1:54–55.
22. Robert Seager II, *and Tyler, Too,* 50–51.
23. *Va. Hist. Rev.,* July 1850, p. 150.
24. TJ to TJR, Nov. 24, 1808, L&B, 12:197–98.
25. *LCB,* 62, No. 81.
26. TJ to JP, Jan. 20, 1763, *PTJ,* 1:7.
27. *ATJ,* 5.
28. TJ to JP, Jan. 20, 1763, *PTJ,* 1:8.
29. *ATJ,* 5.
30. TJ to William Duval, June 14, 1806, mss., LC, No. 27898.
31. TJ to Thomas Turpin, Feb. 5, 1769, *PTJ,* 1:24.
32. TJ, "Notes for the Biography of George Wythe," L&B, 1:169–70.
33. Quoted in Lyon Tyler, *George Wythe,* 74.
34. *ATJ,* 5.

35. Hugh B. Grigsby, *The Virginia Convention of 1776,* 120.
36. Quoted in C. Bowers, *Young Jefferson,* 29.
37. Lyon Tyler, *George Wythe,* 74.

4. "I Was Bold in the Pursuit of Knowledge"

TJ, as Law Student: F.L. Dewey, *TJ, Lawyer,* 9–17; E. Dumbauld, *TJ and the Law,* 3–17; M. Peterson, *TJ and the New Nation,* 16–20; J. Blackburn, *George Wythe of Williamsburg,* 33–39; I.E. Brown, *American Aristides,* 43–108.

Stamp Act Crisis: E.S. and H.M. Morgan, *The Stamp Act Crisis;* R. Middlekauff, *The Glorious Cause,* 76–93; R.R. Beeman, *Patrick Henry,* 31–44; J. Axelrad, *Patrick Henry,* 28–45; I.E. Brown, *American Aristides,* 55–61; W.J. Van Schreeven, et al., eds., *Revolutionary Virginia,* vol. 1; W.S. Randall, *Benedict Arnold,* 48–60; *A Little Revenge,* 196–200; M. Peterson, *TJ and the New Nation,* 16–20; C. Cullen, "New Light on John Marshall's Legal Education."

1. Quoted in M. Kimball, *Jefferson: The Road to Glory,* 73.
2. L&B, 1:169–70.
3. TJ to John Minor, Aug. 30, 1814, L&B, 2:420–21.
4. TJ to JP, Dec. 25, 1762, 5.
5. *ATJ,* 5.
6. TJ to L.H. Girardin, Jan. 15, 1815, L&B, 4:231.
7. *ATJ,* 5.
8. TJ to Bernard Moore, Aug. 30, 1814, L&B, 2:420–21.
9. TJ to Thomas Cooper, Feb. 10, 1814, L&B, 14:85.
10. TJ to Ralph Izard, July 17, 1788, *PTJ,* 13:372.
11. TJ to TJ Randolph, June 14, 1806, L&B, 12:197–8.
12. Ibid.
13. TJ to JP, Jan. 20, 1763, *PTJ,* 1:5.
14. TJ to Thomas Cooper, Feb. 10, 1814, L&B, 14:85.
15. TJ to Horatio G. Spafford, March 17, 1814, L&B, 14:120.
16. TJ to Bernard Moore in TJ to John Minor, Aug. 30, 1814, L&B, 2:420–21.
17. TJ to T. Cooper, Feb. 10, 1814, L&B, 14:85.
18. TJ to William Wirt, Aug. 5, 1815, Ford, *Writings,* 9:476.
19. TJ to JP, Jan. 23, 1764, *PTJ,* 1:15.
20. Ibid.
21. TJ to JP, Dec. 25, 1762, *PTJ,* 4.
22. Ibid.
23. Ibid.
24. Ibid., 4–5.
25. Ibid., 5.
26. Ibid., 6.
27. TJ to JP, July 15, 1763, *PTJ,* 1:9.

28. TJ to JP, Jan. 20, 1763, *PTJ,* 1:7.
29. TJ to JP, July 15, 1763, *PTJ,* 1:11.
30. TJ to JP, Jan. 20, 1763, *PTJ,* 1:7.
31. TJ to JP, July 16, 1763, *PTJ,* 1:9.
32. Ibid., 10.
33. Ibid.
34. TJ to JP, Oct. 7, 1763, *PTJ,* 1:11.
35. Ibid., 12.
36. TJ to William Fleming, [October], 1763, *PTJ,* 1:12.
37. TJ to JP, Jan. 19, 1764, *PTJ,* 1:13–14.
38. Ibid., 14.
39. Ibid.
40. TJ to JP, Jan. 23, 1764, *PTJ,* 1:15.
41. TJ to Wm. Fleming, Mar. 20, 1764, *PTJ,* 1:16.
42. Ibid.
43. TJ to JP, April 9, 1764, *PTJ,* 1:17.
44. Betsy Ambler to Mildred Smith, Richmond, 1781, "Old Virginia Correspondence," *AM,* 84 (1899), 538.
45. *Fee Book,* mss., HEH.
46. F.L. Dewey, *TJ, Lawyer,* 11.
47. C. Cullen, "New Light on Marshall's Legal Education," 348.
48. TJ to Thomas Turpin, Feb. 5, 1769, *PTJ,* 1:24.
49. Quoted in M. Kimball, *Road to Glory,* 116.
50. A. Burnaby, *Travels,* 55.
51. J.P. Kennedy and H.R. McIlwaine, eds. *JHBV,* 1761–65, 302–305.
52. James Maury to John Camm, Dec. 12, 1763, in Ann Maury, *Memoirs of a Huguenot Family,* 418.
53. Ibid., 418–24.
54. Ibid.
55. Ibid.
56. Maury to Camm, Maury Letterbook, mss., *PAPS.*
57. TJ to William Wirt, Aug. 4, 1805, *PMHB,* 34 (1910), 394.
58. Ibid.
59. Quoted in G. Morgan, *Patrick Henry,* 100.
60. *JHBV,* lxiv–lxv.
61. Quoted in R. Beeman, *Patrick Henry,* 37.
62. Edward Carrington to William Wirt, in Henry, *Patrick Henry,* 1:86–87.
63. TJ to Wirt, Aug. 14, 1814, Ford, 9:470.
64. Ibid., 470.
65. Gov. Fauquier to Board of Trade, June 6, 1765, in *JHBV,* 1761–1765, lxvii.
66. TJ to Wirt, Aug. 14, 1814, Ford, 9:468.
67. Fauquier to Board of Trade, *supra.*
68. Ibid.
69. TJ to Wirt, *supra.*

70. Fauquier to Board of Trade, *supra.*
71. Fauquier to Board of Trade, Nov. 3, 1765, in *JHBV,* 1761–1765, lxviii–lxxi.
72. S.H. Hochman, "Personal Financial Biography," 34.
73. TJ, "List of Services," [Sept.] 1800, LC, 219:39161.
74. TJ to GW, March 1, 1779, *PTJ,* 2:235.
75. Randall, I:41.
76. SR, 38–39.
77. TJ, *Garden Book,* 1771, 25.
78. Ibid., 26.
79. Ibid., 1.
80. TJ to TJR, L&B, 12:197–98.
81. George Gilmer to John Morgan, May 11, 1766, *PTJ,* 1:18.
82. TJ to JP, May 25, 1766, *PTJ,* 1:19–20.
83. M. Kimball, *Jefferson: The Road to Glory,* 139.
84. TJ, *Garden Book,* 1766, 1.
85. TJ to TJR.
86. Quoted in H.T. Dickinson, *Bolingbroke,* London, 1970, 298.
87. TJ to JP, May 25, 1766, *PTJ,* 1:19.

5. *"An Untiring Spirit of Investigation"*

Thomas Jefferson, Lawyer: F. Dewey, "TJ's Law Practice," *VMHB,* 85 (1977), 289–303; "New Light on the General Court of Colonial Virginia," *William and Mary Law Review,* 21 (1979), 1–14; *TJ, Lawyer,* 18–113; A.M. Smith, "Virginia Lawyers, 1680—1776: The Birth of an American Profession," unpub. doct. diss., Johns Hopkins U., 1967; John W. Davis, "TJ, Attorney at Law," *Proceedings Virginia State Bar Assoc.,* 38 (1926): 361–77; *TJ and the Law,* 3–143; TJ, *Reports of Cases Determined in the General Court of Virginia;* F.H. Hart, *The Valley of Virginia in the American Revolution;* J.A. Waddell, *Annals of Augusta County, Va., 1726—1871;* D. Malone, 1:113–27; M. Kimball, *Jefferson: The Road to Glory,* 1:89–99; M. Peterson, *TJ and the New Nation,* 20–22; D.J. Mays, *Edmund Pendleton,* vol. 1; H.F. Rankin, "The General Court of Colonial Virginia," *VMHB,* 70 (1962): 142–53; C. Eaton, "A Mirror of the Southern Colonial Lawyer: The Fee Books of Patrick Henry, TJ and Waightsill Avery," *WMQ,* 3rd ser., 8 (1951), 420–34; A.G. Roeber, *Faithful Magistrates and Republican Lawyers,* 73–159.

1. TJ to JP, Feb. 21, 1770, *PTJ,* 1:36.
2. Quoted in H.S. Randall, 1:68.
3. Diary of Isaac Jefferson in *WMQ,* 3rd ser., 10 (1951), 568.
4. Quoted in SR, 46.
5. Quoted in G. Wood, *Radicalism of the American Revolution,* 51.
6. Ibid.
7. Quoted in Dewey, "TJ's Law Practice," 289.

8. TJ to Will Fleming, [Oct. 1763], *PTJ,* 1:13.
9. J.W. Davis, 369.
10. Edmund Randolph, "Essay on the Revolutionary History of Virginia," *VMHB,* 43 (1953), 115.
11. Ibid.
12. Isaac Jefferson, "Memoirs of a Monticello Slave," *WMQ,* 3rd. ser., 10 (1951), 572.
13. Ibid., 572n.
14. Dewey, *TJ, Lawyer,* 26.
15. TJ, *Notes,* 136.
16. TJ to JP, Feb. 21, 1770, *PTJ,* 1:35–36.
17. I. Jefferson, "Memoirs," 572n.
18. E. Randolph, "Essay," 115.
19. Quoted in M. Kimball, *Jefferson: The Road to Glory,* 134.
20. I. Jefferson, "Memoirs," 572.
21. M. Kimball, *Jefferson: The Road to Glory,* 134–35.
22. TJ, *Expense Book,* mss., HEH.
23. *Case Book,* mss., HEH.
24. S. Hochman, "Personal Financial Biography of TJ," 56.
25. E. Randolph, "Essay," 123.
26. *Case Book,* mss., HEH.
27. Enclosure 1 to Governor Fauquier's letter to the Earl of Shelburne, May 20, 1767, in George Reese, ed., *Official Papers of Francis Fauquier,* 1449.
28. F. Dewey, *TJ, Lawyer,* 37.
29. TJ, *Expense Book,* mss., MHS.
30. Ibid.
31. TJ to John Walker, Sept. 3, 1769, *PTJ,* 1:32.
32. F. Dewey, *TJ, Lawyer,* 75.
33. Ibid., 79.
34. Ibid.
35. *Case Book,* mss., HEH.
36. TJ, *Account Book,* mss., MHS.
37. Thomas Lewis to Col. Wm. Preston, Sept. 18, 1769, mss., Preston Davie Collection, Preston Family Papers, VHS.
38. *JHBV, 1770–72,* 35.
39. Ibid., 57.
40. Dewey, *TJ, Lawyer,* 41.
41. Thomas Lewis to Wm. Preston, Aug. 28, 1770, mss., Preston Davie Collection, Preston Family Papers, VHS.
42. TJ, *Account Book,* 1770, mss., MHS.
43. E. Pendleton to Wm. Preston, Mays, *Letters of Pendleton,* 1:64.
44. *Case Book,* HEH, Case 122.
45. Dewey, *TJ, Lawyer,* 89.
46. Quoted in J.W. Davis, 368.

47. TJ to George Wythe, Mar. 1, 1779, *PTJ*, 2:235.
48. Ibid.
49. E. Randolph, "Essay," 123.
50. Quoted in J.W. Davis, 376.
51. TJ, *Account Book*, mss., NYPL.
52. TJ, *Notes*, 24–25.
53. Ibid.
54. Ibid., 19–20.
55. TJ, *LCB*, 102.
56. *Virginia Gazette*, May 20, 1773.
57. TJ, *Fee Book*, HEH.
58. TJ, memo books, 1767–70, mss., LC.
59. TJ to James Fishback, Sept. 27, 1809, L&B, 12:315.
60. TJ to JP, Jan. 23, 1764, *PTJ*, 1:15.

6. "All Men Are Born Free"

TJ and Slavery Cases: TJ, *Reports*, 90–96; TJ, *Notes on the State of Virginia*, 264–68; M. Kimball, *Jefferson: The Road to Glory*, 207–8; D. Malone, 1:121–22; F.M. Brodie, *Thomas Jefferson*, 103–5; E. Dumbauld, *TJ and the Law*, 85–87, 90, 139–43; A. Mapp, *TJ: A Strange Case of Mistaken Identity;* 167–73.

TJ and Divorce: F. Dewey, "TJ and a Williamsburg Scandal," *VMHB*, 89 (1981), 44–63; Dewey, "TJ's Notes on Divorce." *WMQ*, 3rd ser., 39 (1982), 212–23.

TJ's Law Practice: J.M. Hemphill, ed., "Edmund Randolph Assumes TJ's Practice," *VMHB*, 67 (1965), 170–71.

Growing Colonial Crisis: Van Schreeven, *Revolutionary Virginia*, vol. I; J.P. Greene, *The Quest for Power*, 357–98; M. Kimball, *Jefferson: The Road to Glory*, 187–208; D. Malone, 1:128–42.

1. TJ to Thomas Turpin, Feb. 5, 1769, *PTJ*, 1:23–4.
2. TJ, *Garden Book*, Aug. 3, 1767, p. 6.
3. TJ to JP, Feb. 21, 1770, *PTJ*, 1:36.
4. TJ to Turpin, Feb. 5, 1769, *PTJ*, 1:24.
5. Ibid.
6. D. Malone, 1:130.
7. Letter of Theodorick Bland, Sr., June 27, 1765, mss., Bland Papers, 1:27.
8. C.S. Sydnor, *Gentlemen Freeholders*, 53–54.
9. TJ, Resolutions for an Answer to Governor Botetourt's Speech, [May 8, 1769], *PTJ*, 1:26–27.
10. TJ to Wm. Wirt, Aug. 5, 1815, *PTJ*, 1:27n.
11. Quoted in B. Bailyn, *Ideological Origins of the American Republic*, 32, 58n.
12. D. Malone, 1:134.
13. Ibid., 1:135.

14. *JHBV, 1766–69,* April 15, 1769, 174.
15. Benjamin Franklin to Thomas Cushing, June 10, 1771, *PBF,* 18:121–22.
16. *JHBV, 1766–69,* April 15, 1769, 174.
17. BF to Cushing, June 10, 1771, *PBF,* 18:121–22.
18. *JHBV, 1766–69,* April 15, 1769, 214.
19. Ibid., 218.
20. Ibid., May 18, 1769, xxxix–xlii.
21. Quoted in D.S. Freeman, *George Washington,* 3:214.
22. GW to Mason, Apr. 5, 1769, quoted in ibid., 214.
23. Van Schreeven, 1:72.
24. Ibid.
25. TJ to John Taylor, May 28, 1816, L&B, 11:327.
26. Quoted in F. Dewey, *TJ, Lawyer,* 48.
27. James Parker to Charles Steuart, Oct. 20, 1769, MS5025, f. 215, National Library of Scotland, Edinburgh, Scotland. Microfilm copy at Colonial Williamsburg Foundation.
28. Quoted in SR, 43.
29. TJ to JP, Feb. 21, 1770, *PTJ,* 1:34–35.
30. TJ, *Reports,* 94.
31. James Ogilvie to TJ, Mar. 28, 1770, *PTJ,* 1:38.
32. TJ to Ogilvie, Feb. 20, 1771, *PTJ,* 1:62–63.
33. Quoted in Dumbauld, 83.
34. TJ, *Notes,* 157.
35. Ibid., 158.
36. Ibid., 159.
37. E. Dumbauld, 79.
38. TJ to Danbury Baptist Association, Jan. 1, 1802, quoted in Dumbauld, *Bill of Rights,* 106.
39. TJ, *Garden Book,* May 28, 1767, 5.
40. Ibid., 6.
41. TJ, *LCB,* no. 175, 83.
42. Horace, Epode II, ed. C.E. Bennett, 365–69.
43. Wm. Byrd to Lord Egmont, quoted in Beverly B. Munford, *Virginia's Attitude Toward Slavery and Secession,* 16–17.
44. *ATJ,* 5.
45. TJ to Edward Coles, August 25, 1814, L&B, 9:477.
46. *ATJ,* 5.
47. *Virginia Gazette,* Sept. 7, 1769.
48. Isaac Jefferson, "Memoir," 574.
49. TJ, *Account Book,* mss., NYPL.
50. Ibid.
51. Wood, *Radicalism of the American Revolution,* 50–51.
52. TJ, *Notes,* 137.
53. TJ, *Case Book,* HEH, Case no. 345.

54. Baskett, *Acts of Assembly,* 309.
55. Dumbauld, 84.
56. TJ, *Reports,* 95–108.
57. Ibid., 96.

7. "The Pursuit of Happiness"

TJ Marries: D. Malone, 1:153–65; M. Kimball, *Jefferson: The Road to Glory,* 1:166–86; F. Brodie, *TJ: An Intimate Portrait,* 86–100; *PTJ,* vol. 1.

Building Monticello: J. McLaughlin, *Jefferson and Monticello;* TJ, Account Books and *Garden Book.*

TJ's Law Practice: F. Dewey, "TJ's Notes on Divorce," 213; *TJ, Lawyer,* 57–72.

1. F. Dewey, "TJ's Notes on Divorce," 218.
2. TJ, *Garden Book,* July 27, 1769, p. 15.
3. TJ, *Account Book,* 1769, mss., MHS.
4. TJ, *Garden Book,* July 27, 1769, p. 15.
5. Mary Walker Lewis to TJ, April 14, 1770, *PTJ,* 1:40.
6. J. McLaughlin, *Jefferson and Monticello,* 58.
7. TJ to James Ogilvie, Feb. 20, 1771, *PTJ,* 1:63.
8. TJ to Robert Skipwith, Aug. 3, 1773, *PTJ,* 1:76.
9. TJ to JP, February 21, 1770, *PTJ,* 1:34.
10. Ibid., Dec. 25, 1762, *PTJ,* 1:5.
11. SR, 43.
12. Randall, 1:63.
13. TJ to R. Skipwith, Aug. 3, 1773, *PTJ,* 1:78.
14. TJ to J. Ogilvie, Feb. 20, 1771, *PTJ,* 1:63.
15. Quoted in M. Peterson, *TJ and the New Nation,* 39.
16. TJ to Thomas Adams, Feb. 20, 1771, *PTJ,* 1:62.
17. Quoted in T. Fleming, *The Man from Monticello,* 9.
18. Ibid.
19. Mrs. Drummond to TJ, March 12, [1771], *PTJ,* 1:66.
20. R. Skipwith to TJ, Sept. 20, 1771, *PTJ,* 1:84.
21. TJ to Thomas Adams, June 1, 1771, *PTJ,* 1:71–72.
22. *PTJ,* 1:86–87.
23. J. Wayles to TJ, October 20, 1772, *PTJ,* 1:95–96.
24. Quoted in Dewey, *TJ, Lawyer,* 59.
25. James Parker to Charles Steuart, May 3, 1771, mss., Steuart Papers, VSL, MS5015, f. 265.
26. TJ, *Reports,* quoted in Dewey, *TJ, Lawyer,* 59.
27. Parker to Steuart, May 25, 1772, mss., Steuart Papers, VSL, MS5027, f. 159.
28. William Lee Letter Book, 1769–71, mss., VSL.
29. Dewey, 62.

30. TJ, *Account Book,* 1772, LC.
31. F. Dewey, "TJ's Notes on Divorce," 213.
32. Ibid., 213n.
33. TJ, *Reports,* 1:94–95.
34. Quoted in Dewey, *TJ, Lawyer,* 65.
35. TJ to Trumbull, Feb. 15, [1789], *PTJ,* 14:561.
36. TJ to R. Skipwith, Aug. 3, 1773, *PTJ,* 1:79.
37. Douglas Wilson, in TJ, *LCB,* 156.
38. F. Dewey, "TJ's Notes," 219.
39. Ibid., 216.
40. Ibid., 216–17.
41. Ibid., 217–19.
42. TJ, quoted in Dewey, *TJ, Lawyer,* 69.
43. Quoted in ibid., 63.
44. Mrs. Eustace to Thomas Burke, Nov. 4, 1773, in Dewey, *TJ, Lawyer,* 64.
45. TJ to Wm. Wirt, quoted in Dewey, *TJ, Lawyer,* 63.

8. "God Gave Us Liberty"

Growing Crisis in Va.: W.J. Van Schreeven, ed., *Revolutionary Virginia,* 1:88–104; J.P. Kennedy, ed., *JHBV,* 1770–72, 1773–75; P. Maier, *From Resistance to Revolution.*

TJ and House of Burgesses: D. Malone, 1:169–80; M. Peterson, *TJ and the New Nation,* 69–73; M. Kimball, *Jefferson: The Road to Glory,* 229–37; A. Mapp, *TJ: A Strange Case of Mistaken Identity,* 73–80; *PTJ,* 1:103–16.

1. *ATJ,* 7.
2. Ibid.
3. J.P. Kennedy, ed. *JHBV,* 1770–72, xxxi.
4. TJ to Thomas Adams, Feb. 20, 1771, *PTJ,* 1:61–62.
5. TJ to T. Adams, June 1, 1771, *PTJ,* 1:71–72.
6. Ibid.
7. *ATJ,* 6.
8. Richard Bland to T. Adams, Aug. 1, 1771, Van Schreeven, 1:85.
9. *WMQ,* 1st ser., 5 (1896), 155–56.
10. *JHBV,* 1770–72, 283–84.
11. H.T. Catterall, 1:4–5.
12. Quoted in P. Maier, *From Resistance to Revolution,* 222–25.
13. Van Schreeven, 1:91.
14. L&B, 1:10–11.
15. Van Schreeven, 1:91.
16. *JHBV,* 1773–76, x.
17. TJ, *Garden Book,* 47.
18. W.S. Simpson, "Dabney Carr," 12.

19. Randall, 1:84.
20. TJ, *Garden Book*, 55.
21. Horace, *Odes*, 1634.
22. *LCB*, 82.
23. Ibid.
24. *Virginia Gazette*, October 1772.
25. J. Wayles to TJ, Oct. 20, 1772, *PTJ*, 1:95.
26. Alexander McCaul to TJ, July 8, 1772, *PTJ*, 1:93.
27. Quoted in M. Peterson, 40.
28. *ATJ*, 5.
29. Madison Hemings quoted in "Life Among the Lowly, No. 1," Pike County, O., *Republican*, Mar. 13, 1873.
30. TJ, *Account Book*, 1774, quoted in D. Malone, 1:165.
31. P. Mazzei, *Memoirs*, 192.
32. TJ, *Garden Book*, 51–52.
33. Ibid., 52.
34. Quoted in F. Dewey, *TJ, Lawyer*, 99.
35. Quoted in M. Boatner, *Encyclopedia of the American Revolution*, 99.
36. Quoted in M. Peterson, 70.
37. L&B, 1:11.
38. Van Schreeven, 1:95.
39. Ibid., 94–95.
40. *ATJ*, 9.
41. *JHBV*, 1773–76, 132.
42. *ATJ*, 8–9.
43. James Parker to Charles Steuart, *Magazine of History*, 3 (1906), 153.
44. GW to Geo. Wm. Fairfax, June 10, 1774, *WW*, 2:224.
45. Van Schreeven, 1:97.
46. Quoted in Mapp, 78.
47. *Va. State Papers*, 8:53, mss., VSL.
48. *ATJ*, 9–10.
49. Van Schreeven, 1:103.
50. Ibid., 104.
51. TJ, Petition of George Mason, [June 1774] *PTJ*, 1:112–15.
52. TJ, *Notes on the State of Virginia*, 62.
53. Ibid., 62.

9. "Let Those Flatter Who Fear"

Civil Strife in Virginia: J. Boucher, *"Reminiscences of An American Loyalist"*; Van Schreeven, 1:101–57; J. Ferling, *The First of Men: A Life of George Washington*, 86–110; D. Malone, 1:180–96; B. Bailyn, *Ideological Origins of the American Revolution;* M.

Peterson, *TJ and the New Nation,* 69–77; P. Maier, *From Resistance to Revolution,* 238–81; *PTJ,* 1:116–149.

TJ as Writer of Revolution: Garry Wills, *Inventing America;* M. Peterson, *TJ and the New Nation,* 33–79; D.W. Carrithers, "Montesquieu, Jefferson, and the Fundamentals of Eighteenth-Century Republican Theory"; M. Kimball, *Jefferson: The Road to Glory,* 209–28;

1. TJ, Agreement with John Randolph, [April] 11, 1771, *PTJ,* 1:66.
2. Van Schreeven, 1:204.
3. [J. Randolph,] "A Plea for Moderation," in Van Schreeven, 1:213.
4. J. Boucher, in *View of the Causes and Consequences of the American Revolution,* London, 1797.
5. Ibid.
6. J. Ferling, 100.
7. F. Dewey, *TJ, Lawyer,* 110.
8. TJ, L&B, 12:355.
9. TJ and John Walker, *PTJ,* 1:116.
10. R.H. Lee to S. Adams, June 23, 1774, in Ballagh, *Letters,* 1:112.
11. Van Schreeven, 1:259.
12. *PTJ,* 1:117–18.
13. Ibid., 119–20.
14. Quoted in M. Peterson, *TJ and New Nation,* 47.
15. Ibid., 45–46.
16. Ibid., 50.
17. TJ, *Reports,* 1:96.
18. G. Chinard, ed., *Commonplace Book of TJ,* [1774].
19. TJ to JA, April 11, 1823, in L.J. Cappon, ed., *The Adams-Jefferson Letters,* 591.
20. Quoted in Mapp, 63.
21. Ibid.
22. Ibid.
23. Henry Home, Lord Kames, *Essays on the Principles of Moral and Natural Religion,* Edinburgh, 1751, 78.
24. Adam Smith, quoted in Gary Wills, *Inventing America,* 238.
25. Quoted in Peterson, 48.
26. Ibid.
27. John Locke, *Two Treatises on Government,* 2:141.
28. Abraham Stanyan, *Grecian History,* quoted in Kimball, 245.
29. D.W. Carrithers, 161.
30. Quoted in Carrithers, 161.
31. TJ to R. Skipwith, Aug. 3, 1771, *PTJ,* 1:79.
32. Montesquieu, *Spirit of the Law,* 250.
33. Ibid.
34. Ibid.
35. Ibid., 11.

36. TJ, "Instructions to the Virginia Delegates," in *PTJ,* 1:121–35.
37. *ATJ,* 10.
38. Ibid., 11.
39. E. Randolph, *History of Virginia,* 205.
40. TJ, undated memorandum, 1809, in *PTJ,* 1:669–71.
41. Ibid., 1:671.
42. *PTJ,* 1:143.
43. Ibid.
44. Quoted in Malone, 1:204.
45. Galloway to Wm. Franklin, Mar. 26, 1775, *New Jersey Archives,* 1st ser. 10:579.
46. BF to Jonathan Shipley, Mar. 10, 1774, *PBF,* 21:138.

10. "I Speak the Sentiments of America"

Second Continental Congress: R. Coupland, *Quebec Act; LDC,* vols. 1–4; *PTJ,* 1:174–523; M. Peterson, *TJ and the New Nation,* 79–107; D. Malone, 1:197–231.

Revolution in Virginia: Van Schreeven, vols. 2 and 3; C.S. Sydnor, *Gentlemen Freeholders,* 112–19; J.E. Selby, *Revolution in Virginia,* 14–99; Axelrad, *Patrick Henry,* 101–62; Jensen, *Founding of a Nation,* 643–45; R.D. Beeman, "The American Revolution," in M. Peterson, ed. *TJ, A Reference Biography,* 25–46.

1. TJ, *Garden Book,* 66.
2. P.V. Fithian, *Journal and Letters, 1767–1774,* 279.
3. *PTJ,* 1:116.
4. D. Ramsey, quoted in Edmund Morgan, *A Sense of Power,* 4.
5. Van Schreeven, 2:363.
6. Ibid., 2:242–43.
7. Ibid., 2:298–300.
8. Ibid., 2:368.
9. Ibid.
10. Quoted in K. Cmiel, 50–52.
11. Axelrad, *Patrick Henry,* 106.
12. Ibid., 107.
13. R.H. Lee, *Memoirs,* 1:101.
14. Van Schreeven, 2:368–69.
15. Ibid., 369.
16. Axelrad, 50–52.
17. Van Schreeven, 2:369.
18. Ibid., 375.
19. *PTJ,* 1:159.
20. Van Schreeven, 2:383–84; 387.
21. Ibid., 2:377–78.
22. J. Ferling, *First of Men,* 333.

23. Quoted in F. Brodie, 87.
24. TJ, *Account Book,* May 7, 1775, mss., HEH.
25. TJ to Wm. Small, May 7, 1775, *PTJ,* 1:165–67.
26. Van Schreeven, 3:103–104.
27. R.H. Lee to F.L. Lee, May 21, 1775, in *LDC,* 3:366–67.
28. *Virginia Gazette,* June 3, 1775.
29. Van Schreeven, 3:12.
30. Ibid., 13.
31. Ibid.
32. Ibid.
33. Ibid., 15.
34. Ibid., 14–15.
35. Ibid., 17.
36. Ibid., 17–18.
37. Ibid., 18.
38. Ibid.
39. Ibid., 19.
40. TJ to D'Ivernois, Feb. 6, 1795, L&B, 7:2.
41. BF to David Hartley, May 8, 1775, *PBF,* 22:34.
42. Samuel Ward to Henry Ward, June 22, 1775, *LDC,* 1:535.
43. JA to Timothy Pickering, Aug. 6, 1822, in JA, *Works,* 2:512.
44. Quoted by JA, *Works,* 2:422.
45. JA to AA, June 23, 1775, *LDC,* 1:537.
46. For detailed description of battle, see Randall, *A Little Revenge,* 341–46.
47. JA to James Warren, *LDC,* 1:545.
48. TJ to Francis Eppes, July 4, 1775, *PTJ,* 1:184–85.
49. William Livingston to Wm. Alexander, July 4, 1775, quoted in *PTJ,* 1:189.
50. John Dickinson in *PTJ,* 1:205.
51. *ATJ,* 13–14.
52. *PTJ,* 1:199–203.
53. JA to TJ, Nov. 12, 1813, *Works,* 10:78–80.
54. JA to AA, June 17, 1775, *LDC,* 1:497.
55. JA to Timothy Pickering, Aug. 6, 1822; C.F. Adams, *Works,* 2:512–14, n. 1.
56. TJ to John Randolph, Aug. 25, 1775, *PTJ,* 1:240–42.
57. TJ, *Garden Book,* Sept. 21, 1775, 66.
58. Ibid., 68.
59. TJ to John Randolph, Nov. 29, 1775, *PTJ,* 1:268–69.
60. TJ, *Notes,* 13.
61. *PTJ,* 1:276–77.
62. TJ to F. Eppes, Oct. 10, 1775, *PTJ,* 1:247.
63. TJ to JP, Oct. 31, 1775, *PTJ,* 1:251.
64. TJ to Eppes, Nov. 7, 1775, *PTJ,* 1:252.
65. Quoted in Van Schreeven, 5:8.
66. Ibid.

67. Ibid., 5:8–9.
68. Ibid., 5:9.
69. Ibid.
70. Ibid., 5:124–25.
71. Ibid., 5:12.
72. Ibid., 5:139–40.
73. Ibid., 5:12.
74. Ibid.
75. JP to TJ, Nov. 11, 1775, *PTJ*, 1:258.
76. Van Schreeven, 5:13.
77. Ibid., 5:15.
78. Ibid.
79. Ibid., 5:16–17.
80. T. Nelson to TJ, Feb. 4, 1776, *PTJ*, 1:286.
81. TJ to T. Cooper, Oct. 27, 1808, L&B, 12:180.
82. *PTJ*, 1:277–84.
83. TJ to J. Randolph, June 1776, *PTJ*, 1:408.
84. SR, 21–22.
85. *PTJ*, 1:259n.
86. TJ to J. Randolph, Nov. 29, 1775, *PTJ*, 1:268–70.
87. Thomas Paine, *Common Sense*, Phila., 1776.
88. TJ to Nelson, May 16, 1776, *PTJ*, 1:292.
89. JP to TJ, Apr. 6, 1776, *PTJ*, 1:287.
90. J. McClurg to TJ, Apr. 6, 1776, *PTJ*, 1:287.
91. Van Schreeven, 5:3.
92. Van Schreeven, 5:47.
93. Ibid., 7:1.
94. Ibid., 7:3.
95. Ibid., 7:4.
96. Ibid.
97. Ibid., 7:143–144n.
98. Ibid., 7:143.
99. TJ to T. Nelson, May 16, 1776, *PTJ*, 1:292.
100. TJ to Ellen W. Coolidge, Nov. 14, 1825, L&B, 18:349.
101. TJ to T. Nelson, May 16, 1776, *PTJ*, 1:292.
102. TJ to JP, May 17, 1776, *PTJ*, 1:293–94.
103. BJ to Josiah Quincy, Apr. 15, 1776, *LDC*, 3:529.
104. Quoted in M. Peterson, 86.

11. "An Expression of the American Mind"

Declaration of Independence: R. Hamowy, "Jefferson and the Scottish Enlightenment"; P.C. Hoffer, "The Declaration of Independence as a Bill in Equity," R.E.

Luker, "Garry Wills and the New Debate Over the Declaration of Independence"; D.W. Carrithers, "Montesquieu, Jefferson, and the Fundamentals of Eighteenth-Century Republican Theory"; R. Ginsburg, "Suppose that Jefferson's Rough Draft of the Declaration of Independence is a Work of Political Philosophy?"; W.S. Howell, "The Declaration of Independence"; E. Dumbauld, "Independence Under International Law"; E. Gittelman, "Jefferson's 'Slave Narrative' "; S.N. Katz, "TJ and the Right to Property"; J.P. Boyd, *The Declaration of Independence: The Evolution of the Text;* C. Becker, *The Declaration of Independence;* D. Malone, *Jefferson the Virginian,* 215–31; M.D. Peterson, *TJ and the New Nation;* 85–97; *ATJ,* 14–29; *PTJ,* 1:298–328; 413–33. A. Koch, *Philosophy of Jefferson.*

The Virginia Constitution: *PTJ,* 1:328–86; B. Tarter, ed., *Revolutionary Virginia,* vol. 7, part 1, 25–42; 270–79; Selby, *The Revolution in Virginia,* 55–79.

1. *PTJ,* 1:290–91.
2. *PTJ,* 1:309.
3. Ibid., 309–10.
4. Ibid., 310–13.
5. Quoted in Carl Becker, *Declaration of Independence,* 135–36.
6. Ibid., 136.
7. TJ to T. Nelson, May 16, 1776, *PTJ,* 1:292.
8. *PTJ,* 1:366–68.
9. C. Sydnor, *Gentlemen Freeholders,* 123.
10. *ATJ,* 38.
11. *PTJ,* 1:356–64.
12. Ibid., 364n.
13. Ibid., 1:333.
14. W. Fleming to TJ, June 22, 1776, *PTJ,* 1:406.
15. E. Randolph to TJ, June 23, 1776, *PTJ,* 1:407.
16. Quoted in M. Peterson, 86.
17. JA, *Autobiography,* in Adams' Papers, 3:383.
18. JA to AA, May 17, 1776, *LDC,* 4:17–18.
19. JA to Mercy Otis Warren, Apr. 6, 1776, quoted in J. Ferling, *JA,* 153.
20. JA to P. Henry, June 3, 1776, *LDC,* 4:122.
21. Diary of Robt. Treat Paine, quoted in P.H. Smith, 298.
22. TJ to JA, Oct. 28, 1813, in Cappon, ed., *The Adams-Jefferson Letters,* 387–92.
23. JA in *Diary* and *Autobiography,* 3:335–37.
24. *PTJ,* 1:413–27.
25. Quoted in M. Peterson, 90.
26. Ibid.
27. Quoted in W.S. Howell, "Declaration of Independence," 225–26.
28. Ibid.
29. *PTJ,* 1:298.
30. Becker, *Declaration of Independence,* 136.
31. Ibid., 171.

32. Ibid., 172.
33. Ibid., 147.
34. Ibid., 171–72.
35. Ibid., 172.
36. *PTJ,* 1:317–18.
37. C. Van Doren, *Benjamin Franklin,* 551.
38. TJ to R.H. Lee, July 8, 1776, *PTJ,* 1:456.
39. TJ to JP, July 30, 1776, *PTJ,* 1:483.
40. *ATJ,* 21.
41. G. Wythe to TJ, July 27, 1776, *PTJ,* 1:476–77.
42. TJ to W. Fleming, July 1, 1776, *PTJ,* 1:411–12.
43. W. Fleming to TJ, Sept. 27, 1776, *PTJ,* 1:474.
44. E. Pendleton to TJ, July 22, 1776, *PTJ,* 1:472.
45. TJ to E. Pendleton, [ca. June 30, 1776], *PTJ,* 1:408.
46. TJ to F. Eppes, July 23, 1776, *PTJ,* 1:473.
47. TJ to J. Hancock, Oct. 11, 1776, *PTJ,* 1:524.
48. TJ to F. Eppes, July 23, 1776, *PTJ,* 1:471–72.
49. TJ to R.H. Lee, July 29, 1776, *PTJ,* 1:477.
50. E. Pendleton to TJ, July 22, 1776, *PTJ,* 1:472.

12. "With a Single Eye to Reason"

Legislative Revolution: *PTJ,* 1:525–68; E. Dumbauld, *TJ and the Law,* 132–56; C.R. Keim, "Primogeniture and Entail in Colonial Virginia"; M. Peterson, *TJ and the New Nation,* 107–65; D. Malone, *TJ and His Times,* 1:235–97; R. Isaac, *Transformation of Virginia;* N.E. Cunningham, *In Pursuit of Reason,* 54–63;

Conflict over Religious Freedom: M. Peterson, 133–44; R. Isaac, 133–44; J.H. Smylie, "Jefferson's Statutes for Religious Freedom: The Hanover Presbyterian Memorials, 1776–1786"; T.E. Buckley, *Church and State in Revolutionary Virginia, 1776–1787,* 17–62; M.D. Peterson and R.C. Vaughan, *The Va. Statute for Religious Freedom.*

Jefferson and Landholding: *PTJ,* 1:564–76; C.W. Alvord, *The Mississippi Valley in British Politics;* Abernethy, *Western Lands,* 123–35, 162–68; Rudolf Freund, "John Adams and Thomas Jefferson on the Nature of Landholding in America"; *Land Economics,* 24 (1948), 107–19; W.S. Lester, *The Transylvania Colony;* A.M. Lewis, "Jefferson and Virginia's Pioneers," *MVHR,* 34 (1948), 551–88; F.J. Turner, "Western State-Making in the Revolutionary Era," *AHR,* 1 (1895), 70–87, 251–69; D.M. Friedenberg, *Life, Liberty and the Pursuit of Land,* 143–212; A.J. Wall, "The Story of the Convention Army, 1777–1783."

1. G. Wythe to TJ, Oct. 28, 1776, *PTJ,* 1:585.
2. J. Hancock to TJ, Sept. 30, 1776, *PTJ,* 1:522n.
3. TJ to Hancock, Oct. 11, 1776, *PTJ,* 1:522n.
4. *PTJ,* 1:605.

5. *PTJ,* 2:331.
6. *ATJ,* 44.
7. Ibid., 38.
8. Ibid., 38–39; TJ to JA, Oct. 28, 1813, Cappon, 389.
9. Quoted in R. Isaacs, *Transformation,* 279.
10. *ATJ,* 41.
11. TJ, *Notes,* 159.
12. *ATJ,* 41.
13. TJ to JA, Oct. 28, 1813, Cappon, 389.
14. TJ, *Notes,* 158.
15. Isaacs, 281.
16. *ATJ,* 41.
17. *PTJ,* 2:546.
18. Lunenberg County Petition, Nov. 3, 1779, *Va. Religious Petitions,* quoted in Isaacs, 282.
19. Ibid., 283.
20. Ibid.
21. Quoted in Peterson, 144.
22. Ibid.
23. Isaacs, 283–84.
24. JM to TJ, Jan. 22, 1786, *PTJ,* 9:196.
25. TJ to JM, Dec. 16, 1786, *PTJ,* 10:603–4.
26. *ATJ,* 1, 44–45.
27. TJ to John Tyler, June 17, 1812, L&B, 13:167.
28. *ATJ,* 44.
29. Ibid., 45.
30. E. Pendleton to TJ, Aug. 10, 1776, *PTJ,* 1:490.
31. *PTJ,* 2:492–93.
32. *ATJ,* 46.
33. *PTJ,* 2:497.
34. TJ to JM, Dec. 15, 1786, *PTJ,* 10:604.
35. L&B, 2:211n.
36. TJ to G. Wythe, Nov. 1, 1778, *PTJ,* 2:230.
37. *ATJ,* 46.
38. TJ, *Notes on Virginia,* 162–63.
39. Quoted in Peterson, 153.
40. TJ, *Notes,* 138.
41. *ATJ,* 51.
42. TJ, *Notes,* 42.
43. L&B, 5:377–79.
44. *ATJ,* 51.
45. Quoted in Peterson, 154.
46. TJ to G. Wythe, Aug. 13, 1786, *PTJ,* 10:243–45.
47. D. Malone, 1:280.

48. TJ to Wythe, Aug. 13, 1786, *PTJ,* 10:243–45.
49. *PTJ,* 2:526–33.
50. Ibid., 2:538–39.
51. *ATJ,* 39.
52. *PTJ,* 2:526–27.
53. JM to TJ, Jan. 22, 1786, *PTJ,* 9:137.
54. TJ to Rittenhouse, July 19, 1778, *PTJ,* 2:202.
55. E. Pendleton to TJ, May 11, 1779, *PTJ,* 2:266.
56. TJ to P. Henry, Mar. 27, 1779, *PTJ,* 2:237.
57. Maj. Gen. Wm. Phillips to TJ, Apr. 11, 1779, *PTJ,* 2:252.
58. Quoted in D. Malone, 1:298.
59. TJ to Gen. Phillips, June 25, 1779, *PTJ,* 3:15.
60. TJ to JP, June 3, 1779, *PTJ,* 2:279.
61. JP to TJ, June 2, 1779, *PTJ,* 2:278.

13. "It Is Not in My Power to Do Anything"

Governor Jefferson: M. Peterson, *TJ and the New Nation,* 166–240; D. Malone, 1:301–29; N.C. Cunningham, *In Pursuit of Reason,* 64–75; *PTJ,* Vols. 2–4; Billings et al, *Virginia;* R. Stuart, *Halfway Pacifist,* 7–17; W. Goering, "Lovers of Peace and Order"; J.G. Hamilton, *The Pacifism of TJ;* J.G. Simcoe, *Military Journal,* 88–129; M.A. Lewis, "Jefferson and Virginia's Pioneers, 1774–1781"; D.E. Reynolds, "Ammunition Supply in Revolutionary Va."; TJ in *PTJ,* 4:254–423.

Invasion of Virginia: J. Selby, *Revolution in Virginia.;* W.S. Randall, *Benedict Arnold: Patriot and Traitor,* 581–84; Malone, 1:330–51; F. Mackenzie, Diary; F.R. Lassiter, "Arnold's Invasion of Va."; *PTJ,* Vols. 5–6, J.G. Simcoe, *Military Journal;* V. Dabney, *Virginia,* 139–54; Royster, *Revolutionary People at War,* 255–94; I.N. Arnold, *Life of Benedict Arnold,* 295–354; M.J. Wright, "Lafayette's Campaign in Va."; W.M. Wallace, *Traitorous Hero,* 260–83.

1. TJ to R.H. Lee, July 17, 1779, *PTJ,* 3:39.
2. TJ, quoted in Peterson, 170.
3. TJ to R.H. Lee, June 17, 1779, *PTJ,* 2:298.
4. Ibid.
5. Virginia Constitution, *PTJ,* 1:380.
6. Wm. Davies to TJ, Feb. 1, 1781, *PTJ,* 4:493–94.
7. TJ to ML, Mar. 10, 1781, *PTJ,* 5:113.
8. Peterson, 560.
9. Ibid.
10. Peterson, 124.
11. TJ to B. Harrison, Sept., 1779, quoted in Malone, 1:315.
12. Peterson, 172.
13. Ibid.

14. Henry Hamilton, July 6, 1781, in *George Rogers Clark Papers, 1771–1781,* 174–207.
15. Peterson, 179.
16. *PTJ,* 2:292–94.
17. Peterson, 179.
18. Ibid., 177.
19. Wm. Phillips to TJ, July 5, 1779, *PTJ,* 3:25–28.
20. TJ to Phillips, July 22, 1779, *PTJ,* 3:44–49.
21. GW to TJ, Aug. 6, 1779, *PTJ,* 3:61.
22. TJ to GW, quoted in Peterson, 182.
23. Ibid., 183.
24. Madison to TJ, Mar. 27, 1780, *PTJ,* 3:335.
25. TJ to Nathaniel Greene, Apr. 1, 1781, *PTJ,* 4:312.
26. TJ to H. Gates, Sept. 3, 1780, *PTJ,* 3:588.
27. TJ to La Luzerne, Aug. 31, 1780, *PTJ,* 3:588.
28. TJ to D'Anmours, Nov. 30, 1780, *Notes,* xiiin.
29. JP to TJ, Dec. 9, 1780, *PTJ,* 4:192.
30. GW to TJ, Dec. 9, 1780, *PTJ,* 4:195.
31. Quoted in Peterson, 206.
32. Ibid., 207.
33. Benedict Arnold to Henry Clinton, Jan. 23, 1781, mss., Clinton Papers, WLCL.
34. Ibid.
35. TJ to GW, Jan. 10, 1781, *PTJ,* 4:333–35.
36. Randall, *Benedict Arnold,* 582.
37. TJ to David Jameson, Apr. 16, 1781, *PTJ,* 5:468.
38. MWJ to Eleanor Madison, Aug. 8, 1780, *PTJ,* 3:532.
39. TJ to JM, May 20, 1782, *PTJ,* 6:186.
40. TJ to Lafayette, May 14, 1781, *PTJ,* 5:644.
41. TJ, Diary of Arnold's Invasion, *PTJ,* 4:260–61; L&B, 1:79.
42. TJ to GW, May 28, 1781, *PTJ,* 6:33.
43. GW to TJ, June 8, 1781, *PTJ,* 6:83.

14. "I Tremble for My Country"

Retires from Governorship: Randall, *Life of TJ,* 1:330–410; D. Malone, 1:373–423; N. Cunningham, *In Pursuit of Reason,* 76–100; *PTJ,* vols. 4–8; M. Peterson, *TJ and the New Nation,* 241–96; *ATJ,* 53–63; F. Brodie, *TJ: An Intimate History,* 215–32.

1. Randall, 1:338.
2. TJ to Wm. Gordon, July 16, 1788, *PTJ,* 13:363–64.
3. TJ, Diary, in *PTJ,* 4:265.
4. TJ to ML, Aug. 4, 1781, *PTJ,* 6:112.

5. Lafayette to Luzerne, Aug. 4, 1781, *AHR,* 20:665.
6. BF to TJ, July 15, 1782, *PTJ,* 6:194.
7. TJ to E. Randolph, Sept. 16, 1781, *PTJ,* 6:118.
8. *ATJ,* 63.
9. TJ, *Notes,* 200–2; 163.
10. TJ to Monroe, Oct. 5, 1781, *PTJ,* 6:127.
11. TJ to GW, Oct. 28, 1781, *PTJ,* 6:129.
12. TJ to Monroe, May 20, 1782, *PTJ,* 6:185.
13. TJ to G. Nicholas, July 28, 1781, *PTJ,* 6:105.
14. E. Randolph to TJ, Sept. 16, 1781, *PTJ,* 6:128.
15. *PTJ,* 6:136n.
16. TJ to Monroe, May 20, 1782, *PTJ,* 6:185.
17. Betsy Ambler to Mildred Smith, 1781, in "An Old Virginia Correspondence," *AM,* 84 (Oct. 1899), 538.
18. *Journal of the House of Delegates,* Dec. 17, 1781.
19. TJ to Isaac Zane, Dec. 24, 1781, *PTJ,* 6:143.
20. TJ to JM, Dec. 8, 1784, *PTJ,* 7:558.
21. Chastellux, *Travels,* 2:41–46.
22. Tyler to TJ, May 16, 1782, *PTJ,* 6:184.
23. Monroe to TJ, May 11, 1782, *PTJ,* 6:183.
24. TJ to Monroe, May 20, 1782, *PTJ,* 6:186.
25. Randall, 1:382.
26. *PTJ,* 6:196.
27. SR, 63.
28. TJ, *Garden Book,* June 25, 1782.
29. TJ to Elizabeth Wayles Eppes, [October 3, 1782], *PTJ,* 6:198.
30. TJ to Chastellux, Nov. 26, 1782, *PTJ,* 6:203.
31. Randall, 1:383.
32. TJ to Chastellux, Nov. 26, 1782, *PTJ,* 6:203.
33. TJ, *Garden Book,* Sept. 23, 1782.
34. TJ to Chastellux, Nov. 26, 1782.
35. JM, Madison Papers, 5:268–69.
36. TJ to R.R. Livingston, Nov. 26, 1782, *PTJ,* 6:206.
37. TJ to Chastellux, Nov. 26, 1782, *PTJ,* 6:203.
38. TJ to JM, Jan. 31, 1783, *PTJ,* 6:225.
39. TJ to Livingston, Feb. 7, 1783, *PTJ,* 6:229.
40. TJ to JM, Feb. 14, 1783, *PTJ,* 6:241.
41. TJ to E. Randolph, Feb. 15, 1783, *PTJ,* 6:249.
42. JM to TJ, Feb. 11, 1783, *PTJ,* 6:235.
43. TJ to E. Randolph, Feb. 15, 1783, *PTJ,* 6:247.
44. *PTJ,* 6:221.
45. TJ to Randolph, Feb. 15, 1783, *PTJ,* 6:247–48.
46. TJ to Abner Nash, Mar. 11, 1783, *PTJ,* 6:255.
47. TJ, *Notes,* 210.

48. TJ to MWJ, Nov. 28, 1783, J. Bear, ed., *Family Letters of TJ*, 19.
49. TJ to Peter Carr, Dec. 11, 1783, *PTJ*, 6:379–80.
50. TJ to Barbé-Marbois, Dec. 5, 1783, *PTJ*, 6:374.
51. TJ to MWJ, Dec. 22, 1783, *PTJ*, 6:416–17.
52. TJ to JM, Dec. 11, 1783, *PTJ*, 6:381–82.
53. TJ, *Notes*, 161.
54. TJ to JM, Dec. 11, 1783, *PTJ*, 6:381.
55. *WW*, 27:285–86n.
56. *WW*, 27:284–85.
57. TJ to B. Harrison, Dec. 24, 1783, *PTJ*, 6:419.
58. Ibid.
59. TJ to JM, Jan. 1, 1784, *PTJ*, 6:438.
60. TJ to Harrison, Dec. 24, 1783, *PTJ*, 6:419.
61. *ATJ*, 60–61.
62. Ibid., 60.
63. *PTJ*, 7:82n.
64. TJ to MWJ, Jan. 15, 1784, *PTJ*, 7:466; to Short, Mar. 1, 1784, *PTJ*, 6:570.
65. G.K. van Hogendorp to TJ, [Apr. 6, 1:784], *PTJ*, 7:81.
66. TJ to van Hogendorp, May 4, 1784, *PTJ*, 7:208–9.
67. TJ to Harrison, Jan. 16, 1784, *PTJ*, 6:468–69.
68. TJ, *Notes on Coinage*, [Mar.—May 1784], *PTJ*, 7:176.
69. *PTJ*, 7:608.
70. TJ to JM, Apr. 25, 1784, *PTJ*, 7:118.
71. TJ to Jean Nicolas Démeunier, "Observations on Démeunier's Manuscript," June, 1786, *PTJ*, 10:58.
72. Ibid., June 26, 1786, *PTJ*, 10:63.
73. TJ to E. Pendleton, May 25, 1784, *PTJ*, 7:292.
74. Hector St. John de Crèvecoeur to TJ, July 15, 1784, *PTJ*, 7:74.
75. Diary of Ezra Stiles, [June 8, 1784], *PTJ*, 7:303.
76. TJ to Chastellux, Sept. 2, 1785, *PTJ*, 8:468.
77. TJ to E. Gerry, July 2, 1784, *PTJ*, 7:358.
78. MWJ to Eliza Trist, after Aug. 24, 1785, *PTJ*, 8:437.
79. Ibid.

15. "I Do Love This People"

TJ in Paris: M. Kimball, *Jefferson: The Scene of Europe*, 3–158; D. Malone, 2:3–49; M. Peterson, *TJ and the New Nation*, 297–346; *PTJ*, vols. 7–10; A. Mapp, *TJ: A Strange Case of Mistaken Identity*, 207–33; H. Rice, *TJ's Paris;* J. Ferling, *John Adams*, 186–275; F. Brodie, *TJ: An Intimate Biography*, 233–51; N. Cunningham, *In Pursuit of Reason*, 90–100; *ATJ*, 63–72; C.L. Lopez, *Mon Cher Papa;* E. Dumbauld, *TJ: American Tourist;* C. Fohlen, *Thomas Jefferson*, 43–65; E. Wright, *Franklin of Philadelphia*, 256–338. R.R. Palmer, "The Dubious Democrat: TJ in Bourbon, France."

1. TJ to Monroe, Mar. 18, 1785, *PTJ,* 8:43.
2. Thomas Lee Shippen quoted in *PTJ,* 13:164n.
3. Peterson, 389.
4. JA to TJ, *Works,* 3:298.
5. *Letters,* C.F. Adams, ed., 1:193.
6. Ibid.
7. MWJ to Eliza Trist, after Aug. 24, 1784, *PTJ,* 8:437.
8. TJ to JM, quoted in Rice, 77.
9. Ibid., 78.
10. Ibid.
11. Peterson, 300.
12. MWJ to Eliza Trist, after Aug. 24, 1784, *PTJ,* 8:436–39.
13. JA to J. Warren, Aug. 27, 1784, *PTJ,* 7:382n.
14. A. Lee to JA, quoted in Brodie, 239.
15. JA to A. Lee, Jan. 31, 1785, quoted in Page Smith, *John Adams,* 2:616.
16. TJ to JM, Feb. 14, 1783, *PTJ,* 6:241.
17. TJ to ML, Apr. 2, 1790, *PTJ,* 16:293.
18. TJ to Giovanni Fabbroni, June 8, 1778, *PTJ,* 2:195.
19. M. Peterson, *TJ and the New Nation,* 332.
20. BF to Margaret Hewson, Jan. 12, 1777, Smyth, *Writings,* 7:10.
21. BF to Jane Mecom, Dec. 8, 1776, C. Van Doren, *Letters of BF and Jane Mecom,* 168.
22. BF to Sarah Franklin Bache, June 3, 1779, Smyth, *Writings,* 7:347.
23. Quoted in C.A. Lopez, *The Private Franklin,* 233.
24. C.F. Adams, ed., *Works of JA,* 1:660.
25. JA, Diary, 3:189.
26. BF to Richard Bache, June 2, 1779, Smyth, *Writings,* 7:345.
27. BF to Richard and Sarah Bache, May 14, 1781.
28. BF to Sarah Bache, June 27, 1780.
29. "Dialogue between Franklin and the Gout," Smyth, *Writings,* 8:154.
30. Trumbull, *Autobiography,* Aug. 13, 1786, 116.
31. JA to TJ, Jan. 22, 1825, Cappon, ed., *Adams-Jefferson Letters,* 606.
32. Francis L. Humphreys, *David Humphreys,* 1:i.
33. JA to John Jay, Apr. 13, 1785, in Malone, 2:21.
34. Peterson, 304.
35. *PTJ,* 7:478.
36. TJ to Monroe, Dec. 10, 1784, *PTJ,* 7:562–65.
37. TJ to Monroe, June 17, 1785, *ATJ,* 8:231.
38. *PTJ,* 7:486.
39. *ATJ,* 65.
40. Peterson, 306.
41. BF to Richard Price, Feb. 6, 1780, Smyth, *Writings,* 8:9.
42. Malone, 2:9.
43. Rice, *TJ's Paris,* 39–41.

44. TJ to Monroe, Mar. 18, 1785, *PTJ,* 8:43.
45. Hester Lynch Piazzi, *Observations and Reflections Made in the Course of a Journey through France,* London, 1789, 1:22.
46. Quoted in Rice, 29.
47. TJ to AA, Sept. 26, 1785, *PTJ,* 8:548.
48. Dr. James Currie to TJ, Nov. 20, 1784, *PTJ,* 7:538–39.
49. TJ to F. Eppes, Jan. 13, 1785, *PTJ,* 7:601.
50. TJ to Elizabeth Eppes, Dec. 14, 1786, *PTJ,* 10:594.
51. TJ to Elizabeth Blair Thompson, Jan. 19, 1787, *PTJ,* 11:59.
52. TJ to Eliza Trist, Aug. 18, 1785, *PTJ,* 8:404.
53. TJ to Charles Bellini, Sept. 30, 1785, *PTJ,* 8:568–69.
54. TJ to Jay, Oct. 8, 1787, *PTJ,* 12:214–16.
55. TJ to GW, May 2, 1788, *PTJ,* 13:128.
56. *PTJ,* 13:269.
57. TJ to JM, Jan. 30, 1787, *PTJ,* 11:95–96.
58. TJ to Eliza Trist, Aug. 18, 1785, *PTJ,* 8:404.
59. TJ to Banister, Oct. 15, 1785, *PTJ,* 8:636.
60. TJ to Bellini, Sept. 30, 1785, *PTJ,* 8:568–69.
61. BF to Cabanis, Smyth, *Writings,* 7:375.
62. TJ to Bellini, Sept. 30, 1785, *PTJ,* 8:569.
63. C.A. Lopez, *Mon Cher Papa,* 135.
64. *Letters of Mrs. Adams,* 90.
65. JA, *Works,* 3:135.
66. TJ to AA, July 7, 1785, *PTJ,* 8:265.
67. M. Kimball, 187.
68. Medlin, Dorothy, "TJ, André Morellet, and the French Version of Notes on the State of Virginia," 85–99.
69. AA quoted in Lopez, 257–58.
70. Crèvecoèur to TJ, July 15, 1784, *PTJ,* 7:303.
71. Mme. Brillon to BF, Smyth, *Writings,* 10:426.
72. *Letters of Mrs. Adams,* 241–42.
73. La Duchesse d'Abrantes, *Histoire des Salons de Paris,* 1:361–406.
74. TJ to Monroe, June 17, 1785, *PTJ,* 8:229.
75. TJ to Chastellux, June 7, 1785, *PTJ,* 8:184.
76. JM to TJ, May 11, 1785 and November 15, 1785, *PTJ,* 8:147–48.
77. Thomson to TJ, Nov. 2, 1785, *PTJ,* 9:38.
78. TJ to JM, Sept. 1, 1785, *PTJ,* 8:462.
79. TJ to Dumas, Feb. 2, 1786, *PTJ,* 9:244.
80. Medlin, 85.
81. *ATJ,* 64.
82. JA to TJ, May 22, 1785, *PTJ,* 8:245.
83. Chastellux, *Travels in North America,* 2:606.
84. Quoted in N. Cunningham, *In Pursuit of Reason,* 96.

16. "A Situation Much More Pleasing"

TJ in Europe: E. Dumbauld, *TJ, American Tourist,* M. Kimball, *Jefferson: The Scene of Europe,* 3–198; D. Malone, 2:3–49; N. Cunningham, *In Pursuit of Reason,* 90–100; F. Brodie, *TJ: An Intimate History,* 215–76.

TJ in England: C.R. Ritcheson, "The Fragile Memory: TJ at the Court of George III"; R. Watson, "TJ's Visit to England, 1786"; M. Batey and D. Lambert, *English Garden Tour;* J. Ferling, *John Adams,* 277–91; H.C. Rice, Jr., "Jefferson's Gift of Fossils to the Museum of Natural History in Paris"; B. Fink, "Jefferson's Palatable Pleasures" in *Voyage et Tourisme en Bourgogne,* ed. Baridon and Chevignard, 115–40; J. Brooke, *George III;* Malone, 2:50–63; Kimball, *Jefferson: The Scene of Europe,* 127–158.

1. Burnett, *Letters of Members of the Continental Congress,* 8:25.
2. Ibid.
3. TJ to JJ, May 11, 1785, *PTJ,* 8:146.
4. TJ to JJ, June 17, 1785, *PTJ,* 8:226.
5. TJ, *Account Book,* mss., HEH.
6. Humphreys to GW, in Humphreys, 1:317.
7. *Mémoires du Comte de Cheverny,* 1:82–84.
8. TJ to JM, Jan. 30, 1787, *PTJ,* 11:95–96.
9. Cheverny, *Mémoires,* 83–84.
10. TJ to JJ, June 17, 1785, *PTJ,* 8:226.
11. TJ to JM, June 20, 1787, *PTJ,* 11:480–84.
12. *ATJ,* 91.
13. Ibid., 104–5.
14. *Letters of Mrs. Adams,* 2:93.
15. TJ to JM, Jan. 30, 1787, *PTJ,* 9:94–95.
16. Mme. Helvetius to BF, quoted in Lopez, *Mon Cher Papa,* 299–300.
17. TJ to Samuel Smith, Aug. 28, 1798, L&B, 10:55.
18. TJ to Wm. Smith, Feb. 19, 1791, *PTJ,* 8:157n–158n.
19. TJ to Wm. Short, May 3, 1785, *PTJ,* 8:134.
20. TJ to Eliz. Eppes, Sept. 22, 1785, *PTJ,* 8:540.
21. TJ to Dr. James Currie, Sept. 27, 1785, *PTJ,* 8:558.
22. TJ to AA, June 21, 1785, *PTJ,* 8:239.
23. Ibid., Sept. 4, 1785, *PTJ,* 8:473.
24. TJ to JM, May 25, 1788, *PTJ,* 13:202.
25. TJ to Rev. James Madison, Oct. 2, 1785, *PTJ,* 8:574.
26. TJ to Nicholas Lewis, Sept. 17, 1785, *PTJ,* 12:135.
27. M. Kimball, *The Scene of Europe,* 261.
28. John Ledyard to TJ, July 29, 1787, *PTJ,* 11:638.
29. *ATJ,* 70.
30. Rumsey to Benjamin West, Mar. 20, 1789, mss., LC.
31. Rumsey to Charles Morrow, Mar. 27, 1789, mss., LC.

32. P. Mazzei, *Memoirs,* 293.
33. TJ to Nathanael Greene, Jan. 12, 1785, Malone, 2:27.
34. Ibid., 28.
35. JJ to TJ, quoted in Peterson, 311.
36. Ibid., 312.
37. Ibid., 313.
38. Ibid., 314.
39. Ibid.
40. Bailyn, "Boyd's Jefferson: Notes for a Sketch," *NEQ,* 33(1960), 380–401.
41. *ATJ,* 67.
42. TJ to Francis Hopkinson, Jan. 13, 1785, *PTJ,* 7:602–3.
43. TJ, *Notes on the State of Virginia,* 64–65.
44. Ibid., 55.
45. Ibid., 54.
46. Ibid., 200–2.
47. L&B, 10:332.
48. JA to TJ, Feb. 21, 1786, *PTJ,* 9:295.
49. TJ to JJ, Mar. 12, 1786, *PTJ,* 9:325–26.
50. TJ, Account Book, Mar. 11, 1786, mss., HEH.
51. TJ to AA, quoted in R. Watson, "TJ's Visit to England," 4.
52. Peterson, 309–10.
53. *ATJ,* 66.
54. TJ and JA to JJ, Apr. 25, 1786, *PTJ,* 9:406–7.
55. *ATJ,* 66.
56. C.R. Ritcheson, "Fragile Memory," 7–8.
57. TJ to W.S. Smith, Sept. 28, 1787, *PTJ,* 12:193.
58. TJ to Alexander McCaul, Apr. 19, 1786, *PTJ,* 9:388.
59. TJ to Wm. Gordon, July 16, 1788, *PTJ,* 13:363–64.
60. TJ to McCaul, Apr. 19, 1786, *PTJ,* 9:389.
61. TJ and JA to JJ, Apr. 25, 1786, *PTJ,* 9:406.
62. Ibid., 9:375n.
63. TJ to JP, May 4, 1786, *PTJ,* 9:444–46.
64. TJ to JJ, Apr. 23, 1786, *PTJ,* 9:402.
65. TJ to W.T. Franklin, May 7, 1786, *PTJ,* 9:466–67.
66. TJ to Wm. Carmichael, L&B, 6:380.
67. Quoted in R. Watson, 5.
68. L&B, 6:145.
69. TJ to C. Thomson, Apr. 22, 1786, *PTJ,* 9:400.
70. TJ to Thomas Cooper, Sept. 10, 1814, L&B, 14:180–82.
71. R. Watson, 9.
72. Quoted in ibid., 6.
73. Ibid., 10.
74. *Letters of Mrs. Adams,* 287.
75. J. Murray, quoted in M. Batey and D. Lambert, 13.

76. Anon., *London and Its Environs,* 1761, 2:115.
77. TJ, *Garden Book,* 111.
78. Ibid.
79. Ibid., 111–12.
80. Ibid., 112.
81. Ibid.
82. Ibid.
83. Ibid.
84. Ibid.
85. Ibid., 112–13.
86. Ibid.
87. JA, Diary, 3:394–97.
88. PTJ, 9:374n.
89. Quoted in Kimball, 150.
90. TJ, *Garden Book,* 113.
91. Ibid., 114.
92. Ibid.
93. JA quoted in *PTJ,* 9:374n.
94. JA, Diary, 3:397.
95. TJ to JP, May 4, 1786, *PTJ,* 9:445.
96. Eliz. Cometti, ed., "Mr. Jefferson Prepares an Itinerary," 100.
97. TJ to JA, Nov. 15, 1816, Cappon, 495–99.

17. "My Head and My Heart"

TJ and Maria Cosway: *PTJ,* 10; Rice, *TJ's Paris;* M. Kimball, *Jefferson: The Scene of Europe,* 150–83; F. Brodie, *TJ: An Intimate History,* 252–76; D. Malone, 2:67–81; T. Fleming, *Man from Monticello,* 132–42; N. Cunningham, *Pursuit of Reason,* 103–6; M. Peterson, *TJ and the New Nation,* 347–49; L.H. Butterfield and H.C. Rice, Jr., "Jefferson's Earliest Note to Maria Cosway"; *PTJ,* Vol. 10; C.B. Van Pelt, "TJ and MC"; Robert Davidoff, "Man of Letters," *TJ: A Reference Biography;* Olivier Choppin de Janvry, "TJ au desért de Retz" in *Voyage et Tourisme en Bourgogne,* ed. Baridon and Chevignard, 144–55.

1. Chastellux, *Travels,* 1:46–51.
2. Hart and Biddle, *Houdon,* 208–9.
3. L&B, 6:774–75.
4. Quoted in Kimball, *Jefferson: The Scene of Europe,* 70.
5. Ibid., 2:71.
6. Wm. Short to TJ, July 28, 1784, *PTJ,* 7:385–86.
7. TJ to James Buchanan and Wm. Hay, Aug. 13, 1785, *PTJ,* 8:366.
8. TJ to Dr. Currie, Jan. 28, 1786, *PTJ,* 9:240.
9. Mazzei, *Memoirs,* 296.

10. L&B, 8:18–19.
11. *Letters of Mrs. Adams,* 250–51.
12. Lyman Butterfield, ed., *Adams Papers,* 4:41n.
13. Smyth, x:467.
14. TJ to Mme. d'Enville, Apr. 2, 1790, *PTJ,* 16:290–91.
15. *Letters of Mrs. Adams,* 215.
16. M. Kimball, *Jefferson: The Scene of Europe,* 65–66.
17. Ibid., 66.
18. TJ to MC, Oct. 12, 1786, *PTJ,* 10:444.
19. Malone, 2:69.
20. Rice, *TJ's Paris,* 19.
21. Wm. Hazlitt, *Conversations with James Northcote,* 99.
22. G.C. Williamson, *Richard Cosway,* 52.
23. Quoted in F. Brodie, *TJ: An Intimate History,* 254.
24. Quoted in ibid.
25. Hazlitt, 99.
26. *London Morning Chronicle,* quoted in Walpole, *Anecdotes of Painting in England,* 5:127–28.
27. TJ to AA, Aug. 9, 1786, *PTJ,* 10:202.
28. TJ to JJ, Jan. 11, 1789, *PTJ,* 14:430.
29. MC to R. Cosway, Mar. 22, 1815, quoted in Brodie, 255.
30. Ibid.
31. Cunningham, *Lives of the Most Eminent British Painters,* 5:10.
32. TJ to MC, Oct. 12, 1786, *PTJ,* 10:445.
33. Ibid., 446.
34. Ibid., 445.
35. Ibid., 446.
36. TJ to AA, Aug. 9, 1786, *PTJ,* 10:203.
37. TJ to Wm. Smith, Aug. 10, 1786, *PTJ,* 10:213.
38. TJ to MC, Oct. 12, 1786, *PTJ,* 10:446.
39. Ibid., 445.
40. Ibid., 446.
41. TJ, mss., LC, *TJ Papers,* 234:41823–37.
42. L&B, 18:430.
43. Ibid., 443.
44. MC to TJ, Feb. 15, 1787, *PTJ,* 9:148–49.
45. TJ to MC, Oct. 12, 1786, *PTJ,* 10:450.
46. TJ to WS, Oct. 22, 1786, *PTJ,* 10:478.
47. Randall, 1:456.
48. MC to TJ, Sept. 20, 1786, *PTJ,* 10:393–94.
49. MC to TJ, Oct. 5, 1786, *PTJ,* 10:433.
50. TJ to MC, Oct. 12, 1786, *PTJ,* 10:448.
51. TJ to MC, Oct. 5, 1786, *PTJ,* 10:431–32.
52. TJ to MC, Oct. 12, 1786, *PTJ,* 10:448.

53. MC to TJ, Oct. 5, 1786, *PTJ,* 10:433.
54. TJ to MC, Oct. 12, 1786, *PTJ,* 10:443–44.
55. Ibid.
56. Ibid.
57. Ibid., 445.
58. Ibid., 446.
59. Ibid., 447.
60. Ibid., 451–52.
61. TJ to Trumbull, Feb. 23, 1787, *PTJ,* 11:181.
62. Ibid., 13:546.
63. MC to TJ, Oct. 30, 1786, *PTJ,* 10:494–95.
64. Ibid., Nov. 17, 1786, *PTJ,* 10:538–39.
65. TJ to MC, Nov. 19, 1786, *PTJ,* 10:542.
66. Ibid., Nov. 29, 1786, *PTJ,* 10:555.
67. Ibid., Dec. 24, 1786, *PTJ,* 10:627.

18. "A Master of My Own Secret"

TJ's Southern Tour: *PTJ,* 11; M. Kimball, *Jefferson: The Scene of Europe,* 184–201; E. Dumbauld, *TJ, American Tourist,* 83–109; D. Malone, 2:112–30; E. Cometti, ed. "Mr. Jefferson Prepares an Itinerary"; J.-F. Bazin and P. Dupuy, "Bicentennial of TJ's Trip to Burgundy"; J. Boyd, *TJ Among the Antiquities of Southern France in 1787,* G. Galtier, *La Viticulture de l'Europe Occidentale à la veille de la Révolution Française;* L. King, "America's First Connoisseur"; R. Lawrence, Sr., ed., *Jefferson and Wine,* 48–55; 107–20; Max Terrier, "Carriages of Jefferson in Europe," *Carriage Journal,* 14(1976), 59–62; B. McEwan, *TJ, Farmer,* 66–71; F. Shuffleton, "Travelling in the Republic of Letters," in *Voyage et Tourisme en Bourgogne,* ed., Baridon and Chevignard, 1–16; T.J. McCormick, "Virginia's Gallic Godfather," *Arts in Virginia,* 4(1964), 2–13.

1. TJ to J. Rutledge, June 19, 1788, *PTJ,* 13:263.
2. TJ to JJ, May 4, 1787, *PTJ,* 11:338–43.
3. List of services, 1800, LC, 39161.
4. TJ to JM, Jan. 30, 1787, *ATJ,* 11:96.
5. TJ to WS, Mar. 15, 1787, *PTJ,* 11:214.
6. "Jefferson's Hints to Americans Travelling in Europe," *PTJ,* 13:264–76.
7. TJ to ML, Apr. 11, 1787, *PTJ,* 11:283.
8. TJ, "Hints," *PTJ,* 13:268.
9. TJ, "Notes of a Tour into the Southern Parts of France, etc.", *PTJ,* 11:415.
10. TJ, Expense Account, mss., HEH.
11. TJ, "Notes," *PTJ,* 11:415.
12. TJ to ML, Apr. 11, 1787, *PTJ,* 11:285.
13. TJ to WS, Mar. 15, 1787, *PTJ,* 11:214.

14. TJ, "Notes," *PTJ,* 11:417.
15. TJ to WS, Mar. 15, 1787, *PTJ,* 11:214–15.
16. J.-F. Bazin and P. Dupuy, 10; TJ to Parent, Mar. 13, 1787, *PTJ,* 11:211–14.
17. TJ to WS, Mar. 15, 1787, *PTJ,* 11:215.
18. Rice, *TJ's Paris,* 94.
19. TJ to MT, Mar. 20, 1787, *PTJ,* 11:226.
20. TJ, "Notes," *PTJ,* 11:420.
21. TJ to WS, Mar. 15, 1787, *PTJ,* 11:215.
22. TJ, "Notes," *PTJ,* 11:421.
23. TJ to MT, Mar. 20, 1787, *PTJ,* 11:226.
24. TJ, "Notes," *PTJ,* 11:420.
25. TJ, "Hints," *PTJ,* 13:273.
26. TJ to MT, Mar. 20, 1787, *PTJ,* 11:226.
27. TJ to JJ, May 4, 1787, *PTJ,* 11:338.
28. TJ, "Notes," *PTJ,* 11:425.
29. TJ to WS, Mar. 27, 1787, *PTJ,* 11:247.
30. TJ to John Rutledge, Jr., June 19, 1788, *PTJ,* 13:263.
31. TJ to WS, Mar. 27, 1787, *PTJ,* 11:247.
32. Ibid., 248.
33. TJ to WS, Mar. 28, 1787, *PTJ,* 11:252.
34. Ibid., 11:254.
35. TJ to JJ, May 4, 1787, *PTJ,* 11:338.
36. TJ, "Notes," *PTJ,* 11:428.
37. TJ, "Hints," *PTJ,* 13:273, June 19, 1788.
38. TJ to P. Mazzei, May 6, 1787, *PTJ,* 11:254.
39. TJ to Mme. de Tott, Apr. 5, 1787, *PTJ,* 11:271.
40. TJ to JJ, May 4, 1787, *PTJ,* 11:338.
41. TJ, *Notes on the State of Virginia,* 42.
42. L&B, 12:204–5.
43. TJ, "Notes," *PTJ,* 11:430–31.
44. TJ, "Hints," *PTJ,* 13:271.
45. Ibid.
46. G. Galtier, *Viticulture,* 19n.
47. TJ, "Hints," *PTJ,* 13:272.
48. TJ to MC, July 1, 1787, *PTJ,* 11:519–20.
49. TJ, "Notes," *PTJ,* 11:439.
50. TJ to E. Rutledge, July 14, 1787, *PTJ,* 11:587–88.
51. TJ, "Hints," *PTJ,* 13:272.
52. TJ to MC, July 1, 1787, *PTJ,* 11:519–20.
53. TJ, "Notes," *PTJ,* 11:440–41.
54. TJ, "Hints," *PTJ,* 13:271.
55. TJ to MJR, May 5, 1787.
56. L&B, 17:202.
57. TJ, "Hints," *PTJ,* 13:271.

58. TJ to WS, May 21, 1787, *PTJ,* 11:372.
59. TJ, "Notes," *PTJ,* 11:458.
60. TJ to John Banister, June 19, 1787, *PTJ,* 11:477.
61. TJ to MC, July 1, 1787, *PTJ,* 11:519–20.
62. TJ to Trumbull, Nov. 13, 1787, *PTJ,* 12:358.
63. MC to TJ, Dec. 1, 1787, *PTJ,* 12:387.
64. MC to TJ, Dec. 7, 1787, *PTJ,* 12:403.
65. TJ to Anne Willing Bingham, Feb. 7, 1787, *PTJ,* 11:123.
66. MC to TJ, Apr. 29, 1788, *PTJ,* 13:115.
67. MC to TJ, Mar. 6, 1788, *PTJ,* 12:646.
68. *PTJ,* 13:264.
69. Mary Jefferson to TJ, quoted in SR, 104.
70. AA to TJ, Sept. 10, 1787, *PTJ,* 12:112.
71. TJ to Eliz. Eppes, July 28, 1787, *PTJ,* 11:634.
72. AA to TJ, June 17, 1787, *PTJ,* 11:503.
73. Ellen Randolph, quoted in D. Malone, 4:498.
74. TJ to JJ, quoted in Peterson, 356.
75. TJ to GW, Aug. 14, 1787, *PTJ,* 12:37–8.
76. TJ to JJ, Sept. 22, 1787, *PTJ,* 12:166.
77. TJ to Carrington, Jan. 16, 1787, *PTJ,* 11:48–49.
78. TJ to AA, Feb. 22, 1787, *PTJ,* 11:174.
79. TJ to Carrington, Jan. 16, 1787, *PTJ,* 11:49.
80. TJ to JA, Nov. 13, 1787, *PTJ,* 12:350.
81. TJ to W.S. Smith, Nov. 13, 1787, *PTJ,* 12:356.
82. TJ to JA, Nov. 13, 1787, *PTJ,* 12:350.
83. TJ to GW, May 2, 1788, *PTJ,* 13:124–28.
84. TJ to W.S. Smith, Feb. 2, 1788, *PTJ,* 12:558–59.
85. TJ to John Rutledge, Jr., Feb. 2, 1788, *PTJ,* 12:556–57.
86. TJ to JM, July 31, 1788, *PTJ,* 13:440–43.
87. TJ to Humphreys, Mar. 18, 1789, *PTJ,* 14:676–79.
88. TJ to JJ, May 4, 1788, *PTJ,* 13:133–37.
89. Gouverneur Morris, *Diary,* 63.
90. TJ to JJ, July 19, 1789, *PTJ,* 15:284–90.
91. TJ to ML, July 6, 1789, *PTJ,* 15:250.
92. TJ to JM, Sept. 6, 1789, *PTJ,* 15:392–97.
93. TJ to John Jay, June 19, 1789, *PTJ,* 15:222.
94. "A Fourth of July Tribute," July 4, 1789, *PTJ,* 15:239–40.
95. *ATJ,* 100.
96. *ATJ,* 108.
97. TJ to JJ, July 19, 1789, *PTJ,* 15:288–89.
98. TJ to JM, July 22, 1789, *PTJ,* 15:299–300.
99. TJ to JJ, Aug. 5, 1789, *PTJ,* 15:334.
100. ML to TJ, Aug. 25, 1789, *PTJ,* 15:354.
101. *ATJ,* 108.

102. MC to TJ, Dec. 23, 1788, *PTJ,* 14:372.
103. TJ to MC, Jan. 14, 1789, *PTJ,* 14:446.
104. TJ to Mme. de Corny, Oct. 14, 1789, *PTJ.*
105. TJ to MC, Oct. 14, 1789, *PTJ,* 15:521.
106. *ATJ,* 111.

19. "The Blessings of Self-Government"

TJ, Secretary of State: M. Peterson, *TJ and the New Nation,* 390–517; K.R. Bowling, *Politics in the First Congress*; N. Cunningham, *The Jeffersonian Republicans* and *In Pursuit of Reason,* 131–94; D. Malone, 2–3; *PTJ,* vols. 16–20; L&B, vols. 7–12; A. Mapp, 1:275–422; F. Brodie, *TJ: An Intimate History,* 319–75; J. Tagg, *B.F. Bache and the Philadelphia Aurora;* D.H. Stewart, *Opposition Press of the Federalist Era;* P.S. Foner, *Democratic-Republican Societies*; R.A. Hendrickson, *Rise and Fall of AH;* S. Watts, *The Republic Reborn;* J. Ferling, *John Adams,* 295–413; J. Charles, *Origins of the American Party System;* W.M. Chambers, *Political Parties in a New Nation;* R. Hofstadter, "Idea of a Party System"; M.P. Ryan, "Party Formation in the U.S. Congress, 1789–1796"; K.R. Bowling, "Dinner at Jefferson's."

TJ, Vice President: S.G. Kurtz, *Presidency of JA;* B. Spivak, *Jefferson's English Crisis;* Palmer, *Stoddert's War;* A. Koch and H. Ammon, "The Virginia and Kentucky Resolutions."

1. TJ to WS, Dec. 15, 1789, *PTJ,* 16:28.
2. TJ to GW, Dec. 15, 1789, *PTJ,* 16:34–35.
3. Randall, 1:522.
4. JM to TJ, Oct. 8, 1789, *PTJ,* 15:509–10.
5. TJ to GW, Dec. 15, 1789, *PTJ,* 16:34–35.
6. TJ, "Holy Cause of Freedom," Feb. 12, 1790, *PTJ,* 16:178–79.
7. Nathaniel Cutting, Diary, Oct. 12, 1789.
8. TJ to Mme. de Corney, Apr. 2, 1790, *PTJ,* 16:290.
9. TJ to TMR, Sr., Feb. 4, 1790, *PTJ,* 16:154–55.
10. TJ to MJR, Apr. 4, 1790, *PTJ,* 16:300.
11. TJ, 16:653–54.
12. Wm. Maclay, Diary, *PTJ,* 16:381n.
13. Ibid., 296.
14. TJ to E. Rutledge, July 4, 1790, *PTJ,* 16:600–01.
15. TJ to GW, July 12, 1790, *PTJ,* 17:108–11.
16. Ibid.
17. Ibid.
18. Ibid., 110.
19. G. Beckwith to Dorchester, *PAH,* 6:496.
20. TJ to WS, Apr. 6, 1790, *PTJ,* 16:316.
21. N. Cunningham, *In Pursuit of Reason,* 139.

22. TJ to GW, Sept. 9, 1792, L&B, 7:137.
23. TJ to GW, [Sept. 14, 1790], *PTJ*, 17:461–63, 466.
24. TJ to Congress, Dec. 30, 1790, quoted in Cunningham, 163; Maclay, Diary, 353.
25. DePauw, ed. *Documentary History of the First Federal Congress*, 2:114–15.
26. TJ to GW, Sept. 9, 1792, *PTJ*.
27. AH, "Report to the President," Jan. 28, 1791, *PAH*, 7:570–607.
28. Peterson, 422.
29. Ibid.
30. TJ, "Report to the President," Jan. 18, 1791, *PTJ*, 18:570.
31. TJ, "Report to the President," Feb. 15, 1791, *PTJ*, 19:275–80.
32. AH, quoted in N. Cunningham, *In Pursuit of Reason*, 166–7.
33. TJ to Jonathan B. Smith, Apr. 26, 1791, *PTJ*, 2:290.
34. TJ to JM, May 9, 1791, *PTJ*, 20:293.
35. Robert Troup to AH, June 15, 1791, *PAH*, 8:478.
36. Cunningham, *In Pursuit of Reason*, 169–70.
37. TJ to Freneau, Feb. 28, 1791, *PTJ*, 19:351.
38. Cunningham, 170.
39. TJ to GW, Sept. 9, 1792, L&B, 7:145.
40. "An American," *Gazette of the U.S.*, Aug. 4, 1792, *PAH*, 12:159.
41. *PAH*, 12:125.
42. TJ, Anas, May 23, 1793, L&B, 1:274.
43. TJ, quoted in Cunningham, 172.
44. AH to GW, quoted in ibid.
45. GW, quoted in ibid.
46. GW to TJ, Apr. 23, 1792, *WW*, 32, 128–34.
47. GW to AH, Aug. 26, 1792, ibid.
48. TJ to GW, Sept. 9, 1792, L&B, 7:137–38.
49. [Catullus] in Cunningham, 175.
50. JM to GW, June 20, 1792, *PJM*, 14:321.
51. TJ to GW, May 23, 1792, L&B, 6:487–95.
52. B. Rush to Aaron Burr, Sept. 24, 1792, L.H. Butterfield, ed., *Letters of B. Rush*, 1:623.
53. TJ to TMR, Nov. 16, 1792, L&B, 7:179.
54. TJ to Thomas Pinckney, Nov. 8, 1792, L&B, 7:177.
55. TJ to MJR, *Family Letters*, 110.
56. TJ to WS, Jan. 3, 1793, L&B, 7:203.
57. TJ to G. Morris, Dec. 30, 1792, L&B, 7:175.
58. TJ to WS, Jan. 3, 1793, L&B, 7:203.
59. Ibid.
60. TJ to GW, Apr. 7, 1793, L&B, 7:275.
61. GW to AH, *PAH*, 14:326–7.
62. Ford, 7:407–08.
63. GW to TJ, J. A. Carroll and M. W. Ashworth, *GW: First in Peace*, 73–75.

64. TJ to JM, May 19, 1793, L&B, 7:337–38.
65. TJ quoted in Cunningham, 183.
66. TJ, Anas, July 5, 1793, L&B, 1:280–81.
67. TJ to G. Morris, June 13, 1793, L&B, 7:407–08.
68. JM to TJ, June 10, 1793, Hant, ed. *Writings of Madison,* 6:127n.
69. TJ, quoted in Cunningham, 186.
70. TJ to JM, June 30, 1793, L&B, 7:420–22.
71. TJ to JM, July 7, 1793, L&B, 7:436.
72. GW to TJ, July 11, 1793, *WW,* 33:4.
73. TJ to GW, Aug. 11, 1793, L&B, 7:471.
74. TJ to Angelica Church, Nov. 27, 1793, L&B, 8:79.
75. TJ to MJR, Dec. 15, 1793, *Family Letters,* 127.
76. TJ to Cooper, Oct. 27, 1808, L&B, 12:180.
77. Lucy Paradise to TJ, Mar. 2, 1790, *PTJ,* 16:198.
78. MC to TJ, Apr. 6, 1790, *PTJ,* 16:312.
79. TJ to MC, June 23, 1790, *PTJ,* 16:550–51.
80. Horace Walpole to Mary Berry, June, 1791, *Works,* 11:285.
81. MC to R. Cosway, Mar. 1, 1793, quoted in F. Brodie, 330.
82. MC to TJ, Nov. 13, 1794, quoted in ibid.
83. TJ to MC, Sept. 8, 1795, quoted in Brodie, 366.
84. TJ to JM, Apr. 27, 1795, L&B, 9:302–3.
85. TJ to MJE, Mar. 3, 1802, *Family Letters,* 219.
86. TJ to JA, May 27, 1795, Cappon, *Adams-Jefferson Letters,* 258.
87. TJ to Archibald Stuart, Jan. 3, 1796, L&B, 8:212–14.
88. TJ to GW, Sept. 12, 1795, mss., LC.
89. TJ to Carlo Bellini, Sept. 30, 1785, *PTJ,* 8:569.
90. TJ to MT, Mar. 20, 1787, *PTJ,* 11:226.
91. TJ to John Brown, Apr. 5, 1797, mss., *TJ Papers,* LC.
92. Duc de La Rochefoucauld-Liancourt, *Travels,* 2:70.
93. Cunningham, 200.
94. TJ to GW, May 14, 1794, L&B, 8:150.
95. TJ to E. Randolph, Sept. 7, 1794, L&B, 8:152.
96. TJ to Tench Coxe, Sept. 10, 1795, L&B, 8:190.
97. TJ to JM, Apr. 27, 1795, L&B, 8:169–70.
98. TJ to JM, Feb. 7, 1796, *PJM,* LC.
99. JM to Monroe, Feb. 26, 1796, *Letters,* 2:83.
100. John Beckley to Wm. Irvine, Sept. 15, 1796, mss., Irvine Papers, HSP.
101. Cunningham, 201.
102. Ibid.
103. Adet to T. Pickering, Oct. 27, 1796, *American State Papers,* 1:576–77.
104. TJ to JM, *WJM,* 6:296–302.
105. JA, May 15, 1797, *Annals of Congress,* 5th Congress, 1st Session, 56.
106. TJ to Mazzei, Apr. 24, 1796, L&B, 8:237–40.

107. TJ to Burr, June 17, 1797, M. J. Kline and J. W. Ryan, eds., *Political Correspondence and Public Papers,* 1:298–300.
108. *Virginia Gazette,* May 24, 1797.
109. TJ to Peregrine Fitzhugh, June 4, 1797, L&B, 8:302.
110. Cunningham, *Circular Letters of Congressmen to Their Constituents,* 1:67–71.
111. TJ to Monroe, Sept. 7, 1797, Jefferson Papers, LC, microfilm series I, 17483–84, reel 21.
112. TJ to JM, Mar. 21, 1798, L&B, 8:386.
113. Samuel J. Cabell, circular letter, Apr. 6, 1798, in Cunningham, *Circular Letters,* 1:119.
114. TJ to P. Carr, Apr. 12, 1798, L&B, 8:405.
115. TJ to JM, Apr. 6, 1798, L&B, 8:403.
116. TJ to JM, Apr. 26, 1798, L&B, 8:411–12.
117. TJ to John Taylor, June 1, 1798, L&B, 8:432.
118. TJ to Stevens Thomson Mason, Oct. 11, 1798, L&B, 8:450.
119. TJ to JM, Feb. 5, 1799, L&B, 9:34.

20. *"The Empire of Liberty"*

President Jefferson: N. Cunningham, *Jeffersonian Republicans in Power. The Process of Government Under Jefferson,* and *In Pursuit of Reason,* 238–321; D. Malone, vols. 4 & 5; L&B, vols. 9–12; M. Peterson, *TJ and the New Nation,* 652–921; D.R. Hickey, *War of 1812,* 3–21; R.W. Tucker and D.C. Hendrickson, *Empire of Liberty;* Carl E. Prince, "The Passing of the Aristocracy, 1801–1805," *JAH,* 57 (1971), 563–75; A. Koch and H. Ammon, "The Virginia and Kentucky Resolutions"; D. Sisson, *Revolution of 1800;* C.O. Lerche, "Jefferson and the Election of 1800"; H. Adams, *History of the U.S. During the Administration of TJ;* M. Grodzins, "Political Parties and the Crisis of Succession in the U.S.: The Case of 1800"; P.F. Boller, Jr. *Presidential Campaigns,* 10–32. J. Larus, "Pell-Mell Along the Potomac"; C.F. Hobson, ed., *Papers of John Marshall,* 6:477–97.

Lewis and Clark Expedition: Malone, 5:142–211; Peterson, *TJ and the New Nation,* 762–64; P.R. Cutright, *Lewis and Clark: Pioneering Naturalists;* B. DeVoto, *Journals of Lewis and Clark;* K. Tobin-Schlesinger, "Jefferson to Lewis: The Study of Nature in the West"; E.G. Chuinard, "TJ and the Corps of Discovery."

Marbury v. Madison: J.M. O'Fallon, "Marbury"; A.R. Amar, "Marbury, Section 13, and the Original Jurisdiction of the Supreme Court"; E. McCaughey, "Marbury v. Madison: Have We Missed the Real Meaning?"

Black Snake Affair: S. Mills, ed., *Trial of Cyrus Dean,* 19–23; K. Kenny, "The Black Snake"; Mary P. Ryan, "Party Formation in the U.S. Congress, 1789 to 1796: A Quantitative Analysis"; R.K. Faulkner, "John Marshall and the Burr Trial."

Sally Hemings Affair: F.M. Brodie, *TJ: An Intimate History;* V. Dabney, *The Jefferson Scandals: A Rebuttal;* M. Durey, *With the Hammer of Truth;* C.A. Jellison, "That Scoundrel Callender," *VMHB,* 64(1959), 295–306.

1. TJ to JM, Feb. 5, 1799, L&B, 9:34.
2. *JHD*, 1797, 55–58.
3. TJ to JM, June 7, 1798, L&B, 7:266–67.
4. TJ to William G. Munford, June 18, 1799, mss., Teachers College Library, Columbia University.
5. W.C. Nicholas to J. Breckenridge, Oct. 10, 1798, mss., Breckinridge Papers, LC.
6. Quoted in A. Koch and H. Ammon, "The Virginia and Kentucky Resolutions," 145.
7. Taylor to JM, March 4, 1799, Rives Papers, LC.
8. Koch, 164.
9. Governor Davie to Iredell, June 17, 1799, G.J. McRee, ed., *Life and Correspondence of James Iredell*, 2:57.
10. TJ to JM, Aug. 23, 1799, mss., Rives Papers, LC.
11. L&B, 7:439.
12. TJ to Monroe, July 15, 1803.
13. TJ to Monroe, Feb. 11, 1799, L&B, 9:36.
14. TJ to Philip Norborne Nicholas, Apr. 7, 1800, L&B, 9:128; [T. Cooper], *Political Arithmetic*, n.p., 1798, 5.
15. TJ to E. Gerry, Jan. 26, 1799, L&B, 9:17–19.
16. Philadelphia *Aurora*, Oct. 4, 1800.
17. Syrett, *PAH*, 21:279.
18. Philadelphia *Aurora*, Oct. 23, 1795.
19. Richmond *Recorder*, Nov. 3, 1802.
20. TJ to JM, April 26, 1;798, quoted in Durey, 106.
21. James Haw, *Stormy Patriot: The Life of Samuel Chase*, Baltimore: Md. Historical Society, 1980, 202–3.
22. Rev. Timothy Dwight, *The Duty of Americans at the Present Crisis*, New Haven, 1798, 20–21.
23. *Gazette of the U.S.*, Sept. 10, 1800.
24. *The Voice of Warning*, 8.
25. Ibid., 22–23.
26. [Rev. Wm. Linn,] *Serious Considerations on the Election of a President*, [New York, 1800] 4–6.
27. Joseph Bloomfield, *To the People of New Jersey*, Sept. 30, 1800, broadside, HSP.
28. Quoted in P. F. Boller, Jr., *Presidential Campaigns*, 11.
29. *Connecticut Courant*, Sept. 15, 1800.
30. Quoted in Boller, 17–18.
31. [John Beckley], *Address to the People of the U.S.*, Phila., 1800.
32. Wm. Austin, *An Address to the Voters for Electors of President*, Richmond, Va., May 26, 1800, LC.
33. AH to JJ, May 7, 1800, *PAH*, 24:465.
34. AH to Theodore Sedgwick, May 4, 1800, *PAH*, 24:453.
35. N. Cunningham, *Jeffersonian Republicans*, 8–9.

36. *Baltimore American,* quoted in Boller, 13.
37. TJ to Monroe, Dec. 20, 1800, *Monroe Papers,* LC.
38. Burr to John Taylor, Oct. 23, 1800, *Papers of Burr,* 1:451.
39. Burr to TJ, Dec. 23, 1800, ibid., 1:473–4.
40. AH to Oliver Wolcott, Jr., Dec. 16, 1800, *PAH,* 25:257.
41. TJ to JM, Feb. 18, 1801, L&B, 9:182.
42. TJ, L&B, 12:136.
43. *Alexandria Times,* Mar. 6, 1801.
44. A. Gallatin to his wife, Jan. 15, 1801, H. Adams, *Life of Gallatin,* 254.
45. Mrs. Samuel Harrison Smith to Susan B. Smith, Mar. 4, 1801, in Gaillard Hunt, ed., *The First Forty Years of Washington Society,* 25.
46. TJ, *First Inaugural Address,* Mar. 4, 1801, TJP, LC.
47. TJ to Monroe, Mar. 7, 1801, L&B, 9:203.
48. *American Citizen,* New York City, June 5, 1801; *American Mercury,* Hartford, July 30, 1801.
49. TJ, Anas, May 15, 1801, L&B, 1:365.
50. Levi Lincoln, ibid., 365–66.
51. TJ to Wilson Cary Nicholas, June 11, 1801, L&B, 9:264–65.
52. TJ to John Dickinson, June 21, 1801, TJP, LC.
53. TJ to B. Rush, Mar. 24, 1801, L&B, 9:230–31.
54. TJ, L&B, 9:204.
55. *American Citizen,* New York, June 5, 1801.
56. TJ to Elias Shipman, July 12, 1801, L&B, 9:272–73.
57. Quoted in C. Prince, 572.
58. TJ to T.M. Randolph, Jr., Nov. 16, 1801, TJP, LC.
59. TJ to Gallatin, Sept. 18, 1801, H. Adams, *Writings of Gallatin,* 1:55.
60. TJ to Uriah Tracy, Jan. 26, 1811, L&B, 9:185.
61. Sen. Wm. Plumer to Jeremiah Smith, Dec. 9, 1802, *Wm. Plumer Papers,* LC.
62. TJ, Rules of Etiquette, [1803], L&B, 10:47.
63. Plumer to Jeremiah Smith, Dec. 9, 1802, supra.
64. Samuel Latham Mitchill to his wife, Feb. 10, 1802, mss., *Mitchill Papers,* Museum of City of New York.
65. TJ to David R. Williams, Jan. 31, 1806, in Cunningham, *Process of Government,* 41–44.
66. Dickinson W. Adams, ed., *Jefferson Extracts from the Gospels,* 12–16.
67. TJ to Short, Oct. 31, 1819, Adams, 389.
68. TJ to JA, Oct. 12, 1813, Cappon, ed., *Adams-Jefferson Letters,* 381–83.
69. Quoted in V. Dabney, *The Jefferson Scandals,* 9.
70. Richmond *Examiner,* June 23, 1802.
71. Philadelphia *Aurora,* Aug. 25, 1802.
72. Richmond *Recorder,* Sept. 1, 1802.
73. Richmond *Examiner,* Sept. 1, 1802.
74. LC, mss., 155, 27117–21.
75. D. Malone, 1:450n.

76. TJ to Robert Smith, Sec. of the Navy, July 1, 1805, quoted in D. Malone, 1:448n.
77. William Burwell Memoir, mss., LC.
78. TJ, draft, First Annual Message, Dec. 8, 1801, TJP, LC.
79. Michael Leib to Lydia Leib, Dec. 9, 1801, Leib-Harrison Papers, HSP.
80. TJ to Nathaniel Macon, May 14, 1801, mss., TJP, LC.
81. L&B, 3:438–9.
82. Ibid., 4:100.
83. TJ, 5th Annual Message, Dec. 3, 1805, *Messages and Papers,* 1:373.
84. TJ to Benjamin Latrobe, quoted in R. Stuart, *Halfway Pacifist,* 36.
85. TJ to the Earl of Buchan, July 10, 1803, L&B, 10:401.
86. TJ to Madame de Stael, July 16, 1807, ibid., 11:282.
87. Robt. Smith to Edw. Preble, July 13, 1803, Knox, ed., *Naval Documents,* 2:474–77.
88. James Fisk, Jan. 25, 1812, *Annals of Congress,* 12–1, 841.
89. Samuel McKee, Jan. 18, 1812, *Annals of Congress,* 12–1, 968–69.
90. TJ to JM, Sept. 6, 1789, *PTJ,* 15:392–97.
91. TJ to R. R. Livingston, Apr. 18, 1802, L&B, 9:364–65.
92. TJ to Pinckney, Nov. 27, 1802, *WJM,* 6:462.
93. JM to Livingston, Dec. 16, 1802, in I. Brant, *James Madison,* 4:99.
94. TJ, Second Annual Message, Dec. 15, 1802, L&B, 9:409.
95. TJ to Monroe, Jan. 13, 1803, L&B, 9:418–19.
96. *Annals of Congress,* 7th Congress, 2nd Session, Jan. 12, 1803, 370–74.
97. *National Intelligencer,* July 4, 1803.
98. TJ to John Dickinson, Aug. 9, 1803, L&B, 10:29.
99. TJ to John Breckenridge, Aug. 12, 1803, L&B, 10:7.
100. TJ, Third Annual Message Dec. 20, 1803, Oct. 17, 1803, L&B, 10:37.
101. P. R. Cutright, *Lewis and Clark, Pioneering Naturalists,* 13–14.
102. Quoted in ibid., 8.
103. Quoted in E. T. Martin, *TJ, Scientist,* 6.
104. Bernard de Voto, ed., *Journals of Lewis and Clark,* xvii.
105. Ibid.
106. Quoted in Martin, *Scientist,* 6.
107. Ibid.
108. TJ to Elbridge Gerry, Mar. 3, 1804, L&B, 10:73.
109. Jacob Crowninshield to Barnabas Bidwell, Feb. 26, 1804, mss., Taft Coll., MHS.
110. TJ to MJE, Mar. 3, 1804, *Family Letters,* 258.
111. TJ to JP, June 25, 1804, L&B, 11:31–32.
112. L&B, 8:346.
113. Litchfield (Conn.) *Witness,* April 30, 1806.
114. TJ to Nicholas, June 13, 1809, L&B, 12:289.
115. TJ to Thomas Seymour, Feb. 11, 1807, L&B, 11:155.
116. Ibid.

117. TJ to Edward Carrington, Jan. 16, 1787, *PTJ,* 11:49.
118. TJ to Walter Jones, Jan. 2, 1814, L&B, 14:46.
119. TJ to Thomas Lomax, Jan. 11, 1806, Malone, 5:95.
120. Monroe to JM, July 1, 1804, *Monroe Papers,* LC, reel 3.
121. Monroe to JM, Oct. 18, 1805, ibid., reel 11.
122. TJ, quoted in D. Hickey, "Monroe-Pinckney Treaty," 85–87.
123. Nicholson to Gallatin, July 14, 1807, *Gallatin Papers,* reel 14.
124. Anthony Merry to Lord Harrowby, Aug. 6, 1804, *Papers of Burr,* 2:291.
125. J. H. Daveiss to TJ, Jan. 10, 1806, TJP, LC.
126. Gen. James Wilkinson to TJ, Oct. 21, 1806, T. P. Abernathy, *Burr Conspiracy,* 150–52.
127. TJ, special message to Congress, Jan. 22, 1807, L&B, 10:346.
128. TJ to Wm. B. Giles, Apr. 20, 1807, L&B, 10:383–84.
129. TJ to George Hay, June 17, 1807, L&B, 10:400–01.
130. TJ to Hay, June 20, 1807, L&B, 10:404.
131. D. Robertson, *Reports of the Trials of Burr,* 2:445.
132. TJ to DuPont, July 14, 1807, L&B, 10:461–62.
133. TJ to Wm. Duane, July 20, 1787, L&B, 1:471.
134. TJ to JM, Aug. 16, 1807, L&B, 10:476–77.
135. TJ to JM, Sept. 1, 1807, L&B, 10:489.
136. Smith to TJ, Oct. 19, 1807, TJP, LC.
137. TJ to Robt. Williams, Nov. 1, 1787, L&B, 11:378–79.
138. TJ to TMR, Nov. 30, 1787, TJP, LC.
139. TJ, special message, Dec. 18, 1807, L&B, 10:530–31.
140. Geo. C. Campbell, Jan. 22, 1808, quoted in Cunningham, *Circular Letters,* 2:525–26.
141. Chilton, Williamson, *Vermont in Quandary, 1763–1825,* 245.
142. A. Addis and C. P. Van Ness, to Gallatin, *Governor and Council,* 5:472.
143. D. Malone, 5:587.
144. Hance, Dawn. "The Early Militia in Rutland." *Vermont History,* 14(1984), 3.
145. Samuel Mills, ed., *Trial of Cyrus B. Dean,* 16.
146. Governor and Council, 5:477.
147. Thomas Leiper, Aug. 28, 1807, TJP, LC.
148. TJ to Gallatin, Aug. 11, 1808, in Jefferson, *Works,* (ed. Ford), 11:41.
149. TJ to Messrs. Thomas, Ellicot and others, Nov. 13, 1807, TJP, LC.

21. "A Fire Bell in the Night"

Sage of Monticello: D. Malone, vol. 6; M. Peterson, *TJ and the New Nation,* 922–1009; N. Cunningham, *In Pursuit of Reason,* 322–49; F. Brodie, *TJ: An Intimate History,* 570–636; Cappon, ed., *Adams-Jefferson Letters;* Betts and Bear, eds., *Family Letters;* H.S. Randall, vol. 3; A. Mapp, vol. 2:191–360.

1. TJ to Charles Thomson, Dec. 25, 1808, L&B, 12:218.
2. TJ to Comte Diadati, Mar. 28, 1807, L&B, 11:182.
3. TJ to Martha Jefferson Randolph, Feb. 27, 1809, *Family Letters,* 386.
4. TJ to JA, Jan. 21, 1812, Cappon, 292.
5. TJ to MC, Dec. 27, 1820, SR, 374.
6. Ellen W. Coolidge, 1850, in SR, 345.
7. TJ to Tadeusz Kosciuszko, Feb. 26, 1810, L&B, 12:369–70.
8. TJ to JA, Jan. 21, 1812, Cappon, 290.
9. TJ to JA, June 19, 1815, Cappon, 443.
10. TJ to John W. Eppes, June 24, 1813, L&B, 11:297.
11. TJ to S. Roane, Sept. 6, 1819, L&B, 12:135.
12. TJ to Sam. J. Cabell, Dec. 28, 1822, mss., TJP, LC.
13. TJ to JA, Nov. 15, 1813, Cappon, 397.
14. Richmond *Enquirer,* Nov. 16, 1824.
15. Geo. Ticknor to Wm. H. Prescott, Dec. 16, 1824, *Life of Ticknor,* 1:348.
16. TJ to Wm. Fitzhugh Gordon, Jan. 1, 1826, in Bear, "The Last Few Days," MACH, 32 (1974), 63.
17. Ibid.
18. MJR to Ellen Coolidge, Apr. 5, 1826, Malone, *TJ and His Times,* 6:473.
19. L&B, 10:362–72.
20. TJR to TJ, Jan. 31, 1826, *Family Letters,* 467.
21. TJ to TJR, Feb. 11, 1826, *Family Letters,* 470.
22. Cabell to TJ, Feb. 10, 1826.
23. TJ to Cabell, Feb. 3, 1826.
24. *ATJ,* 51.
25. TJ to John Holmes, Apr. 22, 1820, mss., TJP, LC.
26. Ibid.
27. TJ, *Notes on the State of Virginia,* 258.
28. TJ to Samuel Kercheval, July 12, 1816, *Life and Selected Writings of TJ,* 674.
29. TJ, *Notes,* 256.
30. TJ to Edw. Coles, Aug. 31, 1814, mss., MHS, 7th Ser., 1:200–2.
31. TJ, quoted in Cunningham, *Pursuit of Reason,* 328.
32. TJ, *Farm Book.*
33. TJ, June 24, 1826, to *National Intelligencer.*
34. Robley Dunglison, "The Autobiographical Ana," Samuel X. Radbill, ed., *TAPS* 53 (1963), 34.
35. SR, 428.
36. Page Smith, *John Adams,* 2:1136–37.
37. Mss., TJP, LC.

Bibliography

(Key to abbreviations is found on page 596.)

Abrams, Rochonne. "Meriwether Lewis: The Logistical Imagination." *MHSB* 36 (1980): 228–40.

———. "Meriwether Lewis: Two Years with Jefferson, the Mentor." *MHSB* (1979): 3–18.

Ackerman, James S. "TJ" in *The Villa: Form and Ideology of Country Houses.* Princeton: Princeton Univ. Press, 1990, 185–211.

Adams, C. F., ed., *Works of John Adams,* 10 vols., Boston: Little, Brown, 1860–65.

Adams, Dickinson W., ed. *Jefferson's Extracts from the Gospels: "The Philosophy, Life and Morals of Jesus."* Princeton: Princeton Univ. Press, 1983.

Adams, Henry. *History of the United States during the Administrations of TJ.* New York: Library of America, 1986.

Adams, John. *Diary and Autobiography.* 4 vols. ed., L.H. Butterfield: Cambridge, Mass.: Harvard Univ. Press, 1961.

Adams, Randolph G. *Political Ideas of the American Revolution.* New York: Barnes & Noble, 1922.

Adams, William Howard. *The Eye of Thomas Jefferson.* Washington, D.C.: Washington National Gallery of Art, 1976.

———. *Jefferson and the Arts: An Extended View.* Washington, D.C.: Washington National Gallery of Art, 1976.

———. *Jefferson's Fine Arts Library.* Charlottesville, Univ. Press of Virginia, 1976.

Akers, Charles W. *Abigail Adams: An American Woman.* Boston: Little, Brown, 1980.

Aldridge, Alfred Owen. *Franklin and His French Contemporaries.* New York: New York Univ. Press, 1957.

Alengry, Franck. *Turgot.* Paris: Charles-Lavanzelle, 1942.

Alexander, Edward P. "Jefferson and Kosciuszko: Friends of Liberty and of Man." *PMHB* 92 (1968): 87–103.

Allen, John Logan. *Passage through the Garden.* Urbana: Univ. of Illinois Press, 1975.

Amar, Akhit Reed. "*Marbury,* Section 13, and the Original Jurisdiction of the Supreme Court." *Univ. of Chicago Law Review,* 56 (1989): 443–99.

Anderson, Jefferson Randolph. "Tuckahoe and the Tuckahoe Randolphs." *VMHB* 45 (1927): 55–86.

Anderson, Philip J. "William Linn, 1752–1808: American Revolutionary and Anti-Jeffersonian." *Journal of Presbyterian History* 55 (1977): 381–94.

Andrews, Robert Hardy. "How the CIA Was Born." *Mankind* 5 (1975): 14–15, 68.

Andrews, Stuart. "Joseph Priestley and American Independence." *HT* (Great Britain) 29 (1979): 221–29.

Anonymous. *Free Enquiry.* Windsor, Vermont, 1808.

Appleby, Joyce. *Capitalism and a New Social Order: The Republican Vision of the 1790s.* New York: New York Univ. Press, 1984.

———. "The Changing Prospect of the Family Farm in the Early National Period." *Working Papers for the Regional Economic History Research Center* 4 (1981): 1–25.

———. "Commercial Farming and the 'Agrarian Myth' in the Early Republic." *JAH* 68 (1982): 833–49.

———. "Jefferson: A Political Reappraisal." *Democracy* 3 (1983): 139–45.

———. "The Radical *Double-Entendre* in the Right to Self-Government," in *Origins of Anglo-American Radicalism,* ed. Margaret Jacob and James R. Jacob, Boston: Allen and Unwin, 1984, 275–83.

———. "Jefferson?" *WMQ,* 3rd ser., 39 (1982): 287–304.

Arkes, Hadley. *Beyond the Constitution.* Princeton: Princeton Univ. Press, 1990.

Arnold, I.N. *Life of Benedict Arnold.* Chicago: Jansen, McClurg, 1880.

Arrowood, Charles F. *Thomas Jefferson and Education in a Republic.* New York: McGraw-Hill, 1930.

Ashworth, John. "The Jeffersonians: Classical Republicans or Liberal Capitalists?" *Journal of American Studies* 18 (1984): 425–35.

Axelrad, Jacob. *Patrick Henry: The Voice of Freedom.* Westport: Greenwood, 1947.

Badinter, Elisabeth. *Condorcet (1743–1794).* Paris: Fayard, 1988.

Bailyn, Bernard. *The Ideological Origins of the American Revolution.* Cambridge, Mass.: Harvard Univ. Press, 1967.

———. "Boyd's Jefferson: Notes for a Sketch," *NEQ* 38 (1960): 380–400.

———. *Pamphlets of the American Revolution, 1750–1776.* Cambridge, Mass.: Harvard Univ. Press, 1965.

———. "Politics and Social Structure in Virginia," in *Seventeenth-Century America: Essays in Colonial History,* ed. James Morton Smith. Chapel Hill: Univ. of North Carolina Press, 1959, 108–10.

Banes, Ruth A. "The Exemplary Self: Autobiography in Eighteenth Century America." *Biography* 5 (1982): 226–39.

Banner, James M., Jr. *To the Hartford Convention.* New York: Knopf, 1970.

Banning, Lance. *Jefferson, Nationalism, and the Enlightenment.* New York: Braziller, 1975.

———. *The Jeffersonian Persuasion: Evolution of a Party Ideology.* Ithaca, N.Y.: Cornell Univ. Press, 1978.

———. "Jeffersonian Ideology Revisited: Liberal and Classical Ideas in the New American Republic." *WMQ,* 3rd ser., 43 (1986): 3–19.

Baridon, Michel, and Bernard Chevignard. *Voyage et tourisme en Bourgogne à l'époque de Jefferson.* Dijon: Éditions universitaires de Dijon, 1988.

Barlow, Joel. *Advice to the Privileged Orders in the Social States of Europe . . . (1792).* Ithaca, N.Y.: Cornell Univ. Press, 1956.

Barnes, Howard A. "The Idea that Caused a War: Horace Bushnell Versus Thomas Jefferson." *Journal of Church and State* 16 (1974): 73–83.

Barnet, Richard J. *Rockets Red Glare: War, Politics and the Presidency.* New York: Simon and Schuster, 1990.

Baron, Robert C., ed. *The Garden and Farm Books of Thomas Jefferson.* Charlottesville, Va.: Univ. Press of Virginia, 1987.

Baron, Sherry. "Thomas Jefferson: Scientist as Politician." *Synthesis* 3 (1975): 6–21.

Barrett, Clifton W. "The Struggle to Create a University." *VQR* 49 (1973): 494–506.

Batey, Mavis, and David Lambert. *The English Garden Tour.* London: John Murray, 1990.

Baxter, W.T. "Accounting in Colonial America." In *Studies in the History of Accounting,* ed. A.C. Littleton and B.S. Yamey. London, 1956.

Bayard, Jean E. *Le Quartier latin hier et aujourdhui.* Paris: Editions Roman Nouveau, 1924.

Bazin, Jean-François, and Pierre Duprey. *Le bicentenaire du Voyage de Jefferson en Bourgogne.* Dijon, France: Conseil régional de Bourgogne, 1986.

Bear, James A., Jr. *Jefferson's Memoranda Books,* 1767–1826. 2 vols. Princeton, 1987.

———. "The Hemings Family of Monticello." *VC* 29 (1979): 78–87.

———. *Jefferson at Monticello.* Charlottesville: Univ. Press of Virginia, 1967.

Becker, Carl L. *The Declaration of Independence: A Study in the History of Political Ideas.* New York: Harcourt Brace, 1922.

Bedini, Silvio A. *Thomas Jefferson: Statesman of Science.* New York: Macmillan, 1990.

———. *Thomas Jefferson and His Copying Machines.* Charlottesville, Va: Univ. Press of Virginia, 1984.

———. "The Scientific Instruments of the Lewis and Clark Expedition." *Great Plains Quarterly* 4 (1984): 54–69.

Beeman, Richard R. *Evolution of the Southern Backcountry: A Case Study of Lunenburg County, Va., 1746–1832.* Philadelphia: Univ. of Pennsylvania Press, 1984.

———. *Patrick Henry, A Biography.* New York: McGraw-Hill, 1974.

Beers, H.P. *The French in North America.* Baton Rouge: Louisiana State Univ. Press, 1957.

Beiswanger, William L. "The Temple in the Garden: Thomas Jefferson's Vision Landscape." *Eighteenth-Century Life* 8 (1983): 170–88.

Behr, Edward. *Algerian Problem.* New York: Norton, 1962.

Beloff, Max. "The Sally Hemings Affair: A 'Founding Father'." *Encounter* 43 (1974): 52–56.

Benson, C. Randolph. *Thomas Jefferson as Social Scientist.* Rutherford, N.J.: Fairleigh Dickinson Univ. Press, 1971.

Berger, Raoul. "Jefferson V. Marshall in the Burr Case." *American Bar Association Journal* 60 (1974): 702–706.

Berman, Eleanor D. *Thomas Jefferson Among the Arts.* New York: Philosophical Library, 1947.

Bernstein, Richard. "Jefferson and His Slave: A Relationship in Doubt." *NYT,* Nov. 22, 1987: 42.

Betts, Edwin M., ed. *Thomas Jefferson's Farm Book.* Charlottesville, Va.: Univ. Press of Virginia, 1987.

Betts, Edwin M., and Hazlehurst B. Perkins. *Thomas Jefferson's Flower Garden at Monticello,* 3rd ed. Charlottesville, Va.: Univ. Press of Virginia, 1986.

Betts, Edwin M., and James A. Bear, Jr., eds. *The Family Letters of Thomas Jefferson.* Charlottesville, Va.: Univ. Press of Virginia, 1986.

Billings, Warren M., John E. Selby, and Thad W. Tate. *Colonial Virginia: A History.* White Plains, N.Y.: KTO Press, 1986.

Blackburn, Joyce. *George Wythe of Williamsburg.* New York: Harper & Row, 1975.

Boatner, Mark M. *Encyclopedia of the American Revolution,* New York: David McKay, 1966.

Boller, Paul F., Jr. *Presidential Campaigns.* New York: Oxford Univ. Press, 1984.

Bolles, Albert S. *The Financial History of the U.S. from 1774 to 1789.* 4th ed., New York: D. Appleton & Co., 1896.

Bolster, William J. "The Impact of Jefferson's Embargo on Coastal Commerce." *Log of Mystic Seaport* 37 (1986): 111–23.

Boorstin, Daniel J. *Lost World of Thomas Jefferson.* New York: Henry Holt, 1948.

Bottorff, William K. *Thomas Jefferson.* Boston: Twayne, 1979.

———. "Mr. Jefferson Tours New England." *New-England Galaxy* 20 (1979): 3–7.

Boucher, Francois. *American Footprints in Paris.* New York: George H. Doran, 1921.

Boucher, Jonathan. *Reminiscences of an American Loyalist, 1738–1789.* Boston: Riverside Press, 1925.

Bowen, Catherine Drinker. *John Adams and the American Revolution.* Boston: Little, Brown, 1949.

Bowers, Claude G. *Jefferson and Hamilton: A Struggle for Democracy in America.* Boston, Houghton Mifflin, 1925.

———. *Jefferson in Power: the Death Struggle of the Federalists.* Boston: Houghton Mifflin 1936.

———. *The Young Jefferson, 1743–1789.* Boston: Houghton Mifflin, 1945.

———. "Jefferson and Civil Liberty." *AM* 191 (1953): 52–58.

Bowling, Kenneth R. *Politics in the First Congress, 1789–1791.* New York: Garland, 1990.

———, ed. *Journal of William Maclay.* New York: Albert & Charles Boni, 1927.

———. "Dinner at Jefferson's: A Note on Jacob E. Cooke's 'The Compromise of 1790.' " *WMQ,* 3rd ser., 28 (1971): 607–28.

Boyd, Julian P. *The Declaration of Independence: The Evolution of the Text.* Princeton: Princeton Univ. Press, 1945.

———. "The Declaration of Independence: The Mystery of the Lost Original." *PMHB* 100 (1976): 438–67.

———. "Jefferson's Expression of the American Mind." *VQR* 50 (1974): 538–62.

———. "Jefferson's Final Testament of Faith." *NYT Magazine,* April 10, 1949: 11, 33–35, 37, 39.

———. "Jefferson's French Baggage, Crated and Uncrated." *MHSB* 83 (1971): 16–27.

———. "The Relevance of Thomas Jefferson for the Twentieth Century." *AS* 22 (1952): 61–76.

———. "Subversive of What?" *AM* 182 (1948), 19–23.

———. "Thomas Jefferson and the Police State," *North Carolina Historical Review* 15 (1948): 233–53.

———. "Thomas Jefferson's 'Empire of Liberty.' " *VQR,* (1948): 538–54.

Breen, T.H. *Tobacco Culture: The Mentality of the Great Tobacco Planters on the Eve of the Revolution.* Princeton: Princeton Univ. Press, 1985.

Breitweiser, Mitchell R. "Jefferson's Prospect." *Prospects* 10 (1985): 315–51.

Brick, Blanche H. "Changing Concepts of Equal Educational Opportunity: Views of Thomas Jefferson, Horace Mann and John Dewey." Unpub. doct. diss., Texas A&M, 1984.

Bridenbaugh, Carl. *Myths and Realities: Societies of the Colonial South.* Baton Rouge: Louisiana State Univ. Press, 1952.

Brodie, Fawn M. *Thomas Jefferson: An Intimate History.* New York: Norton, 1974.

———. "The Great Jefferson Taboo." *AH* 23 (1972): 49–57, 97–100.

———. "Thomas Jefferson's Unknown Grandchildren." *AH* 27 (1976): 28–33, 94–99.

Brooke, John. *King George III*, New York: McGraw-Hill, 1972.

Brown, C. Allan. *Poplar Forest: The Mathematics of an Ideal Villa.* Charlottesville, Va.: Univ. Press of Virginia, 1990.

Brown, Imogene E. *American Aristides, A Biography of George Wythe,* Rutherford, N.J.: Fairleigh Dickinson Univer. Press, 1981.

Brown, Ralph A. *Presidency of John Adams.* Lawrence, Kan.: Univ. of Kansas Press, 1975.

———. *The Presidency of Thomas Jefferson.* Lawrence, Kan: Univ. of Kansas Press, 1976.

Brown, Roger H. *The Republic in Peril: 1812.* New York: Norton, 1971.

Bryson, W. Hamilton. *Legal Education in Virginia, 1779–1979.* Charlottesville, Va: Univ. Press of Virginia, 1982.

———. ed. *Virginia Law Reporters Before 1880,* Charlottesville, Va.: Univ. Press of Virginia, 1977.

Buel, Richard. *Securing the Revolution: Ideology in American Politics, 1789–1815.* Ithaca: Cornell Univ. Press, 1972.

Bullock, Helen D. *My Head and My Heart: A Little Chronicle of Thomas Jefferson and Maria Cosway.* New York: Putnam's, 1945.

———. "A Dissertation on Education," *Papers of the Albemarle County Historical Society* II, (1941–1942): 36–40.

———. "The Papers of Thomas Jefferson." *American Archivist* 4 (1941): 238–49.

Burg, B.R. "The Rhetoric of Miscegenation: Thomas Jefferson, Sally Hemings and Their Historians." *Phylon* 47 (1986): 128–38.

Burke, Edmund. *A Philosophical Enquiry into the Sublime and the Beautiful.* London, 1757.

Burnaby, Andrew. *Travels through Middle Settlements in North America,* 1759–1760. London, 1775, repr. N.Y., 1904.

Burnett, Edmund C., ed. *Letters of Members of the Continental Congress.* Washington, D.C.: Carnegie Institution of Washington, 1921–36, 8 vols.

Burr, Aaron. Correspondence with General James Wilkinson. Mss., Newberry Library, Chicago.

———. File, mss. Library of Congress.

———. Papers. New-York Historical Society.

Burroughs, Raymond Darwin. ed., *Natural History of the Lewis & Clark Expedition.* East Lansing: Michigan State Univ. Press., 1961.

Butterfield, Lyman H., and Howard C. Rice. "Jefferson's Earliest Note to Maria Cosway with Some New Facts and Conjectures on His Broken Wrist." *WMQ,* 3rd ser., 5 (1948): 26–33, 620–21.

Cabell, Nathan F., ed. *Early History of the University of Virginia.* Richmond: J.W. Randolph, 1856.

Cabell, William C. Notes. Alderman Library, University of Va.

Caldwell, Lynton K. *The Administrative Theories of Hamilton & Jefferson,* 2nd ed. New York: Holmes and Meier, 1988.

Cappon, Lester J., ed. *The Adams-Jefferson Letters.* Chapel Hill, N.C.: Univ. of North Carolina Press, 1988.

Carnahan, Frances. "Dinner with Thomas Jefferson." *EAL* 16 (1985): 22–27.

Carrithers, David W. "Montesquieu, Jefferson, and the Fundamentals of Eighteenth-Century Republican Theory." *French-American Review* 6 (1982): 160–68.

Carson, David A. "Congress in Jefferson's Foreign Policy, 1801–1809." Unpub. doct. diss., Texas Christian U., 1983.

———. "Jefferson, Congress, and the Question of Leadership in the Tripolitan War." *VMHB* 94 (1986): 409–24.

———. "That Ground Called Quiddism: John Randolph's War With the Jefferson Administration." *Journal of American Studies* 20 (1986): 71–92.

Castiello, Kathleen R. "The Italian Sculptors of the United States Capitol: 1806–1834." Unpub. doct. diss., U. of Michigan, 1975.

Ceram, C.W. "Mr. Jefferson's 'Dig.' " *AHI* 6 (1971): 38–41.

Chambers, William N. *Political Parties in a New Nation.* New York: Oxford Univ. Press, 1963.

Chapin, Bradley. "Colonial and Revolutionary Origins of the American Law of Treason," *WMQ* 17 (1960): 3–21.

Charles, Joseph. *The Origins of the American Party System.* New York: Harper & Row, 1961.

Chastellux, Marquis de. *Travels in North America, 1780, 1781.* Chapel Hill: Univ. of North Carolina Press, 1963.

Cheatham, Edgar. "Reunion at Monticello." *EAL* 8 (1977): 40–43.

Chinard, Gilbert. *Thomas Jefferson, the Apostle of Americanism.* Boston: Little, Brown, 1929.

———. *Letters of Lafayette and Jefferson.* Baltimore: Johns Hopkins Univ. Press, 1929.

———. *The Literary Bible of Thomas Jefferson.* Baltimore: Johns Hopkins Univ. Press, 1928.

Chitwood, Oliver Perry. *John Tyler, Champion of the Old South.* New York: Appleton-Century, 1939.

———. *Justice in Colonial Virginia.* Baltimore: Johns Hopkins Press, 1905.

Chuinard, E.G. "Thomas Jefferson and the Corps of Discovery: Could He Have Done More?" *American West* 12 (1975): 4–13.

Clark, J.C.D. *The Dynamics of Change: The Crisis of the 1750's and English Party Systems.* Cambridge: Harvard Univ. Press, 1982.

Clark, Kenneth. "Thomas Jefferson and the Italian Renaissance." *VQR* 48 (1972): 519–31.

Clinton, Robert Lawry. *Marbury v. Madison and Judicial Review.* Lawrence, Kan.: Univ. Press of Kansas, 1989.

Clive, John, and Bernard Bailyn. "England's Cultural Provinces: Scotland and America." *WMQ,* 3rd ser., 11 (1954): 200–13.

Cmiel, Kenneth. *Democratic Eloquence: The Fight for Popular Speech in Nineteenth-Century America.* New York: Wm. Morrow, 1990.

Coats, Peter. *Great Gardens of Britain.* New York: Putnam's, 1967.

Coatsworth, John H. "American Trade with European Colonies in the Caribbean and South America, 1790–1812." *WMQ,* 3rd ser., 24 (1967): 243–66.

Cohen, J. Bernard, ed. *Thomas Jefferson and the Sciences.* New York: Arno, 1980.

Cole, John. "The Library of Congress in the Nineteenth Century: An Informal Account." *Journal of Library History* 9 (1974): 222–40.

Commager, Henry Steele. *Jefferson, Nationalism and the Enlightenment.* New York: Braziller, 1975.

———. "Jefferson and the Book-Burners" *AH* 9 (1958): 65–68.

Cometti, Elizabeth, ed. "Mr. Jefferson Prepares an Itinerary." *JSH* 12 (1946): 89–106.

Conant, James B. *Thomas Jefferson and the Development of American Public Education.* Berkeley: Univ. of California Press, 1963.

Constant, Caroline. *The Palladio Guide.* Princeton: Princeton Architectural Press, 1985.

Conway, John J. *Footprints of Famous Americans in Paris.* New York, 1912.

Cooke, Jacob E. "The Collaboration of Tench Coxe and Thomas Jefferson." *PMHB* 100 (1976): 468–90.

Corwin, Edward S. *French Policy and the American alliance of 1778.* Princeton: Princeton Univ. Press, 1916. Repr. 1970.

———. *The President: Office and Powers, 1787–1948* New York: New York University Press, 1948.

Coupland, Reginald. *Quebec Act, A Study in Statesmanship.* Oxford: Clarendon Press, 1968.

Cousins, Norman, ed. *In God We Trust: The Religious Beliefs and Ideas of the American Founding Fathers.* New York: Harper & Row, 1958.

Coves, Elliott, ed. *History of the Lewis and Clark Expedition.* 3 vols. New York: Dover, 1965.

Cox, James M. "Jefferson's Autobiography: Recovering Literature's Lost Ground." *Southern Review* 14 (1978): 633–52.

Crackel, Theodore J. *Mr. Jefferson's Army: Political and Social Reform of the Military Establishment, 1801–1809.* New York: New York Univ. Press, 1987.

———. "The Founding of West Point: Jefferson and the Politics of Security." *Armed Forces,* 7 (1981): 529–43.

———. "Jefferson, Politics, and the Army: An Examination of the Military Peace Establishment Act of 1802." *Journal of the Early Republic* 2 (1982): 21–38.

Crader, Diana C. "The Zooarchaeology of the Storehouse and the Dry Well at Monticello." *American Antiquity* 49 (1984): 542–58.

Cranston, Maurice. "Should We Celebrate the French Revolution?" *American Spectator* 22 (1989): 15–17.

Crevecoeur, J. Hector St. John de. *Letters from An American Farmer and Sketches of Eighteenth-Century America.* New Haven: Yale Univ. Press, 1925.

Cripe, Helen. *Thomas Jefferson and Music.* Charlottesville: Univ. Press of Virginia, 1974.

Crocker, Lester. "Interpreting the Enlightenment: A Political Approach." *Journal of the History of Ideas* 46 (1985): 211–30.

Crout, Robert R. "Vergennes, the United States, and History: Or, What's in a Name?" *Consortium on Revolutionary Europe 1750–1850 Proceedings,* (1981): 103–110.

Crow, Jeffrey J., and Merrill Peterson. *Thomas Jefferson and the American Revolution.* Charlottesville, Va: Univ. Press of Virginia, 1976.

Crow, Thomas E. *Painters and Public Life in Eighteenth Century Paris.* New Haven: Yale Univ. Press, 1985.

Cullen, Charles T. "New Light on John Marshall's Legal Education and Admission to the Bar." *American Journal of Legal History. WMQ* 3rd ser., 13 (1956): 40–52.

———. "Thomas Jefferson: Writings on the Constitution." *This Constitution* 13 (1986): 27–33.

Cunliffe, Marcus. "Thomas Jefferson and the Dangers of the Past." *WLB* 6 (1982): 96–107.

Cunningham, Noble E., Jr. *The Jeffersonian Republicans: the Formation of Party Organization. 1789–1801.* Chapel Hill: Univ. of North Carolina Press, 1957.

———. *The Jeffersonian Republicans in Power: Party Operations, 1801–1809.* Chapel Hill: Univ. of North Carolina Press, 1963.

———. *The Process of Government Under Jefferson.* Princeton: Princeton Univ. Press, 1978.

———. *In Pursuit of Reason: The Life of Thomas Jefferson.* New York: Ballantine, 1987.

———. *The Image of Thomas Jefferson in the Public Eye: Portraits for the People 1800–1809.* Charlottesville: Univ. Press of Virginia, 1981.

———. *The United States in 1800: Henry Adams Revisited.* Charlottesville: Univ. Press of Virginia, 1988.

———. "John Beckley: An Early American Party Manager." *WMQ,* 3rd ser. 13 (1956): 40–52.

Cuthbert, Norma B. "Poplar Forest: Jefferson's Legacy to His Grandson." *HLQ* 6 (1943): 333–56.

Cutright, Paul R. *Lewis and Clark: Pioneering Naturalists.* Urbana: Univ. of Illinois Press, 1969.

Dabney, Virginius. *The Jefferson Scandals: A Rebuttal.* New York: Dodd, Mead, 1981.

———. *Mr. Jefferson's University: A History.* Charlottesville: Univ. Press of Virginia, 1981.

———. "The Monticello Scandals: History and Fiction." *VC* 29 (1979): 52–61.

———. *Virginia: The New Dominion.* Garden City, N.Y.: Doubleday, 1971.

Daiker, Virginia. "The Capitol of Jefferson and Latrobe." *QJLC* 32 (1975): 25–32.

Dakin, Douglas. *Turgot and the Ancien Régime in France.* New York: Octagon, 1965.

Daniels, Jonathan. *Ordeal of Ambition: Jefferson, Hamilton, Burr.* Garden City, N.Y.: Doubleday, 1970.

———. *The Randolphs of Virginia.* Garden City, N.Y.: Doubleday, 1972.

Daniloff, Ruth. "A Cipher's the Key to the Treasure in Them Thar Hills." *Smithsonian,* 1981, 12 (1): 126–44.

Darnton, Robert. "What Was Revolutionary about the French Revolution?" *New York Review of Books,* January 19, 1989, pp. 1–6, 10.

———. *The Great Cat Massacre.* New York: Basic Books, 1984.

———. *Literary Underground of the Old Regime,* Cambridge, Mass.: Harvard Univ. Press, 1982.

———. *Mesmerism and the End of the Enlightenment in France.* Cambridge, Mass.: Harvard Univ. Press, 1968.

Daveiss, Joseph H. *A View of the President's Conduct Concerning the Conspiracy of 1806.* Frankfort, Ky., 1806.

Davis, David Brion. *The Problem of Slavery in Western Culture.* Ithaca, N.Y.: Cornell Univ. Press, 1966.

Davis, John W. "Thomas Jefferson, Attorney-at-Law." *Proceedings, Virginia State Bar Association* 38 (1926): 361–77.

Davis, Richard Beale. *Intellectual Life in the Colonial South, 1585–1763.* 3 vols. Knoxville: Univ. of Tenn. Press, 1978.

———. *Intellectual Life in Jefferson's Virginia, 1790–1830.* Chapel Hill: Univ. of North Carolina Press, 1964.

Davis, Richard Beale, ed. *Correspondence of Thomas Jefferson and Francis Walker Gilmer, 1814–1826.* Columbia, S.C.: Univ. of So. Carolina Press, 1946.

Davis, Robert R., Jr. "Pell-Mell: Jeffersonian Etiquette and Protocol." *Historian* 43 (1981): 509–29.

Dawidoff, Robert. "The Fox in the Henhouse: Jefferson and Slavery." *Reviews in American History.* 6 (1978): 503–11.

D'Elia, Donald J. *Jefferson, Rush, and the Limits of Philosophical Friendship. PAPS* 117 (1973): 333–43.

DeVoto, Bernard. ed., *Journals of Lewis and Clark.* Boston: Houghton Mifflin, 1953.

Dewey, Donald O. *Marshall Versus Jefferson: The Political Background of Marbury v. Madison.* New York: Knopf, 1970.

Dewey, Frank L. *Thomas Jefferson, Lawyer.* Charlottesville: Univ. Press of Virginia, 1986.

———. "New Light on the General Court of Colonial Virginia." *William & Mary Law Review.* 21 (1979): 1–14.

———. "The Waterson-Madison Episode: An Incident in Thomas Jefferson's Law Practice." *VMHB* 90 (1982): 165–76.

———. "Thomas Jefferson and a Williamsburg Scandal: The Case of Blair v. Blair." *VMHB* 89 (1981): 44–63.

———. "Thomas Jefferson's Law Practice." *VMHB* 85 (1977): 289–301.

———. *Thomas Jefferson's Law Practice: The Norfolk Anti-Inoculation Riots.* *VMHB* 91 (1983): 39–53.

———. *Thomas Jefferson's Notes on Divorce.* *WMQ* 39 (1982): 212–23.

Dickinson, H.T. *Bolingbroke.* London: Constable, 1970.

Diderot, Denis. *Salons.* Oxford: Clarendon Press, 1975.

Diggins, John Patrick. "Comrades and Citizens: New Mythologies in American Historiography." *AHR* 90 (1985): 614–38.

———. "Slavery, Race and Equality: Jefferson and the Pathos of the Enlightenment," *American Quarterly* 28 (1976): 206–28.

Dill, Alonzo T. *George Wythe, Teacher of Liberty,* Williamsburg, Va.: Independence Bicentennial Commission, 1979.

Dillon, Raymond. *Meriwether Lewis.* New York: Coward-McCann, 1965.

Donnelly, Marian C. "Jefferson's Observatory Design." *Journal of the Society of Architectural History* 36 (1977): 33–35.

Dorman, Robert L. "Thomas Jefferson's Letter to the Indians: Fate of a Frontier Artifact." *Chronicles of Oklahoma* 63 (1985–86): 340–59.

Dos Passos, John. *The Head and Heart of Thomas Jefferson.* Garden City, N.Y.: Doubleday, 1954.

———. *The Shackles of Power: Three Jeffersonian Decades.* Garden City, N.Y.: Doubleday, 1966.

Drago, George. *Jefferson's Louisiana: Politics and the Clash of Legal Traditions.* Cambridge, Mass.: Harvard Univ. Press, 1975.

Dull, Jonathan R. *A Diplomatic History of the American Revolution.* New Haven: Yale Univ. Press, 1984.

———. *Franklin the Diplomat: The French Mission.* Philadelphia: American Philosophical Society, 1982.

Dumbauld, Edward. *The Declaration of Independence and What It Means Today.* Norman, Okla.: Univ. of Oklahoma Press, 1950.

———. "Independence Under International Law." *American Journal of International Law.* 70 (1976): 425–31.

———. *Thomas Jefferson, American Tourist.* Norman, Okla.: Univ. of Okla. Press, 1946.

———. *Thomas Jefferson and the Law.* Norman, Okla.: Univ. of Okla. Press, 1978.

Dunn, John. *Locke.* New York: Oxford Univ. Press, 1984.

———. *The Political Thought of John Locke: An Historical Account of the "Two Treaties of Government".* Cambridge, Eng.: Cambridge Univ. Press, 1969.

———. "The Politics of Locke in England and America in the Eighteenth Century," in *John Locke: Problems and Perspectives,* ed. John W. Yalton. Cambridge, Eng.: Cambridge Univ. Press, 1969.

Durey, Michael. *"With the Hammer of Truth": James Thomson Callender and America's Early National Heroes.* Charlottesville: Univ. Press of Virginia, 1990.

Duroselle, Jean-Baptiste. *La France et les Etats-Unis des origines a nos jours.* Paris: Seuil, 1976.

Eaton, C. "A Mirror of the Southern Colonial Lawyer: The Fee Books of Patrick

Henry, Thomas Jefferson and Waightsill Avery." *WMQ*, 3rd ser., 8 (1951): 420–34.

Eckenrode, Hamilton J. *The Randolphs: The Story of a Virginia Family*. Indianapolis: Bobbs, Merrill, 1946.

Edward, C. *Jefferson, Sullivan, and the Moose. AHI* 9 (1974): 18–19.

Edwards, Everett. E., ed. *Jefferson and Agriculture: A Sourcebook*. Washington: U.S. Department of Agriculture, 1976, repr. of 1943 ed.

Edwards, Rem B. *A Return to Moral and Religious Philosophy in Early America*. Washington, D.C.: Univ. Press of America, 1982.

Egan, Clifford L. *Neither Peace Nor War: Franco-American Relations, 1803–1812*. Baton Rouge: Louisiana State Univ. Press, 1983.

Egnal, Marc. "Origins of the Revolution in Virginia: A Reinterpretation." *WMQ*, 3rd ser., 37 (1980): 401–28.

Eide, Ingvard Henry. *American Odyssey: The Journey of Lewis and Clark*. Chicago: Rand McNally, 1979.

Ellis, Richard E. *The Jeffersonian Crisis: Courts and Politics in the Young Republic*. New York: Oxford Univ. Press, 1971.

Ernst, Joseph Albert. "The Currency Act Repeal Movement: A Study of Imperial Politics and Revolutionary Crisis, 1764–1767." *WMQ*, 3rd ser., 25 (1968): 177–211.

Evans, Emory G. "Planter Indebtedness and the Coming of the Revolution in Virginia." *WMQ*, 3rd ser., 19 (1962): 511–33.

———. "Private Indebtedness and the Revolution in Virginia, 1776–1796." *WMQ*, 3rd ser., 28 (1971): 349–74.

———. "The Rise and Decline of the Virginia Aristocracy in the Eighteenth Century: The Nelsons," in *The Old Dominion*, ed. Garrett D. Ruttman. Charlottesville: Univ. Press of Virginia, 1964, 62–78.

Farrand, Daniel, et al. "Supplement to the Vermont Centinel." Burlington, Vt.: 1809.

Farrand, Max. "If James Madison Had Had a Sense of Humor." *PMHB* 62 (1938): 134–35.

Faulkner, Robert K. "John Marshall and the Burr Trial." *JAH*, 53 (1966): 247–58.

Ferguson, E. James. *The Power of the Purse: A History of American Public Finance, 1776–1790*. Chapel Hill: Univ. of North Carolina Press, 1961.

Ferguson, Henry N. *The Man Who Saved Monticello. AHI* 14 (1980): 20–27.

Ferguson, Robert A. " 'Mysterious Obligations': Jefferson's Notes on the State of Virginia." *American Literature*, 1980. 52 (3): 381–406.

Ferling, John. *John Adams, A Life*. Knoxville: Univ. of Tenn. Press, 1992.

———. *The First of Men*. Knoxville; Univ. of Tenn. Press, 1987.

Fischer, David Hackett. *The Revolution of American Conservatism: The Federalist Party in the Era of Jeffersonian Democracy*, New York: Harper & Row, 1965.

Fisher, George (pseud.) *The American Instructor: or, Young Man's Best Companion*, 9th ed. Philadelphia, 1748.

Fitch, James Marston. "The Lawn: America's Greatest Architectural Achievement." *AHM* 35 (1984): 49–64.

Fitch, Noel Ripley. *Literary Cafes of Paris.* Washington, D.C.: Starrhill Press, 1989.

Fithian, Philip Vickers. *Journals and Letters, 1767–1774,* ed. J.R. Williams, Princeton: Princeton Univ. Press, 1900.

———. *Journals and Letters, 1773–1774.* Williamsburg: Colonial Williamsburg, 1957.

———. *Journals and Letters, 1775–1776.* Princeton: Princeton Univ. Press, 1934.

Fleming, Thomas. *The Man from Monticello: An Intimate Life of Thomas Jefferson.* New York: William Morrow, 1969.

———. "Monticello's Long Career—From Riches to Rags to Riches." *Smithsonian* 4 (1973): 62–69.

Flores, Dan L. *Jefferson and Southwest Exploration.* Norman: Univ. of Oklahoma Press, 1984.

———. "Rendezvous at Spanish Bluff: Jefferson's Red River Exploration." *Red River Valley History Review.* 1979 4 (2): 4–26.

Fohlen, Claude. "The Impact of the American Revolution Abroad," in *U.S. Library of Congress Symposia on the American Revolution Papers.* Washington, D.C.: Library of Congress, 1976.

———. "Jefferson et L'Achat de la Louisiane." *Histoire* (France), 1978 (5): 75–77.

———. *Thomas Jefferson.* Nancy: Presses Universitaires de Nancy, 1992.

Folley, William E. Rice. "Visiting the President: An Exercise in Jeffersonian Indian Diplomacy." *American West* 16 (1979): 4–15, 56.

Foner, Philip S. ed. *Democratic-Republican Societies, 1790–1800.* Westport: Greenwood Press, 1976.

———. *Tom Paine and Revolutionary America.* New York, 1976.

Fontaine, John. *Journal.* ed. E.P. Alexander. Williamsburg: Colonial Williamsburg Foundation, 1972.

Franklin, Benjamin. *Bagatelles From Passy.* New York: Eakins Press, 1967.

———. *The Complete Poor Richard's Almanacks,* intro., Whitfield J. Bill, Jr., 2 vols. Barre, Mass.: 1970.

Freehling, William W. "The Founding Fathers and Slavery." *AHR* 71 (1972): 81–93.

Friedenberg, Daniel M. *Life, Liberty and the Pursuit of Land: The Plunders of America.* Buffalo: Prometheus, 1992.

Frisch, Morton J. "Hamilton's Report on Manufactures and Political Philosophy." *Publius* 8 (1978): 129–39.

Frye, Melinda Y. *Thomas Jefferson and Wine in Early America* (microform). San Francisco: Wine Institute of America, 1976.

Gaines, William H., Jr. *Thomas Mann Randolph, Jefferson's Son-in-Law,* Baton Rouge: Louisiana State Univ. Press, 1966.

Galtier, Gaston. "La Viticulture de l'Europe occidentale à la veille de la Révolution française, d'après les notes de voyage de Thomas Jefferson." *Societé Languedocienne de Géographie* 3 (1968): 7–72.

Gay, Peter. *The Enlightenment: An Interpretation* 2 vols. N.Y.: Knopf, 1966–69.

Gerbi, Antonello. *The Dispute of the New World: The History of a Polemic, 1750–1900* trans. Jeremy Moyle, Pittsburgh: Univ. of Pittsburgh Press, 1973.

Gifford, James M. *Slavery and Jeffersonian Virginia.* Urbana, Ill.: Univ. of Ill. Press, 1973.

Gifford, Prosser. *The Treaty of Paris in a Changing State System.* Lanham, Md.: Univ. Press of America, 1985.

Gillies, Paul S. "Adjusting to Union: An Assessment of Statehood, 1791–1816." In *A More Perfect Union: Vermont Becomes a State, 1777–1816.* ed. Michael Sherman. Montpelier: Vermont Historical Society, 1991.

Gilreath, James, and Douglas L. Wilson, eds. *Thomas Jefferson's Library: A Catalog with the Entries in His Own Order.* Washington, D.C.: Library of Congress, 1989.

Ginsberg, Robert. "Suppose That Jefferson's Rough Draft of the Declaration of Independence is a Work of Political Philosophy." *Eighteenth Century Theory and Interpretation* 25 (1984): 25–43.

Gipson, Lawrence Henry. *The British Empire Before the American Revolution* 15 vols. New York: Knopf, 1966–1970.

———. "Virginia Planter Debts Before the American Revolution." *VMHB* 69 (1961): 259–77.

Gittleman, Edwin. "Jefferson's 'Slave Narrative': The Declaration of Independence as a Literary Text." *EAL* 8 (1974): 239–56.

Goering, Wynn. "Lovers of Peace and Order." *Mennonite Life* 40 (1985): 11–15.

Golladay, V. Dennis. "Jefferson's 'Malignant Neighbor,' John Nicholas, Jr." *VMHB* 86 (1978): 306–19.

Goodman, Paul. "Social Status of Party Leadership, 1797–1804." *WMQ*, 3rd ser., 25 (1968): 465–74.

Goodman, Warren H. "The Origins of the War of 1812: A Survey of Changing Interpretations." *MVHR* 28 (1941): 171–86.

Green, Constance M. *Washington, Village and Capital, 1800–1878.* Princeton: Princeton Univ. Press, 1962.

Greene, Jack P. *The Quest for Power: the Lower Houses of Assembly in the Southern Royal Colonies.* Williamsburg: Institute of Early American History and Culture, 1963.

———, and J.R. Pole, eds. *Colonial British America: Essays in the New History of the Early Modern Era.* Baltimore, Johns Hopkins Univ. Press, 1984.

———. "The Currency Act of 1764 in Imperial-Colonial Relations, 1764–1776." *WMQ*, 3rd ser., 18 (1961): 485–518.

Griswold, A. Whitney. "The Agrarian Democracy of Thomas Jefferson." *American Political Science Review.* (1946): 657–81.

Grodzins, Morton. "Political Parties and the Crisis of Succession in the United States: The Case of 1800," in *Political Parties and Political Development,* ed. Joseph La Palombara and Myron Werner. Princeton: Princeton Univ. Press, 1966.

Guinness, Desmond. "Thomas Jefferson: Visionary Architect." *Horizon,* 22 (1979): 51–55.

Gunderson, Joan R., and Gwen V. Gampel. "Married Women's Legal Status in Eighteenth-Century New York and Virginia." *WMQ,* 3rd ser., 39 (1982): 114–34.

Hamilton, J.G. de Roulhac. "The Pacifism of Thomas Jefferson." *VQR* (1955): 607–20.

Hammond, Bray. *Banks and Politics in America from the Revolution to the Civil War.* Princeton: Princeton Univ. Press, 1957.

Hamowy, Ronald. "Jefferson and the Scottish Enlightenment." *WMQ,* 3rd ser., 36 (1979): 503–23.

Harnsberger, Douglas. "In Delorme's Manner. . . ." *APT Bulletin* (Canada) 13 (1981): 2–8.

Hart, F.H. *The Valley of Virginia in the American Revolution.* Chapel Hill, Univ. of North Carolina Press, 1942.

Haskins, Caryl P. "Mr. Jefferson's Sacred Gardens." *VQR* (1967): 529–44.

———. "Law Versus Politics in the Early Years of the Marshall Court." *University of Pennsylvania Law Review* 130 (1981): 1–27.

Hawke, David Freeman. *Those Tremendous Mountains,* New York: Norton, 1980.

[Hay, George]. *Hortensius. An Essay on the Liberty of the Press.* Philadelphia, 1799.

———. *An Essay on the Liberty of the Press . . .* From Libellers, Richmond, 1803.

Healey, Robert M. "Jefferson on Judaism and the Jews." *AJH* 73 (1984): 359–74.

———. *Jeffersonian Religion in Public Education.* New Haven: Yale Univ. Press, 1962.

Hellenbrand, Harold. *The Unfinished Revolution: Education and Politics in the Thought of Thomas Jefferson.* Newark, Del.: Univ. of Delaware Press, 1990.

———. "Not 'to Destroy but to Fulfil': Jefferson, Indians, and Republican Dispensation." *Eighteenth-Century Studies* 18 (1985): 523–49.

———. "Roads to Happiness: Rhetorical and Philosophical Design in Jefferson's Notes on the State of Virginia." *EAL* 20 (1985): 3–23.

Hemings, Madison. "Life Among the Lowly, No. 1." Pike County (Ohio) *Republican,* March 13, 1873. Repr. in Brodie, Fawn. *Thomas Jefferson: An Intimate History.* 637–45.

Hemphell, John M. II. "John Wayles Rates His Neighbors." *VMHB* 66 (1958): 302–6.

———, ed. "Edmund Randolph Assumes Thomas Jefferson's Practice." *VMHB* 67 (1959): 170–71.

Henderson, Archibald. *Dr. Thomas Walker of the Loyal Company of Virginia.* Worcester, Mass.: American Antiquarian Society, 1951.

Henderson, H. James. *Jefferson, Nationalism, and the Enlightenment.* New York: Braziller, 1975.

Henderson, John C. *Thomas Jefferson's Views on Public Education.* New York: Putnam's, 1890.

Henderson, Phillip, G. "Marshall Versus Jefferson: Politics and the Federal Judiciary in the Early Republic." *Michigan Journal of Political Science* 2 (1983): 42–66.

Hendrickson, Robert A. *The Rise and Fall of AH.* New York: Van Nostrand Reinhold, 1981.

Henrich, Joseph George. "Thomas Paine's Short Career as a Naval Architect, August-October, 1807." *American Neptune* 34 (1974): 123–34.

Herndon, G. Melvin. "Keeping an Eye on the British: William Tatham and the Chesapeake Affair." *VC* 22 (1972): 30–39.

Hickey, Donald R. *War of 1812.* Urbana, Ill.: Univ. of Ill. Press, 1989.

———. "Timothy Pickering and the Haitian Slave Revolt: A Letter to Thomas Jefferson in 1806." *EIHC* 120 (1984): 149–63.

Higgenbotham, Don. *War of American Independence.* Boston: Northeastern Univ. Press, 1983.

Hindle, Brooke. *The Pursuit of Science in Revolutionary America.* Chapel Hill: Univ. of North Carolina Press, 1956.

Hobson, Charles F. "Recovery of British Debts in the Federal Circuit Court of Virginia, 1790 to 1797." *VMHB* 92 (1984): 176–200.

Hochman, Steven H. "Thomas Jefferson: A Personal Financial Biography." Unpub. doct. diss., University of Virginia, 1987.

Hockman, Daniel M. "The Dawson Brothers and the Virginia Commissariat, 1743–1760," Unpub. doct. diss., University of Illinois, 1975.

Hoeveler, J. David, Jr. "Thomas Jefferson and the American 'Provincial' Mind." *Modern Age* 25 (1981): 271–80.

Hoffmann, John. "Queries Regarding the Western Rivers: An Unpublished Letter from Thomas Jefferson to the Geographer of the United States." *Journal of the Illinois State Historical Society* 75 (1982): 15–28.

Hofstadter, Richard. *The Idea of a Party System: The Rise of Legitimate Opposition in the United States, 1780–1840.* Berkeley: Univ. of California Press, 1969.

Honeywell, Roy J. *The Educational Work of Thomas Jefferson,* Cambridge, Mass.: Harvard Univ. Press, 1931.

Horrocks, Thomas. "Thomas Jefferson and the Great Claw." *VC* 35 (1985): 70–79.

Horsman, Reginald. *Diplomacy of the New Republic, 1776–1815.* Arlington, Ill.: Harlan Davidson, 1985.

———. "Thomas Jefferson and the Ordinance of 1784." *Illinois Historical Journal* 79 (1986): 99–112.

Hoskins, Janina. "A Lesson Which All Our Countrymen Should Study: Jefferson Views Poland." *QJLC* 33 (1976): 29–46.

Howell, Wilbur Samuel. "The Declaration of Independence: Some Adventures with America's Political Masterpiece." *Quarterly Journal of Speech* 62 (1976): 221–33.

Howell, Wilbur Samuel, ed. *Jefferson's Parliamentary Writings.* Princeton: Princeton Univ. Press, 1988.

Hulliung, Mark. *Citizen Machiavelli.* Princeton: Princeton Univ. Press, 1983.

———. *Montesquieu and the Old Regime.* Berkeley: Univ. of California Press, 1976.

Hume, Edgar E. *Lafayette and the Society of the Cincinnati.* Baltimore: Johns Hopkins Univ. Press, 1934.

Humphreys, Francis L. *Life and Times of David Humphreys,* 2 vols, New York: G.P. Putnam's Sons, 1917.

Huntley, William B. "Jefferson's Public and Private Religion." *SAQ* 79 (1980): 286–301.

Isaac, Rhys. *Transformation of Virginia, 1740–1790.* Chapel Hill: Univ. of North Carolina Press, 1982.

Jackson, Donald. *A Year at Monticello: 1795.* Golden, Colorado: Fulcrum, Inc., 1989.

———. *Letters of the Lewis and Clark Expedition.* Urbana: Univ. of Ill. Press, 1962.

———. *Jefferson, Meriwether Lewis and the Reduction of the United States Army. PAPS* 124 (1980): 91–96.

———. *Thomas Jefferson and the Stony Mountains: Exploring the West from Monticello.* Urbana, Ill.: Univ. of Illinois Press, 1981.

———. "On the Death of Meriwether Lewis's Servant." *WMQ,* 3rd ser., 21 (1964): 445–48.

Jacobs, Victor. "Was Thomas Jefferson Really Very Bright?" *Manuscripts* 34 (1982): 21–24.

Jayne, Allen. *The Religious and Moral Wisdom of Thomas Jefferson: An Anthology.* New York: Vantage Press, 1984.

[Jefferson, Isaac.] *Memoirs of a Monticello Slave.* ed. Rayford W. Logan. Charlottesville: Univ. Press of Virginia, 1951.

[Jefferson,] Israel. "Reminisinces," in "Life among the Lowly, No. 3," Pike County (Ohio) *Republican,* Dec. 25, 1873.

Jefferson, Thomas. *The Adams-Jefferson Letters: the Complete Correspondence between Thomas Jefferson and Abigail and John Adams* ed. Lester J. Cappon. Chapel Hill: Univ. of North Carolina Press, 1959.

———. *Autobiography.* ed. Adrienne Koch and William Peden. New York: Modern Library, 1944.

———. *Extracts from the Gospels,* ed. Dickinson W. Adams. Princeton: Princeton Univ. Press, 1983.

———. *Family Letters,* ed. Edwin M. Batts and James A. Bear, Jr. Columbia, Mo.: Univ. of Missouri Press, 1966.

———. *Farm Book,* ed. Edwin M. Betts, Philadelphia: American Philosophical Society, 1953. Repr. 1987.

———. *Garden Book, 1766–1824* ed. Edwin M. Betts. Philadelphia: American Philosophical Society, 1944.

———. *The Jefferson Bible,* Charlottesville, Va.: Univ. Press of Virginia, 1988.

———. *Jefferson-Kosciuszko Correspondence,* ed. Bodgan Grzelonski. Warsaw: Interpress, 1978.

———. *Life and Selected Writings of TJ,* ed. Adrienne Koch and William H. Peden. New York: Modern Library, 1944.

———. *Literary Commonplace Book,* ed. Douglas L. Wilson. Princeton: Princeton Univ. Press, 1989.

———. *Message from the President of the United States Communicating Discoveries.* Washington, D.C.: 1806.

———. *Notes on the State of Virginia,* ed. William Peden. Chapel Hill, N.C.: Univ. of North Carolina Press, 1954.

———. *The Papers of Thomas Jefferson* 25 vols. to date. Princeton: Princeton Univ. Press, 1950.

———. *Parliamentary Writings,* ed. Wilbur S. Howell. Princeton: Princeton Univ. Press, 1988.

———. *Reports of Cases Determined in the General Court of Virginia.* Buffalo: William S. Hein, 1981. Repr. of 1829 ed.

———. *Writings,* selected and ed. Merrill D. Peterson. New York: Library of America, 1984.

———. "Some Letters of Jefferson." *Southern Bivouac* of Louisville, Ky., New Series, 2 (1887): 632–38, 752–59.

———. *Thomas Jefferson and His Unknown Brother,* ed. Bernard Mayo. Charlottesville: Univ. Press of Virginia, 1981.

———. *Works.* ed. Paul Leicester Ford. 10 vols. New York, 1892–99.

———. *Writings,* ed. A.A. Lipscomb and A.E. Bergh. 20 vols. Washington, D.C.: TJ Memorial Assn. of the U.S., 1903.

"Thomas Jefferson's Advice to the Cherokees." *Journal of Cherokee Studies,* 4 (1979): 64–66.

Jefferson, Peter. *Account Book.* Mss., HEH.

Jellison, Charles A. "James Thomson Callender: Human Nature in a Hideous Form." *VC* 29 (1979): 62–69.

Jennings, Walter Wilson. *The American Embargo, 1807–1809.* Iowa City, Iowa: Univ. of Iowa Press, 1921.

Jensen, Merrill. *The Founding of a Nation: A History of the American Revolution, 1763–1776.* New York: Knopf, 1968.

———. *The New Nation.* New York: Knopf, 1950.

Johansen, Bruce Eliott. *Franklin, Jefferson and American Indians: A Study in the Cross-cultural Communication of Ideas.* Seattle, Wash.: Univ. of Washington Press, 1979.

Johnson, James W. *The Formation of English Neo-Classical Thought,* Princeton: Princeton Univ. Press, 1967.

Johnson, Ludwell H., III., "Sharper Than a Serpent's Tooth, Thomas Jefferson and His Alma Mater. *VMHB,* 99 (1991): 145–162.

Johnstone, Robert M., Jr. *Jefferson and the Presidency: Leadership in the Young Republic.* Ithaca, N.Y.: Cornell Univ. Press, 1978.

Jones, Alice Hanson. *Wealth of a Nation to Be: The American Colonies on the Eve of the Revolution.* New York: Columbia Univ. Press, 1980.

Jones, Howard Mumford. *America and French Culture, 1750–1848.* Chapel Hill: Univ. of North Carolina Press, 1927.

Jordan, Daniel P. *Political Leadership in Jefferson's Virginia,* Charlottesville: Univ. Press of Virginia, 1983.

———. *A Richmond Reader, 1733–1983.* Chapel Hill, Univ. of North Carolina Press, 1983.

Jordan, Winthrop D. *White Over Black: American Attitudes Toward the Negro, 1550–1812.* Chapel Hill, N.C.: Univ. of North Carolina Press, 1968.

Jurden, D.A. "A Historiography of American Deism." *American Benedictine Review* 25 (1974): 108–22.

Kammen, Michael. *Dimensions of a New Identity: The 1973 Jefferson Lectures in the Humanities.* New York: Norton, 1974.

———. "The Founding Fathers: In Search of Fame and Identity." *RAH* 3 (1975): 196–205.

Kaplan, Lawrence S. *Jefferson and France: An Essay on Politics and Political Ideas.* New Haven, 1980 repr. of 1967 ed.

———. "Reflections on Jefferson as a Francophile." *SAQ* 79 (80): 38–50.

———. "Toward Isolationism: The Jeffersonian Republicans and the Franco-American Alliance of 1778." *Historical Reflections,* (Canada) 3 (1976): 69–81.

Katz, Stanley N. "Thomas Jefferson and the Right to Property in Revolutionary America." *Journal of Law and Economics* 19 (1976): 467–88.

Keats, John. *Eminent Domain: The Louisiana Purchase and the Making of America.* New York: Charterhouse, 1973.

Keim, C. Ray. "Primogeniture and Entail in Colonial Virginia." *WMQ,* 3rd ser., 25 (1968): 545–86.

Kelso, William M. "Mulberry Row: Slave Life at Thomas Jefferson's Monticello." *Archaelogy.* 39 (1986), 28–35.

Kennedy, J.P. and H.R. McIlwaine, eds., *Journal of the House of Burgesses of Virginia,* 13 vols. Richmond: 1905–1915.

Kenny, M. Kate. "The Black Snake Affair: Implications for the Present." *Chittenden County (Vermont) Historical Society Bulletin* 22 (1987): 1–4.

Keohane, Nannerl O. *Philosophy and the State in France: The Renaissance to the Enlightenment.* Princeton: Princeton Univ. Press, 1980.

Kessler, Sanford. "Locke's Influence on Jefferson's 'Bill for Establishing Religious Freedom.' " *Journal of Church and State* 25 (1983): 231–52.

Ketcham, Ralph. *Presidents above Party: The First American Presidency, 1789–1829.* Chapel Hill: Univ. of North Carolina Press, 1984.

———. "The Transatlantic Background of Thomas Jefferson's Ideas of Executive Power." *Studies in 18th Century Culture* 11 (1982): 163–80.

Kett, Joseph F. "Education," in *TJ, A Reference Biography.* ed. M.D. Peterson, New York: Charles Scribner's Sons, 1986.

Kimball, Fiske. *Thomas Jefferson, Architect,* 2nd. ed. New York: Da Capo, 1968.

———. "In Search of Jefferson's Birthplace." *VMHB* 51 (1943), 313–25.

Kimball, Marie. *Jefferson: The Road to Glory, 1743–1776.* New York: Coward: McCann, 1943.

———. *Jefferson: The Scene of Europe, 1784–1789.* New York, Coward-McCann, 1950.

———. *Jefferson, War and Peace, 1776–1784.* New York: Coward-McCann, 1947.

———. *Thomas Jefferson's Cook Book.* Charlottesville: 1976.

———. "Europe Comes to Jefferson." *American-German Review.* 15 (1949), 15–17, 30.

———. "Jefferson in Paris." *North American Review.* 248 (1939): 73–86.

King, Lisa. "America's First Connoisseur." *Wine Spectator* 15 (1991): 24–33.

Kite, Elizabeth Sarah. *L'Enfant and Washington, 1791–1792.* Baltimore: Johns Hopkins, 1929.

Klingelhofer, Herbert E. "Abolish the Navy!" *Manuscripts,* 1981. 33 (4): 277–84.

Kloppenberg, James T. "The Virtues of Liberalism: Christianity, Republicanism, and Ethics in Early American Political Discourse." *JAH* 74 (1987): 9–33.

Knudson, Jerry W. "Jefferson the Father of Slave Children? One View of the Book Reviewers." *JH* 3 (1976): 56–58.

———. "The Jeffersonian Assault on the Federalist Judiciary, 1802–1805: Political Forces and Press Reaction." *AJLH* 14 (1970): 55–75.

———. "The Myth of Black Sally." *NHB* 32 (1969): 15–22.

———. "Political Journalism in the Age of Jefferson." *JH* 1 (1970): 20–23.

Koch, Adrienne. *The American Enlightenment.* New York: G. Braziller, 1965.

———. *Jefferson and Madison: The Great Collaboration.* New York: Knopf, 1964.

———. *The Philosophy of Thomas Jefferson.* New York: Columbia Univ. Press, 1943.

Koch, Adrienne, and Harry Ammon. "The Virginia and Kentucky Resolutions: An Episode in Jefferson's and Madison's Defense of Civil Liberties." *WMQ,* ed ser., 5 (1948): 141–76.

Koch, Adrienne and William Peden, eds. *Life and Selected Writings of Thomas Jefferson.* New York: Modern Library, 1944, repr. 1972.

Koenig, Louis W. "Consensus Politics, 1800–1805." *AH* 18 (1967): 4–7, 74–80.

Kramnick, Isaac F. *Bolingbroke and His Circle.* Cambridge, Mass.: Harvard Univ. Press, 1968.

Kukla, Jon. "Flirtation and Feux D'Artifices: Mr. Jefferson, Mrs. Cosway, and Fireworks." *VC* 26 (1976): 52–63.

Kuper, Theodore F. "Jefferson and Italy: Vital Contacts Between Two Great Peoples." *Atlantica.* 15 (1933): 8–10, 37.

Kurtz, Stephen G. *The Presidency of John Adams.* Philadelphia: Univ. of Pennsylvania Press, 1957.

Labaree, Leonard, and William Wilcox, et al., eds. *Papers of Benjamin Franklin.* 28 vols. to date. New Haven: Yale Univ. Press, 1959 to date.

Ladenson, Alex. "I Cannot Live Without books: Thomas Jefferson, Bibliophile." *WLB* 52 (1978): 624–31.

Lafayette, Marquis de. *Letters of Lafayette and Jefferson,* ed. Gilbert Chinard. Baltimore: Johns Hopkins Univ. Press, 1929.

Landers, H.L. *The Virginia Campaign and the Blockade and Siege of Yorktown.* Washington, D.C.: U.S. Army War College, 1931.

Lang, H. Jack. "Last Letters From the Valiant." *Manuscripts* 28 (1976): 195–201.

Langhorne, Elizabeth. "A Black Family at Monticello." *Magazine of Albemarle County History* 43 (1985): 1–16.

———. *Monticello: A Family Story.* Chapel Hill: Algonquin Books, 1987.

La Rochefoucauld Liancourt, Duc de. *Travels Through the U.S. of North America.* 2 vols. London: R. Phillips, 1799.

Larson, Martin A. *Jefferson, Magnificent Populist.* Washington, D.C.: Robert D. Luce, 1981.

Larus, Joel. "Pell-Mell Along the Potomac," *WMQ,* 3rd ser., 17 (1960): 349–57.

Laslett, Peter, ed. *John Locke, Two Treatises of Government,* Cambridge: Cambridge Univ. Press, 1960.

Lassiter, Francis R. "Arnold's Invasion of Virginia," *Sewanee Review,* 9 (1901): 78–93, 185–203.

Laub, C.H. "Revolutionary Virginia and the Crown Lands," 1775–1783. *WMQ,* 2nd ser., 11 (1931): 304–14.

Lawrence, R. de Trevelle, III. *Jefferson and Wine.* The Plains, Va.: Vinifera Wine Growers Assn., 1976.

Lehmann, Karl. *TJ, American Humanist.* New York: Macmillan, 1947.

Leiner, Frederick C. "The 'Whimsical Phylosophic President' and his Gunboats." *American Neptune* 43 (1983): 245–66.

Lerche, Charles O., Jr. "Jefferson and the Election of 1800: A Case Study in the Political Smear." *WMQ,* 3rd ser., 5 (1948): 467–91.

Levy, Leonard W. *Jefferson and Civil Liberties,* Cambridge, Mass.: Harvard Univ. Press, 1963.

———. "Liberty and the First Amendment, 1790–1800," *AHR* 68 (1962): 22–37.

Lewis, Clayton W. "Style in Jefferson's Notes on the State of Virginia." *Southern Review* 14 (1978): 668–76.

Lewis, Jan. *The Pursuit of Happiness: Family and Values in Jefferson's Virginia.* New York: Cambridge Univ. Press, 1983.

Lewis, Marc Anthony. "Jefferson and Virginia's Pioneers, 1774–1781." *MVHR* 34 (1948): 551–88.

Lewis, Warren H. *The Splendid Century: Life in the France of Louis XIV.* Garden City, N.Y.: Doubleday, 1957.

Llewellyn, Robert. "A New View of Monticello." *HP* 35 (1983): 48–51.

———. *Thomas Jefferson's Monticello.* Charlottesville: Univ. Press of Virginia, 1983.

Lomask, Milton. *Aaron Burr.* 2 vols. New York: Farrar, Strauss and Giroux, 1979.

Lopez, Claude Anne. *Mon Cher Papa: Franklin and the Ladies of Paris.* New Haven: Yale Univ. Press, 1990.

Lopez, Claude-Anne and Eugenia W. Herbert. *The Private Franklin: The Man and His Family.* New York: Norton, 1975.

Lough, J. *An Introduction to Eighteenth Century France.* London: Longmans, 1988.
———. *France on the Eve of Revolution: British Travellers' Observations, 1763–1788.* London: Croom Helm, 1987.
Luker, Ralph E. "Garry Wills and the New Debate Over the Declaration of Independence." *VQR* 56 (1980): 244–66.
Luttrell, Clifton B. "Thomas Jefferson on Money and Banking: Disciple of David Hume and Forerunner of Some Modern Monetary Views." *History of Political Economy* 7 (1975): 156–73.
Lynn, Kenneth S. "Falsifying Jefferson." *Commentary* 66 (1978): 66–71.
Mabbutt, Fred R. "The New Guardians: Education and Technology." *Colorado Quarterly* 24 (1975): 155–71.
Maclay, William. *Diary.* ed. Kenneth B. Bowling. Baltimore: Johns Hopkins Univ. Press, 1988.
Madison, James. *Papers.* 17 vols. to date. Chicago: Univ. of Chicago Press, 1962.
Mahon, John K. *The American Militia: Decade of Decision, 1789–1799.* Gainesville: Univ. of Florida Press, 1960.
Mahon, Terrence. "Virginia Reaction to British Policy, 1763–1776." Unpub. doct. diss., U. Wisconsin, 1960.
Main, Jackson Turner. "The One Hundred." *WMQ,* 3rd ser., 11 (1954): 354–84.
———. *Political Parties Before the Constitution.* Chapel Hill: Univ. of North Carolina Press, 1973.
———. *The Soverign States, 1775–1783.* Chapel Hill: Univ. of North Carolina Press, 1973.
Malone, Dumas. *Thomas Jefferson and His Times.* 6 vols. Boston: Little, Brown, 1948–81.
———. "At Home with Thomas Jefferson." *NYT,* July 1, 1956. pp. 8, 18–19.
———. "The Fear of Ideas." *AS* 21 (1952): 413–422.
———. *The Fry and Jefferson Map of Virginia and Maryland,* Princeton, Princeton Univ. Press: 1950.
———. "Jefferson Goes to School in Williamsburg." *VQR* (1957): 481–96.
———. "Monticello." *Horizon.* 26 (1983): 53–60.
———. "Mr. Jefferson and the Living Generation." *AS* 41 (1972): 587–98.
———. "Mr. Jefferson's Private Life." *PAPS* 84 (1974): 65–72.
———. "A Note on Evidence: The Personal History of Madison Hemings."
———. "The Return of a Virginian." *VQR.* (1951): 528–43.
Manarin, Louis H., and Clifford Dowdey. *History of Henrico County.* Charlottesville, Va.: Univ. Press of Virginia, 1984.
Mannix, Richard. "Gallatin, Jefferson, and the Embargo of 1808." *Diplomatic History* 3 (1979): 151–72.
Mapp, Alf J., Jr. *Thomas Jefferson: A Strange Case of Mistaken Identity.* Lanham, Md.: Madison, 1987.
———. *Thomas Jefferson: Passionate Pilgrim.* Lanham, Md.: Madison, 1991.
Marambaud, Pierre. "William Byrd I: A Young Virginia Planter in the 1670s." *VMHB* 81 (1973): 131–50.

Marchione, Margherita. "Philip Mazzei and the Last King of Poland." *Italian Americana* 4 (1978): 185–99.

Marks, Frederick W. III. "Foreign Affairs: A Winning Issue in the Campaign for Ratification of the United States Constitution." *Political Science Quarterly* 86 (1971): 444–69.

Marsh, Philip M. "The Jefferson-Madison Vacation." *PMHB* 71 (1947): 70–72.

Marshall, John. Papers. ed. Charles F. Hobson. 6 vols. to date. Chapel Hill, N.C.: Univ. of North Carolina Press, 1974.

Martin, Edwin T. *Thomas Jefferson, Scientist.* New York: Henry Schuman, 1952.

Matthews, Richard K. *The Radical Politics of Thomas Jefferson: A Revisionist View.* Lawrence, Kan.: Univ. Press of Kansas, 1984.

Maury, Ann. *Memoirs of a Huguenot Family.* New York: John S. Taylor, 1838.

Maury, James. "A Dissertation on Education in a Formal Letter from James Maury to Robert Jackson, July 17, 1762," ed. Helen D. Bullock, *Papers of the Albemarle County Historical Society* 2 (1941–42): 32–60.

———. *To Christians of Every Denomination Among Us, Especially Those of the Established Church.* Annapolis: printed by Anne Catharine Green, 1771.

May, Henry F. *The Enlightenment in America.* New York: Oxford Univ. Press, 1976.

Mayo, Bernard. *Another Peppercorn for Mr. Jefferson.* Charlottesville: Thomas Jefferson Memorial Foundation, 1976.

———, ed. *Jefferson Himself.* New York, 1942, repr. 1970.

Mays, David J. *Edmund Pendleton, 1721–1803.* 2 vols. Cambridge, Mass.: Harvard Univ. Press, 1952.

———. *Selected Writings and Correspondence,* ed. Margherita Marchione, et al. Prato, Italy: Edizioni del Palazzo, 1983.

McAllister, Elaine. "The Marquis de Condorcet and Thomas Jefferson: Revolutionary Proposals for Civic Education in the Eighteenth Century." Unpubl. doct. diss, Georgia State Univ., 1982.

McCabe, Carol. "Mr. Jefferson's Garden." *EAL* 14 (1982): 44–49.

McCollester, Charles W. "The 'Spitting' Lyon of Vermont." *New England Galaxy* 17 (1975): 25–34.

McColley, Robert. *Slavery and Jeffersonian Virginia.* Urbana: Univ. of Illinois Press, 1964.

McCoy, Drew. *The Elusive Republic: Political Economy in Jeffersonian America.* Chapel Hill: Univ. of North Carolina Press, 1980.

McCusker, John J. *Money and Exchange in Europe and America, 1600–1775: A Handbook.* Chapel Hill: Univ. of North Carolina Press, 1978.

McDonald, Forrest. *The Presidency of Thomas Jefferson.* Lawrence, Kan.: Univ. of Kansas Press, 1976.

McLaughlin, Jack. *Jefferson and Monticello.* New York: Henry Holt, 1988.

McLoughlin, William G. "Thomas Jefferson and the Beginning of Cherokee Nationalism, 1806 to 1809." *WMQ,* 3rd ser., 32 (1975): 547–80.

Medlin, Dorothy. "Thomas Jefferson, André Morellet, and the French Version of Notes on the State of Virginia." *WMQ* 35 (1978): 85–99.

Merrill, Boynton, Jr. *Jefferson's Nephews: A Frontier Tragedy.* Princeton: Princeton Univ. Press, 1976.

Meschutt, David. "Gilbert Stuart's Portraits of Thomas Jefferson." *American Art Journal* 13 (1981): 2–16.

———. "The Adams-Jefferson Portrait Exchange." *American Art Journal* 14 (1982): 47–54.

Midgley, Louis. *The Brodie Connection: Thomas Jefferson and Joseph Smith. Brigham Young University Studies* 20 (1979): 59–67.

Miller, Charles A. *Jefferson and Nature: An Interpretation.* Baltimore: Johns Hopkins Univ. Press, 1988.

Miller, John Chester. *The Wolf by the Ears: Thomas Jefferson and Slavery.* New York: Free Press, 1977.

Miller, Sue Freeman. "Mr. Jefferson's Passion: His Grove at Monticello." *HP* 32 (1980): 32–35.

Minor, Benjamin. "Memoir of the Author," in George Wythe, *Decisions on Cases in Virginia,* Richmond: 1852.

Mirkin, Harris G. "Rebellion, Revolution, and the Constitution: Thomas Jefferson's Theory of Civil Disobedience." *AS* 13 (1972): 61–74.

Mitchell, Robert D. *Commercialism and Frontier: Perspectives on the Early Shenandoah Valley.* Charlottesville: Univ. Press of Virginia, 1977.

Moore, John Hammond. "That 'Commodious' Annex to Jefferson's Rotunda: Was it Really a National Mausoleum?" *VC* 29 (1980): 114–23.

———. *Albemarle: Jefferson's County,* 1727–1976. Charlottesville: Univ. Press of Virginia, 1976.

Mordecai, Samuel. *Richmond in By-Gone Days.* New York: Arno Press, 1975.

Moreau de Saint-Méry, *American Journey, 1793–1798,* Kenneth and Anna M. Roberts, tr. and ed. Garden City: Doubleday, 1947.

Morgan, Edmund S. *American Slavery, American Freedom.* New York: Norton, 1975.

———. *The Meaning of Independence: John Adams, George Washington, Thomas Jefferson.* Charlottesville: Univ. Press of Virginia, 1976.

Morgan, Edmund S., and Helen M. *The Stamp Act Crisis, Prologue to Revolution.* Chapel Hill: Univ. of North Carolina Press, 1953.

Morpugo, J.F. *Their Majesties' Royal College: William and Mary in the Seventeenth and Eighteenth Centuries.* Washington, D.C.: 1976.

Morris, Gouvernour. *A Diary of the French Revolution.* ed. Beatrix Cary Davenport. Boston: Houghton Mifflin, 1939.

Morris, Richard B. "The Legal Profession in America on the Eve of the Revolution," in Harry W. Jones, ed., *Political Separatism and Legal Continuity.* Chicago: Univ. of Chicago Press, 1976.

———. *The Peacemakers, the Great Powers and American Independence.* New York: Harper & Row, 1965.

Morse, Genevieve Forbes. "Captain Jack Jouette." *DAR* 115 (1981): 700–703.

Mossiker, Frances. *Madame de Sevigne: A Life and Letters.* New York: Knopf, 1983.

Mott, Frank L. *Jefferson and the Press.* Baton Rouge, La.: Louisiana State Univ. Press, 1943.

Moulton, Gary E. *Journals of the Lewis and Clark Expedition.* Lincoln, Neb.: Univ. of Nebraska Press, 1986.

Moulton, Sherman R. "A Vermont Treason Trial," Vermont Bar Association, *Proceedings* 24 (1935): 121–41,

Mullin, G.W. *Flight and Rebellion: Slave Resistance in Eighteenth-Century Virginia.* New York: Oxford Univ. Press, 1972.

Murphy, Paul L. *Marshall v. Jefferson: The Political Background of Marbury v. Madison. Borzoi Series in United States Constitutional History.* New York: Knopf, 1970.

Nettels, Curtis P. *The Emergence of a National Economy, 1775–1815.* New York: Holt, Rinehart and Winston, 1962.

Nichols, Frederick D. *Thomas Jefferson's Architectural Drawings,* 4th ed. Charlottesville: Univ. Press of Virginia, 1984.

Nichols, Frederick D., and Ralph E. Griswold. *Thomas Jefferson, Landscape Architect.* Charlottesville: Univ. Press of Virginia, 1978.

Nichols, Roy Franklin. *The Invention of the American Political Parties.* New York: Macmillan, 1967.

Noonan, John T. *Persons and Masks of the Law: Cardozo, Holmes, Jefferson, and Wythe as Makers of the Masks.* New York: Farrar, Straus and Giroux, 1976.

Oakes, James. *Slavery and Freedom: An Interpretation of the Old South.* New York: Knopf, 1990.

O'Brien, Charles F. "The Religious Issue in the Presidential Campaign of 1800." *EIHC* 107 (1971): 82–93.

O'Fallon, James M. *"Marbury." Stanford Law Review,* 44 (1992): 219–60.

Padover, Saul K. *Jefferson.* New York: Harcourt, Brace, 1942.

———. *Thomas Jefferson and the Foundations of American Freedom.* Princeton: D. Van Nostrand, 1965.

———. "Thomas Jefferson and the Election of 1800." *Lithopininian* 7 (1972): 8–14.

Padover, ed. *Thomas Jefferson and the National Capital.* Preface by Harold L. Ickes. Washington, D.C.: Government Printing Office, 1946.

Page, John. "Autobiography." *VQR* 3 (1850): 142–51.

Palmer, Michael A. *Stoddert's War: Naval Operations During the Quasi-War with France, 1798–1801.* Columbia, S.C.: Univ. of South Carolina Press, 1987.

Palmer, Robert Roswell. *The Age of the Democratic Revolution.* Princeton: Princeton Univ. Press, 1959.

———. *The World of the French Revolution,* New York: Harper & Row, 1971.

———. "The Dubious Democrat: Thomas Jefferson in Bourbon, France." *Political Science Quarterly* (1967): 388–404.

Pancake, John S. *Jefferson and Hamilton.* Woodbury, N.Y.: Barron's Educational Series, 1974.

Parker, Harold T. *The Cult of Antiquity and the French Revolutionaries.* Chicago: Univ. of Chicago Press, 1937.

Parker, Iola B. "Whiskey Creek Keeps Running, But Only With Water." *Smithsonian* 5 (1974): 82–89.

Peckham, H.H. *The Colonial Wars, 1689–1762,* Chicago: Univ. of Chicago Press, 1964.

Peden, William H. *Twilight at Monticello.* Boston: Houghton Mifflin, 1973.

Peden, William H., ed. *Notes on the State of Virginia.* Chapel Hill: Univ. of North Carolina Press: 1954.

Peeler, David P. "Thomas Jefferson's Nursery of Republican Patriots: Virginia." *Journal of Church and State* 28 (1986): 70–93.

Perkins, Bradford. *The First Rapprochement: England and the United States, 1795–1805,* Philadelphia: Univ. of Pennsylvania Press, 1955.

Peterson, Merrill D. *Adams and Jefferson: A Revolutionary Dialogue.* Athens, Georgia: Univ. of Georgia Press, 1976.

———. *Jefferson and Madison and the Making of Constitutions.* Charlottesville, Va.: Univ. Press of Virginia, 1987.

———. *The Jefferson Image in the American Mind.* New York: Oxford Univ. Press, 1960.

———. *Thomas Jefferson and the American Revolution.* Williamsburg: Virginia Independent Bicentennial Commission, 1976.

———. *Thomas Jefferson and the Beginnings of American Citizenship.* Charlottesville: Univ. Press of Virginia, 1981.

———. *Thomas Jefferson and the New Nation.* New York: Oxford Univ. Press, 1970.

———. *Thomas Jefferson: A Reference Biography.* New York: Scribner's, 1986.

———, ed. *Visitors to Monticello,* Charlottesville: Univ. Press of Virginia, 1989.

———. *Virginia Statute of Religious Freedom.* Cambridge, Eng.: Cambridge Univ. Press, 1988.

———. "Dumas Malone: The Completion of a Monument." *VQR* 58 (1982): 26–31.

———. "Mr. Jefferson's 'Sovereignty of the Living Generation.'" *VQR* 52 (1976): 437–47.

———. "Process and Personality in Jefferson's Administration." *Reviews in American History* 7 (1979): 189–98.

———. "Thomas Jefferson and the Constitution." *This Constitution* (1986): 12–17.

Peterson, Merrill D., and Robert C. Vaughan, eds. *The Virginia Statute for Religious Freedom.* New York: Cambridge Univ. Press, 1988.

Pfeffer, Leo. "The Revolution in Virginia," in *Church, State and Freedom.* Boston: Beacon Press, 1953, 93–102.

Pierson, Hamilton W. *Jefferson at Monticello.* New York: 1862. repr. and ed. by James A. Bear, Jr., Charlottesville: Univ. Press of Virginia, 1967.

Phillipson, Nicholas, "The Scottish Enlightenment," in *The Enlightenment in National Context,* ed. Roy Porter and Mikulas Teich. Cambridge, Eng.: Cambridge Univ. Press, 1981.

Plumer, William. *William Plumer's Memorandum of Proceedings in the United States Senate, 1803–1807.* New York: Macmillan, 1923.

Pole, J.R. "Enlightenment and the Politics of American Nature," in *The Enlightenment in National Context,* ed. Roy Porter and Mikulas Teich. Cambridge, Eng.: Cambridge Univ. Press, 1981.

———. *Political Representation in England and the Origins of the American Republic.* London: Macmillan, 1966.

Post, David M. "Jeffersonian Revisions of Locke's Education, Property Rights, and Liberty." *Journal of the History of Ideas* 47 (1986): 147–59.

Prager, Frank B. "Trends and Developments in American Patent Law from Jefferson to Clifford (1790–1870)." *AJLH,* 6 (1962): 45–63.

Preyer, Kathryn. "Crime, the Criminal Law and Reform in Post-Revolutionary Virginia." *Law and History Review* 1 (1983): 53–85.

Price, Jacob M. *Capital and Credit in British Overseas Trade.* Cambridge, Mass.: Harvard Univ. Press, 1980.

Pula, James S. "The American Will of Thaddeus Kosciuszko." *Polish American Studies* 34 (1977): 16–25.

Pulley, Judith. "The Bittersweet Friendship of Thomas Jefferson and Abigail Adams." *EIHC* 108 (1972): 193–216.

Quinby, Lee. "Thomas Jefferson: The Virtue of Aesthetics and the Aesthetics of Virtue." *AHR* 87 (1982): 337–56.

Rakove, Jack N. *Beginnings of National Politics.* New York: Knopf, 1975.

———. *Interpreting the Constitution.* Boston: Northeastern Univ. Press, 1990.

Randall, Henry S. *Life of Thomas Jefferson.* 3 vols. New York, 1858. repr. New York: Da Capo Press, 1972.

Randall, Willard Sterne. *A Little Revenge: Benjamin Franklin at War with His Son.* Boston: Little, Brown, 1984, repr. New York: Quill/Morrow, 1991.

———. *Benedict Arnold, Patriot and Traitor,* New York: William Morrow, 1990.

Randolph, Edmund. *Essays on the Revolutionary History of Virginia, 1774–1782, VMHB* 43 (1935): 115, 122–23.

Randolph, Sarah N. *The Domestic Life of Thomas Jefferson.* New York: Harper and Brothers, 1871.

Rankin, H.F. "The General Court of Colonial Virginia." *VMHB,* 70 (1962): 142–53.

Ratzlaff, Robert K. "The Evolution of A Gentleman-Politician: John Rutledge, Jr., of South Carolina." *Midwest Quarterly* 27 (1985): 77–95.

Reardon, John J. *Edmund Randolph: A Biography.* New York: Macmillan, 1975.

———. *Peyton Randolph, 1721–1775, One Who Presided.* Durham: Carolina Academic Press, 1982.

Regis, Pamela. *Describing Early America.* DeKalb, Ill.: Northern Ill. Univ. Press, 1992.

Reinhold, Meyer. *Classica Americana: The Greek and Roman Heritage in the U.S.* Detroit: Wayne State Univ. Press, 1984.

Reynolds, Donald E. "Ammunition Supply in Revolutionary Virginia." *VMHB,* 73 (1965): 56–74.

Ribodeau, François. *La Déstinée sécrete de la Fayette ou le méssianisme révolutionnaire.* Paris: Robert Laffont, 1972.

Riccards, Michael P. "Philip Mazzei: The Jeffersonian as Internationalist." *Italian Americana* 6 (1980): 210–21.

Rice, Howard C., Jr., *L'Hôtel de Langéac, Jefferson's Paris Residence.* Charlottesville: Thomas Jefferson Memorial Foundation, 1947.

———. *Thomas Jefferson's Paris.* Princeton: Princeton Univ. Press, 1976.

———. "Jefferson's Gift of Fossils to the Museum of Natural History in Paris." *PAPS,* 95 (1951): 597–627.

Rice, Howard C., Jr., and Anne S.K. Brown, trans. and eds. *The American Campaigns of Rochambeau's Army, 1780–83.* 2 vols. Princeton: Princeton Univ. Press, 1972.

Rich, Bennett M. *The Presidents and Civil Disorder.* Westport: Greenwood, 1990.

Richardson, E.P. "A Life Drawing of Jefferson by John Trumbull." *Maryland Historical Magazine* 70 (1975): 363–71.

Richardson, William D. "Thomas Jefferson and Race: The Declaration and Notes on the State of Virginia." *Polity* 16 (1984): 447–66.

Riedesel, Baroness von. *Baroness von Riedesel and the American Revolution.* trans. and ed. Marvin L. Brown, Jr. Chapel Hill: Univ. of North Carolina Press, 1965.

Risjord, Norman K. "The Compromise of 1790: New Evidence on the Dinner Table Bargain." *WMQ,* 3rd ser., 33 (1976): 309–14.

Ritcheson, Charles R. "The Fragile Memory: Thomas Jefferson at the Court of George III." *Eighteenth-Century Life* 6 (1981): 1–16.

Robbins, Caroline. *The Eighteenth-Century Commonwealthman.* Cambridge, Mass.: Harvard Univ. Press, 1959.

Robbins, Jan C. "Jefferson and the Press: The Resolution of an Anomaly." *Journalism Quarterly* 48 (1971): 421–30.

Robertson, David, ed. *Reports on the Trial of Colonel Aaron Burr.* Philadelphia: Hopkins & Earle, 1808.

Robertson, Samuel Arndt. "Thomas Jefferson and the Eighteenth-Century Landscape Garden Movement in England." Unpub. doct. diss., Yale, 1974.

Rodrigues, Leda B. "Jose Joaquim da Maia e TJ." *Revista do Instituto Historico e Geographico Brasiliero.* 33 (1981): 53–70.

Roeber, A.G. *Faithful Magistrates and Republican Lawyers: Creators of Virginia Legal Culture, 1680–1810.* Chapel Hill: Univ. of North Carolina Press, 1981.

———. "Authority, Law and Custom: The Rituals of Court Day in Tidewater Virginia, 1720 to 1750," *WMQ,* 3d ser., 37 (1980): 29–52.

Ronda, James P. *Lewis and Clark among the Indians.* Lincoln, Neb.: Univ. of Nebraska Press, 1984.

Roosevelt, T. *Gouverneur Morris.* Boston: Houghton Mifflin, 1970.

Rossiter, Clinton. *Alexander Hamilton and the Constitution.* New York: Harcourt, Brace and World, 1964.

Royster, Charles. *A Revolutionary People at War.* Chapel Hill: Univ. of North Carolina Press, 1979.

———. "A Battle of Memoirs: Light-Horse Harry Lee and Thomas Jefferson." *VC* 31 (1981): 112–27.

Rutland, Robert A. "Madison's Bookish Habits." *QJLC* 37 (1980): 176–91.

Ryan, Mary P. "Party Formation in the United States Congress, 1789 to 1796: A Quantitative Analysis." *WMQ,* 3rd ser., 28 (1971): 523–42.

St. John, Jeffrey. *Forge of Union, Anvil of Liberty.* Ottawa, Ill.: Jameson Books, 1992.

Salviati, Yvette. "La 'Barque sécrete' d'un demi-dieu: Thomas Jefferson dans *La Virginienne.*" *Mythes, Croyances et Réligions dans le Monde Anglo-Saxon*5 (1987): 165–86.

Sanford, Charles B. *Thomas Jefferson and His Library.* Hamden, Conn.: Archon, 1977.

———. *The Religious Life of Thomas Jefferson.* Charlottesville: Univ. Press of Virginia, 1984.

Schackner, Nathan. *Aaron Burr.* New York: Frederick A. Stokes, 1937.

———. *Thomas Jefferson.* New York: Thomas Yoseloff, 1951.

Scharnhorst, Gary. "The Virginian as a Founding Father." *Arizona Quarterly* 40 (1984): 227–41.

Scherr, Arthur. "The 'Republican Experiment' and the Election of 1796 in Virginia." *West Virginia History* 37 (1976): 89–108.

Schmitt, Gary J. "Jefferson and Executive Power: Revisionism and the Revolution of 1800." *Publius* 17 (1987): 7–25.

Schöpf, Johann David. *Travels in the Confederation, 1783–1784,* trans. and ed. Alfred J. Morison. Philadelphia: W.J. Campbell, 1911. 2 vols.

Schultz, Constance B. "Of Bigotry in Politics and Religion: Jefferson's Religion, The Federalist Press, and the Syllabus." *VMHB* 91 (1983): 73–91.

Seager, Robert. *and Tyler too.* New York: McGraw-Hill, 1963.

Sears, Louis M. *Jefferson and the Embargo.* Durham, N.C.: Duke Univ. Press, 1967.

———. "British Industry and the Embargo." Quarterly Journal of Economics. 34 (1919): 88–113.

Seelye, John. "Beyond the Shining Mountains: The Lewis and Clark Expedition as an Enlightenment Epic." *VQR* 63 (1987): 36–53.

Selby, John E. *The Revolution in Virginia.* Williamsburg: Colonial Williamsburg Foundation, 1988.

Shackleton, Robert. *Montesquieu: A Critical Biography.* Oxford: Oxford Univ. Press, 1961.

Shalhope, Robert E. "Thomas Jefferson's Republicanism and Antebellum Southern Thought." *JSH* 42 (1976): 529–56.

———. "Toward a Republican Synthesis: The Emergence of an Understanding of Republicanism in American Historiography." *WMQ,* 3d ser, 2 (Jan. 1972): 49–80.

Sharp, James R. "Adams and Jefferson in the Middle East." *Manuscripts* 33 (1981): 237–40.

Sharp, James Roger. "Unraveling the Mystery of Jefferson's Letter of April 27, 1795." *Journal of the Early Republic* 6 (1986): 411–18.

Shaw, Peter. "Blood Is Thicker Than Irony: Henry Adams 'History.' " *NEQ* 40 (1967): 163–87.

Shawen, Neil McDowell. "The Casting of a Lengthened Shadow: Thomas Jefferson's Role in Determining the Site for a State University in Virginia." Unpub. doct. diss. George Washington Univ., 1980.

———. "Thomas Jefferson and a 'National' University: The Hidden Agenda For Virginia." *VMHB* 92 (1984): 309–35.

Sheehan, Bernard W. "Jefferson and the West." *VQR* 58 (1982): 345–52.

———. *Seeds of Extinction: Jeffersonian Philanthropy and the American Indian.* Chapel Hill: Univ. of North Carolina Press, 1973.

Sheldon, Garrett Ward. *Political Philosophy of Thomas Jefferson.* Baltimore: Johns Hopkins Univ. Press, 1992.

———. "Classical and Modern Influences on American Political Thought: The Political Theories of Thomas Jefferson." Unpub. doct. diss., Rutgers, 1983.

Shenkir, William E., Glenn A. Welsch, and James A. Bear, Jr. "Thomas Jefferson: Management Accountant." *Journal of Accountancy* 133 (1972): 33–47.

Sheps, Arthur. "The American Revolution and the Transformation of English Republicanism." *Historical Reflections* 2 (1975): 3–28.

Sher, Richard B., and Jeffrey R. Smitten, eds. *Scotland and America in the Age of the Enlightenment.* Princeton: Princeton Univ. Press, 1990.

Shonting, Donald Allen. "Romantic Aspects in the Works of Thomas Jefferson." Unpub. doct. diss., Ohio Univ., 1977.

Simcoe, John Graves. *Military Journal.* New York: Bartlett & Wellford, 1844. Repr. 1968.

Simpson, William S. "Dabney Carr: Portrait of a Colonial Patriot." *VC* 23 (1974): 5–13.

Sioussat, St. George L. "The Breakdown of the Royal Management of Lands in the Southern Provinces, 1773–1775." *Agricultural History* 3 (1929): 67–98.

Sisson, Daniel Joseph. *The American Revolution of 1800,* New York: Knopf, 1974.

Smelser, Marshall. *The Congress Sounds the Navy, 1787–1798.* South Bend, Ind.: Univ. of Notre Dame Press, 1959.

Smith, Abigail Adams. *Journal and Correspondence of Miss Adams, Daughter of John Adams,* ed. Caroline DeWindt. New York: Wiley & Putnam, 1841.

Smith, A.M. "Virginia Lawyers, 1680–1776: The Birth of an American Profession." Unpub. doct. diss., Johns Hopkins Univ., Baltimore, 1967.

Smith, Daniel B. *Inside the Great House: Planter Family Life in Eighteenth-Century Chesapeake Society.* Ithaca, N.Y.: Cornell Univ. Press, 1980.

Smith, James Morton. *Freedom's Fetters: The Alien and Sedition Laws and American Civil Liberties.* Ithaca, N.Y.: Cornell Univ. Press, 1956.

Smith, Margaret Bayard. *The First Forty Years of Washington Society.* New York: Scribner's, 1906.

Smith, Paul H. "Time and Temperature: Philadelphia, July 4, 1776." *QJLC* 33 (1976): 294–99.

Smollett, Tobias. *Travels through France and Italy.* London: 1763, repr. Oxford: Oxford Univ. Press, 1981.

Smylie, James H. "Jefferson's Statute for Religious Freedom: The Hanover Presbytery Memorials, 1776–1786." *American Presbyterians* 63 (1985): 355–373.

Smythe, J.F.D. *A Tour of the United States,* 2 vols. London: 1784.

Sobel, Mechal. *The World They Made Together: Black and White Values in Eighteenth-Century Virginia.* Princeton: Princeton Univ. Press, 1987.

Sosin, J.M. *The Aristocracy of the Long Robe: The Origins of Judicial Review in America.* Westport, Conn.: Greenwood Press, 1989.

Souchal, Genevieve. *French 18th-Century Furniture.* New York: Putnam's, 1961.

Sowerby, E. Millicent, comp. *Catalogue of the Library of Thomas Jefferson,* 5 vols. Washington, D.C.: Library of Congress, 1952–59.

Spalding, James C. "Loyalist as Royalist, Patriot as Puritan: The American Revolution as a Repetition of the English Civil Wars." *Church History* 45 (1976): 329–40.

Spivak, Burton. *Jefferson's English Crisis: Commerce, Embargo, and the Republican Revolution.* Charlottesville: Univ. Press of Virginia, 1979.

———. "Republican Dreams and National Interest: The Jeffersonians and American Foreign Policy." *Society for History of American Foreign Relations Newsletter* 12 (1981): 1–21.

Sprague, Marshall. *So Vast, So Beautiful a Land.* Boston: Little, Brown, 1974.

Sprague, Stuart Seely. "Jefferson, Kentucky and the Closing of the Port of New Orleans, 1802–1803." *Register of the Kentucky History Society,*79 (1972): 312–17.

Stanard, Mary N. *Colonial Virginia: Its People and Customs.* Detroit: Singing Tree, 1970.

Stanard, W.G. "Racing in Colonial Virginia," *VMHB* 2 (1894): 293–305.

Starr, Raymond. *Jefferson, Nationalism, and the Enlightenment.* New York: Braziller, 1975.

Stevens, Michael E. "Thomas Jefferson, Indians, and Missing Privy Council Journals." *South Carolina Historical Magazine* 82 (1981): 177–85.

Stevens, William O. *Old Williamsburg and Her Neighbors.* New York: Dodd, Mead, 1938.

Stewart, Donald H. *The Opposition Press of the Federalist Period.* Albany: State Univ. of New York Press, 1969.

Stewart, Robert A. *History of Virginia's Navy of the Revolution.* Richmond: Mitchell and Hotchkiss, 1939.

Stinchcombe, W.C. *The American Revolution and the French Alliance.* Syracuse: Syracuse Univ. Press, 1969.

Stolba, K. Marie. "Music in the Life of Thomas Jefferson." *DAR* 108 (1974): 196–202.

Stourzh, Gerald. *Alexander Hamilton and the Idea of Republican Government.* Stanford: Stanford Univ. Press, 1970.

Stowe, Steven M. "Private Emotions and a Public Man in Early Nineteenth Century Virginia." *HEQ* 27 (1987): 75–81.

Stuart, Reginald C. *The Half-way Pacifist: Thomas Jefferson's View of War.* Toronto: Univ. of Toronto Press, 1978.

Stuart, Reginald Charles. "Encounter with Mars: Thomas Jefferson's View of War." Unpub. doct. diss., University of Florida, 1974.

———. "Thomas Jefferson and the Function of War: Policy or Principle?" *Canadian Journal of History* 11 (1976): 155–71.

———. "Thomas Jefferson and the Origins of War." *Peace and Change* 4 (1976): 22–27.

Suro, Dario. "Jefferson, the Architect." *Americas* 25 (1973): 29–35.

Sydnor, Charles S. *Gentlemen Freeholders: Political Parties in Washington's Virginia.* Chapel Hill: Univ. of North Carolina Press, 1952.

Tagg, James. *Benjamin Franklin Bache and the Philadelphia "Aurora."* Philadelphia: Univ. of Pennsylvania Press, 1991.

Tate, Thad W. *The Negro in Eighteenth-Century Williamsburg.* Williamsburg: Colonial Williamsburg, 1965.

———. "The Coming of the Revolution in Virginia: Britain's Challenge to Virginia's Ruling Class, 1763–1776." *WMQ*, 3d ser., 19 (1962): 323–43.

Thomson, Robert Polk. "The Reform of the College of William and Mary, 1763–1780." *PAPS* 115 (1971): 207–13.

Tobin-Schlesinger, Kathleen. "Jefferson to Lewis: The Study of Nature in the West." *Journal of the West.* 29 (1990): 54–61.

Tocqueville, Alexis de. *The Old Regime and the French Revolution,* trans. Stuart Gilbert. Garden City, N.Y.: Doubleday, 1955.

Torrence, William C. "Henrico County, Virginia: Beginnings of Its Families." *WMQ*, 1st ser., 24 (1915–16): 116–42, 202–10, 262–83.

———. "Thomas and William Branch of Henrico." *WMQ*, 1st ser., 25 (1916–17): 59–70, 107–16.

Trumbull, John. *Autobiography,* ed. Theodore Sizer. New Haven: Yale Univ. Press, 1953.

Tucker, Robert W., and David C. Hendrickson. *Empire of Liberty: The Statecraft of Thomas Jefferson.* New York: Oxford Univ. Press, 1990.

———. *Fall of the First British Empire: Origins of the War of American Independence.* Baltimore: Johns Hopkins Univ. Press, 1982.

Turner, Jesse. "A Phantom Precedent," *American Law Review* 48 (1914): 321–33.

Tully, James. *A Discourse on Property: John Locke and His Adversaries.* Cambridge, Eng: Cambridge Univ. Press, 1980.

Tyack, David B. *George Ticknor and the Boston Brahmins.* Cambridge, Mass.: Harvard Univ. Press, 1967.

Tyler, Lyon Gardimer. "Early Courses and Professors at William and Mary College." *WMQ*, 1st ser., 14 (1905): 71–83.

Vance, Joseph C. "Thomas Jefferson Randolph." Unpub. doct. diss., Univ. of Virginia, 1957.

VanDoren, Carl C. *Benjamin Franklin.* New York: Garden City Publishing Co., 1938.

VanMeter, Suzanne. *"A Noble Bargain: The Louisiana Purchase."* Unpub. doc. diss. Indiana Univ., 1977.

Van Pelt, Charles B. *Thomas Jefferson and Maria Cosway. AH* 22 (1971): 22–29, 102–3.

Van Schreeven, William J., et al. *Revolutionary Virginia: The Road to Independence.* 7 vols. Charlottesville, Va.: Univ. Press of Virginia, 1973–83.

Vaughn, Joseph Lee. *Thomas Jefferson's Rotunda Restored.* Charlottesville: Univ. Press of Virginia, 1981.

Verner, Coolie. "The Maps and Plates Appearing with the Several Editions of Mr. Jefferson's *Notes on the State of Virginia. VMHB* 59 (1951): 21–32.

Waciuma, Wanjohi. *Intervention in Spanish Floridas, 1801–1813: A Study in Jeffersonian Foreign Policy.* Boston: Branden, 1976.

Waddell, Joseph Addison. *Annals of Augusta County, Virginia, 1726–1871.* Bridgewater, Va.: C.J. Carrier Co., 1950.

Wagoner, Jennings L. "Honor and Dishonor at Mr. Jefferson's University: The Antebellum Years." *HEQ* 26 (1986): 155–79.

Wall, A.J. "The Story of the Convention Army." New York Historical Society *Quarterly Bulletin,* 11 (1927): 67–97.

Wall, Charles Coleman, Jr. "Students and the Student Life at the University of Virginia, 1825–1861," Unpub. doct. diss., Univ. of Virginia, 1978.

Wallace, Willard M. *Traitorous Hero: Benedict Arnold.* New York: Harper & Row, 1954.

Walters, Raymond, Jr. *Albert Gallatin: Jeffersonian Financier and Diplomat.* Pittsburgh: Univ. of Pittsburgh Press, 1957.

Washington, George. *Papers,* ed. W.W. Abbot, et al., Colonial Series, 8 vols. to date; Confederation Series, 1 vol. to date. Charlottesville, Va.: Univ. Press of Virginia, 1983.

———. *Journal of the Proceedings of the Presidency.* 1 vol. to date. Charlottesville, Va.: Univ. Press of Virginia, 1993.

Waterman, Julian S. "Thomas Jefferson and Blackstone's Commentaries," *Illinois Law Review* 27 (1933): 629–59.

Watson, Francis J.B. "America's First Universal Man Had a Very Acute Eye." *Smithsonian* 7 (1976): 88–95.

Watson, Ross. "Thomas Jefferson's Visit to England, 1786." *HT* (Great Britain) 27 (1971): 3–13.

Watts, Steven. *The Republican Reborn: War and the Making of Liberal America, 1790–1820.* Baltimore: Johns Hopkins Univ. Press, 1987.

Wayland, J.W. *The Fairfax Line: Thomas Lewis's Journal of 1746.* Newmarket, Va.: Henkel Press, 1925.

Wells, Samuel J. "International Causes of the Treaty of Mount Dexter, 1805." *Journal of Mississippi History* 48 (1896): 177–85.

Wertenbaker, Thomas Jefferson. *Patrician and Plebeian in Virginia.* Charlottesville, Va.: Univ. Press of Virginia, 1910.

———. *The Planters of Colonial Virginia.* Princeton: Princeton Univ. Press, 1922.

Wettstein, Arnold. "Religionless Religion in the Letters and Papers From Monticello." *Religion in Life,* 45 (1976): 152–60.

Weyant, Robert G. "Helvetius and Jefferson: Studies of Human Nature and Government in the Eighteenth Century." *Journal of the History of the Behavioral Science* 9 (1973): 29–41.

Wharton, James. "Jefferson, Expert on Wine." *Commonwealth* 26 (1959): 4, 8, 65–6.

Whipple, A.B.C. *To the Shores of Tripoli: The Birth of the U.S. Navy and Marines.* New York: Wm. Morrow, 1991.

White, Leonard D. *The Jeffersonians: A Study in Administrative History, 1801–29.* New York: Macmillan, 1948.

White, Morton. *Philosophy of the American Revolution.* New York: Oxford Univ. Press, 1978.

White, Patrick C.T. *The Critical Years: American Foreign Policy, 1793–1823.* New York: Wiley, 1970.

Williams, Richard L. "Atop a "Little Mountain" in Virginia, Jefferson Cultivated His Botanical Bent." *Smithsonian* 15 (1984): 68–77.

Williams, T. Harry. "On the Couch at Monticello." *RAH* 2 (1974): 523–29.

———. "Williamsburg—The Old Colonial Capital." *WMQ,* 1st ser., 16 (1907): 1–65.

Wills, Garry. *Inventing America: Jefferson's Declaration of Independence.* New York: Doubleday, 1978.

———. "Jefferson's Other Buildings." *AM* 27 (1993): 80–89.

Wilson, Arthur M. "The Enlightenment Came First to England," in Stephen B. Baxter, ed., *England's Rise to Greatness, 1660–1763.* Berkeley: Univ. of California Press, 1983.

Wilson, Douglas L. "The American Agricola: Jefferson's Agrarianism and the Classical Tradition." *SAQ* 80 (1981): 339–54.

———. "Sowerby Revisited: The Unfinished Catalogue of Thomas Jefferson's Library." *WMQ* 41 (1984): 615–28.

———. *Thomas Jefferson's Early Notebooks. WMQ,* 3rd ser., 42 (1985): 433–52.

Wilson, Douglas L., ed. *Jefferson's Commonplace Book,* Princeton: Princeton Univ. Press, 1989.

———. "Thomas Jefferson's Early Notebooks." *WMQ,* 3rd ser., 42 (1985): 433–52.

———. "What Jefferson and Lincoln Read." *AM* 267 (1991): 51–63.

Windley, Lathan A. "Runaway Slave Advertisements of George Washington and Thomas Jefferson." *Journal of Negro History* 63 (1978): 373–74.

Wolff, Philippe. "Jefferson on Provence and Languedoc." *Proceedings of the Annual Meeting of the Western Society for French History* 3 (1975): 191–205.

Wood, Gordon. *The Creation of the American Republic, 1776–1787.* Chapel Hill: Univ. of North Carolina Press, 1969.

———. *The Radicalism of the American Revolution.* New York: Knopf, 1992.

———. "Rhetoric and Reality in the American Revolution." *WMQ,* 3rd ser., 23 (1966), 3–32.

———. The Bigger the Beast the Better." *AHI* 17 (1982): 30–37.

Woods, Edgar. *Albemarle County in Virginia.* Charlottesville, Va., Univ. Press of Virginia, 1901.

Woods, Mary N. "Thomas Jefferson and the University of Virginia: Planning the Academic Village." *Journal of the Society of Architectural Historians* 44 (1985): 266–83.

Woodson, Minnie Shumate. "Researching to Document the Oral History of the Thomas Woodson Family: Dismantling the Sable Curtain." *Journal of the Afro-American Historical and Genealogy Society* 6 (1985): 3–12.

Woolery, William K. *The Relation of Thomas Jefferson to American Foreign Policy.* Baltimore: Johns Hopkins Univ. Press, 1927.

Wright, Esmond. *Franklin of Philadelphia.* Cambridge, Mass.: Harvard Univ. Press, 1986.

Wright, Louis B. *The First Gentlemen of Virginia.* Charlottesville, Va.: Univ. Press of Virginia, 1940.

Wright, M.J. "Lafayette's Campaign in Virginia." *Southern Historical Assn. Publications* 9 (1905): 234–40, 261–71.

Yoder, Edwin M. "The Sage at Sunset." *VQR* 58 (1958): 32–37.

Young, Alfred. "The Mechanics and the Jeffersonians: New York, 1789–1801," *Labor History* 5 (1964): 247–76.

Young, Arthur. *Travels in France during the Years 1787, 1788 and 1789,* ed. C. Maxwell. Cambridge, Eng.: Cambridge Univ. Press, 1950.

Young, James Sterling. *The Washington Community, 1800–1828.* New York: Harcourt, Brace and World, 1966.

Zimmer, Anne Y. *Jonathan Boucher: Loyalist in Exile.* Detroit: Wayne State Univ. Press, 1978.

Index

About the Author

Willard Sterne Randall is the prizewinning author of thirteen books, including *Ethan Allen: His Life and Times*; *A Little Revenge: Benjamin Franklin at War with His Son*; *Alexander Hamilton: A Life*; *Thomas Jefferson: A Life*; *George Washington: A Life*; and *Benedict Arnold: Patriot and Traitor*. A finalist for the Los Angeles Times Book Prize, he is a professor of history at Champlain College and lives in Burlington, Vermont.

BOOKS BY WILLARD STERNE RANDALL

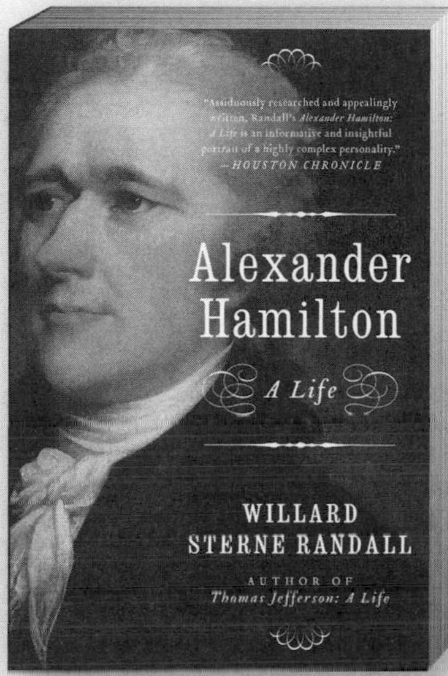

ALEXANDER HAMILTON
A Life

Available in Paperback and eBook

From his less than auspicious start in 1755 on the Caribbean island of Nevis to his untimely death in a duel with his old enemy Aaron Burr in 1804, Alexander Hamilton, despite his short life, left a huge legacy. Orphaned at thirteen and apprenticed in a counting house, the precocious Hamilton learned principles of business that helped him create the American financial system and invent the modern corporation. But first the staunch, intrepid Hamilton served in the American Revolution, acting as General Washington's spymaster. Forging a successful legal career, Hamilton coauthored the Federalist Papers and plunged into politics. Irresistibly attractive to women, he was a man of many gifts, but he could be arrogant and was, at times, a poor judge of character.

"Vividly created. A fine biography." —*Boston Globe*

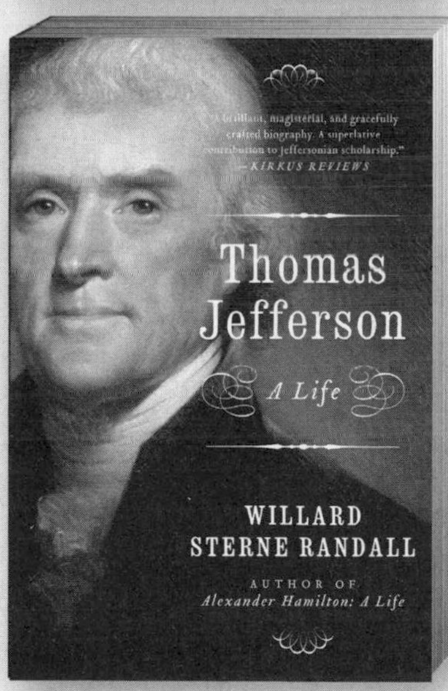

THOMAS JEFFERSON
A Life

Available in Paperback and eBook

The highly praised one-volume biography of Thomas Jefferson that provides illuminating new insights into his public and private life—by the award-winning author of *A Little Revenge: Benjamin Franklin at War with His Son* and *Benedict Arnold: Patriot and Traitor*. Combining firsthand scholarship and material drawn from the Jefferson Papers, Willard Sterne Randall calls on his skills as an investigative journalist to challenge long-held assumptions about the reasoning, motives, and works of this sage, philosopher, politician, and romantic. Revealing Jefferson's inner and outer struggles, this is an examination of both Jefferson's thoughts on slavery and his alleged relations with the slave Sally Hemmings as well as his Revolutionary and diplomatic intrigues.

"Randall's achievement in this flowing meticulously researched narrative is no small matter: He has rescued Jefferson from myth and returned him to us a fully developed person." —*San Francisco Chronicle*